PROJECT APOLLO

By TD Barnes

Title of the Book: Project Apollo

Independently published.

Printed in the United States of America
First Edition

Contents

Glossary

12 Moon Walks, 24 Human visitors, 238,855 miles average distance to Earth.

Glossary for Project Apollo

B.S.: Bachelor of Science degree

CMP: Command Module Pilot

CM: Command Module

CSM: Command and Service Module(s)

GET: Ground Elapsed Time

GH2: Gaseous Hydrogen

LH2: Liquid Hydrogen

LM: Lunar Module

LMP: Lunar Module Pilot

LOX: Liquid Oxygen

LRV: Lunar Rover Vehicle

MET: Mission Elapsed Time

NASA: National Aeronautics and Space Administration

Ph.D.: Doctor of Philosophy degree

S-IB: Saturn IB launch vehicle

S-IVB: Saturn IV-B launch vehicle

SM: Service Module

SPS: Service Propulsion System

Sc.D.: Doctor of Science degree

Components and Experiments

Command Module (CM): The crew's primary living quarters and control center during the mission.

Command and Service Module (CSM): Combined structure including the Command Module (CM) and Service Module (SM), responsible for course corrections, lunar orbit insertions, and Transearth Injection (TEI).

Lunar Module (LM): Used for controlled lunar landing, housed the Ascent and Descent Stages.

Instrument Unit: Provided guidance, navigation, telemetry, and control signals during launch.

Spacecraft/Lunar Module Adapter: The facilitated interface between the launch vehicle and Lunar Module (LM).

Lunar Surface Experiments

Central Station: Nerve center on the lunar surface providing communication, power distribution, and control functions.

Early Apollo Scientific Experiment Package (EASEP): Set of experiments deployed on Apollo 11, powered by solar energy.

Active Seismic Experiments: Used thumpers and explosive charges to generate seismic signals, probing lunar structure.

Heat Flow Experiment: Measured temperature gradients and thermal conductivity near lunar surface.

Lunar Mass Spectrometer: Analyzed lunar atmospheric components and their abundances.

Lunar Seismic Profiling Experiment: Deployed explosive charges to measure lunar structure and seismic activity.

Solar Wind Spectrometer: Analyzed interactions between lunar surface and solar wind.

Laser Ranging Retroreflector: Arrays bounced laser pulses to Earth, enabling precise Earth-Moon distance measurements.

Lunar Surface Magnetometer: Measured Moon's intrinsic magnetic field.

Lunar Ejecta and Meteorites Experiment: Analyzed lunar dust particles and meteorites.

Radioisotope Thermoelectric Generator: Provided continuous electrical power for experiments.

Additional Terms

Mission Elapsed Time (MET): Time measured from solid rocket booster ignition.

Transearth Injection (TEI): Critical maneuver for returning to Earth's orbit.

Preface

In the early 1960s, the Apollo program emerged as a pivotal response to the intense rivalry between the United States and the Soviet Union during the Cold War. This competition, known as the Space Race, saw both nations striving to demonstrate their technological and ideological superiority through achievements in space exploration. The Apollo program followed Project Mercury, which had proven the feasibility of human spaceflight in Earth's orbit. Unlike Mercury's single-astronaut missions, Apollo was designed to accommodate three astronauts, aiming to land humans on the Moon and ensure their safe return.

NASA's Deputy Administrator Hugh L. Dryden introduced the Apollo program in July 1960 during Space Task Group conferences. The program initially envisioned a spacecraft with a separate mission module cabin housing equipment and propulsion systems, distinct from the command module, which contained piloting controls and reentry systems.

NASA launched a competition on August 30, 1960, to explore the feasibility of the proposed spacecraft. On October 25, study contracts were awarded to General Dynamics/Convair, General Electric, and the Glenn L. Martin Company. Concurrently, under Maxime Faget's leadership, NASA conducted internal spacecraft design studies to benchmark industry proposals.

The election of John F. Kennedy as President in November 1960 marked a turning point for the Apollo program. Amid Cold War tensions and concerns about American technological prowess, Kennedy emphasized space exploration as a symbol of national prestige. Initially hesitant about the financial costs, Kennedy's resolve to advance American space capabilities intensified after Yuri Gagarin's orbit of Earth in April 1961.

Kennedy's pivotal decision came on May 25, 1961, when he addressed Congress in a Special Message on Urgent National Needs. He boldly proposed landing a man on the Moon and returning him safely to Earth before the decade's end, setting in motion the ambitious goals of the Apollo program.

NASA faced monumental challenges in achieving Kennedy's lunar landing goal. Despite initial skepticism, Kennedy's directive focused on lunar exploration, surpassing earlier goals of space stations and circumlunar flights. NASA discarded the feasibility study designs and adopted Maxime Faget's integrated command and service module design as the basis for the Apollo spacecraft.

In October 1961, NASA initiated a new competition for spacecraft procurement bids based on Faget's design specifications. On November 28, 1961, North American Aviation emerged as the winner, chosen over Martin despite receiving a higher technical rating due to its established collaboration with NASA.

The Apollo program required unprecedented technological innovation and financial commitment, ultimately costing $25 billion, equivalent to approximately $182 billion in 2023. At its peak, the program employed 400,000 individuals and engaged over 20,000 industrial firms and academic institutions across the United States.

NASA established the Marshall Space Flight Center (MSFC) in Huntsville, Alabama, on July 1, 1960, to support the ambitious lunar missions. MSFC played a crucial role in designing the heavy-lift Saturn launch vehicles necessary for the Apollo missions, highlighting the scale and determination of America's pursuit of Kennedy's lunar vision.

President Kennedy's commitment to landing a man on the Moon before the decade's end, declared on May 25, 1961, marked a defining moment in American space policy. This declaration set the stage for the Apollo program's extraordinary lunar missions and solidified Kennedy's vision of American leadership in space exploration.

The Apollo program began with several uncrewed tests of the Saturn launch vehicle and the Command and Service Modules (CSM) to ensure the reliability and safety of future crewed missions:

Saturn I Launch Vehicle Tests: The Apollo program's initial development involved testing the Saturn I rocket. Between October 1961 and September 1964, ten uncrewed launches validated the rocket's design and capabilities, focusing on structural integrity, propulsion systems, and flight performance.

AS-201 (Apollo-Saturn 201): Launched on February 26, 1966, this mission was the first flight of the Saturn IB launch vehicle and the Block I CSM. It tested the spacecraft's heat shield and other systems, successfully demonstrating the performance of the launch vehicle and the CSM during suborbital flight.

AS-202 (Apollo-Saturn 202): Launched on August 25, 1966, AS-202 further tested the Saturn IB and the Block I CSM. It conducted a more extended suborbital flight, providing additional data on the spacecraft's heat shield, propulsion, and guidance systems.

AS-203 (Apollo-Saturn 203): Launched on July 5, 1966, AS-203 focused on testing the behavior of liquid hydrogen fuel in the Saturn IB's second stage in a zero-gravity environment. This mission was critical for understanding fuel management in space and contributed valuable information for future missions.

Apollo 1: Initially designated AS-204, Apollo 1 was intended to be the first crewed mission of the Apollo program. Tragically, on January 27, 1967, during a pre-flight test, a cabin fire claimed the lives of astronauts Gus Grissom, Ed White, and Roger B. Chaffee. This disaster led to significant design and safety changes in the Apollo spacecraft, emphasizing the importance of crew safety.

NASA's Apollo program, meticulously planned for lunar exploration, adopted a structured seven-step approach to achieve its ultimate goal of a Moon landing. This plan, established on September 20, 1967, outlined each mission phase, culminating in the historic lunar landing:

Apollo 4 and Apollo 6 (A missions) tested the Saturn V launch vehicle using uncrewed Block I versions of the Command and Service Module (CSM) in Earth orbit. These missions validated the Saturn V rocket's capabilities for future crewed missions.

Apollo 5 (B mission) focused on testing the Lunar Module (LM) in Earth orbit. It provided crucial data on the LM's performance and systems ahead of crewed missions.

Apollo 7 (C mission), scheduled for October 1968, was a crewed Earth-orbit flight solely testing the CSM. It aimed to ensure the spacecraft's readiness for more ambitious missions beyond Earth's orbit.

Apollo 8 (D mission), originally planned to test the LM in a low Earth orbit, evolved into a groundbreaking lunar orbital flight in December 1968. Led by Frank Borman, James Lovell, and William Anders, Apollo 8 aimed to orbit the Moon, paving the way for future lunar landings.

Following Apollo 8's success, James McDivitt commanded Apollo 9 (E mission). This mission, in early 1969, tested the LM in an elliptical medium Earth orbit, validating the LM's capabilities and operational readiness for lunar missions.

Apollo 10 (F mission) was designated to test both the CSM and LM in lunar orbit. With Thomas Stafford, John Young, and Eugene Cernan, this mission aimed to simulate all aspects of a lunar landing mission except for the actual touchdown. It served as a critical dress rehearsal for Apollo 11.

Apollo 11 (G mission), the culmination of NASA's efforts, achieved the first manned Moon landing on July 20, 1969. Led by Neil Armstrong, Buzz Aldrin, and Michael Collins, this historic mission fulfilled the final step of NASA's seven-step plan, marking a monumental achievement in human space exploration.

Following the success of Apollo 11, NASA continued with several more lunar missions, each building on the achievements of the previous ones and further expanding our understanding of the Moon:

Apollo 12, launched on November 14, 1969, was the sixth crewed mission in the Apollo program and the second to land on the Moon. Astronauts Charles Conrad and Alan L. Bean conducted extensive lunar surface experiments and retrieved parts of the Surveyor 3 spacecraft, which had landed on the Moon in April 1967.

Apollo 13, launched on April 11, 1970, was intended to be the third crewed mission to land on the Moon. However, an oxygen tank explosion en route to the Moon forced the mission to be aborted. The crew—James A. Lovell, John L. Swigert, and Fred W. Haise—managed to return safely to Earth, demonstrating remarkable ingenuity and resilience in the face of near-disaster.

Apollo 14, launched on January 31, 1971, saw astronauts Alan B. Shepard and Edgar D. Mitchell successfully land on the Moon's Fra Mauro highlands, the original target of the aborted Apollo 13 mission. They conducted two Moonwalks, collecting over 94 pounds of lunar samples and deploying scientific instruments.

Apollo 15, launched on July 26, 1971, was the first of the "J missions," which were capable of a longer stay on the Moon and greater surface mobility. David R. Scott and James B. Irwin explored the Hadley-Apennine region using the Lunar Roving Vehicle, greatly extending the range of their surface activities and collecting a significant amount of geological data.

Apollo 16, launched on April 16, 1972, targeted the Descartes Highlands. Astronauts John W. Young and Charles M. Duke conducted extensive lunar surface exploration and deployed a suite of scientific instruments. This mission provided critical insights into the lunar highlands' geology.

Apollo 17, launched on December 7, 1972, was the final mission of the Apollo program. It featured the first scientist-astronaut, Harrison H. Schmitt, a geologist, who joined Eugene A. Cernan in exploring the Taurus-Littrow region. They conducted three Moonwalks, collected a record 243 pounds of lunar samples, and left a legacy of scientific knowledge that continues to inform lunar research today.

Each phase of the Apollo program, from initial tests of launch vehicles and spacecraft in Earth orbit to the ultimate lunar landings, demonstrated NASA's meticulous planning, technological prowess, and the courage of astronauts who ventured into the unknown depths of space.

The Space Evolution

The evolution of space exploration has indeed seen a significant shift with the emergence of commercial astronauts alongside traditional government-sponsored astronauts. Historically, governmental space agencies like NASA or military programs exclusively selected and trained astronauts, focusing on missions crucial to national interests and scientific advancement.

While the term "astronaut" historically referred strictly to individuals trained by government space agencies, it has now expanded to include these commercially trained professionals. This broader definition encompasses individuals from diverse backgrounds—scientists, engineers, pilots, and even private citizens—selected and trained to conduct missions in space.

In the annals of aerospace history, the journey toward crewed space exploration was paved by the daring exploits and technological strides of many. Since the inception of the pioneering X-1 in 1946, which famously broke the sound barrier with Chuck Yeager at the helm, the United States has spearheaded a lineage of experimental aircraft aimed at testing the boundaries of flight.

Following the X-1's groundbreaking achievement, subsequent X-planes, such as the X-2 and X-15, pushed these boundaries even further. The Bell X-2, designed to explore speeds beyond Mach 3 and altitudes over 100,000 feet, wrestled with high-speed stability and thermal protection challenges. Meanwhile, the X-15, a rocket-powered marvel, soared to the edge of space, exceeding Mach 6 and reaching altitudes surpassing 350,000 feet. These feats expanded our understanding of hypersonic flight and reentry dynamics and earned their pilots the prestigious astronaut wings, marking pivotal strides toward manned spaceflight.

Among these pioneering pilots were icons like Neil Armstrong and Joseph Walker, whose missions aboard the X-15 provided crucial data that would later inform early spaceflight endeavors. Later, in the afternoon, joining NASA, Armstrong immortalized himself as the first human to set foot on the Moon during the historic Apollo 11 mission. Following his X-15 achievements, Walker continued to shape aerospace research with his contributions to space exploration.

In parallel, the Man in Space Soonest project, initiated by the USAF in 1958, aimed ambitiously at achieving crewed spaceflight ahead of the Soviet Union. Although the project was short-lived, its roster of distinguished test pilots, including Armstrong, Walker, and others, laid the groundwork for future achievements in space exploration. Armstrong, in particular, would etch his name in history with his lunar landing, forever cementing the United States' place in the cosmos.

On July 10, 2024, the last surviving X-15 pilot, Joe Engle passed away, marking the end of an era defined by bravery and technological breakthroughs. Their collective efforts, from breaking the sound barrier to setting foot on the Moon, shaped the trajectory of space exploration, underscoring the collaborative efforts between government and private sectors to expand humanity's reach beyond Earth.

The Mercury Seven, NASA's first group of astronauts selected in April 1959 for Project Mercury, stand as pillars of American space exploration history. These seven pioneers were chosen for their courage, intellect, and physical endurance, essential traits for pushing the boundaries of human exploration beyond Earth's atmosphere.

Scott Carpenter, Gordon Cooper, John Glenn, Gus Grissom, Wally Schirra, Alan Shepard, and Deke Slayton each contributed uniquely to the Mercury program and beyond. Scott Carpenter piloted Aurora 7, Gordon Cooper commanded Faith 7, and later flew on Gemini 5, while John Glenn's historic flight aboard Friendship 7 made him the first American to orbit the Earth. Tragically, Gus Grissom lost his life in the Apollo 1 fire, but not before flying Liberty Bell 7 and Gemini 3. Wally Schirra, who piloted Sigma 7, holds the distinction of flying in all three NASA spacecraft types: Mercury, Gemini (6A), and Apollo (7). Alan Shepard made history as the first American in space aboard Freedom 7, and later commanded Apollo 14, becoming the fifth person to walk on the Moon. Deke Slayton, medically disqualified from Mercury, flew on the Apollo-Soyuz Test Project in 1975.

The Mercury Seven laid the foundation for America's human spaceflight program. Their pioneering missions demonstrated the technical prowess needed for orbital flight and advanced spacecraft technology, paving the way for future missions in the Gemini, Apollo, and Space Shuttle programs.

In 1960, amidst the escalating space race between the United States and the Soviet Union, pivotal selections were made to shape the future of space exploration for both nations.

On March 7, the Soviet Union announced the selection of Air Force Group 1, comprising twenty experienced jet pilots from the Soviet Air Force. These individuals, including Yuri Gagarin, Valentina Tereshkova, and Gherman Titov, formed the initial cadre of Soviet cosmonauts. They were chosen for their exceptional piloting skills and underwent rigorous training to prepare for the challenges of space travel. As of 2024, Boris Volynov is the sole surviving member of this historic group.

In April of the same year, the United States secretly selected the first group of astronauts for the Dyna-Soar program, a pioneering initiative to develop a reusable spaceplane capable of both orbital flight and strategic military missions. The group included notable figures like Neil Armstrong, who later gained fame as the first person to walk on the Moon during the Apollo 11 mission, and William H. Dana. Armstrong and Dana departed the program in the summer of 1962, marking the conclusion of their involvement in Dyna-Soar.

William H. Dana, the last surviving member of Dyna-Soar Group 1, passed away in 2014. Their selection and training represented America's early efforts to match and surpass Soviet achievements in space exploration, laying the groundwork for subsequent manned space programs such as Gemini and Apollo. On both sides of the Cold War divide, these early pioneers propelled humanity toward new frontiers in the cosmos, forever altering the course of history and inspiring future generations of space explorers.

In 1962, pivotal selections in both the Soviet Union and the United States marked significant milestones in the global space race, reflecting rapid advancements in astronautics and setting the stage for historic achievements in space exploration.

On March 12, 1962, the Soviet Union expanded its cosmonaut training program by adding five civilian women with parachuting expertise. Among them, Valentina Tereshkova emerged as a trailblazer by becoming the first woman to fly in space aboard Vostok 6 in 1963. Her historic mission underscored Soviet leadership in gender equality in astronautics. Tatyana Kuznetsova, the youngest member at age 20, joined Valentina Ponomaryova, Irina Solovyova, and Zhanna Yorkina in contributing to the advancement of women in space exploration, despite not flying in space themselves.

Meanwhile, on September 17, 1962, NASA announced its second group of astronauts, famously known as the "Next Nine" or the "Nifty Nine." This group, designated as NASA Group 2, included Neil Armstrong, who later made history as the first person to walk on the Moon during the Apollo 11 mission. Other members—Frank Borman, Pete Conrad, Jim Lovell, Jim McDivitt, Elliot See, Tom Stafford, Ed White, and John Young—played pivotal roles in both the

Gemini and Apollo programs. Their achievements ranged from lunar walks to commanding groundbreaking space missions.

Tragically, Elliot See lost his life in a flight accident during preparations for Gemini 9, and Ed White perished in the Apollo 1 launchpad fire, highlighting the risks inherent in pushing the boundaries of space exploration.

Additionally, on September 19, 1962, Albert Crews joined the Dyna-Soar program, contributing to the advanced spaceplane missions aimed at orbital and strategic military roles. Although the Dyna-Soar program was eventually canceled, it provided crucial insights and technologies that influenced future space endeavors.

The 1963 selections marked significant advancements in the Soviet Union's and the United States' space capabilities. The Soviet Union and the United States intensified their efforts in space exploration by expanding their astronaut and cosmonaut groups, ushering in a new era of ambitious missions and technological advancements.

On January 10, 1963, the Soviet Union selected its second group of cosmonauts from the Air Force, known as Air Force Group 2. This cohort included Yuri Artyukhin, Eduard Buinovski, Lev Dyomin, Georgy Dobrovolsky, and others—comprising fifteen highly skilled individuals. These cosmonauts played crucial roles in various Soviet space missions and programs throughout the 1960s and 1970s, contributing significantly to the nation's space exploration efforts.

Meanwhile, on October 17, 1963, NASA announced its third group of astronauts, famously known as "The Fourteen." This group, designated as NASA Group 3, included notable figures such as Buzz Aldrin, William Anders, Charles Bassett, Alan Bean, and Eugene Cernan, among others. The Fourteen made substantial contributions to both the Gemini and Apollo programs, marking key milestones in American space exploration:

Buzz Aldrin, Alan Bean, Eugene Cernan, and David Scott walked on the Moon during various Apollo missions, demonstrating America's capability to achieve lunar exploration goals.

Five members—Aldrin, Cernan, Collins, Gordon, and Scott—also participated in the Gemini missions, conducting crucial tests and experiments in Earth's orbit.

Eugene Cernan became the only astronaut from this group to fly to the Moon twice, commanding Apollo 10 and Apollo 17, the final Apollo lunar landing mission.

Alan Bean further commanded the Skylab 3 mission, contributing to the success of America's first space station.

Tragically, the Apollo 1 fire claimed the life of Roger Chaffee, while Charles Bassett, Theodore Freeman, and Clifton Williams lost their lives in accidents involving NASA T-38 jet trainers. These losses underscored the risks inherent in pushing the boundaries of space exploration.

In 1964, the Soviet Union made significant expansions to its cosmonaut groups, reflecting their strategic advancements in space exploration:

On January 25, 1964, Georgi Beregovoi (1921–1995) was added to the Soviet Air Force cosmonaut group. Beregovoi would later make history as the commander of Soyuz 3 in 1968, becoming the first Soviet cosmonaut to fly alone in space. His mission marked a significant achievement in Soviet spaceflight capabilities.

On May 26, 1964, the Soviet Union selected the Voskhod Group, also known as Medical Group 1. This group included Vladimir Benderov, Georgy Katys, Vasili Lazarev, Boris Polyakov, Aleksei Sorokin, and Boris Yegorov. Boris Yegorov notably became the first physician to venture into space when he flew aboard Voskhod 1 in October 1964. This mission was the first to carry more than one crew member and showcased Soviet expertise in manned space missions.

On June 11, 1964, Konstantin Feoktistov (1926–2009) was selected as part of the Civilian Specialist Group 1. Feoktistov served as the flight engineer on Voskhod 1, contributing crucial expertise to the mission's success. Voskhod 1 demonstrated Soviet capabilities in multi-crew spaceflights, advancing their achievements in space exploration.

These selections in 1964 underscored the Soviet Union's commitment to expanding and diversifying its cosmonaut corps, focusing on specialized skills and capabilities required for their ambitious space missions.

In 1965, both the Soviet Union and the United States made significant selections in their astronaut and cosmonaut groups, reflecting their continued advancements in space exploration:

On June 1, 1965, three civilian journalists—Yaroslav Golovanov, Yuri Letunov, and Mikhail Rebrov—were selected for cosmonaut training to fly on a Voskhod mission. However, when the Voskhod program was canceled, Golovanov and Letunov were dismissed. Rebrov remained associated with the Soviet space program as a journalist until 1974, contributing to the documentation and public understanding of Soviet space achievements.

Also on June 1, 1965, three physicians—Yevgeni Illyin, Aleksandr Kiselyov, and Yuri Senkevich—were selected for long-duration Voskhod flights. These missions were ultimately canceled to prioritize the Soviet Moon program. Consequently, all three physicians were dismissed from the cosmonaut corps at the beginning of the following year, despite their training and readiness for spaceflight.

On June 28, 1965, NASA selected its fourth group of astronauts, famously known as "The Scientists." This group included Owen Garriott, Edward Gibson, Duane Graveline, Joseph Kerwin, Curt Michel, and Harrison Schmitt. Notable achievements from this group include:d

Harrison Schmitt walked on the Moon during Apollo 17.

Owen Garriott flew on Skylab and later on Space Shuttle flight STS-9, becoming the first amateur radio operator to operate from orbit.

Edward Gibson and Joseph Kerwin also flew on Skylab missions, contributing to extended stays in space and scientific research.

On October 28, 1965, the Soviet Union selected Air Force Group 3, comprising many cosmonauts. This group was intended to participate in various Soyuz programs, including military applications and lunar missions. However, due to program cancellations and adjustments, only a few from this group had the opportunity actually to fly in space. The Soviet space program evolved significantly during this period, focusing on orbital missions and space station programs.

In November 1965, the US Air Force selected Group 1 for the Manned Orbiting Laboratory (MOL) program. Members included Michael J. Adams, Albert H. Crews Jr., John L. Finley, Richard E. Lawyer, Lachlan Macleay, Francis G. Neubeck, James M. Taylor, and Richard H. Truly. Richard Truly later transferred to NASA after the cancellation of the MOL program and flew on the Space Shuttle. He became the NASA Administrator in 1989, marking a significant leadership role in American space exploration.

In 1966, the United States began the Apollo Project to put 14 astronauts on the Moon's surface.

The Launch Vehicles

Before the inception of the Apollo program, Dr. Wernher von Braun and his dedicated team of rocket engineers were deeply immersed in visionary projects aimed at conquering the vast frontier of space. Central to their plans were the development of formidable launch vehicles, the Saturn and the even more ambitious Nova series. These rockets were envisaged to be colossal in scale, capable of carrying monumental payloads into the depths of space.

The Marshall Space Flight Center's first Saturn I vehicle, SA-1, lifts off from Cape Canaveral, Florida, on October 27, 1961. This early configuration, Saturn I Block I, 162 feet tall and weighing 460 tons, consisted of the eight H-1 engines S-I stage and the dummy second stage (S-IV

Von Braun's pivotal role shifted dramatically when he transitioned from military service in the Army to a leadership position at NASA's Marshall Space Flight Center. As director, he steered rocketry and space exploration toward unprecedented heights.

Initially, the Apollo program was conceived around a direct ascent strategy. This audacious plan entailed launching a three-person crew aboard the Apollo command and service module directly toward the lunar surface, perched atop a massive descent rocket stage. Such an endeavor demanded the capabilities of a Nova-class launcher capable of hoisting payloads exceeding 180,000 pounds (82,000 kg).

SA-1 before launch

However, a critical decision made on June 11, 1962, marked a pivotal shift in Apollo's trajectory. NASA opted for lunar orbit rendezvous (LOR) instead, a decision that revolutionized the mission architecture. This strategic pivot enabled the Marshall Space Flight Center to develop the Saturn V rocket, supplanting the need for the Nova series and setting the stage for the epic lunar missions that would follow.

NASA adopted a systematic approach to standardizing its mission planning. Each spacecraft-launch vehicle combination was assigned a specific series number: AS-10x for Saturn I, AS-20x for Saturn IB, and AS-50x for Saturn V. This departure from sequential numbering, as seen in Project Gemini, reflected NASA's meticulous planning and readiness to embark on the historic human flights that defined the Apollo era.

Following the lessons learned from the Mercury program, which necessitated a robust launch escape system (LES) for crew safety, the Apollo program similarly required thorough testing. To qualify this critical system, NASA needed a specialized rocket larger than the Mercury-era Little Joe. Thus, General Dynamics/Convair developed the Little Joe II, designed explicitly for qualification flight testing. After a successful qualification test flight in August 1963, four additional test flights of the LES (A-001 through A-004) were conducted at White Sands Missile Range between May 1964 and January 1966.

Meanwhile, the development of the Saturn rocket family played a pivotal role in shaping Apollo's capabilities. The Saturn I, NASA's first heavy-lift launch vehicle, was initially intended for low Earth orbit tests of partially equipped Command and Service Modules (CSMs). Its first stage, powered by eight Rocketdyne H-1 engines burning RP-1 and liquid oxygen (LOX), generated a thrust of 1,500,000 pounds-force (6,670 kN). The second stage, S-IV, employed six Pratt & Whitney RL-10 engines fueled by liquid hydrogen, each producing 90,000 pounds-force (400 kN) of thrust. The S-V third stage was flown passively on four Saturn I launches.

The Saturn I launch vehicle was initially designed to carry crewed Command Module flights into low Earth orbit. However, its payload capacity was constrained to 20,000 pounds (9,100 kg), which needed to be increased to lift even a partially fueled service module. To address this limitation, there were considerations for developing a lightweight retrorocket module to enable deorbiting, but these plans proved impractical and were eventually abandoned.

Instead, the uprated Saturn IB emerged as the preferred choice for crewed Earth orbit tests. This modified version could launch the Command Module paired with a half-fueled Service Module, aligning better with operational requirements. Consequently, Saturn I's role was refocused primarily on advancing Saturn launch vehicle development, conducting tests with boilerplate Command and Service Modules, and deploying three micrometeoroid satellites crucial for Apollo missions.

On October 27, 1961, the Saturn-Apollo 1 (SA-1) mission marked a historic milestone as the inaugural flight of the Saturn I space launch vehicle, serving as the pioneering mission of the American Apollo program. Launching from Cape Canaveral, Florida, SA-1 represented a quantum leap in size and power compared to its predecessors. Standing three times taller and requiring six times more fuel than the earlier Juno I rocket that launched Explorer 1, America's first satellite in 1958, the Saturn I boasted ten times the thrust capacity.

NASA's approach to the SA-1 mission reflected the cautious testing methodologies of the time. It opted against all-up testing, where entire systems are validated in a single launch. Instead, each stage of the rocket was slated for separate test flights. For SA-1, only the S-I first stage was live. It aimed to evaluate the structural integrity of the launch vehicle during a suborbital flight, utilizing a nose cone inherited from a Jupiter rocket.

Preparations for SA-1 were meticulous and novel. It marked the first occasion a rocket stage had been transported to Cape Canaveral by barge, signaling the feasibility for future Saturn rockets. Despite challenges encountered during its voyage, such as grounding incidents due to inaccurate nautical charts and minor damage from a bridge collision, the first and dummy upper stages arrived successfully on August 15, 1961.

At Pad 34, the booster erection proceeded smoothly five days later, albeit slightly behind schedule amidst manual testing procedures. Unlike today's automated protocols, testing involved manually activating switches in the control center to gauge the rocket's responses.

On the day of launch, October 27, 1961, at 12:30 p.m. EST, RP-1 propellant began flowing into the rocket's tanks, initially exceeding requirements by 3% to allow for easy drainage if needed. Liquid oxygen followed at 3:00 a.m. the next day, filling stages in stages from 10% to 97%, before topping off gradually to ensure leak checks were effective.

Despite minor delays caused by adverse weather conditions, SA-1 launched only an hour past its scheduled time. Initial assessments had given the rocket a 75% chance of liftoff and a mere 30% likelihood of completing a nominal flight without incident. Even during nominal conditions, potential damage was anticipated, reflecting cautious optimism in the face of unprecedented technological challenges.

While deemed successful, the launch itself bore a peculiar note: witnesses noted the sound was notably subdued compared to expectations set by Atlas rocket launches. Atmospheric conditions were later identified as a factor in attenuating sound propagation, particularly notable in comparisons drawn between Cape Canaveral and the Redstone Arsenal.

The flight trajectory achieved remarkable precision, reaching an altitude of 136.5 km and impacting 345.7 km downrange in the Atlantic Ocean. However, the only technical glitch noted was the premature cutoff of engines, attributed to a 400 kg surplus of liquid oxygen and a 410 kg deficit of RP-1—despite SA-1 being loaded at 83% of its propellant capacity for safety during this test flight.

SA-1 thus inaugurated the Saturn I's journey, laying crucial groundwork for subsequent missions in the Apollo program and NASA's broader ambitions in space exploration.

SA-2: On April 25, 1962, the Saturn-Apollo 2 (SA-2) mission marked a critical milestone as the second flight of the Saturn I launch vehicle and the inaugural mission of Project Highwater within the American Apollo program. Launching from Cape Canaveral, Florida, SA-2 built upon the foundational successes of its predecessor, SA-1, while introducing new elements crucial to advancing space exploration.

The preparations for SA-2 commenced on February 27, 1962, with the arrival of the second Saturn I launch vehicle at Cape Canaveral. This mission saw notable enhancements over SA-1, including additional baffles in the propellant tanks to minimize fuel sloshing—a key lesson learned from previous testing. Despite encountering minor technical issues during launch preparations, such as a detected leak and operational glitches in guidance systems and service structures, none necessitated a delay beyond the scheduled April 25 launch date.

SA-2 lifted off precisely at 14:00:34 UTC from Launch Complex 34, following a brief 30-minute hold due to a vessel entering the flight safety zone downrange. The rocket carried 619,000 pounds (281,000 kg) of propellant, optimized to approximately 83% of its maximum capacity.

During its brief yet impactful flight, the H-1 engines powered the rocket to an altitude of 35 miles (56 km), achieving a maximum velocity of 3,750 miles per hour (6,040 km/h; 1,680 m/s). Officials initiated a terminate command approximately 2 minutes and 40 seconds after liftoff, triggering charges that led to the rocket's controlled destruction at 65.4 miles (105.3 km).

The primary objectives of SA-2 mirrored those of SA-1, focusing on validating the Saturn I rocket's propulsion performance, structural design, aerodynamics, and the efficacy of its guidance and control systems. Notably, SA-2 also served as the inaugural mission for Project Highwater, designed to study Earth's ionosphere and observe phenomena such as noctilucent clouds and the behavior of ice in space. The mission involved intentional releases of ballast water from the upper stages—approximately 190,000 pounds

(86,000 kg)—to simulate future payload masses.

Upon termination, the dynamite charges split the second stage longitudinally, instantly dispersing its water load, while primacord charges punctured the third stage, gradually releasing its water over several seconds. Ground cameras captured the resulting water cloud, which ascended to a maximum altitude of 100 miles (161 km), generating lightning-like effects observed as a "synthetic thunderstorm" in space—a phenomenon noted by Dr. Wernher von Braun.

SA-2 achieved all its objectives, marking another significant step forward in NASA's journey toward lunar exploration and expanding humanity's understanding of space.

SA-3: On November 16, 1962, Saturn-Apollo 3 (SA-3) marked a significant milestone as the third flight of the Saturn I launch vehicle, following its predecessors SA-1 and SA-2. This mission, integral to the American Apollo program, aimed to test further and validate key components crucial for future space exploration efforts.

Preparations for SA-3 began with the delivery of the Saturn I launch vehicle components to Cape Canaveral on September 19, 1962. However, the initial assembly was delayed until September 21 due to adverse weather conditions caused by a tropical depression. The dummy second and third stages (S-IV and S-V) and payload were integrated into the booster by September 24. Following assembly, ballast water was loaded into the dummy stages on October 31, and RP-1 fuel was loaded on November 14.

The launch, scheduled for November 16, 1962, from Launch Complex 34 at Cape Canaveral, experienced a 45-minute delay due to a power failure in ground support equipment. Nevertheless, SA-3 successfully lifted off at 17:45:02 UTC, carrying a full load of approximately 750,000 pounds (340,000 kg) of propellant—a significant increase compared to previous flights.

During its ascent, SA-3's eight H-1 engines performed admirably, with the four inner engines shutting down at 2 minutes 21.66 seconds after launch, followed by the outer engines at 2 minutes 29.09 seconds. This allowed the rocket to achieve a maximum velocity of 4,046 miles per hour (6,511 km/h). The vehicle continued its trajectory until officials initiated a terminate command at 4 minutes 52 seconds after launch, leading to the controlled destruction of the dummy upper stages at an altitude of 103.91 miles (167.22 km).

The primary objectives of SA-3 aligned closely with those of its predecessors. They focused on testing and validating the first-stage booster (S-I) and its H-1 engines, as well as assessing ground support equipment and overall vehicle performance in flight. Additionally, SA-3 continued the experiments of Project Highwater, involving the intentional release of ballast water from the second and third stages to study Earth's ionosphere and other atmospheric phenomena.

For Project Highwater, SA-3's dummy upper stages were filled with 192,528 pounds (87,329 kg) of water, which was released upon command to generate observable effects in space. The resulting ice cloud, tracked by ground cameras and aircraft, provided valuable data on atmospheric interactions and was visible for several seconds, confirming the experiment's success.

SA-3 also incorporated ten special tests on advancing technologies and procedures for future Apollo missions. These included propulsion tests with full propellant loads, the first use of retrorockets on Apollo hardware for stage separation simulations, and testing new telemetry and instrumentation systems crucial for automated spacecraft operations.

Despite minor telemetry issues during flight, SA-3 achieved all engineering goals and validated critical systems for the Apollo program's future missions. The mission's success underscored NASA's growing confidence in the Saturn I launch vehicle and laid essential groundwork for subsequent advancements in space exploration.

SA-4: Saturn-Apollo 4 (SA-4) marked a pivotal milestone in the early phases of the Apollo Program. It was the fourth launch of the Saturn I launch vehicle. This mission was crucial for testing the capabilities and structural integrity of the rocket's S-I first stage, a necessary step before advancing to more complex orbital missions.

One of the primary objectives of SA-4 was to simulate the scenario of an engine failure mid-flight. Approximately 100 seconds after liftoff, one of the engines was deliberately shut down to assess the rocket's ability to redistribute fuel among the remaining engines. This critical test was a precursor to similar maneuvers successfully executed during later Apollo missions using the larger Saturn V rockets, notably on Apollo 6 and the dramatic Apollo 13 mission.

In preparation for future flights, SA-4 also enhanced its dummy second stage, incorporating the aerodynamic design elements of the actual second stage. This included installing vent ducts, fairings, and dummy camera pods alongside antennas tailored for the rocket's Block II configuration.

Notable operational achievements and challenges characterized the launch itself. Despite experiencing the longest series of holds in any previous mission, totaling 120 minutes, SA-4 proceeded smoothly through its initial phase. Following the planned engine shutdown, the rocket demonstrated flawless performance, validating the efficacy of its fuel rerouting system. Contrary to concerns, the shutdown engine did not suffer heat-related disintegration due to a lack of cooling propellant, a reassuring outcome affirming the robustness of the clustered engine design.

SA-4 achieved a peak altitude of 129 kilometers and attained a maximum velocity of 5,906 kilometers per hour. Additionally, the mission included a crucial test of retrorockets designed for future stage separation maneuvers, although stage separation was intended on something other than this occasion. Instead, the retrorockets' successful firing confirmed their readiness for upcoming missions where separation would be pivotal.

In summary, Saturn-Apollo 4 exemplified meticulous testing and pivotal advancements essential to the Apollo Program's pursuit of lunar exploration, setting a solid foundation for subsequent missions that would ultimately lead humanity to the Moon.

SA-5: Saturn-Apollo 5 (SA-5) marked a pivotal advancement in the Apollo Program. President Kennedy hailed it as a crucial step toward asserting U.S. leadership in space following the Soviet Union's early successes with Sputnik. Launched in 1964, SA-5 introduced significant upgrades that set the stage for future lunar missions.

Saturn SA-4 on the pad

The key innovation of SA-5 was the integration of the Block II Saturn I configuration, featuring a two-stage system.

The first stage, now enlarged to accommodate its full 340,000 kilograms of propellant, boasted eight upgraded engines generating 188,000 pounds of thrust each. This enhancement was complemented by adding eight fins, enhancing stability during flight. Notably, SA-5 carried a Jupiter nosecone rather than a full Apollo spacecraft, emphasizing its role as a developmental mission rather than a crewed flight.

The second stage represented a leap forward, powered by six engines burning liquid hydrogen—a design (RL10) initially intended for the Centaur upper stage and delivered to the launch site via the unique Aero Spacelines Pregnant Guppy aircraft. This stage marked the first operational use of liquid hydrogen propulsion in the Apollo Program, showcasing advanced technology critical for future deep space missions.

Critical to the mission's success was the positioning of the guidance and control computer above the second stage, a configuration mirroring future Saturn V flights. This Instrument Unit autonomously managed the rocket's ascent, compensating dynamically for atmospheric conditions and thrust fluctuations—a precursor to the complex guidance systems of later lunar missions.

SA-5 achieved another milestone by becoming the first orbital mission of the Apollo Program. Its powerful first and second stages propelled it into an elliptical orbit, demonstrating the capability to place substantial payloads into space. This achievement was underscored by President Kennedy's acknowledgment in a speech shortly before his tragic assassination, highlighting SA-5 as a pivotal step toward U.S. space dominance.

The launch itself was challenging. A technical issue during the initial fill of liquid oxygen necessitated a brief postponement, highlighting spaceflight preparation's intricacies and rigorous protocols. However, once launched on January 29, 1964, SA-5 performed flawlessly. It transmitted extensive telemetry data and was tracked by telescopes and cameras throughout its flight. Stage separation, captured by multiple cameras, proceeded smoothly, validating the functionality of retrorockets and ullage systems critical for future missions.

Dr. Wernher von Braun explains the Saturn Launch System to President John F. Kennedy — at Cape Canaveral, Florida. NASA Deputy Administrator Robert Seamans is to the left of von Braun. On 16 November 1963, six days before the President's assassination.

Upon reaching orbit, the second stage briefly became the largest satellite in orbit at the time, weighing nearly 17,000 kilograms—a feat that underscored America's capacity to develop and deploy large-scale launch vehicles comparable to those of its Cold War rival.

SA-6: AS-101, also known as SA-6, marked a pivotal moment in the Saturn I launch series and the Apollo Program, lifting off from Cape Kennedy Air Force Station Space Launch Complex 37B on May 28, 1964. This mission was significant as it carried the first boilerplate Apollo spacecraft, BP-13, into low Earth orbit, initiating a crucial testing phase for the Apollo Command and Service Module (CSM) and its Launch Escape System (LES) tower.

The launch was challenging. Initially scheduled for earlier attempts, the first launch was scrubbed due to liquid oxygen contamination from a damaged wire mesh screen. In contrast, the second attempt faced a setback when the rocket's guidance system overheated. Finally, on the third attempt, AS-101 successfully soared into the skies.

The ascent proceeded nominally until 116.9 seconds after liftoff when an unexpected anomaly occurred—engine number eight of the Saturn I's first stage shut down prematurely. This unplanned event triggered an automatic compensation by the guidance system, extending the burn of the remaining seven engines by 2.7 seconds beyond the intended duration. Despite this setback, the first stage separated as planned, allowing the second stage to ignite seamlessly. Shortly after that, the Launch Escape System was jettisoned as scheduled, followed by the release of film cameras from the first stage, documenting the separation process.

Throughout its flight, AS-101 performed admirably, with the second stage cutting off slightly earlier than predicted after transmitting valuable data for four orbits. The boilerplate Apollo spacecraft, weighing 17,000 pounds and resembling the CSM in size and shape, continued its mission in a stable orbit for 54 orbits, providing extensive telemetry until its batteries depleted.

Post-flight analysis revealed that stripped teeth caused engine number eight's premature shutdown on one of the turbopump gears—a mechanical issue promptly addressed by engineers. Fortunately, this isolated incident did not delay subsequent launches, as improvements to the gear design were already in progress.

SA-7: AS-102, also known as SA-7, marked another crucial milestone in the Saturn I launch series and the Apollo Program, lifting off from Cape Kennedy, Florida on September 18, 1964. This mission aimed to replicate the success of AS-101 by carrying the boilerplate Apollo spacecraft, BP-15, into low Earth orbit, continuing vital testing for the Apollo Command and Service Module (CSM) and its systems.

This photograph depicts an intense moment during the SA-6 launch at the Firing Room. Dr. von Braun, Director of the Marshall Space Flight Center (MSFC) is at center; to his left is Dr. George Mueller, Associate Director for Manned Space Flight; and far right is Dr. Eberhard Rees, Director for Research and Development, MSFC. The SA-6, the sixth flight of the Saturn 1 vehicle, launched a S-IV stage (a second stage) and an Apollo boilerplate spacecraft.

The objectives of AS-102 closely mirrored those of its predecessor, AS-101. BP-15, similar to BP-13 flown on AS-101, featured instrumentation on one of its simulated reaction control system thrusters to monitor launch temperatures and vibrations. Notably, AS-102 introduced the ST-124 programmable guidance computer aboard the Saturn rocket, a significant upgrade from the preprogrammed "black box" used in earlier launches. This advancement allowed for in-flight reprogramming, enhancing flexibility in response to any unexpected behaviors during the mission.

Preparation for AS-102 encountered challenges, including the discovery of a small crack in engine number six in early July. This necessitated the removal of the engine and the inspection of all eight engines, a meticulous process involving numerous connections that caused a delay of approximately two weeks.

Further setbacks followed due to Hurricanes Cleo and Dora, extending the launch timeline.

Apollo Command Module boiler plate #15 checked out prior to SA-7 launch on September 18, 1964

When AS-102 finally launched on September 18, it demonstrated robust performance. The first stage burned for 147.7 seconds before separation, followed swiftly by the ignition of the second stage and the planned jettison of the Launch Escape System (LES). The second stage burned until +621.1 seconds, achieving an orbit with parameters of 212.66 by 226.50 kilometers.

Throughout its flight, BP-15 transmitted telemetry data for five orbits, successfully meeting all mission objectives. However, the recovery of eight film-camera pods, which were intended to document stage separation but landed unexpectedly far downrange, posed a challenge exacerbated by Hurricane Gladys. Despite this, two pods eventually washed ashore two months later, their film miraculously intact despite being encrusted with barnacles.

AS-103, also known as SA-9, was conducted on February 16, 1965, and represented a significant milestone in the Apollo program and space exploration history. This mission marked the third orbital flight test of a boilerplate Apollo spacecraft. It was the inaugural flight of the Pegasus micrometeoroid detection satellite, demonstrating NASA's increasing capabilities with the Saturn I launch vehicle.

Objectives: AS-103 had a comprehensive set of objectives, totaling 12. Two main objectives were centered around the Pegasus satellite: first, to demonstrate the operational functionality of its mechanical, structural, and electronic systems, and second, to evaluate the sampling of meteoroid data in near-Earth orbit. The launch trajectory for AS-103 was specifically designed to place the Pegasus satellite into its intended orbit, differing significantly from previous missions like AS-101 and AS-102.

Launch: The launch configuration included an S-I first stage, an S-IV second stage, and an instrument unit. The spacecraft itself consisted of a boilerplate command and service module (BP-16), a launch escape system, and a service module/launch vehicle adapter. The Pegasus 1 satellite was housed within the service module, attached to the S-IV stage. The launch occurred from Cape Kennedy Launch Complex 37B at 9:37:03 a.m. EST, following a hold of 1 hour and 7 minutes due to a power failure in the Eastern Test Range flight safety computer. Another built-in hold of 30 minutes allowed for a battery check on the Pegasus satellite.

The launch proceeded smoothly, with the spacecraft successfully inserted into orbit approximately 10.5 minutes after liftoff. During launch, the launch escape system and subsequently the command module were jettisoned as planned. The Pegasus satellite, weighing about 3,980 pounds (1,810 kg) and measuring 208 by 84 by 95 inches (5.3 by 2.1 by 2.4 m) with deployed wings spanning 96 feet (29 m), was deployed into an orbit with a perigee of 307.8 miles (495.4 km), an apogee of 461.9 miles (743.4 km), and an orbital inclination of 31.76°.

Results: AS-103 achieved all of its mission objectives successfully. The trajectory and space-fixed velocity closely matched the planned parameters. The shroud separating the Apollo spacecraft from the Pegasus satellite occurred approximately 804 seconds after liftoff, followed by the deployment of the satellite's meteoroid detection panel wings one minute later. The Pegasus A satellite operated nominally until decommissioned on August 29, 1968, with a predicted useful lifetime of 1188 days. Despite minor malfunctions encountered during the mission, both in the launch vehicle and the satellite, AS-103 contributed significantly to NASA's understanding of orbital operations and micrometeoroid detection.

The spacecraft remained in orbit until July 10, 1985, when it re-entered the atmosphere and safely landed in the ocean, concluding a successful mission that paved the way for further advancements in space technology and exploration.

AS-104, also known as SA-8, was an important mission in the Saturn I series. It served as the fourth orbital test of a boilerplate Apollo spacecraft and the second flight of the Pegasus micrometeoroid detection satellite. Launched on May 25, 1965, this mission continued to expand NASA's capabilities and test new technologies critical for the Apollo program.

Objectives: The primary objective of AS-104 was to demonstrate the Saturn launch vehicle's iterative guidance mode and evaluate its system accuracy. Similar to its predecessor, AS-103, this mission aimed to refine the launch vehicle's capabilities regarding guidance and control.

The payload configuration for AS-104 mirrored that of AS-103, with notable differences. Specifically, a single reaction control engine assembly was mounted on the boilerplate service module (BP-26). This assembly was instrumented to gather additional data on launch environment temperatures. Unlike the engines used in AS-101, two of the four in this configuration were of a prototype design, highlighting ongoing refinements in engine technology.

Launch: AS-104 marked a significant milestone as the first nighttime launch in the Saturn I series. A built-in 35-minute hold ensured that the launch occurred within the optimal window.

The launch occurred from Cape Kennedy Launch Complex 37B at 2:35:01 a.m. EST (07:35:01 GMT) on May 25, 1965. The mission proceeded nominally, with the payload successfully inserted into orbit approximately 10.6 minutes after liftoff. The total mass placed in orbit, including the spacecraft, Pegasus B satellite, adapter, instrument unit, and S-IV stage, amounted to 34,113 pounds (15,473 kg). The orbit achieved had a perigee of 314.0 miles (505 km), an apogee of 464.1 miles (747 km), and an orbital inclination of 31.78°. The Pegasus 2 satellite, weighing 1397 kilograms (3080 pounds), was stowed inside the boilerplate's service module and remained attached to the S-IV stage.

Results: AS-104 encountered minor malfunctions in the S-I stage propulsion system, yet all mission objectives were successfully achieved. The actual trajectory closely matched the predicted path, facilitating the spacecraft's separation 806 seconds after liftoff. This mission contributed valuable data to NASA's ongoing efforts to refine the Saturn launch vehicle and its associated technologies, paving the way for future advancements in space exploration and the Apollo program.

AS-105 was the fifth and final orbital flight of a boilerplate Apollo spacecraft and the third and final launch of a Pegasus micrometeoroid detection satellite. It was launched by SA-10, the tenth and final Saturn I rocket, in 1965.

Overview

AS-105 was an Apollo boilerplate spacecraft; boilerplate BP-9A was used for the flight. The spacecraft reentered on November 22, 1975.[2] The Saturn launch vehicle (SA-10) was similar to those of missions AS-103 and AS-104. As on the previous mission, the boilerplate service module was equipped with a test installation of a reaction control engine package.

The primary flight objective was to continue demonstrating the launch vehicle's iterative guidance mode and evaluating system accuracy.

Launch

AS-105 was launched from Cape Kennedy Launch Complex 37B at 08:00 EST (13:00 GMT) on July 30, 1965, on the last Saturn I rocket, SA-10. A planned thirty-minute hold ensured that the launch time coincided with the opening of the Pegasus launch window. The launch was normal and the payload was inserted into orbit approximately 10.7 minutes after lift-off. The total mass placed in orbit, including the spacecraft, Pegasus spacecraft, adapter, instrument unit, and S-IV stage, was 34,438 pounds (15,621 kg).

The spacecraft was separated 812 seconds after lift-off, and the separation and ejection system operated as planned. The Pegasus 3 spacecraft, which was attached to the S-IV stage of the Saturn I and stowed inside the boilerplate service module, was deployed 40 seconds after command initiation, at 872 seconds. Pegasus 3 was a 1423.6 kilogram (3138.6 pounds) micrometeoroid detection satellite bolted to the S-IV.

In the early stages of Project Apollo, the sequencing and naming of the initial uncrewed Apollo-Saturn (AS) missions presented some initial confusion. Originally, AS-204 was slated to be the first crewed mission, following three uncrewed test flights. Tragically, AS-204 was later redesignated as Apollo 1 after a fatal fire claimed the lives of its crew during a routine test and training session on the launch pad.

Following this setback, NASA resumed its uncrewed Apollo missions to rigorously test the Saturn V launch vehicle and the Lunar Module. These missions, designated Apollo 4, Apollo 5, and Apollo 6, aimed to validate crucial systems and procedures for the upcoming crewed missions. As a result of the fire, the first crewed Apollo mission was postponed until Apollo 7.

Interestingly, the initial three uncrewed flights were never formally given "Apollo" designations, though there was brief contemplation of renaming AS-201, AS-202, and AS-203 as Apollo 1-A, Apollo 2, and Apollo 3, respectively. However, this idea did not materialize into official designations.

Meanwhile, the Saturn IB rocket, an upgraded version derived from the Saturn I, played a pivotal role in the Apollo program. By replacing the S-IV second stage with the more powerful S-IVB, which would later serve as the third stage for the Saturn V, NASA significantly boosted its payload capacity to approximately 46,000 pounds (21,000 kg). This enhancement allowed for critical maneuvers such as orbiting a Command Module with a partially fueled Service Module or deploying a fully loaded Lunar Module.

During this phase, NASA conducted several pivotal tests, including two suborbital trials of the Apollo Block I Command and Service Module, a development test of the S-IVB stage, and a successful test of the Lunar Module. The triumph of the Lunar Module test prompted NASA to cancel a second uncrewed flight, underscoring confidence in the vehicle's readiness for lunar missions.

AS-201

In February 1966, NASA embarked on a pivotal milestone in its Apollo program with the AS-201 mission. This mission marked the inaugural flight of the Saturn-IB rocket, a crucial step toward fulfilling President John F. Kennedy's ambitious lunar landing goal by the decade's end.

The AS-201 mission adopted an innovative "all-up" approach, a comprehensive test philosophy that validated all Apollo Command and Service Module (CSM) components in a single, uncrewed flight. The towering Saturn-IB, propelled by its formidable first stage generating 1.6 million pounds of thrust from eight upgraded H-1 engines, lifted off from Kennedy Space Center's Launch Complex 34. Its mission: to rigorously assess the integrated system's capabilities and readiness for future crewed lunar missions.

On February 26, 1966, the AS-201 mission finally commenced from Launch Pad 34 following several delays. As the Saturn-IB rocket cleared the tower, control transitioned from Cape Kennedy Air Force Station to NASA's Mission Control Center at the Manned Spacecraft Center in Houston, overseen by Flight Director Glynn S. Lunney and his engineers. This critical shift marked the beginning of meticulous monitoring of every aspect of the mission.

AS-201 launch

Recovery swimmers from the prime recovery ship the U.S.S. Boxer (LPH-4) preparing the AS-201 Command Module (CM) for retrieval.

The Saturn-IB performed admirably, with both stages executing their roles flawlessly, propelling the Apollo Command and Service Module (CSM) to a peak altitude of 303 miles on a precise suborbital trajectory. A strategically placed camera inside the first stage captured key moments: the dramatic separation of stages, firing of the S-IVB stage's ullage motors to settle propellants, ignition of the J-2 engine, and jettison of the Launch Escape System (LES).

The U.S.S. Boxer approaches the CM for retrieval.

After separation from the rocket, the Apollo spacecraft's Service Propulsion System (SPS) engine initiated a crucial burn to adjust its reentry velocity. However, the ingestion of helium into the propellant lines caused the engine to deliver lower thrust than anticipated during this critical 104-second burn. A subsequent 10-second burn to test the engine's restart capability was similarly affected by this issue.

Despite these propulsion challenges, the spacecraft executed a turnaround maneuver as planned, and the Command Module (CM) separated from the Service Module (SM) to initiate its reentry sequence. Although a fault in the electrical power system caused a loss of control, resulting in an unintended roll during reentry, the CM's heat shield performed impeccably, shielding the capsule from the intense heat of atmospheric reentry.

Following a nail-biting descent, the spacecraft deployed its parachutes precisely as scheduled and splashed down in the south Atlantic Ocean, approximately 200 miles west of Ascension Island—a deviation of 46 miles from the intended target.

Following the successful recovery of the Command Module (CM) from the AS-201 mission by swimmers aboard the U.S.S. Boxer (LPH-4), the capsule was hoisted aboard the carrier approximately three hours after liftoff. The U.S.S. Boxer then sailed to Norfolk, Virginia, reaching port on March 6th. From there, the CM was transported to the North American Aviation (NAA) plant in Downey, California, where it underwent thorough postflight inspections.

In 1968, CM-009 was utilized for land impact tests at the Manned Spacecraft Center (MSC), contributing further to NASA's ongoing evaluations of spacecraft performance and durability.

At the time of the AS-201 mission, NASA had planned two additional uncrewed Apollo Saturn-IB flights to certify both the rocket and spacecraft for manned missions. These plans were tragically altered on January 27, 1967, during a pre-launch test for Apollo 1, resulting in the loss of astronauts Gus Grissom, Ed White, and Roger B. Chaffee in a fire aboard the spacecraft. This devastating event profoundly impacted NASA's approach to crew safety and mission readiness, prompting extensive changes in spacecraft design and safety protocols before resuming crewed Apollo missions.

In 1966, amidst the tense backdrop of the Cold War, significant strides were being made in the realm of space exploration on both sides. On April 4th, NASA introduced Group 5, a cohort of astronauts who would come to be known as the "Original Nineteen," a term coined by the esteemed astronaut John Young. Among them were Vance Brand, John S. Bull, Gerald Carr, Charles Duke, Joseph Engle, Ronald Evans, Edward Givens, Fred Haise, James Irwin, Don Lind, Jack Lousma, Ken Mattingly, Bruce McCandless II, Edgar Mitchell, William Pogue, Stuart Roosa, Jack Swigert, Paul Weitz, and Alfred Worden.

This diverse assembly of astronauts represented the next wave of explorers destined to leave an enduring imprint on the annals of space history. Roughly half of these astronauts would go on to play pivotal roles in the Apollo missions, while others made significant contributions to projects like Skylab and the nascent Space Shuttle program. Their collective efforts underscored NASA's ambitious push to conquer new frontiers beyond Earth's bounds.

Notably, Vance Brand's involvement extended beyond Apollo missions; he also participated in the groundbreaking Apollo-Soyuz Test Project in 1975. This historic mission, which saw American and Soviet spacecraft docking in space, briefly thawed Cold War tensions and exemplified the potential for cooperative ventures in space exploration.

On April 4th, NASA introduced Group 5, affectionately termed the "Original Nineteen" by veteran astronaut John Young. This esteemed cohort included Vance Brand, Gerald Carr, Charles Duke, Fred Haise, James Irwin, Ken Mattingly, Edgar Mitchell, Stuart Roosa, and others, whose names would become synonymous with NASA's ambitious missions. About half of these astronauts would go on to play integral roles in the Apollo missions, while others contributed significantly to Skylab and the burgeoning Space Shuttle programs.

Vance Brand, a notable member of Group 5, achieved another milestone in space cooperation by participating in the American leg of the Apollo-Soyuz Test Project in 1975. This mission marked a historic moment of détente, showcasing joint efforts in space exploration during a tense era of geopolitical rivalry.

However, Group 5 also faced tragedy. Edward Givens tragically lost his life in a car accident in 1967, and John Bull resigned from the Astronaut Corps in 1968 due to health reasons. Despite these setbacks, the group boasted several achievements: Joseph Engle stood out as the only astronaut to have earned his astronaut wings before selection, while Brand, Haise, Mattingly, Lousma, and Weitz distinguished themselves by flying missions across both Apollo and Shuttle programs.

Meanwhile, in the Soviet Union, May 23rd marked the selection of Civilian Specialist Group 2, a cadre including Sergei Anokhin, Georgi Grechko, Valeri Kubasov, and others renowned for their expertise and readiness to advance Soviet space missions. This selection highlighted the USSR's dedication to enhancing scientific and operational capabilities in space.

On June 30th, the United States Air Force's Manned Orbiting Laboratory (MOL) program continued to influence American space ambitions by selecting Group 2. Comprising future NASA stalwarts like Robert Crippen, Gordon Fullerton, and others, this group underwent rigorous training for orbital missions under military auspices. Following the program's cancellation, all members transferred to NASA, where they played pivotal roles during the Shuttle era.

Simultaneously in the Soviet Union, September marked the initiation of cosmonaut training for the Military Cosmonaut Group under the Soyuz 7K-VI Zvezda program. Participants like Pavel Popovich and Gennadiy Kolesnikov prepared rigorously for missions designed to push the boundaries of Soviet space capabilities. However, the program faced an abrupt end in December 1967, underscoring the challenges inherent in pioneering space technologies during the intense geopolitical climate of the Cold War.

AS-202

In January 1966, components of the Saturn IB rocket destined for the Apollo AS-202 mission began arriving at Cape Kennedy Air Force Station in Florida, now known as Cape Canaveral Space Force Station. Just six days after the Apollo AS-201 mission launched from Launch Pad 34, workers swiftly erected the rocket's first stage on March 4, followed by the addition of the second stage on March 10. Over the next month, engineers meticulously tested the rocket to ensure it was meticulously prepared for its upcoming mission.

Launch of AS-202.

Meanwhile, the Apollo spacecraft, bearing serial number 011, arrived at NASA's Kennedy Space Center (KSC) in April 1966. Housed in the Manned Spacecraft Operations Building, the spacecraft was configured nearly identically to a crewed version, complete with a fully operational life support system. The primary distinction was the absence of crew couches onboard CM-011.

For the AS-202 mission, scheduled objectives included rigorous testing of the Service Module's (SM) two power-generating fuel cells and validation of the Service Propulsion System (SPS) main engine's critical multiple restart capability. Extensive vacuum chamber and other testing protocols were completed to ensure every component functioned flawlessly under demanding space conditions.

On July 2, ground crews transported the spacecraft to Launch Pad 34 and meticulously stacked it atop the Saturn IB rocket. A comprehensive countdown demonstration on August 8 verified the readiness of both rocket and spacecraft, marking a crucial milestone as NASA continued its relentless pursuit of lunar exploration.

On the afternoon of August 25, 1966, the Apollo AS-202 mission commenced with a dramatic liftoff from Launch Pad 34, propelled by the mighty Saturn IB rocket's eight H-1 engines. Standing 225 feet tall, the rocket ascended under the watchful eyes of a dedicated team of flight controllers at NASA's Mission Control Center, led by Flight Director John D. Hodge at the Manned Spacecraft Center, now known as NASA's Johnson Space Center in Houston.

The first stage engines burned fiercely for just over two minutes, propelling the rocket to an altitude of 36 miles. As the initial stage concluded, the second stage separated smoothly and ignited its own engine, an event documented by two ejectable and recoverable camera pods positioned atop the first stage. Approximately seven and a half minutes later, at 138 miles, the second stage engine ceased its burn, and the Apollo spacecraft detached from the rocket.

Moments later, the Service Module's (SM) Service Propulsion System (SPS) engine fired up, its thrust lasting slightly over three and a half minutes. This pushed the spacecraft to its peak altitude of 710 miles approximately 41 minutes post-launch, setting the stage for a planned splashdown in the Pacific Ocean.

As the spacecraft began its descent toward Earth, the SPS engine reignited for 90 seconds, enhancing its reentry velocity, thereby testing the robustness of the heat shield. Demonstrating its versatility, the SPS engine performed two additional brief burns to showcase its multiple restart capability, affirming crucial engineering milestones for future manned lunar missions.

One hour and 11 minutes after liftoff, the Command Module (CM) separated from the Service Module (SM) and oriented its heat shield to face the direction of reentry. At approximately 400,000 feet (about 75 miles), the capsule encountered the initial layers of Earth's atmosphere at an astonishing speed of 19,440 miles per hour. Utilizing its guidance system, the CM executed a maneuver known as a double-skip reentry: first descending to around 40 miles altitude, then utilizing its lift capabilities to briefly ascend back to nearly 50 miles before resuming its final descent. This strategic approach helped mitigate the physical stresses on the capsule.

During this harrowing descent, the heat shield endured temperatures reaching about 1,500 degrees Celsius (2,732 degrees Fahrenheit), yet the cabin's interior remained comfortable at 21 degrees Celsius (70 degrees Fahrenheit). As the CM descended to 24,000 feet, two drogue parachutes deployed to stabilize and slow its descent, followed by three main parachutes at 10,000 feet.

Splashdown occurred in the Pacific Ocean southeast of Wake Island after a total flight duration of one hour and 33 minutes. However, the landing point was 235 miles short of the intended target area, a discrepancy attributed to the CM's lower-than-predicted lift-to-drag ratio.

The U.S.S. Hornet (CV-12), serving as the primary recovery vessel, embarked on an eight and a half hour journey to reach the capsule and successfully retrieve it from the ocean, concluding a critical mission milestone in NASA's Apollo program.

Following its recovery, CM-011 was transported aboard the U.S.S. Hornet to Long Beach Naval Station in California, arriving on September 2. Workers safely offloaded the capsule to a portside hangar, where hazardous propellants were drained from the vehicle. On September 7, CM-011 was transported by truck to its manufacturer, the North American Aviation plant in Downey, California. At the plant, extensive postflight testing and analysis were conducted to assess the spacecraft's performance and structural integrity.

After completing these inspections, CM-011 returned to the Manned Spacecraft Center (MSC), now NASA's Johnson Space Center, as its mission continued. The spacecraft played a significant role at the U.S. pavilion during the World Expo in Montreal from April to October 1967, showcasing America's pioneering efforts in space exploration.

Following its stint at the Expo, engineers at MSC modified CM-011 for a critical drop test on October 31, 1968. This test evaluated whether a space-flown Apollo spacecraft could withstand a land landing, crucial for emergency scenarios. Designated CM-011A for this test, the capsule was dropped at a 23-degree angle, resulting in a crack at the base.

In the 1970s, NASA entrusted CM-011A to the Smithsonian Institution's National Air and Space Museum in Washington, D.C., where it was periodically displayed, including a notable exhibit in 2002. In June 2004, CM-011A was transported overland to its current home at the U.S.S. Hornet Sea, Air, and Space Museum in Alameda, California. Technicians painstakingly restored the capsule to its post-drop test condition, preserving an iconic piece of space exploration history for future generations to admire and study.

On October 19, 1966, NASA, buoyed by the successes of three Saturn IB missions, including two with operational Apollo spacecraft, announced a historic next step: the AS-204 mission, utilizing spacecraft 012, would mark the first crewed Apollo flight, destined for Earth orbit. This pivotal mission could last up to 14 days, showcasing NASA's progress toward President Kennedy's lunar landing goal.

The prime crew for AS-204 was assigned as follows: Virgil I. Grissom as command pilot, Edward H. White as senior pilot, and Roger B. Chaffee as pilot. A backup crew was also designated, comprising astronauts James A. McDivitt, David R. Scott, and Russell L. Schweickart. While a specific launch date had not been set, NASA anticipated the mission to take place in February 1967, heralding an exciting new era in space exploration.

Tragically, these plans took a devastating turn on January 27, 1967.

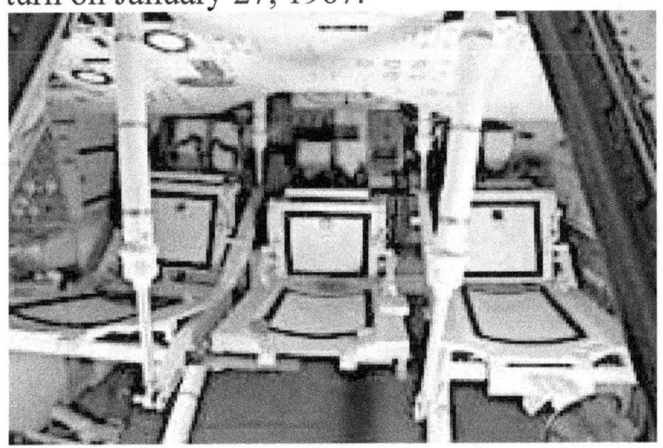

AS-202 CM-011 interior

Recovery teams lifting CM-011 aboard the prime recovery ship U.S.S. Hornet.

AS-203

The AS-203 mission marked a pivotal moment in the Apollo program, primarily focusing on the Saturn IB rocket designated SA-203. This variant represented a significant upgrade over its predecessor, the Saturn I, which had been pivotal in NASA's early spaceflight endeavors under the guidance of rocket engineer Wernher von Braun at the Marshall Space Flight Center in Huntsville, Alabama.

At the heart of the Saturn IB was its first stage, the S-IB, constructed by Chrysler. This stage featured a cluster of eight tanks, each 1.8 meters in diameter, containing liquid oxygen (LOX) and RP-1 derived from the Redstone rocket's proven technology. Central to the stage was a larger 2.7-meter LOX tank adapted from the Jupiter rocket. The S-IB underwent significant modifications and refinements, resulting in a weight reduction of 9,000 kilograms compared to its predecessor. Eight improved Rocketdyne H-1 engines were powering the S-IB, collectively generating 7,120 kilonewtons of thrust, an enhancement over previous iterations.

The most substantial advancement in the transition to the Saturn IB was introducing its second stage, the S-IVB, manufactured by Douglas Aircraft Company. Unlike the Saturn I's S-IV stage, which employed six Pratt & Whitney RL-10 engines, the S-IVB utilized a single Rocketdyne J-2 engine producing 890 kilonewtons of thrust. This stage employed liquid hydrogen and LOX, a high-energy propellant combination that would prove crucial in future missions.

The Saturn IB's capability to lift approximately 21 metric tons into low Earth orbit made it suitable for testing both the Command/Service Module (CSM) and Lunar Module (LM) in Earth's vicinity. Beyond its primary mission objectives, early Saturn IB launches provided vital flight-testing opportunities for the S-IVB stage, identical to the one destined for use as the third stage in the Saturn V rocket, pivotal for lunar missions. Topping off the S-IVB was the Instrument Unit (IU), a ring-shaped apparatus weighing 2,100 kilograms crucial for guiding the Saturn rocket during flight.

Standing at 43.2 meters tall without its payload, the Apollo-Saturn IB soared to a total height of 68.3 meters with a launch mass of 598 metric tons.

The objectives of the Apollo AS-203 mission were centered around thorough evaluations of the Saturn IB rocket, particularly focusing on the shared S-IVB stage, a critical component also used in the Saturn V Moon rocket. This mission aimed to simulate conditions akin to those encountered during a lunar mission, specifically in Earth parking orbit.

The S-IVB stage, designated S-IVB-203 for this flight, played a central role in these evaluations. One primary objective involved observing the behavior of liquid hydrogen within the S-IVB's fuel tank using onboard TV cameras. This observation was crucial for verifying the efficacy of various tank design elements under weightless conditions, mimicking scenarios encountered during prolonged orbital missions and lunar transits.

Additionally, the AS-203 mission provided an opportunity to test the Rocketdyne J-2 engine's restart capabilities. This involved simulating an in-orbit restart scenario, a critical maneuver necessary for successfully deploying lunar missions using the Saturn V rocket.

Since the Apollo spacecraft was not required for AS-203, the Saturn IB (SA-203) was fitted with a streamlined 1,680-kilogram aerodynamic nose cone. This nose cone remained attached throughout the mission and housed an experiment focused on the behavior of cryogenic liquid nitrogen under weightless conditions. This experiment was vital for future developments in fuel cell systems, as liquid nitrogen was a surrogate for liquid oxygen (LOX), offering similar handling properties and safety considerations.

To optimize the Saturn IB's performance, the S-IVB-203 stage carried a reduced load of LOX—53,800 kilograms instead of the typical 86,600 kilograms. This reduction in LOX significantly shortened the burn time of the stage to approximately four minutes and 50 seconds, ensuring that a substantial amount of liquid hydrogen—critical for propulsion—remained available after achieving orbit. The total estimated mass of the S-IVB stage, nose cone, and associated equipment in orbit, including residual propellants, reached 26,550 kilograms, marking it as the heaviest payload ever orbited by the United States up to that point.

The AS-203 mission required several adaptations to the S-IVB-203 stage, distinct from its counterpart, the S-IVB-500, intended for use in the Saturn V lunar missions. Both versions incorporated an Auxiliary Propulsion System (APS) crucial for controlling orientation during flight. Positioned near the base of the stage, this system on the S-IVB-203 housed two modules containing three 670-newton engines fueled by hypergolic propellants, enclosed within aerodynamic fairings. This setup was scaled down compared to the S-IVB-500's APS, designed to accommodate a larger payload and longer coasting periods in Earth parking orbit, featuring an additional 310-newton ullage engines to stabilize propellants during weightless flight.

For the AS-203 mission, the S-IVB-203 stage utilized a modified approach for ullage control. It employed rear-facing nozzles to expel gaseous oxygen generated by its tank's residual evaporation of approximately 1,700 kilograms of LOX. Combined with continuous venting of excess gaseous hydrogen, this system provided sufficient ullage to meet mission objectives, expected to span approximately four orbits.

Another critical objective was to simulate the J-2 engine restart, a maneuver essential for propelling Apollo out of Earth's orbit toward the Moon. The J-2 engine on the S-IVB-203 was equipped with a chill down and recirculation system akin to that planned for the S-IVB-500 model. This system utilized cryogenic propellants to prechill the engine and associated systems for five minutes before ignition. Although an actual reignition of the J-2 engine was not attempted during AS-203, as it would have necessitated additional LOX and equipment, engineers leveraged ground testing insights to prioritize other mission goals over an in-orbit restart demonstration.

The conclusion of the AS-203 mission involved deliberately overpressurizing the liquid hydrogen tank until the common bulkhead with the LOX tank failed. This test aimed to validate ground-tested failure scenarios under real-world conditions. No retrieval efforts were made for the orbiting hardware following mission completion as planned.

Liftoff of Saturn Mission 203, the second in the uprated Saturn I Development Mission Series, was accomplished from the Kennedy Space Center (KSC) Launch Complex 37 at 10:53 a.m., 07/05/1966. KSC, FL.

The preparations for the Apollo AS-203 mission at Cape Kennedy unfolded methodically, beginning with delivering the S-IVB-203 second stage on April 6, 1966. This crucial component was soon followed by the arrival of the Saturn IB's first stage, S-IB-3, which reached the site by barge on April 12. Erected on the newly refurbished Pad B at Launch Complex 37, this stage supported the final Saturn I launch nearly nine months earlier. Meanwhile, Launch Complex 34, occupied since March 4 by the Saturn IB intended for AS-202, was unavailable for the AS-203 mission.

On April 21, the S-IVB stage was integrated into the assembly, with the Instrument Unit (IU) and nose cone added later the same day. Powering up operations for the S-IB stage commenced on April 25, beginning an extensive series of pre-launch tests.

Following the successful flight readiness test on June 27, preparations for the countdown demonstration test (CDDT) began, focusing initially on RP-1 tanking operations for the S-IB stage. The first phase of the CDDT was completed on June 29, followed by the second phase on July 1. With these critical tests accomplished, the countdown resumed at T-11 hours 30 minutes on July 4, scheduled for a launch at 9:00 AM EDT the following morning.

The countdown proceeded smoothly until July 5, 8:45 AM EDT, when a hold was initiated at T-15 minutes. This pause allowed ground controllers to address tasks and investigate a transmission issue with one of the two TV cameras designated to monitor the liquid hydrogen inside the S-IVB fuel tank. After nearly an hour and a half without resolution, the decision was made to recycle the countdown to T-15 minutes and proceed with the launch using only one operational camera.

In the Mission Control Center (MCC) at the Manned Spacecraft Center, now NASA's Johnson Space Center in Houston, Flight Director John D. Hodge monitors the progress of the AS-203 mission

Shortly before launch, another brief hold occurred three minutes prior to liftoff to verify radar status at the Bermuda tracking station. Finally, at 10:53:17 AM EDT, amidst careful monitoring and anticipation, AS-203 lifted off from LC-37B, marking a significant step in advancing the Apollo program's capabilities and preparations for future lunar missions.

During the AS-203 mission, the Saturn IB performed impressively, launching SA-203 into a successful orbital trajectory with notable achievements. The launch was characterized by a faster acceleration than previous flights, reaching Mach 1 at an altitude of 6.67 kilometers just 51.6 seconds after liftoff — 12.9 seconds quicker than SA-201. The S-IB stage operated nearly nominally, with the separation of the S-IVB stage occurring only 0.8 seconds ahead of schedule at 143.4 seconds into the flight.

Overall view of the MCC during the AS-203 flight.

Following stage separation, camera pods deployed from the spent S-IB stage captured the separation, though only one was recovered from the Atlantic Ocean. The S-IVB stage, designated S-IVB-203 for this mission, performed admirably, shutting down its engine just 2.9 seconds earlier than planned after 7 minutes and 13 seconds of powered flight. It achieved an orbit of 185.4 by 189.3 kilometers with an inclination of 32.0°, slightly below the desired 190-kilometer circular orbit.

During the first orbit, the operational TV camera on the S-IVB-203 successfully observed the behavior of liquid hydrogen in its tank, validating predictions about its response to different venting strategies and internal structures. This observation was crucial for understanding the dynamics of cryogenic fluids in space. Subsequently, the simulated restart of the J-2 engine in the following orbit tested and verified the effectiveness of the chill down and recirculation systems, critical components for the future S-IVB-500 model used in Saturn V missions.

The AS-203 mission concluded with the execution of its final test, which unfortunately led to the loss of the S-IVB-203 stage. After successfully completing all primary and secondary objectives over several orbits, the final test involved closing the vents on the liquid hydrogen tank while opening those on the LOX tank. This maneuver was intended to decrease the internal pressure of the LOX tank by allowing the liquid hydrogen to boil off, thereby increasing the pressure differential across the common bulkhead between the two tanks.

Between 1966 and 1967, a select group of military cosmonauts embarked on rigorous training for the ambitious aerospace system known as Project "Spiral." Led by Gherman Titov, this cadre included Anatoly Kuklin, Vasily Lazarev, and other seasoned cosmonauts such as Anatoly Berezovoy, Yuri Romanenko, and Lev Vorobyov. Their mission was to pioneer new frontiers in aerospace technology, positioning them at the forefront of Soviet innovation. However, in 1973, Project "Spiral" was disbanded, reflecting shifting priorities within the Soviet space program.

In January 1967, the global race for space exploration continued to escalate with significant developments on both sides of the Iron Curtain. The Soviet Union expanded its Civilian Specialist Group 2 Supplemental with the addition of Nikolai Rukavishnikov and Vitali Sevastyanov. This group, which also included Aleksei Yeliseyev, bolstered the Soviet space program's expertise, although Sevastyanov, the last surviving member, passed away in 2010.

February witnessed heightened Soviet efforts in crewed lunar programs, organizing cosmonauts into two distinct training groups. Under Vladimir Komarov's leadership, the first group featured iconic figures like Yuri Gagarin, Andrian Nikolayev, Pavel Popovich, and engineer-cosmonauts such as Viktor Gorbatko and Georgi Grechko. The second group, led by Alexei Leonov, included Pavel Belyayev, Boris Volynov, and Oleg Makarov, alongside engineers like Nikolai Rukavishnikov and Yuri Artyukhin. These groups underscored the Soviet Union's unwavering commitment to lunar exploration amidst escalating tensions of the space race.

On May 7th, the Soviet Air Force announced the formation of Group 4, comprising Vladimir Alekseyev, Mikhail Burdayev, and other elite aerospace specialists. This group, rigorously trained in cutting-edge aerospace technology and operations, represented the USSR's strategic military interests in space during the height of the Cold War.

Simultaneously, the Academy of Sciences in the Soviet Union assembled a group on May 22nd, including distinguished scientists and researchers such as Mars Fathulin and Rudolf Gulyayev. Their pivotal roles focused on advancing the scientific objectives of Soviet space missions, emphasizing technological innovations and theoretical advancements.

In the United States, June marked the selection of USAF MOL Group 3, featuring James Abrahamson, Robert Herres, Robert H. Lawrence Jr., and Donald Peterson. Robert Lawrence Jr., notably the first African-American selected as an astronaut, tragically lost his life in a jet accident before the MOL program's cancellation. His pioneering role paved the way for future African-American astronauts like Guion Bluford and Ronald McNair. Peterson later transferred to NASA, where he flew on the Space Shuttle, while James Abrahamson remains the sole surviving member of this distinguished group.

On October 4th, NASA unveiled Group 6, famously dubbed the XS-11 or "The Excess Eleven." This unique cohort of scientist-astronauts included Joseph Allen, Philip Chapman, and others who would leave an indelible mark on the agency's space endeavors. Unlike their predecessors, these individuals brought a wealth of scientific expertise to complement their astronaut training.

Initially tasked as support crew members for the final Apollo missions and later as backup crew members for Skylab, Group 6's contributions were pivotal during NASA's transition to the Space Shuttle era. Several members, such as Story Musgrave, distinguished themselves as mission specialists in subsequent Shuttle missions. Musgrave's remarkable flight on STS-80 at the age of 61 earned him the distinction of being the "oldest astronaut" before John Glenn's historic return to space.

Apollo 1

The AS-204 mission, or Apollo 1, was a pivotal milestone in NASA's Apollo program. It was designed primarily as a crewed spacecraft verification test. Scheduled for launch on February 21, 1967, from Cape Kennedy LC-34, the mission aimed to validate the capabilities of the Apollo Command and Service Module, designated as CSM-012, developed by North American Aviation.

This mission marked a crucial step toward the ultimate goal of landing humans on the Moon. With a launch mass of 20,000 kilograms (45,000 lb), the spacecraft was to orbit Earth in a low Earth orbit configuration. Planned orbital parameters included a perigee altitude of 220 kilometers (120 nautical miles), an apogee altitude of 300 kilometers (160 nautical miles), and an inclination of 31 degrees, with an anticipated orbital period of 89.7 minutes.

Tragically, disaster struck during a pre-launch test on January 27, 1967, at 23:31:19 UTC. A fire broke out inside the spacecraft during a routine test on the launch pad, claiming the lives of all three crew members: Gus Grissom, Edward H. White II, and Roger B. Chaffee. This catastrophic event resulted in the loss of valuable astronaut lives and prompted extensive reassessments and safety enhancements across NASA's manned spaceflight programs.

Apollo 1, originally designated AS-204, stood poised as the inaugural crewed mission of NASA's ambitious Apollo program, aimed at achieving the historic feat of landing humans on the lunar surface. Scheduled for launch on February 21, 1967, it was intended as a critical low Earth orbit test of the Apollo command and service module—a pivotal step toward lunar exploration.

Official portrait of prime and backup crews for AS-204, as of April 1, 1966. The backup crew (standing) of McDivitt (center), Scott (left) and Schweickart were replaced by Schirra, Eisele and Cunningham in December 1966.

Donald Kent "Deke" Slayton, a pivotal figure in American aerospace history, began his journey as a pilot and engineer with the U.S. Army Air Forces during World War II, where he flew missions across Europe and the Pacific. Following the war, Slayton pursued his passion for aeronautical engineering, earning a Bachelor of Science degree from the University of Minnesota in 1949. His career trajectory led him to Boeing, where he honed his skills before joining the Minnesota Air National Guard.

Transitioning to the United States Air Force, Slayton furthered his expertise at the U.S. Air Force Test Pilot School in 1955, solidifying his reputation as a distinguished test pilot. His achievements caught the attention of NASA, and in 1959, Slayton was selected as one of the elite Mercury Seven astronauts, pioneers in America's fledgling manned spaceflight program.

Astronaut Donald K. “Deke” Slayton.

Initially slated to pilot one of the early orbital missions, Slayton encountered a setback when diagnosed with atrial fibrillation in 1962, temporarily grounding him. Undeterred, he remained integral to NASA's operations. He served as the agency's first Chief of the Astronaut Office and Director of Flight Crew Operations, pivotal roles in crew assignments and mission planning.

In 1972, after a decade-long hiatus from spaceflight due to his medical condition, Slayton received medical clearance to return to active duty. In 1975, he made history as the docking module pilot for the Apollo-Soyuz Test Project, a groundbreaking mission symbolizing international cooperation in space exploration.

Beyond his illustrious astronaut career, Slayton contributed significantly to the development of the Space Shuttle, leveraging his experience and technical acumen to shape NASA's future initiatives. His tenure at the agency spanned until 1982, marking decades of dedicated service and visionary leadership.

Tragically, Slayton's life was cut short by brain cancer on June 13, 1993, leaving behind a legacy of courage, innovation, and unwavering commitment to advancing humanity's reach into the cosmos.

In January 1966, Deke Slayton, the Director of Flight Crew Operations, assembled the initial crew for the Apollo 1 mission. The crew consisted of Gus Grissom as Command Pilot, Ed White as Senior Pilot, and Donn F. Eisele as Pilot. However, Eisele faced setbacks when he dislocated his shoulder twice during training aboard the KC-135 weightlessness aircraft. This necessitated surgery on January 27, rendering him medically unfit to continue with the mission.

Deke Slayton decided to replace Eisele with Roger B. Chaffee on the crew in response. NASA officially announced this revised crew lineup on March 21, 1966. Meanwhile, James McDivitt, David Scott, and Russell Schweickart were designated as the backup crew, poised to step in if needed.

In October 1966, NASA announced plans to equip the Apollo 1 mission with a pioneering addition: a miniature television camera. This camera was intended to broadcast live from within the command module during the mission and provide crucial visuals for flight controllers to monitor the spacecraft's instrument panel in real time—a significant advancement in mission monitoring and public engagement. This innovation marked the beginning of a tradition where television cameras would accompany all subsequent crewed Apollo missions, offering unprecedented views of space exploration to audiences worldwide.

Astronauts (left to right) Gus Grissom, Ed White, and Roger Chaffee, pose in front of Launch Complex 34 which is housing their Saturn 1 launch vehicle. The astronauts died ten days later in a fire on the launch pad.

Portrait of the Apollo 1 prime crew for first manned Apollo space flight. L-R:: Edward H. White II, Virgil I. "Gus" Grissom, and Roger B. Chaffee

Regarding the mission insignia, Gus Grissom's crew received initial approval in June 1966 to design a patch named Apollo 1. However, this approval was temporarily withdrawn pending a final decision on the mission's designation, which was ultimately resolved after the tragic fire. The crew-crafted design featured a central motif depicting the command and service module flying over the southeastern United States, with Florida prominently displayed as the launch site. In the background, the Moon loomed, symbolizing the ultimate goal of the Apollo program.

Surrounding this central image was a yellow border adorned with the mission and astronaut names, complemented by a secondary border featuring stars and stripes, highlighted with gold trim. The artwork for this emblem was skillfully executed by Allen Stevens, an employee of North American Aviation, showcasing the collaborative effort and artistic vision of the Apollo 1 crew. This insignia stood as a poignant tribute to the mission's objectives and the aspirations of those who embarked on the journey toward lunar exploration.

The preparation and testing phase of the Apollo 1 mission were marked by a series of challenges and critical evaluations, reflecting the immense complexity of the Apollo command and service module (CSM). Appointed in October 1963, Joseph F. Shea assumed the role of Apollo Spacecraft Program Office (ASPO) manager, tasked with overseeing the design and construction of both the CSM and the Lunar Module (LM).

By August 19, 1966, during a spacecraft review meeting with Shea, the Apollo 1 crew raised concerns about the significant presence of flammable materials within the cabin. Nylon netting and Velcro, while convenient for securing tools and equipment, posed a serious fire hazard in the pressurized pure oxygen environment of the spacecraft. In response to the crew's apprehensions, Shea received a parody crew portrait depicting them with heads bowed and hands clasped in prayer, bearing the inscription: "It isn't that we don't trust you, Joe, but this time we've decided to go over your head."

Prompted by these concerns, Shea directed his team to instruct North American Aviation (NAA), the prime contractor, to remove the flammable materials from the spacecraft. However, Shea did not personally oversee the execution of this directive.

The Apollo 1 crew expressed their concerns about their spacecraft's problems by presenting this parody of their crew portrait to ASPO manager Joseph Shea on August 19, 1966.

Despite these precautionary measures, challenges persisted. The spacecraft, designated CM-012, was delivered to Kennedy Space Center (KSC) on August 26, 1966, under a conditional Certificate of Flight Worthiness, signifying 113 significant engineering changes yet to be completed. Additionally, 623 engineering change orders were issued and addressed after delivery.

Intensive testing followed at KSC, including altitude chamber tests and combined system evaluations. Issues surfaced during these tests, including a flawed design in the environmental control unit (ECU) within the command module, necessitating its return for redesign and rework after leaking coolant upon its initial deployment.

Despite these setbacks, by December 30, 1966, the reassembled spacecraft completed altitude chamber testing with the backup crew led by Walter Schirra. The crew expressed satisfaction with the spacecraft's condition and performance, signaling readiness for the upcoming mission.

On January 6, 1967, CM-012 was mated to its Saturn IB launch vehicle on pad 34, marking the final preparations before launch. Throughout this process, the crew, led by Gus Grissom, maintained a steadfast commitment to their mission despite acknowledging the inherent risks of space exploration.

Command module 012, labeled Apollo One, arrives at Kennedy Space Center on August 26,

Grissom's reflections on these risks underscored the sobering reality of spaceflight. He acknowledged the diligent efforts to mitigate risks while recognizing the inevitability of potential failures. His pragmatic approach reflected the ethos of NASA's pioneering efforts, emphasizing thorough preparation, rigorous training, and unwavering dedication to pushing the boundaries of human exploration.

Tragically, the mission never left the ground. Disaster struck during a routine launch rehearsal test at Cape Kennedy Air Force Station Launch Complex 34 on January 27.

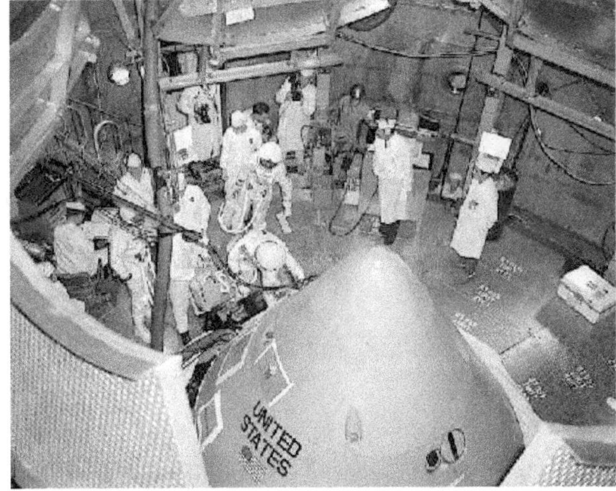

Apollo 1 crewmen enter their spacecraft in the altitude chamber at Kennedy Space Center, October 18, 1966.

The crew, consisting of seasoned astronauts Lt. Colonel Virgil Ivan "Gus" Grissom (USAF), Lt. Colonel Edward Higgins White, II (USAF), and Lt. Commander Roger Bruce Chaffee (USN), faced a fateful setback during a routine practice session. As they prepared inside the command module for what would have been the first crewed Apollo flight, a fire erupted suddenly, engulfing the spacecraft and claiming their lives. This tragic incident not only resulted in the loss of these courageous astronauts but also brought the entire Apollo program to an abrupt halt.

In tribute to their sacrifice, NASA officially named the mission Apollo 1, forever honoring the crew's memory and emphasizing their enduring contribution to the space program's relentless pursuit of lunar exploration.

Virgil Ivan "Gus" Grissom, an integral figure in America's early space endeavors, was among the select few chosen in the pioneering 1959 astronaut group. His career marked significant milestones in the nation's space exploration efforts, beginning with his role as the pilot of MR-4, the second and final suborbital flight of Project Mercury. He later commanded Gemini 3, the first two-person space mission, showcasing his exceptional piloting skills and leadership qualities.

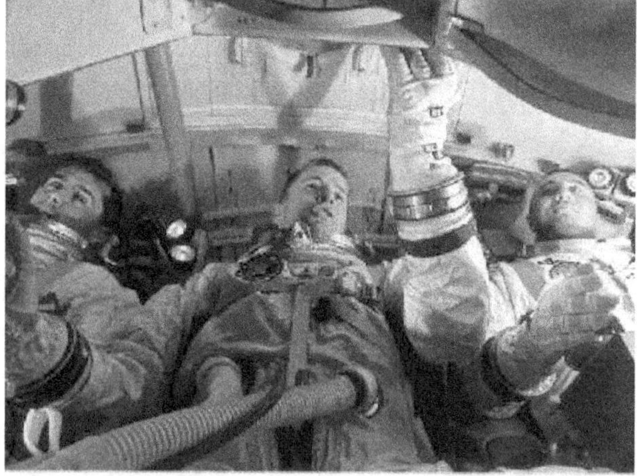

Chaffee, White, and Grissom training in a simulator of their command module cabin, January

Virgil Ivan "Gus" Grissom: A Pioneer's Journey

Born on 3 April 1926 in Mitchell, Indiana, Grissom brought a deep-rooted passion for aviation and exploration to his illustrious career. Educated in mechanical engineering at Purdue University, where he earned his B.S. in 1950, Grissom's technical acumen and dedication propelled him into the forefront of NASA's astronaut corps.

Tragically, on the day of the Apollo 1 fire, Grissom was 40 years old, serving as the command pilot for what was to be a pivotal mission in the lunar landing program. His backup, Captain Walter Marty "Wally" Schirra (USN), stood ready to support Grissom in the endeavor.

Gus Grissom's legacy endures through his contributions to space exploration and the enduring lessons learned from the Apollo 1 tragedy. His unwavering commitment to pushing the boundaries of human achievement continues to inspire generations of astronauts and space enthusiasts alike.

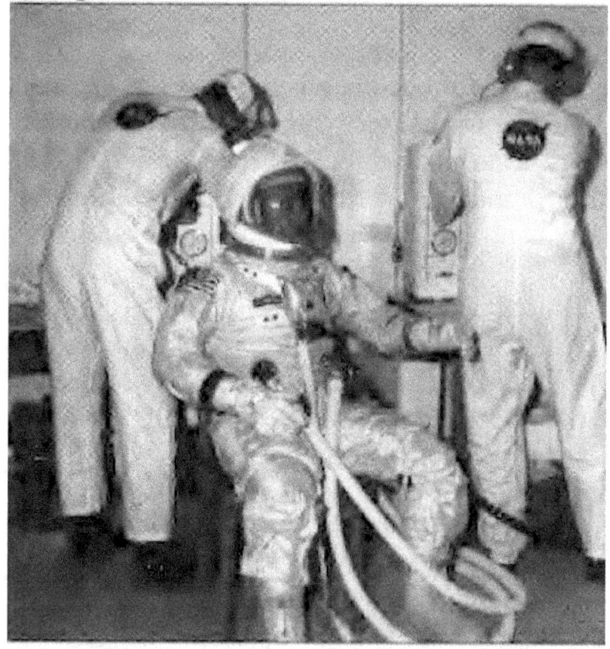

Grissom being checked out in Apollo 1 pressure suit

Edward Higgins White, II: A Spacewalking Pioneer

Edward Higgins White, II, a distinguished astronaut selected in 1962, made history as the first American to step into the vastness of space during the Gemini 4 mission. Born on 14 November 1930 in San Antonio, Texas, White exemplified courage and innovation in pursuing space exploration.

Educated at the U.S. Military Academy at West Point, where he earned his B.S. in 1952, White's career path was steeped in aeronautical engineering, culminating in an M.S. from the University of Michigan in 1959. His selection as an astronaut reflected his exceptional capabilities and commitment to advancing the boundaries of human spaceflight.

Tragically, White was only 36 years old on the day of the Apollo 1 fire, where he served as senior pilot alongside Grissom and pilot Roger Bruce Chaffee. His contributions to the Gemini program and his pioneering spacewalk set a precedent for future missions, demonstrating the indomitable spirit and dedication of those who dare to explore beyond Earth's confines.

Roger Bruce Chaffee: A Promising Career Cut Short

Roger Bruce Chaffee, born on 15 February 1935 in Grand Rapids, Michigan, embarked on his journey into space exploration with unwavering determination and a passion for engineering. Educated at Purdue University, he earned his B.S. in aeronautical engineering in 1957, laying the foundation for a career marked by technical expertise and dedication.

Selected as an astronaut in 1963, Chaffee was training for his first spaceflight as part of the Apollo program. On the day of the Apollo 1 fire, at 31 years old, he represented a new generation of astronauts poised to push the boundaries of space exploration.

Tragically, alongside his esteemed colleagues Gus Grissom and Edward Higgins White, II, Chaffee's promising career was cut short during a routine practice session within the Apollo command module. His backup, Ronnie Walter "Walt" Cunningham, stood ready to support Chaffee in his inaugural space mission.

The Apollo 1 Tragedy: The Sequence of Events

On January 27, 1967, tragedy struck during a critical phase known as the Plugs Out Integrated Test for the Apollo 1 mission. This pivotal test was designed to simulate the actual conditions of a launch as closely as possible, validating the space vehicle systems and operational procedures essential for a successful space mission.

The crew of Apollo 1—Command Pilot Gus Grissom, Senior Pilot Ed White, and Pilot Roger B. Chaffee—were preparing for a crucial "plugs-out" test on Launch Pad 34 at Cape Kennedy Air Force Station. This test was pivotal as it aimed to verify that the spacecraft could function autonomously on internal power, disconnected from all external cables and umbilicals, a prerequisite for the scheduled February 21 launch.

Beginning at 1:00 pm EST (1800 GMT), Grissom, Chaffee, and White entered the command module fully suited and strapped into their seats, connected to the spacecraft's oxygen and communication systems. Almost immediately, Grissom detected an unusual odor inside his suit, likening it to "sour buttermilk." Concerned, the countdown was briefly halted at 1:20 pm to investigate the source of the odor. Despite thorough checks, no cause was identified, and the countdown resumed at 2:42 pm.

Three minutes later, preparations to seal the hatch began. The hatch consisted of a removable inner hatch, a hinged outer hatch integral to the spacecraft's heat shield, and an outer hatch cover part of the boost protective cover. However, the outer hatch cover was only partially latched due to a slight distortion caused by cabling running underneath to facilitate the simulated internal power, as the spacecraft's fuel cell reactants were not yet loaded for this test. Once sealed, the cabin atmosphere was replaced with pure oxygen at a pressure of 16.7 psi (115 kPa), slightly higher than atmospheric pressure.

The spacecraft's inertial measurement unit and biomedical sensors monitored the astronauts' movement during the test. Issues with the communications loop led to difficulties in communication between the crew, the Operations and Checkout Building, and the control room at Complex 34. Frustrated by these communication glitches, Grissom remarked on the challenges of achieving a lunar mission if basic communication hurdles couldn't be overcome.

Grissom, Chaffee, and White during Apollo 1 training (NASA S66-49181).

As the test progressed, further delays occurred at 5:40 pm to troubleshoot the communication problems. Despite these setbacks, all countdown procedures up to the simulated internal power transfer were completed by 6:20 pm. The countdown then paused at T minus 10 minutes at 6:30 pm, awaiting the resolution of the remaining issues.

This plugs-out test underscored the meticulous preparation and rigorous testing required for crewed space missions, highlighting both the technical complexities and the operational challenges that NASA and its astronauts faced in the ambitious pursuit of lunar exploration.

In the tense minutes leading up to the tragic fire on board Apollo 1, meticulous troubleshooting efforts dominated the scene as engineers grappled with persistent communication glitches. Beginning with the T-10-minute hold at 23:20 GMT, routine procedures to resolve these technical challenges unfolded amidst an otherwise seemingly normal operational environment.

The Block I hatch, as used on Apollo 1, consisted of two pieces, and required pressure inside the cabin to be no greater than atmospheric to open. A third outer layer, the boost protective hatch cover,

From 23:20 to approximately 23:30 GMT, all systems other than communications functioned nominally, with no discernible indicators of the impending catastrophe. Engineers focused on rectifying the malfunctioning microphone, ensuring clear and reliable lines of communication with the crew inside the command module.

At 23:30:14 GMT, voice transmissions ceased abruptly, marking the onset of a critical period that would alter the course of the mission. Over the subsequent 50 seconds, data streams from biomedical sensors, the live microphone, the guidance and navigation system, and the environmental control system hinted at activity within the spacecraft. This activity, however, remained enigmatic, lacking a definitive correlation to the tragic events that followed.

Specifically, the biomedical data revealed that the senior pilot, Edward Higgins White, II, initially exhibited minimal activity until around 23:30:21 GMT, where a slight uptick in pulse and respiratory rates was observed. Moments later, at 23:30:30 GMT and 23:30:39 GMT, brief muscular activity was registered on the electrocardiogram, indicating movement within the module. Notably, these responses did not suggest a state of alarm or distress.

By 23:30:45 GMT, all biomedical parameters had reverted to baseline levels indicative of rest, setting a somber backdrop against the impending tragedy. The events that unfolded in the minutes after that would forever change the trajectory of NASA's Apollo program, prompting a profound reevaluation of safety protocols and spacecraft design that would shape future missions to come.

Apollo 1 commander Grissom (l.) inspects the CM during a visit to North American Aviation in 1966 (NASA S66-40760).

In the fateful minutes preceding the fire aboard Apollo 1, critical data from the command module revealed subtle yet significant indicators of crew activity. Beginning around 23:30 GMT, the command pilot's live microphone captured distinct brushing and tapping noises reminiscent of earlier movements observed during test phases. These sounds persisted until 23:30:58.6 GMT, marking a period of crucial actions within the spacecraft.

Meanwhile, the guidance and navigation system, designed to monitor module stability, detected minor movements starting at 23:30:24 GMT. This initial activity intensified between 23:30:39 GMT and 23:30:44 GMT, indicative of more pronounced shifts within the command module. These observations underscored ongoing efforts by the crew to address issues or prepare for forthcoming procedures.

At 23:31:00 GMT, additional movements were noted until data transmission abruptly ceased with the onset of the fire. Concurrently, increased oxygen flow rates to the crew suits signaled continued activity, reflecting adjustments or preparations underway within the confined space.

During the Plugs Out Integrated Test of Apollo 1, the astronauts encountered a critical series of events that would forever alter the course of the Apollo program. As the crew conducted their simulated mission within the command module, their spacesuits experienced minor leaks, a known issue that varied based on their positions inside the spacecraft.

At approximately 23:30:24 GMT, sensors began registering an unusual increase in air flow, coinciding with reports from the crew of a movement that appeared to exacerbate this issue. Data logs confirmed a gradual rise in flow rate, peaking just before 23:31 GMT when a significant voltage spike on AC Bus 2 was recorded at 23:30:54.8 GMT. Concurrently, other monitored parameters exhibited anomalous behavior, signaling a growing disturbance.

Then, at 23:31:04.7 GMT, the crew transmitted their first distress call, signaling a fire erupting within the command module. According to emergency procedures, the senior pilot, positioned centrally, opened the hatch while securing his harness. Witnesses watching the televised feed of the module noted movements suggesting the senior pilot's attempt to reach for the inner hatch handle.

However, post-incident examination revealed his harness buckle remained fastened, indicating he had begun the standard hatch-opening protocol amidst the escalating crisis.

Following the detection of a fire within Apollo 1's command module, data from the Guidance and Navigation System indicated significant activity inside the spacecraft. This activity corresponded to movements by the crew, likely in response to the fire's proximity or as they initiated emergency evacuation procedures as per their training protocols documented in Apollo 1 crew training.

On Adjustable Level 8, adjacent to the command module, personnel swiftly responded upon receiving the fire report. The pad leader promptly issued orders to commence crew egress procedures, directing technicians toward the White Room enclosure surrounding the hatch, where the crew would exit in an emergency.

Then, at precisely 23:31:19 GMT, a catastrophic rupture occurred within the command module.

By 23:31:22.4 GMT, just three seconds after the rupture, all voice and data transmission from the Apollo 1 spacecraft had ceased entirely. Witnesses monitoring the television feed of the hatch window described a harrowing scene: flames initially spread from the left to the right side of the command module, swiftly engulfing the entire visible area.

The rupture allowed flames and gases to escape forcefully, spreading into the space between the command module's pressure vessel and heat shield through access hatches. This rapid dissemination ignited combustible materials, posing an immediate danger to personnel on levels A-8 and A-7 of the service structure surrounding the launch pad. These intense flames not only threatened the safety of nearby personnel but also severely hindered initial rescue efforts.

The eruption of fire and the alarming sounds of structural rupture led many personnel on the launch pad to fear an imminent explosion of the Apollo 1 command module. In response, all personnel on level A-8 swiftly evacuated for their safety. However, their evacuation was quickly followed by a courageous return to the scene to initiate rescue operations.

Apollo 1 crew members inspect equipment before fire (NASA S66-40472).

Moving swiftly onto the swing arm from the umbilical tower, several personnel retrieved fire extinguishers and hastened back along the swing arm to the White Room. Their immediate objective was to commence rescue efforts and assist any crew members inside the module. Meanwhile, others located fire extinguishers from various points across the service structure, joining forces to combat the spreading flames and support firefighting efforts.

The Apollo 1 command module was equipped with three hatches. The outermost hatch, the Boost Protective Cover (BPC) hatch, formed part of the launch cover system designed to protect the module during liftoff. This hatch was designed to be jettisoned before orbital operations, facilitating access and deployment of spacecraft systems.

The ablative hatch's middle hatch served as the outer hatch once the Boost Protective Cover (BPC) was jettisoned following launch. This hatch was designed to protect the module's exterior during the intense heat of reentry into the Earth's atmosphere.

Apollo 1 after the fire

Inside the pressure vessel of the command module was the inner hatch, which sealed the crew compartment. This inner hatch was intended to be the first to be opened by the astronauts during a planned or emergency egress.

On the day of the tragic fire, the outer BPC hatch was in place but not fully latched due to distortion caused by temporary wire bundles installed for testing. In contrast, the middle and inner hatches were securely latched after the crew entered the module. Despite the BPC hatch not being fully secured, a specially designed tool was needed to provide a handhold for lifting it away from the command module.

Amidst the unfolding emergency, the White Room—a crucial area around the hatch where personnel could assist crew ingress and egress—rapidly filled with dense, dark smoke emanating from the interior of the command module and secondary fires spreading across Level A-8 of the service structure.

During the chaotic aftermath of the fire aboard Apollo 1, some personnel managed to locate and don functional gas masks, while others struggled to find operable masks or make theirs functional. Unfortunately, even the operational masks proved inadequate for the dense smoke that filled the environment, as they were primarily designed for toxic atmospheres rather than dense smoke conditions.

Visibility inside the White Room, where crucial rescue operations were underway, was severely limited. The smoke was so thick that visibility was reduced to mere inches, forcing personnel to rely predominantly on tactile senses rather than sight.

Within the White Room, a specialized hatch removal tool was available. After extinguishing a small fire near the BPC hatch and locating the tool, the pad leader and an assistant proceeded to remove the BPC hatch. Despite the hatch not being latched, its removal proved challenging due to the smoke-filled conditions.

Upon successfully removing the hatch, the personnel involved were forced to evacuate the White Room due to the escalating smoke, which made it impossible to remain inside safely.

After evacuating the smoke-filled White Room, the team passed the hatch removal tool to others who joined the effort. Five individuals participated in opening the three hatches, making multiple trips into the White Room, and back out to breathe.

The middle hatch was removed with comparatively less difficulty than the Boost Protective Cover (BPC) hatch. Moving to the inner hatch, they unlatched it and attempted to raise it from its supports to lower it to the command module floor. However, they encountered difficulties as the hatch could not be fully lowered and instead had to be pushed to the side.

Upon opening the inner hatch, they were met with intense heat and smoke billowing from within the command module. This discovery underscored the severity of the fire and the challenges faced in accessing and rescuing any survivors.

Once the pad leader confirmed that all hatches were open, he quickly exited the White Room and moved along the swing arm a short distance. There, he put on his headset and promptly reported this critical development. Analysis of voice recordings later determined that this report was made approximately 5 minutes and 27 seconds after the initial report of the fire.

The pad leader recalled that his report followed the opening of the inner hatch by no more than 30 seconds. Based on this timeline, it was deduced that all hatches were successfully opened, and the two outer hatches were removed approximately five minutes after the fire was first reported, around 23:36 GMT.

Based on medical analysis from autopsy reports, it was determined that the likelihood of resuscitation diminished rapidly once consciousness was lost, occurring approximately 15 to 30 seconds after the first signs of suit failure. By 23:36 GMT, medical experts concluded that resuscitation efforts were no longer viable. The primary cause of unconsciousness leading to death was cerebral hypoxia resulting from cardiac arrest induced by myocardial hypoxia. Environmental conditions such as temperature, pressure, and concentrations of gases like carbon monoxide, carbon dioxide, and oxygen fluctuated rapidly, complicating the physiological and metabolic responses within the crew members.

Visibility inside the command module was severely impaired despite the dimly illuminated lights. Initially, the crew members were not visible to the personnel outside. Despite efforts from the rescue team who had opened the hatches, they could not locate the crew members.

Throughout this critical period, additional personnel on Level A-8 were combating secondary fires, intensifying concerns that the fires might spread to ignite the launch escape tower positioned above the command module, posing a significant threat to the launch complex.

According to log records, emergency services were alerted shortly after the fire was reported, with fire apparatus and personnel dispatched around 23:32 GMT. The doctor overseeing the test from the blockhouse near the pad responded immediately, moving toward the umbilical tower base. The precise arrival time of firefighters on Level A-8 remains uncertain, underscoring the chaotic and urgent nature of the response efforts amidst the unfolding tragedy.

The personnel who had opened the hatches confirmed unanimously that all hatches were fully open before any firefighters were observed on Level A-8 or within the White Room. However, when the first firefighters reached Level A-8, they reported that while the outer hatches were open, the inner hatch remained inside the command module when they arrived, indicating their arrival was after 23:36 GMT.

Based on estimates, it took the firefighters approximately seven to eight minutes to travel from the fire station to the launch complex and ascend the elevator to Level A-8. Thus, their estimated arrival time was shortly before 23:40 GMT.

Upon their arrival, firefighters faced challenging conditions, with dense smoke obscuring visibility. They could vaguely discern the positions of the crew couches and crew members through the haze. Initial efforts to extract the senior pilot from the command module were unsuccessful.

Subsequent inspections revealed significant details about the crew's positions and conditions. The command pilot's couch, positioned at a "170-degree" angle, was essentially horizontal, with released foot restraints and harness and oxygen hoses connected to the suit. His helmet visor was closed and locked, and he was found lying supine on the aft bulkhead or floor of the module, with his head beneath the headrest and feet on his couch. A fragment of his suit material found outside the command module suggested his suit had failed before the rupture at 23:31:19.4 GMT, allowing debris to be carried out through the rupture.

The senior pilot's couch, positioned at a "96-degree" angle, had the back portion horizontal and the lower portion raised, indicating a different posture within the module than the command pilot.

After the fire, the investigation into the conditions inside the command module revealed critical details about the crew's final moments. Though burned through, the command pilot's shoulder straps and lap belts were found with the buckle still securing them, indicating he had not attempted to release them. His suit's oxygen outlet hose was connected, but the inlet hose was disconnected. His helmet visor was closed and locked, and all electrical connections remained intact. He was discovered lying transversely across the command module, just below the hatchway.

In contrast, the senior pilot's couch (the rightmost couch) was in a "264-degree" position, with the back portion horizontal and the lower portion dropping toward the floor. His restraints were disconnected, and all hoses and electrical connections were intact. Similar to the command pilot, his helmet visor was closed and locked, and he was found supine on his couch.

From these findings, it was deduced that the command pilot likely left his couch to escape the initial fire. In contrast, the senior pilot remained in his designated position as per emergency procedures, attempting to open the hatch until his restraints burned through. He intended to maintain communications until the hatch could be opened, as planned. However, due to a slightly higher pressure inside the command module than outside, opening the inner hatch became impossible due to the resultant force on the hatch. This failure of the pressure relief system, exacerbated by the fire, made opening the inner hatch unfeasible until after the cabin ruptured.

The intense and widespread fire that followed, coupled with rapidly increasing carbon monoxide levels, further hindered any attempt at egress. It remains uncertain whether the crew managed to move the inner hatch handle, as its opening from the White Room also shifts the handle within the module to an unlatched position.

Upon the arrival of firefighters, the pad leader, affected by smoke inhalation, was relieved and conveyed the grim situation to medical personnel without immediately disclosing it over communication channels. The doctors arrived at the White Room around 23:45 GMT, shortly after the firefighters. Medical support was officially reported as available by approximately 23:43 GMT, underscoring the rapid response efforts despite the challenging conditions.

In the White Room of the spacecraft, three doctors solemnly entered, confronted by the aftermath of intense heat, smoke, and thermal burns that had tragically claimed the lives of the crew. Their initial assessment, hampered by the lingering fumes and smoke within the command module, revealed the impossibility of immediate crew extraction. With no breathing apparatus, they swiftly directed firefighters to cease any premature removal attempts.

Once the command module had been sufficiently ventilated, the doctors returned equipped to extract the crew. Yet, what they encountered was a poignant testament to the severity of the incident: the crew's suits had fused extensively with melted nylon from the spacecraft, rendering extraction perilous and potentially disruptive to the investigation.

In deference to the integrity of the accident inquiry, it was decided to suspend extraction efforts momentarily. Instead, meticulous photographs of the command module were taken, capturing the scene with the crew still aboard, ensuring that critical evidence remained undisturbed. Only after this documentation was complete did the doctors and their team cautiously resume extraction procedures, commencing around 00:30 GMT on 28 January.

The removal of the crew was a meticulous process that spanned approximately 90 minutes, concluding approximately seven and a half hours after the tragic accident.

The fire's progression is chronologically detailed. It is believed to have originated in the lower forward section of the left equipment bay, situated to the left of the command pilot and notably below his seat. The fire unfolded in distinct stages, each marked by significant developments. Within 15 seconds of the first report of fire at 23:31:19 GMT, the command module's cabin ruptured amid a rapid temperature surge and a sharp increase in internal pressure.

During the initial stage, flames swiftly propagated from the ignition point, coursing through designed debris traps to prevent objects from falling into critical equipment areas during tests and missions. Concurrently, nearby Velcro strips ignited, contributing to the fire's early propagation. Notably, the intensity of the blaze did not peak until approximately 23:31:12 GMT, suggesting a gradual buildup likely due to the initial zone containing minimal combustible materials.

The gradual pressure escalation during this phase was likely tempered by the aluminum structure of the command module, which absorbed and dissipated much of the heat, influencing the fire's progression and the ensuing operational challenges faced by the recovery and investigative teams.

The initial flames, originating from the lower forward section, ascended vertically before spreading across the cabin ceiling. Within this confined space, the debris traps contained combustible materials and acted as conduits for firebrands of molten nylon, further fueling the fire's progression. These burning fragments scattered throughout the cabin, intensifying the spread of flames.

By 23:31:12 GMT, the fire had breached its initial confines, forming a formidable wall of flames along the left side of the module. This barrier thwarted the command pilot, positioned in the left couch, from accessing the valve crucial for venting the module to the outside atmosphere—a pivotal step in emergency egress protocols. However, even if the valve had been successfully opened, its venting capacity was inadequate to counteract the rapid pressure escalation driven by the intense fire.

Analyses suggested that activating the valve would have only marginally delayed the rupture of the command module, prolonging its structural integrity by less than a second amidst the overwhelming forces unleashed by the blaze. This sobering assessment underscored the harrowing conditions within the stricken spacecraft during those fateful moments.

The command module of Apollo 1 was engineered to withstand internal pressures of approximately 13 pounds per square inch above external pressures without compromising its structural integrity. However, during the fire, data recordings indicated that this design limit was exceeded toward the latter part of the initial fire stage. The critical rupture occurred precisely at approximately 23:31:19 GMT, at the juncture where the floor or aft bulkhead of the module met the wall—an area diametrically opposed to the fire's origin.

In the tense moments preceding the rupture, around 23:31:16.8 GMT, the crew began the final communication. Within seconds, as the module breached, this communication abruptly ceased at 23:31:21.8 GMT. Shortly after that, telemetry data was lost at 23:31:22.4 GMT, marking the tragic conclusion of Apollo 1's ill-fated mission.

The rupture of the command module signaled the onset of its brief second stage of fire. This phase was marked by intensified conflagration, driven by the rapid expulsion of gases through the breached pressure vessel. The forceful convection currents created by this outflow dispersed burning fragments, igniting fires across the crew compartment. This heightened state of combustion peaked around 23:31:25 GMT, culminating in a swirling vortex of flames that spread rapidly throughout the module.

Subsequent examination of the module and crew suits revealed clear evidence that the fire had spread from the left side toward the rupture site. Signs of the fire's intensity included burst and charred aluminum tubing within the oxygen and coolant systems, predominantly located at floor level.

The fire's third and final stage was characterized by the swift production of concentrated carbon monoxide, a byproduct of incomplete combustion in the dwindling oxygen-depleted atmosphere. As pressure within the command module dissipated and flames engulfed the entire compartment, the oxygen levels plummeted to levels insufficient to sustain further combustion. This sequence marked the tragic conclusion of Apollo 1's mission, underscoring the catastrophic consequences of the fire within the confined space of the spacecraft.

Following the intense combustion stages, the fire's third phase brought a stark transformation within Apollo 1's command module. Unlike earlier stages characterized by relatively smokeless flames, this phase generated heavy smoke, depositing significant amounts of soot on interior spacecraft surfaces as temperatures cooled.

The duration of this third stage was notably brief, spanning only a few seconds after the crew had been safely removed. It was swiftly evident that the oxygen supply within the module had rapidly diminished, rendering the atmosphere lethal shortly after 23:31:30 GMT. This marked a critical point, just five seconds after the spacecraft was inspected by NASA, North American Aviation management, and Apollo 204 Review Board members, who assessed switch positions and overall module conditions.

The inspection conducted by experts, including consultants, focused on the exterior of Spacecraft 012, evaluating critical elements amidst the aftermath of the tragic fire.

Following the initial catastrophic stages, the fire inside Apollo 1's command module was swiftly extinguished due to a rapid depletion of oxygen. However, a concentrated and intense fire persisted in the vicinity of the environmental control unit. This unit, situated in the left equipment bay near the suspected origin of the fire, sustained damage to its oxygen and water/glycol lines. These compromised lines continued to supply oxygen and fuel, sustaining the localized blaze.

The persistent fire caused significant damage, melting sections of the aft bulkhead and scorching adjacent portions of the inner surface of the command module's heat shield.

Immediately following the tragic accident, stringent security measures were implemented at Launch Complex 34, where Apollo 1 was housed. The site was secured to preserve vital evidence. Before any disturbance to the scene, meticulous external and internal photographs were meticulously captured.

After the initial documentation of the command module's condition with a series of close-up stereo photographs, meticulous steps were taken to preserve the integrity of the evidence. The crew couches were carefully removed, and a specialized false floor featuring removable 18-inch transparent squares was installed. This innovative setup allowed for comprehensive access to the entire interior of the command module without disturbing critical evidence.

A thorough inspection of the spacecraft's interior followed, ensuring a detailed assessment of all systems and components. Subsequently, the investigating board overseeing the Apollo 1 incident drafted and approved a comprehensive disassembly plan. Command module 014 was transported to NASA's Kennedy Space Center (KSC) on 1 February 1967, housed in the Pyrotechnics Installation Building. It served as a testbed for refining disassembly techniques for selected components before extracting them from command module 012.

By 7 February 1967, the disassembly plan was fully operational and meticulously executed, with each component's removal documented through extensive photography. Approximately 5,000 detailed photographs were captured throughout the step-by-step disassembly process, serving as crucial visual records for the investigation. These efforts underscored the meticulous and systematic approach to uncovering the circumstances surrounding the tragic event aboard Apollo 1.

All critical interfaces, including electrical connectors, tubing joints, and physical mounts of components, underwent rigorous inspection and documentation phases. These processes occurred immediately before, during, and after the meticulous disassembly of the command module. Each component removed from the spacecraft underwent careful tagging, sealing in sterile plastic containers, and transportation under strict security protocols to bonded storage facilities.

On 17 February 1967, following extensive wiring tests and disassembly progress, the Board determined that the command module could be relocated without compromising the integrity of evidence. It was transferred to the Pyrotechnics Installation Building at NASA's Kennedy Space Center (KSC), where more favorable working conditions enhanced efficiency and precision.

Under these improved conditions, a structured work schedule of two eight-hour shifts per day, six days a week, was adopted to maintain pace with the rigorous analysis and disassembly tasks. The only exception to this schedule was a dedicated three-day period, during which three eight-hour daily shifts were employed. This intensive effort was focused on removing the aft heat shield, relocating the command module to a more accessible workstation, and carefully dismantling the crew compartment heat shield.

These meticulous procedures and scheduling decisions exemplified the methodical approach taken by investigators and technicians alike, ensuring thorough examination and preservation of evidence crucial to understanding the Apollo 1 tragedy.

Following the tragic fire that claimed the lives of Gus Grissom, Ed White, and Roger B. Chaffee during the Apollo 1 mission rehearsal, NASA swiftly launched comprehensive investigations into the incident. An Accident

Review Board was convened immediately to delve into the root causes of the fire. Simultaneously, both houses of the United States Congress initiated their inquiries to oversee NASA's investigative process.

The investigation revealed that the fire's ignition source was electrical, exacerbated by highly combustible nylon material and the spacecraft's pressurized pure oxygen atmosphere. This combination created a rapidly spreading blaze within the command module's confined cabin. Tragically, the plug door hatch design, intended to seal the cabin, prevented swift rescue efforts as it could not be opened against the internal pressure.

Compounding the tragedy was the realization that the test, conducted with an unfueled rocket, was not considered hazardous. This oversight led to inadequate emergency preparedness measures, contributing to the inability to respond quickly to the emergency.

During the congressional investigations, Senator Walter Mondale brought to light an internal NASA document called the Phillips Report. This document highlighted issues with North American Aviation, the prime contractor for the Apollo spacecraft. Its revelation caused embarrassment for NASA Administrator James E. Webb, who had not been aware of its existence. The disclosure sparked controversy and scrutiny over NASA's management of the Apollo program.

Despite the congressional scrutiny and displeasure over NASA's information handling, both committees ultimately concluded that the issues raised in the Phillips Report did not directly contribute to the tragic accident. The investigations, however, spurred critical reforms in spacecraft safety protocols and oversight, ensuring that future missions would be conducted with greater vigilance and preparedness.

Following the in-flight failure of the Gemini 8 mission in March 1966, NASA underwent a significant procedural shift under the guidance of Deputy Administrator Robert Seamans. Management Instruction 8621.1 was enacted in April that year, marking a pivotal moment in NASA's approach to mission failure investigations. Drawing from established military aircraft accident protocols, this directive empowered the Deputy Administrator to lead independent investigations into major failures, supplementing the responsibilities of program officials.

The tragic fire during a pre-flight test of Apollo 1 on January 27, 1967, prompted an immediate and thorough response. NASA Administrator James E. Webb swiftly sought President Lyndon B. Johnson's approval to adhere to NASA's established investigation procedures. Webb pledged transparency in identifying causes and committed to keeping Congressional leaders informed.

Under Seamans' direction, the Apollo 204 Review Board was formed, chaired by Floyd L. Thompson of Langley Research Center. The board included notable figures such as astronaut Frank Borman and spacecraft designer Maxime Faget. The team's mandate was clear: comprehensively investigate every incident aspect. They began by impounding all Apollo 1 hardware and software, restricting access to authorized personnel only.

Central to their investigation was the meticulous disassembly of Command Module CM-012, a process guided by previously tested procedures used on an identical module, CM-014. This meticulous approach extended to examining every component and system within the module, supported by detailed stereo-photographic documentation. The astronauts' autopsy results and witness testimony were also pivotal in uncovering the sequence of events leading to the tragic outcome.

Seamans regularly updated Administrator Webb throughout the investigation, ensuring transparency and accountability at every stage. The board's findings culminated in a comprehensive final report released on April 5, 1967, which detailed the causes of the Apollo 1 incident and recommended corrective actions to prevent future tragedies.

Based on the findings of the review board, the Apollo 1 tragedy was attributed to a combination of critical factors that led to the fire and the subsequent loss of the crew:

The board determined that the ignition source likely originated from vulnerable wiring carrying spacecraft power and plumbing carrying a combustible and corrosive coolant. These components, crucial for the spacecraft's operation, were susceptible to failure under certain conditions.

The spacecraft's interior was maintained in a pure oxygen atmosphere at higher than atmospheric pressure. While this environment was necessary for the mission, it proved highly combustible in the event of an ignition source, such as the one identified.

The cabin featured a hatch cover that could not be quickly removed under high-pressure conditions. This hindered the crew's ability to swiftly escape or be rescued in an emergency.

Extensive combustible materials were distributed throughout the cabin. These materials, essential for spacecraft operations, contributed significantly to the rapid spread of the fire once ignited.

The review board highlighted deficiencies in emergency preparedness, including the lack of effective rescue or medical assistance protocols and mechanisms for crew escape in an emergency. These shortcomings compounded the severity of the incident.

The autopsy findings revealed that the primary cause of death for astronauts Grissom, White, and Chaffee was cardiac arrest due to exposure to high concentrations of carbon monoxide. The severe burns suffered by the crew were largely postmortem, as the fire had compromised their suits and oxygen supply, exposing them to the lethal atmosphere within the cabin.

The investigation by the Apollo 204 Review Board pinpointed several critical findings regarding the origin and cause of the fire that claimed the lives of astronauts Grissom, White, and Chaffee during the Apollo 1 mission.

Firstly, the board identified a momentary electrical power failure at precisely 23:30:55 GMT as pivotal. This disruption was accompanied by evidence of electric arcs within the spacecraft's interior, indicating potential electrical faults. While multiple electric arcs were observed, the board could not definitively attribute the fire to a single ignition source.

Based on their analysis, the fire was believed to have been initiated near the floor in the lower left section of the cabin, close to the Environmental Control Unit (ECU). This area was identified as a focal point due to its proximity to critical systems and wiring.

A significant discovery during the investigation implicated a silver-plated copper wire that ran through the ECU area near the center couch. The wire had lost its Teflon insulation and showed signs of abrasion, likely from repeated opening and closing of a small access door nearby. This compromised wiring was positioned near a junction in an ethylene glycol/water cooling line, known to be prone to leaks.

Further investigation revealed that electrolysis of the ethylene glycol solution with the silver anode of the wire could lead to a violent exothermic reaction. In the Command Module's pure oxygen environment, this reaction could ignite the ethylene glycol mixture, contributing to the rapid spread of the fire.

Experiments at the Illinois Institute of Technology corroborated these findings, confirming that silver-plated wires posed a specific hazard in such environments. As a result, NASA's Apollo Spacecraft Program Office (ASPO) swiftly directed North American and Grumman contractors to eliminate silver or silver-coated electrical contacts from areas susceptible to ethylene glycol spills in future Apollo spacecraft.

The plugs-out test conducted on Apollo 1 aimed to simulate the critical launch procedures under conditions meant to replicate those of an actual mission. The spacecraft's cabin was pressurized with pure oxygen at a high level of 16.7 psi (115 kPa), more than five times the partial pressure of oxygen found in standard atmospheric conditions at sea level. While essential for spaceflight, this environment significantly increased the flammability of materials that would not normally ignite in Earth's atmosphere.

NASA chose a high-pressure oxygen atmosphere from successful precedents set in the Mercury and Gemini programs. Before launch, this atmosphere served two primary purposes: to purge nitrogen from the cabin air and replace it with pure oxygen and to secure the plug door hatch cover by maintaining a higher internal pressure than the ambient environment. During the launch itself, the pressure gradually decreased to an in-flight level of 5 psi (34 kPa), ensuring adequate oxygen for the astronauts while mitigating fire risks.

Before the fatal incident, the Apollo 1 crew had successfully tested this procedure in the Operations and Checkout Building's altitude chamber. These tests, executed on October 18 and 19, 1966, involved fully pressurizing the command module with pure oxygen on four occasions, totaling six hours and fifteen minutes. Notably, this duration exceeded the plugs-out test by two and a half hours.

The investigation board highlighted these tests as pivotal moments in their analysis. They underscored the spacecraft's extensive exposure to a high-pressure oxygen environment during these trials, providing critical context for understanding the conditions leading up to the tragic fire on January 27, 1967.

These insights prompted significant reassessments in NASA's safety protocols and spacecraft design strategies, emphasizing the need for robust fire prevention measures and crew protection mechanisms in future missions. The lessons learned from Apollo 1's tests and subsequent investigation profoundly influenced the evolution of space exploration practices, ensuring safer conditions for astronauts as they ventured into the depths of space.

The Apollo 204 Review Board identified a critical issue concerning combustible materials near potential ignition sources within the Apollo 1 spacecraft. Among these materials, the board highlighted the extensive use of Velcro— 34 square feet (3.2 m^2)— installed throughout the cabin, almost resembling carpeting in its coverage.

Velcro, while commonly used for its convenience in securing items in microgravity, was found to be highly flammable in the spacecraft's high-pressure, pure oxygen environment. This posed a significant fire hazard, especially given the rapid combustion characteristics of materials in such conditions.

According to astronaut Buzz Aldrin's account in his book "Men From Earth," concerns regarding the flammability of Velcro were raised as early as August 19. Following these concerns and under the direction of Joseph Shea, head of NASA's Office of Manned Space Flight, efforts were made to remove the flammable material. However, despite these efforts, Velcro was reportedly reinstated before the spacecraft's delivery to Cape Kennedy on August 26.

This sequence of events underscored lapses in oversight and adherence to safety protocols during the spacecraft's preparation phase. Despite known flammability risks and crew objections, the reinstallation of Velcro highlighted systemic challenges in managing safety concerns effectively within the Apollo program.

The review board's findings prompted immediate corrective actions, including stricter material selection criteria and enhanced safety inspections to mitigate fire risks in future spacecraft. These measures reflected NASA's commitment to learning from the Apollo 1 tragedy and implementing robust safety measures to safeguard astronauts and spacecraft integrity during future missions.

The design of the inner hatch cover used in the Apollo 1 spacecraft employed a plug door configuration, which required higher pressure inside the cabin than outside to maintain a secure seal. During normal operations, such as launch preparations, the cabin was pressurized at approximately two psi (14 kPa) above ambient pressure. This pressure difference was sufficient to keep the hatch cover closed until the internal pressure could be safely vented.

In emergencies, the procedure called for astronaut Grissom to open the cabin vent valve first, allowing astronaut White to remove the hatch cover. However, Grissom could not reach the valve positioned to the left during the tragic fire, obstructed by the initial wall of flames that quickly engulfed the cabin.

Compounding this critical issue was the venting system's design limitation. While it could effectively handle the normal pressure venting process, its capacity was inadequate to manage the rapid pressure spike caused by the fire's intense heat, which quickly escalated to 29 psi (200 kPa). This limitation severely hampered any effective venting of the cabin atmosphere during the emergency.

Initially, North American Aviation, the contractor responsible for the Apollo spacecraft, had proposed an outward-opening hatch design with explosive bolts similar to those used in Project Mercury. This design would have allowed for quick emergency egress by blowing the hatch open in an emergency. However, NASA rejected this proposal, citing concerns that the hatch could accidentally open, as had occurred during Gus Grissom's Liberty Bell 7 flight.

Instead, NASA's Manned Spacecraft Center opted for a mechanically operated inward-opening hatch for the Gemini and Apollo programs, considered more reliable for maintaining integrity during space missions. Before the Apollo 1 fire, the astronauts had recommended transitioning to an outward-opening hatch design, already planned for inclusion in the Block II command module.

Donald K. Slayton, in his testimony following the accident, clarified that the astronauts' recommendation for an outward-opening hatch primarily stemmed from its practicality for spacewalks and mission conclusion rather than emergencies. This distinction highlighted the evolving considerations in spacecraft design and safety protocols aimed at ensuring operational efficiency and crew safety throughout all phases of space missions.

Several factors drove the decision to use a pure oxygen atmosphere in the Mercury and subsequent Apollo spacecraft, each aimed at balancing safety, functionality, and operational efficiency.

Initially, NASA considered using a nitrogen/oxygen mixture for the Mercury spacecraft to mitigate fire risks during launch preparations. However, this approach was discarded due to several key considerations. First and foremost, a pure oxygen atmosphere at 5 psi (34 kPa) was found to be comfortably breathable by astronauts, significantly reducing the pressure load on the spacecraft in the vacuum of space. This simplicity in breathing gas management was crucial for minimizing the complexity of life support systems onboard.

Secondly, using nitrogen in combination with in-flight pressure reductions carried the risk of decompression sickness, commonly known as "the bends." This potential health hazard for astronauts further influenced the decision to opt for a pure oxygen environment, eliminating the need for nitrogen.

However, criticisms of the pure oxygen design arose following a serious accident on April 21, 1960, while testing a Mercury cabin/spacesuit atmosphere system in a vacuum chamber. McDonnell Aircraft test pilot G. B. North experienced unconsciousness and serious injury due to nitrogen-rich (oxygen-poor) air leaking into his spacesuit feed. This incident underscored the dangers of improper gas mixtures in confined spacecraft environments.

Despite recommendations from North American Aviation to consider using an oxygen/nitrogen mixture for the Apollo spacecraft, NASA ultimately decided against it. The pure oxygen design was deemed safer, less complicated, and lighter in weight, aligning with the overarching goal of minimizing risks and maximizing mission efficiency.

Criticism of NASA's approach intensified following the Apollo 1 tragedy. In his monograph "Project Apollo: The Tough Decisions," Deputy Administrator Seamans acknowledged that not conducting a fire test on the command module before the plugs-out test was a significant oversight. Astronaut Jim McDivitt, in the BBC documentary series "NASA: Triumph and Tragedy," expressed that NASA had underestimated the implications of a 100% oxygen atmosphere on fire safety. Similar sentiments were echoed by other astronauts in documentaries such as "In the Shadow of the Moon" (2007), highlighting the lessons learned and the evolution of safety practices in manned spaceflight.

Ultimately, NASA's decision-making process surrounding atmospheric composition in spacecraft reflected a complex balance of technological capability, safety considerations, and the evolving understanding of human factors in space exploration.

The aftermath of the Apollo 1 tragedy saw intense scrutiny and investigations from committees in the United States Congress, particularly the Senate Committee on Aeronautical and Space Sciences chaired by Senator Clinton P. Anderson. NASA's top leadership, including Administrator James E. Webb, Manned Space Flight Administrator Dr. George E. Mueller, and Apollo Program Director Maj Gen Samuel C. Phillips, were summoned to testify before Anderson's committee.

During the February 27, 1967 hearings, Senator Walter F. Mondale raised concerns about alleged performance issues at North American Aviation, the prime contractor for the Apollo spacecraft. Webb and other NASA officials initially denied knowledge of any significant problems. However, it was revealed that Phillips had conducted a critical "tiger team" investigation in late 1965, highlighting quality deficiencies, schedule delays, and cost overruns in the Apollo CSM and Saturn V second-stage contracts with North America. This report, known as the "Phillips Report," had not been disclosed to Congress until Mondale pressed the issue.

Deputy Administrator Seamans, Administrator Webb, Manned Space Flight Administrator George E. Mueller, and Apollo Program Director Phillips testify before a Senate hearing on the Apollo

Mondale's persistence in questioning NASA's handling of these reports led to tensions and accusations of deception from Congress toward NASA leadership. Webb defended NASA's initial selection of North American as the contractor for Apollo but faced criticism for not fully disclosing the challenges identified by Phillips' team. The Senate committee's final report acknowledged that the Phillips review did not directly contribute to the accident but emphasized the importance of transparency in government-contractor relationships.

Freshman Senators Edward W. Brooke III and Charles H. Percy expressed stronger dissent in their appended views, criticizing NASA for withholding crucial information from Congress. Mondale issued a strongly worded statement condemning NASA's evasiveness and lack of candor in their communications with Congress.

Despite the political fallout, President Lyndon B. Johnson's support for NASA and the Apollo program helped mitigate the potential backlash. Johnson, a staunch advocate of space exploration since its inception, effectively portrayed Apollo as a continuation of President John F. Kennedy's legacy.

Internally at NASA, tensions with North American escalated over assigning blame for the tragedy. Webb intervened by demanding the resignation of Chief Engineer Harrison A. Storms, which led to internal strife within the contractor's leadership. Meanwhile, Joseph Shea, a key figure in the Apollo program, struggled with personal coping mechanisms following the tragedy, prompting concerns about his mental health. Despite attempts to address the situation, Shea's role was eventually reassigned, marking a difficult period for NASA's leadership and its handling of the aftermath.

The astronauts' widows requested that the mission name "Apollo 1" be reserved for the ill-fated flight their husbands never completed. On April 24, 1967, George Mueller, then Associate Administrator for Manned Space Flight, officially announced this change. The mission AS-204, which ended tragically on the ground, would be designated as "Apollo 1 - first manned Apollo Saturn flight – failed on the ground test."

Before Apollo 1, there had been three unmanned Apollo missions: AS-201, AS-202, and AS-203. Only AS-201 and AS-202 carried spacecraft but were not originally designated as "Apollo" missions. In accordance with Mueller's announcement, the next mission, AS-501, the first unmanned Saturn V test flight, was designated Apollo 4. Subsequent missions were numbered sequentially based on flight order. The designations Apollo 2 and Apollo 3 were officially unused.

This pause in crewed flights allowed NASA to address delays in developing the Saturn V rocket and the lunar module. Apollo 4, the first unmanned Saturn V test flight, was successfully launched in November 1967. Meanwhile, the Saturn IB rocket originally intended for Apollo 1 (AS-204) was relocated from Launch Complex 34 to Launch Complex 37B. It was later used to launch Apollo 5 in January 1968, an unmanned Earth orbital test flight of the lunar module (LM-1).

Apollo 6 (AS-502), another unmanned Saturn V test flight, followed in April 1968. Finally, Grissom's backup crew, consisting of Wally Schirra, Don Eisele, and Walter Cunningham, successfully flew the first crewed mission after Apollo 1 as Apollo 7 (AS-205) in October 1968. This mission marked the debut of the Block II Command and Service Module (CSM).

Apollo 4

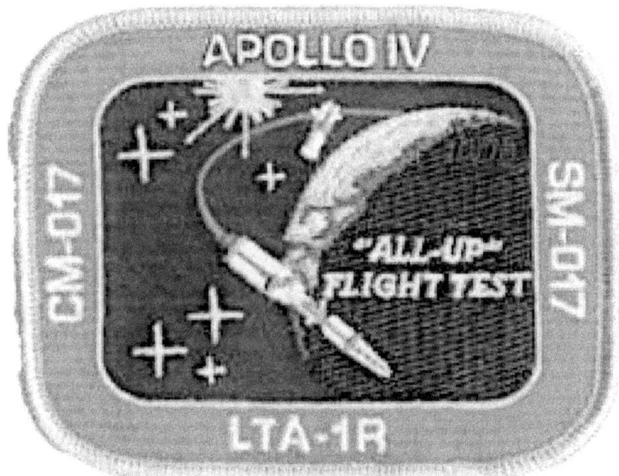

On November 9, 1967, Apollo 4, designated SA-501, heralded a monumental milestone in space exploration as NASA's uncrewed test flight of the Saturn V launch vehicle. This mission, assembled meticulously within Kennedy Space Center's iconic Vehicle Assembly Building, marked the inaugural launch from Launch Complex 39, purpose-built for the Saturn V.

Apollo 4 pioneered the "all-up" approach, where every component of the rocket system, including the S-IC first stage and S-II second stage, undertook their maiden flight together. Notably, it validated the in-flight restart capability of the S-IVB third stage, crucial for lunar missions. The command and service module, a Block I variant modified to incorporate key revisions of the upcoming Block II design, underwent rigorous testing, including a simulated lunar-return heat shield evaluation at high velocities and angles.

Originally slated for an early 1967 launch, delays ensued due to spacecraft component issues and rigorous pre-flight testing protocols intensified following the tragic Apollo 1 fire in January 1967. These setbacks underscored NASA's commitment to safety and thoroughness in mission preparation.

Thus, when Apollo 4 thundered skyward from LC-39, it embodied the culmination of meticulous planning, technological innovation, and America's steadfast commitment to realizing President Kennedy's lunar vision. This pivotal moment propelled the nation closer to achieving the monumental feat of landing on the Moon.

On November 9, 1967, Apollo 4, the first test flight of the Apollo/Saturn V space vehicle, was launched from Kennedy Space Center Launch Complex 39. This was an unmanned test flight intended to prove that the complex Saturn V rocket could perform its requirements. All three stages separated successfully and their engines performed as planned. The third stage also restarted in orbit, which was a requirement for lunar missions. At the end of the flight, the unmanned Apollo spacecraft reentered and proved that it could survive the intense heat generated during a high-speed return from the moon.

Earlier in January 1965, Major General Samuel C. Phillips, the Apollo Program Director, set an ambitious schedule for SA-501, the first test flight of the Saturn V rocket, aiming for launch by January 1967. This timeline allowed little margin for delays, especially with two additional Saturn V launches planned for later in 1967. However, early doubts among Apollo program officials about meeting this deadline proved well-founded when an explosion involving a liquid oxygen line at LC-39 threatened to postpone SA-501 by several weeks.

Significant issues with North American Aviation, the contractor responsible for the S-II stage of the Saturn V, and the Apollo Command and Service Module (CSM) were Compounding these challenges. NASA had encountered persistent problems related to scheduling, cost management, and quality control across both programs. In response to these concerns, General Phillips personally led a thorough investigation at North American's California facility between November and December 1965. This intensive review culminated in a detailed report to his superior, George Mueller, highlighting critical deficiencies in program management that needed urgent rectification.

Technical challenges continued to complicate the progress of the Saturn V rocket leading up to Apollo 4. One significant setback emerged when cracks were discovered in the S-II stage, delaying its testing and approval by NASA. Despite these setbacks, the logistical puzzle of assembling the massive Saturn V at Kennedy Space Center (KSC) began to take shape.

The arrival of key components commenced in August 1966, starting with the S-IVB stage transported by the unique Pregnant Guppy aircraft. This was followed in September by the S-IC first stage, which arrived via barge. These deliveries were pivotal for the assembly process scheduled to take place in the Vehicle Assembly Building (VAB) at KSC.

However, the S-II stage, crucial for the rocket's complete assembly, faced delays beyond its original delivery target by July 1966. To proceed with the assembly timeline, NASA improvised with a temporary "spacer" instead of the S-II stage, ensuring that other stages and components could be integrated as planned. This temporary measure underscored NASA's adaptive approach in managing unforeseen delays while maintaining momentum toward the ambitious launch schedule set for Apollo missions.

In January 1967, tragedy struck NASA's Apollo program with a devastating fire during a critical test of Apollo 1 on the launch pad, claiming the lives of three astronauts. This incident not only dealt a profound emotional blow but also triggered a rigorous reassessment of safety protocols across the agency.

Initially intended as an unmanned mission, SA-501, scheduled to be the first Apollo flight, came under intense scrutiny following the Apollo 1 disaster. NASA's meticulous inspection of the Command and Service Module (CSM) revealed a startling 1,407 discrepancies. These ranged from improperly routed wires to serious concerns such as exposed wiring vulnerable to short circuits. The thoroughness of these inspections underscored the critical need for precision and reliability in spacecraft assembly.

Compounding these challenges, preparations for Apollo 4 faced unexpected setbacks. During inspections, an unanticipated bolt was discovered lodged within one of the J-2 engines, raising significant concerns about assembly oversight and quality control processes. This discovery highlighted the exacting standards required at every stage of the intricate Saturn V rocket assembly process.

By March 1967, a comprehensive review chaired by Major General Samuel C. Phillips revealed a daunting 1,200 issues affecting the Saturn V rocket itself. Technicians and engineers embarked on a concerted effort to address these issues, aiming to rectify at least eighty discrepancies daily in preparation for upcoming missions.

As repairs progressed on the CSM, the focus shifted to integrating the S-II stage into the assembly stack. However, new challenges emerged when hairline cracks were detected in another S-II stage under construction. This discovery necessitated a meticulous inspection and repair process, delaying progress until mid-June.

A pivotal development unfolded amid the ongoing efforts to repair the Command and Service Module (CSM) following the Apollo 1 tragedy. The temporary spacer, integral during the assembly process, was removed to allow for the integration of the S-II stage into the spacecraft stack. However, progress encountered a significant hurdle on May 24 when engineers discovered hairline cracks in another S-II stage under construction. This unexpected finding necessitated the immediate removal of the stage for thorough inspection and subsequent repairs, delaying the assembly process until mid-June.

Despite these setbacks, substantial strides were made in the Apollo program. By May 24, 1967, the reassembled SA-501, now with the CSM reinstated and the inspected S-II stage positioned, was poised for rollout from the Vehicle Assembly Building to Launch Complex 39. This milestone event occurred on August 26, 1967, marking a critical step forward in NASA's preparations for its lunar missions.

Two days later, the Mobile Servicing Structure, vital for spacecraft and launch vehicle access, arrived at the launch site via crawler transport. This deployment represented NASA's first instance of assembling a spacecraft away from its primary launch location. The decision to relocate operations aimed to safeguard equipment and personnel from the challenging environmental conditions of Florida, ensuring optimal conditions for the impending historic mission.

Apollo 4 inside the VAB

On August 26, 1967, the fully assembled Apollo 4 vehicle was transported to Launch Complex 39 (LC-39). This marked a significant milestone as it was the first time a NASA spacecraft had been fully assembled away from its launch site. The Mobile Servicing Structure, crucial for providing access to the launch vehicle and spacecraft, was also transported to LC-39 two days later, utilizing a crawler for transport. This relocation allowed NASA to protect sensitive equipment and personnel from Florida's hot and humid climate, ensuring optimal launch readiness.

The lead-up to Apollo 4's inaugural flight tested NASA's mettle and technical expertise to the fullest extent. Initially planned for a swift turnaround, the countdown demonstration test encountered a series of formidable obstacles. Originally set for September 20, 1967, the start was postponed to September 25, eventually commencing late on September 27 due to unforeseen delays. As October arrived, further technical issues arose, pushing the countdown to within 45 minutes of launch by October 4.

Yet, persistent challenges persisted. A critical computer malfunction on October 9 forced a reset, reverting the countdown to minus 13 hours. Complications with equipment reliability added to the strain, exacerbated by the fatigue plaguing the launch team. Recognizing the need for recuperation, a two-day respite was called to restore readiness.

Ultimately, the exhaustive countdown demonstration test extended far beyond its initial timeframe of just over a week, concluding on October 13 after spanning three arduous weeks. Despite the setbacks, NASA's commitment to meticulous preparation and resilience in overcoming technical hurdles set the stage for the successful launch of Apollo 4, a pivotal milestone in America's journey to the moon.

Amid mounting pressure and media skepticism surrounding Apollo 4's feasibility, NASA Administrator James E. Webb intervened after concerns about the mission's readiness surfaced. In a tense meeting, Webb asserted NASA's commitment to proceeding on its terms despite the setbacks and public scrutiny.

These challenges, while taxing, provided invaluable learning opportunities for the launch crew, refining their procedures and bolstering their readiness for the impending historic mission. Following a rigorous flight readiness review on October 19, Apollo 4 received clearance for launch pending the completion of final tests and adjustments.

However, ongoing concerns about potential leaks in the liquid oxygen tanks' Teflon seal rings and drain valves, exacerbated by prolonged exposure to Florida's sun on the launch pad, prompted Major General Samuel C. Phillips to postpone the launch from its initially targeted date of November 7 to November 9, 1967. This decision underscored NASA's meticulous attention to detail and unwavering commitment to ensuring mission success.

Apollo 4 and its counterpart Apollo 6 played an essential role in certifying the Saturn V launch vehicle, the Apollo spacecraft, and the associated ground systems crucial for upcoming crewed lunar missions. Notably, Apollo 4 marked the maiden flight of the Saturn V rocket, achieving significant milestones for its untested S-IC first stage and S-II second stage, both making their operational debut (the S-IVB stage had flown previously as part of the Saturn IB).

The primary objectives of Apollo 4 were extensive and pivotal. NASA aimed to gather crucial flight data on multiple fronts, validating the structural integrity and compatibility of the Saturn V and Apollo spacecraft under various flight loads. This included studying the performance during critical separation events between stages and assessing the operational capabilities of essential subsystems, particularly the emergency detection systems vital for crew safety.

Another critical goal was to evaluate the Apollo Command Module (CM) heat shield under conditions simulating high-speed reentry from a lunar mission. This test ensured the CM could endure the intense heat and stresses encountered during Earth's atmosphere reentry, safeguarding astronauts returning from the Moon.

Additionally, Apollo 4 aimed to verify the in-space restart capability of the S-IVB stage, a capability essential for maneuvers like lunar orbit insertion and trans-lunar injection during crewed missions.

Equipped with CSM-017, designed under the Block I configuration of the Command and Service Module (CSM), Apollo 4 focused on Earth orbit tests and validation missions early in the Apollo program. Unlike later Block II CSMs used for lunar missions, CSM-017 could not dock with a Lunar Module (LM).

Ultimately, Apollo 4 successfully achieved all its objectives, providing vital data that confirmed the readiness of the Saturn V rocket, Apollo spacecraft, and ground systems for the ambitious goal of landing astronauts on the Moon and returning them safely to Earth.

CSM-017, integral to NASA's Project Apollo, combined two vital components: the Command Module (CM-017) and the Service Module (SM-020). CM-017, NASA's second fully operational command module, initially designated for the ill-fated Apollo 1 mission, underwent a critical reassignment following the tragic fire that claimed CM-012. This disaster prompted extensive redesigns aimed at enhancing crew safety for subsequent missions.

The Service Module, SM-020, initially slated for use with CSM-020 in a Saturn V test flight, saw its role reassigned after SM-017 was damaged in an explosion. This shift underscored NASA's adaptive approach to overcoming setbacks to advance toward lunar exploration goals.

In preparation for manned missions, CSM-017 underwent significant Block II modifications. These enhancements were essential as NASA imposed stringent certification requirements for crewed missions. Upgrades included bringing the heat shield up to Block II standards to endure the intense heat of re-entry. The spacecraft also adopted an upgraded CM-to-SM umbilical connector and improved VHF and S-band antennas typical of Block II design.

CSM-017 is moved into position.

Notably, one of the most crucial modifications addressed the CM's hatch, a safety feature tragically underscored by the Apollo 1 incident. The original hatch design had hindered the astronauts' escape, leading to a comprehensive redesign. Though the redesigned hatch was scheduled for deployment on Apollo 6, its critical components underwent rigorous flight-qualification tests starting with Apollo 4. During this mission, a specially designed test panel simulated the new seals and exterior heat shield, ensuring they met the rigorous demands of space travel.

In its Block I configuration, Apollo 4, utilizing CSM-017, marked an early milestone in testing the Apollo spacecraft in Earth orbit. Unlike later Block II configurations geared for lunar missions, CSM-017 could not dock with a Lunar Module (LM).

Despite being uncrewed, Apollo 4 was equipped with special equipment allowing Mission Control to operate the CSM's systems remotely. Additionally, a camera was installed to automatically capture images from one of the command module's windows during its final orbit.

Notably, since Apollo 4 carried no crew, the command module lacked the usual interior fittings such as couches, controls, and displays that would be essential for astronauts during a crewed mission. These modifications and enhancements ensured that CSM-017 was thoroughly tested and ready to support future crewed Apollo missions to the Moon.

During the Apollo 4 mission, a Lunar Module Test Article (LTA-10R) was included in the payload. It was housed inside the Spacecraft-LM Adapter (SLA-8), positioned on the third stage of the Saturn V rocket throughout the flight.

LTA-10R was a test article to simulate the lunar module's descent and ascent stages. The descent stage was configured similarly to a flight-type module but lacked landing gear. Its fuel and oxidizer tanks were filled with a mixture of water, glycol, and freon to simulate the weight and characteristics of actual propellants. On top of the descent stage mockup was an ascent stage made of aluminum, primarily for ballast purposes and devoid of actual flight systems.

Both the SLA-8 and LTA-10R were extensively instrumented to measure stresses and conditions experienced during launch and orbital ascent aboard the Saturn V. This data was crucial for evaluating the performance and structural integrity of the SLA and lunar module configurations under real flight conditions.

After completing its mission role, LTA-10R would meet its end when the S-IVB stage re-entered the Earth's atmosphere, leading to its destruction as part of the mission's conclusion. This test article played a significant role in validating the design and readiness of the lunar module for future crewed lunar missions within the Apollo program.

Apollo 4 marked a significant milestone as the first flight of the Saturn V rocket, the largest launch vehicle ever attempted. This mission introduced the "all-up" testing concept to the Apollo program, a decision that had far-reaching implications for its development and schedule.

George Mueller, head of the NASA Office of Manned Space Flight, championed the idea of all-up testing, where each stage of the launch vehicle and the spacecraft is fully functional and integrated from the first flight. Drawing on his experience with military missile projects like the Minuteman ICBM, Mueller recognized the efficiency of testing complete systems rather than incrementally testing each component. He believed that all-up testing would expedite the Apollo program's schedule.

In a directive issued in late 1963, Mueller mandated that both the first Saturn IB and the first Saturn V flights be uncrewed, with each stage fully operational and carrying a working spacecraft. Subsequent flights of each rocket type would also follow this pattern: the second flight uncrewed, and the third flight crewed. This approach aimed to validate the entire launch vehicle and spacecraft system in one go, minimizing the number of test flights needed.

This decision was a departure from the incremental testing approach favored by Wernher von Braun's Marshall Space Flight Center team, where new rockets were traditionally tested stage by stage. The all-up testing approach required rigorous preparation and confidence in the reliability of each component from the outset.

While some Apollo program managers initially had reservations about all-up testing, fearing the risks of a comprehensive failure, they ultimately agreed to Mueller's strategy. The alternative of incremental testing would have likely extended the timeline for achieving a crewed lunar landing beyond the ambitious goal set for 1970.

Apollo 4, therefore, not only tested the capabilities of the Saturn V rocket and validated the spacecraft systems but also established a precedent for comprehensive testing methodologies in large-scale space missions.

In the days leading up to Apollo's launch, Kennedy Space Center (KSC) buzzed with anticipation and activity as VIPs from various sectors converged for this historic event. Among them was Wernher von Braun, the visionary rocket engineer, who arrived on November 6th. His presence underscored the significance of the upcoming mission, marked by an exclusive executive dinner and conference that evening.

NASA executives, industry leaders, congressional dignitaries, and diplomats also made their way to KSC, each centering around the pivotal Launch Complex 39. Here, NASA meticulously organized VIP guest lists, ensuring each director could extend personal invitations without overlap. These esteemed guests gathered in uncovered bleachers near the towering Vehicle Assembly Building (VAB), eagerly awaiting the momentous launch.

Meanwhile, KSC itself transformed into a hub of media activity. Press headquarters were established in nearby Cocoa Beach, accommodating accredited journalists who would play a crucial role in disseminating the mission's progress to the world. Tours of KSC were arranged for visiting media representatives, facilitated by a frequent shuttle service from Cocoa Beach.

For KSC employees and their families, the experience was uniquely personal. Positioned near their respective work assignments, they, too, awaited the launch with a blend of professional pride and personal anticipation.

The Apollo 4 launch vehicle (right) is rolled out from the Vehicle Assembly Building (far left) past the Mobile Servicing Structure.

Acknowledging exceptional contributions, NASA honored 43 employees from contractor teams with the "Manned Flight Awareness" distinction. These select individuals were treated to a VIP tour of KSC, participated in social events attended by six astronauts, and received a privileged view of the historic launch.

Throughout this bustling pre-launch atmosphere, NASA ensured extensive telephone facilities at the press site near LC-39, bearing the cost of facilitating seamless media coverage. This attention to detail reflected NASA's commitment to sharing the Apollo mission's achievements with a global audience, cementing KSC as the epicenter of an unforgettable journey into space exploration.

Apollo 4 marked a pivotal moment in space exploration, as the maiden flight of the Saturn V, a colossal launch vehicle that captured global attention. Media outlets struggled to convey the sheer scale of this technological marvel to the public. Descriptions likened its towering height to surpassing the Statue of Liberty and its weight to thirteen times that of a standard reference, such as a "good-sized navy destroyer," a comparison North American made in their media handouts.

The day before the historic launch, key figures gathered for an outdoor press conference, including Mueller, Phillips, von Braun, Deputy Administrator Robert C. Seamans, and Kennedy Space Center Director Kurt Debus. Against the backdrop of the mammoth Saturn V, more than a thousand journalists, including representatives from the Soviet Union, gathered to witness and report on this momentous occasion. The presence of such an international contingent underscored the global significance and interest in Apollo 4's ambitious mission, setting the stage for a new era in space exploration.

On November 6, 1967, at 10:30 pm EST (03:30 November 7 UTC), the meticulous countdown for Apollo 4 commenced, spanning a grueling 56 and a half hours. This intricate sequence kicked off with the critical task of loading propellants. A staggering logistics effort supported the mission, involving 89 trailer-truck loads of liquid oxygen, 28 trailer loads of LH2 (liquid hydrogen), and 27 rail cars filled with RP-1, a highly refined kerosene.

Despite the complexity of the operation, this countdown was notably smooth, with only minor issues encountered along the way. Crucially, the launch was completed on time thanks to strategic built-in holds in the countdown procedure. These scheduled pauses allowed for any accumulated delays to be addressed promptly, ensuring the mission remained on track for its historic liftoff. This strategic approach underscored NASA's meticulous planning and readiness, paving the way for the successful launch of the Saturn V and the groundbreaking Apollo 4 mission.

Apollo 4 on the launch pad

On November 9 at 7:00 am EST (noon UTC), Apollo 4 roared to life with the ignition of its five F-1 engines, unleashing a deafening cacophony across Kennedy Space Center. Despite the launch pads at LC-39 being over five kilometers (three miles) from the Vehicle Assembly Building, the sheer force of the sound exceeded expectations, buffeting structures such as the VAB, Launch Control Center, and nearby press buildings. The intensity was such that dust shook loose from the ceiling of the Launch Control Center, settling over the mission controllers' consoles.

William Donn of Columbia University likened the blast to one of the loudest noises in human history, excluding nuclear explosions. CBS commentator Walter Cronkite and producer Jeff Gralnick, observing from their trailer, felt the pressure so intensely that they feared for the integrity of their observation window as ceiling tiles fell around them. Renowned for his space coverage, Cronkite later described Apollo 4 as one of the most nerve-wracking missions he had witnessed.

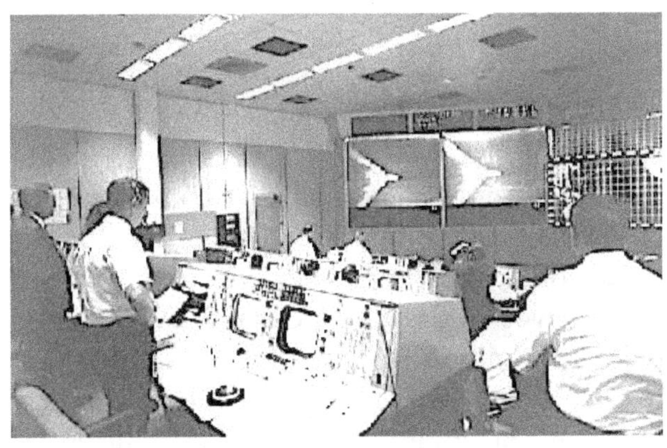

Mission controllers watch Apollo 4 climb to orbit.

The launch successfully placed the S-IVB stage and Command/Service Module (CSM) into a nearly circular orbit of approximately 190 kilometers (100 nautical miles), a crucial step mirroring future lunar missions. A critical simulation of the trans-lunar injection burn was executed following two orbits. This maneuver saw the S-IVB stage undergo its first in-space re-ignition, placing the spacecraft into an elliptical orbit with an apogee reaching 17,218 kilometers (9,297 nautical miles) and a perigee carefully aimed 84.6 kilometers (45.7 nautical miles) below Earth's surface. This trajectory ensured high-speed re-entry for the Command Module (CM) and deliberate destruction of the S-IVB stage after re-entry.

Upon completing this phase, the CSM separated from the S-IVB and adjusted its orbit to reach an apogee of 18,092 kilometers (9,769 nautical miles). Subsequently, the service module engine fired for 281 seconds to increase re-entry speed to 11,168 meters per second (36,639 feet per second) at an altitude of 120 kilometers (400,000 feet), replicating conditions expected during a return journey from the Moon.

The CM touched down approximately 8.6 nautical miles (16 kilometers) northwest of Midway Island in the North Pacific Ocean. The descent was observed from the aircraft carrier USS Bennington, the primary recovery vessel. Within hours, recovery teams had retrieved the spacecraft and one of its parachutes—an Apollo first, retrieved intact for detailed inspection. The CM was transported to Hawaii for deactivation before being sent to North American's Downey, California facility for comprehensive post-flight analysis. This mission validated crucial systems for future lunar endeavors and marked a significant step forward in America's space exploration ambitions.

Apollo 4 carried two motion-picture cameras strategically mounted on the Saturn V to capture pivotal moments during the mission. These cameras were designed to film the separation of the first stage and interstage from the launch vehicle. Once their task was complete, they were ejected and descended to the Atlantic Ocean in pods equipped with parachutes and radio beacons. Recovery efforts successfully retrieved them approximately 870 kilometers (470 nautical miles) downrange from Kennedy Space Center.

One significant consequence of the longer burn times was that the Command Module (CM) re-entered Earth's atmosphere slightly faster and at a shallower angle than initially calculated. This deviation, however, remained well within permissible margins. The anomaly arose not from any fault in the guidance system, which performed admirably but rather from the control of the burn from ground-based commands.

The CM's environmental control system effectively maintained cabin temperatures and pressures within safe ranges throughout the mission. During atmospheric re-entry, cabin temperature increased by only 5.6 °C (10 °F), a testament to the system's reliability and engineering. This ensured that the crew experienced optimal conditions despite the rigorous demands of their journey.

Earth photographed with the command module camera

President Lyndon Johnson captured the global significance of Apollo 4's launch, describing it as an awe-inspiring event visible to the entire world—the debut of the largest rocket ever flown. He emphasized how the launch symbolized America's harnessing of immense power for peaceful space exploration.

Wernher von Braun, the pioneering rocket engineer, praised the mission as a masterful achievement from start to finish, highlighting its flawless execution from liftoff precisely on schedule to the flawless performance of every stage of the Saturn V.

Roger E. Bilstein, in his historical account of the Saturn V, noted that Apollo 4's flawless mission brought immense joy and optimism to the entire NASA community. The mission's success bolstered confidence across the agency, fostering anticipation and belief that astronauts could land on the Moon by mid-1969.

Apollo 6, the subsequent flight of the Saturn V launched on April 4, 1968, faced more challenges than Apollo 4. Issues such as pogo oscillations during the first stage and an early shutdown of the second-stage engine led to deliberations on the necessity of further unmanned tests. Despite these setbacks, NASA proceeded with crewed missions, culminating in the historic Apollo 8 flight—the first to orbit the Moon.

The Command Module (CM) from the Apollo 4 mission, designated CM-017, holds historical significance and has found a place in public display. Initially transferred to the Smithsonian Institution in January 1969, it later moved to the North Carolina Museum of Life and Science in 1978. From there, it was relocated to NASA's Stennis Space Center, where it was exhibited until 2017. Currently, CM-017 is on display at the Infinity Science Center, located at Stennis Space Center in Pearlington, Mississippi, allowing visitors to witness a tangible piece of space exploration history up close.

Apollo 4 represented a pivotal achievement on multiple fronts—technically, managerially, and psychologically—marked by a series of groundbreaking milestones. It marked the debut flight of the Saturn V rocket's first and second stages, with the S-IVB stage having previously flown on Saturn IB launch vehicles. Notably, it was the inaugural launch of the complete Saturn V configuration, showcasing the culmination of years of engineering and testing.

Key among its achievements was the successful restart of the S-IVB stage in orbital flight, demonstrating crucial capabilities for future lunar missions. Apollo 4 also saw the first liftoff from Launch Complex 39 at Kennedy Space Center, a facility purpose-built for launching lunar missions.

Technological advancements were also evident in the first flight test of the Block II command module heatshield, critical for protecting astronauts during re-entry into Earth's atmosphere. Moreover, the mission included the first flight of a simulated lunar module, further refining procedures for future lunar landings.

The flawless execution of these milestones, with minimal technical issues, instilled a profound sense of confidence within NASA. As Administrator James E. Webb's successor, Thomas O. Paine, succinctly summarized, Apollo was clearly "on the way to the moon." This confidence was pivotal as NASA moved forward with ambitious plans to fulfill President John F. Kennedy's vision of landing astronauts on the lunar surface by the decade's end.

Additionally, the command module of Apollo 4 housed an automatic 70 mm film camera that played a significant role in capturing stunning images of Earth. Over a span of two hours and thirteen minutes, as the spacecraft approached and surpassed its highest point (apogee), this camera snapped 755 color photographs. These images were taken through the Command Pilot's forward-looking window at altitudes ranging from 13,510 to 18,092 kilometers (7,295 to 9,769 nautical miles). They represented the highest-altitude color images taken up to that time.

Although these photographs lacked the resolution required for detailed scientific analysis, they held immense value for Earth sciences enthusiasts and researchers alike. They provided a unique perspective on our planet from the vantage point of space, offering a glimpse into its beauty and complexity from a rarely seen distance.

During the Apollo 4 mission, all launch vehicle and spacecraft systems operated flawlessly. As the Saturn V rocket ascended to orbit, its three stages burned slightly longer than anticipated. This resulted in the spacecraft achieving an orbit approximately one kilometer higher than originally planned, a variation well within acceptable limits.

In the space race of the Cold War, in 1968, the Soviet Union continued to strengthen its cadre of specialists with the formation of Civilian Specialist Group 3, which included Vladimir Fartushny, Viktor Patsayev, and Valeri Yazdovsky. These individuals brought diverse expertise to Soviet space missions, contributing significantly to the advancement of their ambitious space exploration programs.

The following year, on August 14th, NASA established Group 7 in the United States. This group, initially trained for the USAF Manned Orbiting Laboratory (MOL) program, transitioned to NASA after the MOL program's cancellation. Among its members were Karol Bobko, Robert Crippen, Gordon Fullerton, and others who would go on to play pivotal roles in the early years of the Space Shuttle program.

Apollo 5

Apollo 5, also known as AS-204, marked a pivotal milestone in the Apollo program as the first uncrewed mission designed to test the Lunar Module (LM), essential for future Moon landings. Scheduled to launch on January 22, 1968, from Cape Kennedy, the mission aimed to validate the LM's performance in space conditions.

Delays plagued the mission's preparation, primarily due to challenges in developing the LM by Grumman Aircraft Industries. Originally intended to fly earlier, the mission faced setbacks, including the tragic Apollo 1 fire that claimed the lives of three astronauts and necessitated changes in mission planning and spacecraft safety protocols.

The launch vehicle, a Saturn IB rocket, eventually lifted off, carrying the LM into space. Despite encountering programming issues that altered the mission's objectives, Apollo 5 achieved its primary goals. It demonstrated critical maneuvers and operations of the LM in the space environment, laying crucial groundwork for subsequent crewed missions.

Preparations for the mission involved meticulous testing and integration efforts at the Kennedy Space Center, where the LM-1 arrived in June 1967. Months were dedicated to rigorous testing and the intricate process of mating the LM with the Saturn IB rocket. Technical challenges persisted, causing further delays before the final countdown commenced on January 21, 1968, culminating in the successful launch the following day.

Apollo 5's Saturn IB on the launchpad

Once the Lunar Module (LM) separated from the S-IVB booster in orbit, Apollo 5 embarked on a series of crucial orbital tests. However, shortly into the mission, a scheduled burn was automatically aborted when the Apollo Guidance Computer detected that the craft was not achieving the planned velocity. This unexpected event prompted a rapid response from Flight Director Gene Kranz and his team at Mission Control in Houston.

Facing this challenge, Kranz and his team swiftly formulated an alternate mission plan on the fly. Despite the setback, the revised objectives successfully met the mission's primary goals of testing the capabilities of the LM-1 under actual space conditions. This adaptability and problem-solving ability demonstrated the agility and expertise of NASA's mission control team during critical moments.

The outcomes of Apollo 5 were sufficiently promising that NASA decided to forego a planned second uncrewed mission dedicated to further LM testing. This decision accelerated NASA's timeline for achieving President Kennedy's ambitious goal of landing astronauts on the Moon by the end of the 1960s.

In 1961, President John F. Kennedy issued a bold challenge to the United States: to land an astronaut on the Moon and safely return them to Earth by the decade's end. This ambitious goal sparked intense deliberation within NASA, the US government's spaceflight agency, on how best to achieve lunar exploration.

By late 1962, NASA settled on a lunar orbit rendezvous approach for its Apollo missions. This strategy involved launching the complete Apollo spacecraft into lunar orbit using the powerful Saturn V launch vehicle's third stage, the S-IVB. Once in lunar orbit, astronauts destined for the Moon would transfer to the lunar excursion module (LEM), later renamed the lunar module (LM). This specialized craft would detach from the Apollo spacecraft's command and service module (CSM) and descend to the lunar surface.

Upon completing their tasks on the Moon, the astronauts would return to the LM, ascend from the lunar surface, and rendezvous with the orbiting CSM. After re-entering the CSM, they would jettison the LM before beginning their journey back to Earth.

In pursuit of this monumental task, NASA invited eleven companies to bid for the construction contract of the LM in 1962. On November 7 that year, NASA announced that Grumman Aircraft Engineering Corporation in Bethpage, New York, had been awarded the contract.

As with Apollo 4, Apollo 5 faced substantial delays primarily due to the Lunar Module (LM) development setbacks, which fell significantly behind schedule. Major General Samuel C. Phillips, the Apollo Program Manager, originally aimed for the uncrewed test flight of LM-1, the first lunar module, to launch by April 1967. NASA had anticipated six months for vehicle checkout and testing, prompting them to request Grumman Aircraft to deliver LM-1 to Kennedy Space Center by September 1966.

However, manufacturing challenges plagued the production of LM-1, causing repeated delays in its delivery. By January 1967, uncertainties persisted regarding the delivery date of LM-1 while preparations for its intended launch vehicle, AS-206, proceeded at Launch Complex 37. The tragic Apollo 1 fire later that month prompted a shift in plans: AS-204, initially designated for Apollo 1, was relocated to Launch Complex 37 to replace AS-206.

This decision was driven by AS-204's comprehensive research and development instrumentation, which made it suitable for the inaugural flight of the LM despite the suspension of crewed missions. NASA's adaptation aimed to leverage AS-204's capabilities to advance the LM program despite the challenging delays and setbacks encountered during its development phase.

Without a functional Lunar Module (LM), Grumman Aircraft constructed a plywood mockup at Launch Complex 37 to facilitate facilities verification, highlighting the urgency and complexity of the LM's development for NASA's Apollo program. On May 12, 1967, George M. Low, Apollo Program Spacecraft Manager, relayed to NASA headquarters that Grumman had committed to delivering LM-1 by June 28, despite acknowledging the formidable challenges ahead.

(November 1967) --- Lunar Module-1 being moved into position for mating with Spacecraft Lunar Module Adapter (SLA)-7 in the Kennedy Space Center's Manned Spacecraft Operations Building. LM-1 and SLA-7 are scheduled to be flown on the Apollo 5 (LM-1/Saturn 204) unmanned space mission.

True to their commitment, LM-1 arrived at Cape Kennedy on June 23 aboard Aero Spacelines' Super Guppy transport aircraft. Within days, the stages of the LM were successfully mated together. Overseeing this critical phase was John J. Williams, a seasoned veteran of launch operations from the Mercury and Gemini programs, who led a team of 400 personnel in meticulously verifying LM-1 against stringent specifications. Under their supervision, Grumman technicians conducted exhaustive testing and modifications to ensure the vehicle's readiness.

However, the journey to readiness was fraught with technical setbacks. Persistent leaks in the LM's ascent stage necessitated debating the stages in August for repairs. Despite efforts to rectify these issues and remate the stages, additional leaks surfaced in September, prompting further demating and repair work. Throughout this challenging period, Grumman technicians diligently addressed each issue, occasionally requiring the removal and repair of various equipment components.

By October, after rigorous efforts and multiple stages of demating and remating, the LM stages were finally reassembled for the next testing and validation phase. These trials underscored the meticulousness and resilience required in preparing the LM-1 for its pivotal role in advancing NASA's lunar exploration ambitions amidst intense technical scrutiny and relentless pursuit of mission success.

By September 6, 1967, Apollo 5 had fallen approximately 39 days behind the original schedule set on July 18. Despite this setback, diligent efforts were underway to address all known issues, except persistent leaks from the propulsion system, which remained a concern.

Progress continued through late 1967, with most mission documents finalized by the end of the year under the guidance of Mission Director William C. Schneider. On November 18, 1967, mission rules were formally issued, marking a critical step toward readiness. The following day, LM-1 was successfully mated to its designated launch vehicle, and comprehensive space vehicle readiness tests were completed by December.

As preparations intensified into early January 1968, NASA Administrator James E. Webb's office announced that the launch of Apollo 5 was scheduled no earlier than January 18, 1968. However, minor issues such as clogged filters prompted slight additional delays. The countdown demonstration test concluded satisfactorily on January 19, paving the way for an abbreviated 22-hour countdown on January 21.

Apollo 5 was specifically designed to validate the operational capabilities of the Lunar Module (LM) by conducting crucial tests during its flight. Among the primary objectives was activating both the ascent and descent engines of the LM. A pivotal "fire in the hole" test was scheduled to verify the ability of the ascent stage engine to ignite while still connected to the descent stage. This procedure simulated scenarios critical to lunar operations and potential abort situations during a lunar landing. The term "fire in the hole" originates from mining, signifying the imminent use of explosives.

In addition to these tests, Apollo 5 aimed to confirm the restart capability of the LM engines after initial use, ensuring reliability during extended missions. Moreover, the mission included testing the Instrument Unit in its configuration for the Saturn V launch vehicle, advancing preparations for future Apollo missions.

LM-1's ascent stage was expected to remain in orbit for about two years before re-entering the atmosphere and disintegrating, and the descent stage for about three weeks.

Apollo 5 launched into orbit atop the Saturn IB rocket, designated SA-204R, originally slated for Apollo 1 before that mission's tragic fire. Arriving at Cape Kennedy in August 1966, SA-204R underwent rigorous inspections following the fire incident, ensuring it was free from any damage or corrosion. With a total ignition weight of 589,413 kilograms (1,299,434 lb), including the spacecraft and propellant, SA-204R was a pivotal launcher for the Apollo program.

LM-1 is delivered by Super Guppy aircraft, June 23, 1967

The space vehicle for Apollo 5 stood 55 meters (180 ft) tall, presenting a robust yet compact appearance due to the absence of a Command and Service Module (CSM) and launch escape system. Instead, the Lunar Module (LM), designated LM-1, was housed within the spacecraft-lunar module adapter (SLA) at the top of the vehicle stack. SLA-7, situated just below the nose cap, featured four panels designed to open upon orbital insertion, enabling the LM to separate and maneuver independently.

LM-1, the inaugural flight-ready Apollo lunar module, was optimized for its test mission objectives. Notably, to reduce weight and streamline operations for the unmanned flight, LM-1 was configured without landing legs. Following structural concerns from testing, LM-1's windows, similar to those intended for LM-5 (destined for Apollo 11), were replaced with aluminum plates to mitigate potential risks during flight.

Despite being unmanned, LM-1 was equipped with a mission programmer to facilitate remote control operations. However, not all systems were fully activated, and consumables were deliberately limited: primary batteries were partially discharged to prevent over-voltage issues, and oxygen tanks for environmental control systems were only partially filled.

On January 22, 1968, Apollo 5 initiated its historic launch from Launch Complex 37B at Cape Kennedy Air Force Station, precisely at 17:48:08 Eastern Standard Time (22:48:08 UTC). The Saturn IB rocket performed flawlessly, propelling the second stage and Lunar Module (LM) into an initial orbit measuring 88 by 120 nautical miles (163 by 222 km).

Shortly after reaching orbit, the spacecraft underwent critical maneuvers: the nose cone was jettisoned, and following a 43-minute, 52-second coast, the LM separated from its adapter, now in a stable orbit of 90 by 120 nautical miles (167 by 222 km).

During the mission's second orbit, the planned 39-second burn of the descent engine commenced but was abruptly halted after only four seconds. The Apollo Guidance Computer intervened, detecting an anomaly where the spacecraft's velocity did not align with expectations. The issue stemmed from a suspected leak in one of the engine's valves, which had not been primed until just before ignition in orbit. This delayed the arrival of propellant to the engine, contributing to the observed velocity lag. Additionally, the fuel tanks were only half full, further impacting performance.

In crewed missions, astronauts would have been equipped to assess such situations and make informed decisions, a capability not available during this unmanned test flight. Although adjustments to the software could have potentially compensated for the delay, this option was not implemented due to procedural oversight.

Gene Kranz, renowned as the flight director for Apollo 5, was pivotal in overseeing the mission's critical operations. With Mission Control under his command, Kranz and his team faced initial communication challenges with the spacecraft. Still, they resolved to proceed with essential engine and "fire-in-the-hole" tests under manual control. These tests were crucial; their omission would have deemed the mission a failure. Despite the obstacles, Kranz's team successfully executed every planned engine burn, demonstrating their expertise and adaptability under pressure.

However, approximately eight hours into the mission, following the completion of the engine burns, the ascent stage encountered a setback when it began to spin uncontrollably. This issue was traced back to a malfunction in the guidance system, highlighting the intricate technical complexities involved in lunar module operations.

The ascent and descent stages were intentionally left in a low orbit as the mission progressed. This strategic decision ensured atmospheric drag would naturally decay their orbits, allowing them to re-enter the Earth's atmosphere. The ascent stage re-entered on January 24 and disintegrated upon re-entry, while the descent stage followed suit on February 12, splashing down in the Pacific Ocean several hundred miles southwest of Guam.

Additionally, simulations indicated that the launch vehicle's S-IVB stage, identified as 1968-007B, re-entered the atmosphere approximately 15.5 hours into the flight, concluding its role in the mission.

Apollo Spacecraft Program Manager George M. Low attributed Apollo 5's success to the robust hardware and the exceptional leadership of Gene Kranz and his flight control teams. Despite encountering challenges during the descent-engine burn, NASA deemed the mission successful in validating the Lunar Module (LM) systems crucial for future lunar missions. This accomplishment led to the cancellation of a planned second uncrewed flight test using LM-2.

Director of Flight Operations Christopher C. Kraft (left) and Manned Spaceflight Center director Robert R. Gilruth in Mission Control during Apollo

Following Apollo 5, the first crewed flight of the Lunar Module occurred during Apollo 9 in March 1969. This mission further tested and validated the LM's capabilities in Earth orbit, paving the way for subsequent missions that would culminate in landing astronauts on the lunar surface as part of the Apollo program's ambitious goals.

Christopher Columbus Kraft Jr. (1924–2019) was a pioneering figure in American aerospace engineering and NASA's early history. He played a pivotal role in shaping NASA's Mission Control Center, influencing its organizational development profoundly. Kraft began his career at NACA before joining NASA in 1958 with the Space Task Group. As NASA's first flight director, he directed operations for key milestones including America's first crewed spaceflight and the first spacewalk. Transitioning into management during the Apollo era, Kraft's strategic insights guided NASA's ambitious lunar missions. He served as director of the Johnson Space Center from 1972 to 1982, leaving a lasting legacy of excellence in human spaceflight.

Kraft works at his console inside the Flight Control area of the Mercury Control Center.

Apollo 6

Apollo 6, designated as Apollo Mission A, represented a pivotal milestone in NASA's Apollo program as an uncrewed Earth orbital flight primarily focused on testing the Command and Service Module (CSM). Operated by NASA, this mission utilized the Apollo CSM-020, manufactured by North American Rockwell, along with the Apollo Lunar Test Article 2R (LTA-2R), with a total launch mass of 36,930 kilograms (81,420 lb).

The mission commenced on April 4, 1968, at 12:00:01 UTC, launching from Kennedy Space Center's LC-39A atop the Saturn V SA-502 rocket. Apollo 6's primary objective was to evaluate the CSM's performance and systems in a space environment, a crucial step toward crewed lunar missions.

Over 9 hours, 57 minutes, and 20 seconds, Apollo 6 completed 3 orbits around Earth, operating within a low-Earth orbit regime. During this period, the spacecraft's performance and various systems were rigorously tested, providing critical data for assessing its readiness for lunar missions. 4,546,918.3 miles, 163 orbits.

After completing its mission objectives, Apollo 6 was successfully recovered by the USS Okinawa on April 4, 1968, at 21:57:21 UTC. The recovery took place in the North Pacific Ocean, specifically at coordinates 27°40′N, 157°55′W, north of Hawaii.

Apollo 6 launch as seen from a launch tower camera

Apollo 6's successful execution provided essential insights and data into the CSM's performance, propulsion systems, and overall operational capabilities crucial for NASA's subsequent crewed lunar missions.

Apollo 6, launched on April 4, 1968, marked a critical milestone in the United States' Apollo Program and the development of the Saturn V launch vehicle. Designated as AS-502, it was the final uncrewed mission in the Apollo series and the second dedicated test of the towering Saturn V rocket. This mission was pivotal in certifying the Saturn V for upcoming crewed lunar missions, paving the way for the historic Apollo 8 mission in December 1968.

Preparations for Apollo 6 began in early 1967 when its components started arriving at the Kennedy Space Center. However, progress was hindered by the extensive testing required for the Saturn V intended for Apollo 4, the rocket's maiden voyage launched in November 1967. Following Apollo 4's success, albeit with delays, preparations for Apollo 6 gained momentum, though the mission itself was rescheduled from March to April 1968 due to remaining technical refinements.

The primary objective of Apollo 6 was to validate the performance of the Saturn V's third stage, known as the S-IVB, in propelling both itself and the Apollo spacecraft toward lunar distances. This critical test aimed to confirm the rocket's capability to carry future astronauts beyond Earth orbit, crucial for subsequent missions aiming to reach and land on the Moon.

Apollo 6's mission profile underscored NASA's rigorous approach to testing and validation, ensuring each component and stage of the Saturn V performed flawlessly under extreme space conditions. Despite encountering technical challenges and delays, Apollo 6 successfully demonstrated the Saturn V's readiness for human spaceflight, setting the stage for the ambitious lunar exploration missions that followed.

The flight plan for Apollo 6 included a crucial maneuver known as a trans-lunar injection, followed by a contingency procedure for a direct return abort utilizing the service module's main engine. This complex operation was scheduled to unfold over approximately 10 hours. However, unforeseen challenges arose when vibrations during flight caused damage to some of the Rocketdyne J-2 engines in the second and third stages. Internal fuel lines ruptured, leading to the premature shutdown of a second-stage engine. Compounding the issue, a second engine in the same stage shut down due to cross-wiring with the first malfunctioning engine.

Despite these setbacks, the onboard guidance system reacted swiftly, compensating for the engine issues by extending the burns of the second and third stages. This adjustment resulted in achieving a parking orbit that, while more elliptical than planned, allowed the mission to continue.

Unfortunately, the damaged third-stage engine failed to reignite for the critical trans-lunar injection maneuver essential for sending the spacecraft toward the Moon. In response, flight controllers replicated the flight profile used successfully in the earlier Apollo 4 mission. This decision enabled the spacecraft to attain a high orbit and execute a high-speed return trajectory.

Apollo 6, the second test flight of the Saturn V launch vehicle, aimed to validate crucial aspects of the Apollo spacecraft's capabilities and readiness for lunar missions. The mission plan involved launching a command and service module (CSM) alongside a Lunar Test Article (LTA), a simulated lunar module (LM) equipped with structural vibration sensors. Unlike subsequent Apollo missions intended for lunar landing, Apollo 6's trajectory was designed to achieve trans-lunar velocities without entering lunar proximity.

The pivotal moment of the mission was the burn of the Saturn V's third stage, the S-IVB, which propelled the spacecraft toward a trajectory extending beyond the Moon's orbit. Following this burn, the CSM was slated to separate from the S-IVB, after which the service module's engine would initiate to decelerate the craft. This maneuver aimed to reduce the spacecraft's apogee to approximately 22,204 kilometers (11,989 nautical miles), simulating a "direct-return" abort scenario. In this critical contingency, the spacecraft would return swiftly to Earth without achieving lunar orbit.

During the return journey, another crucial engine firing was planned to accelerate the spacecraft, replicating the high-speed conditions anticipated during an actual return from the Moon. The re-entry angle was set at approximately -6.5 degrees, with a velocity of

11,100 meters per second (36,500 feet per second), mimicking the rigorous conditions of Earth re-entry following a lunar mission.

Apollo 6 aimed to thoroughly assess the Saturn V launch vehicle's capability to transport the complete Apollo spacecraft to the Moon. This mission focused specifically on testing the structural integrity of the Lunar Module (LM) under the stresses and vibrations generated by the near-full loads encountered during launch.

The Lunar Module Test Article (LTA-2R) being moved for mating with the spacecraft–LM adapter.

Having already qualified the Apollo spacecraft for crewed missions through the successful Apollo 4 flight, which marked the Saturn V's maiden voyage, Apollo 6 sought to validate the launch vehicle fully. The primary objectives included executing all planned mission phases to achieve the initial parking orbit and restarting the S-IVB stage. This crucial maneuver was intended to propel the spacecraft beyond the Moon's orbit, testing the Saturn V's performance under simulated lunar mission conditions.

Apollo 6, designated AS-502, was the second operational flight of the Saturn V launch vehicle and carried several significant payloads and modifications to advance the Apollo program's lunar objectives. The spacecraft included CSM-020, a Block I Command and Service Module (CSM) that incorporated some Block II upgrades but lacked the capability for docking with a Lunar Module, a feature reserved for later Block II versions.

One notable improvement in CSM-020 was its redesigned crew hatch, specifically engineered to enhance operability under lunar return conditions. This new hatch replaced the problematic design criticized by the Apollo 1 investigation board for its difficulty in emergency openings—a flaw tragically highlighted by the Apollo 1 fire on January 27, 1967, which claimed the lives of three astronauts.

llowing damage to its predecessor, SM-017, due to an explosion.

Due to its short duration, Apollo 6's mission did not activate all systems of SM-014. Notably, systems such as the radiators for managing electrical heat and the environmental control system were disconnected.

Kenneth S. Kleinknecht, the Command and Service Module manager at the Manned Spaceflight Center in Houston, expressed satisfaction with CSM-020 upon its arrival at Kennedy Space Center from North American Aviation. Unlike its ill-fated predecessor Apollo 1, which arrived with numerous unresolved issues, CSM-020 had only 23 mostly routine problems, indicating significant progress in spacecraft preparation and safety enhancements.

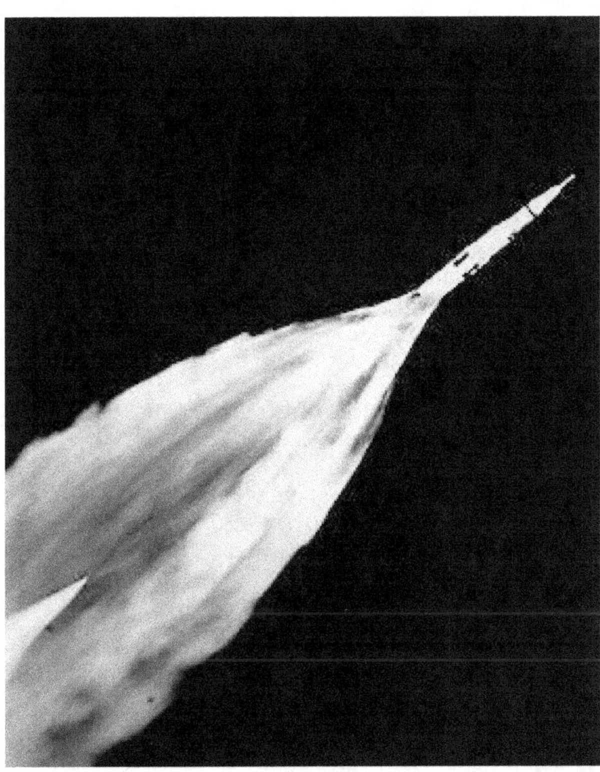

This view of the Apollo 6 launch was taken from a chase plane.

The command module for Apollo 6 was CM-020, equipped with essential mission programming and remote operation capabilities. Its service module, SM-014, differed from the originally intended SM-020, which had been reassigned to Apollo 4 fo

Accompanying CSM-020 on Apollo 6 was the Lunar Test Article (LTA-2R), a simulated lunar module housed within the Spacecraft-Lunar Module Adapter (SLA-9). LTA-2R featured a flight-type descent stage filled with a water-glycol mixture and freon in its oxidizer tanks. Designed without functional flight systems, its ballasted aluminum ascent stage was equipped with instruments to measure vibration, acoustics, and structural integrity throughout the mission.

The preparations for Apollo 6, designated AS-502, involved a series of logistical challenges and adjustments to accommodate the mission's timeline and technical requirements. The S-IC first stage arrived at Kennedy Space Center by barge on March 13, 1967, and was promptly erected in the Vehicle Assembly Building (VAB) four days later. This marked the beginning of assembly operations for the massive Saturn V rocket.

However, the S-II second stage, crucial for the mission, was not yet ready. A dumbbell-shaped spacer was temporarily used in place of the S-II to proceed with testing and assembly. This spacer replicated the dimensions and electrical connections of the actual S-II stage and allowed testing to continue until the arrival of the real S-II stage on May 24. After extensive integration and testing, the S-II was stacked and mated into the rocket assembly on July 7.

Apollo 6 also utilized High Bay 3 of the VAB for the first time. However, it soon became evident that the air conditioning systems in this new facility were inadequate to handle the Florida heat and the heat generated by the equipment and workers. Portable high-capacity units were brought in to maintain suitable working conditions during assembly and testing.

Delays during April were exacerbated by personnel and equipment commitments to the preceding Apollo 4 mission, which took priority. This overlap caused scheduling conflicts and slowed progress on Apollo 6's preparation. Additionally, the Mobile Service Launcher 2, which would support the rocket during launch, faced delays in installing its swing arms, crucial for vehicle stabilization and fueling operations.

Further complicating matters was the delayed arrival of the Command and Service Module (CSM). Originally scheduled for late September, its delivery was pushed back by two months, further compressing the final integration and testing timeline.

Following the successful launch of Apollo 4 on November 9, 1967, the pace of preparations for Apollo 6 accelerated, although challenges persisted with the flight hardware. The Command and Service Module (CSM) was mounted atop the launch vehicle on December 11, 1967, marking a critical step in the assembly process. Subsequently, on February 6, 1968, the fully assembled spacecraft stack was transported to Launch Complex 39A for final preparations.

The rollout to the launch pad was a complex operation that spanned an entire day, complicated by heavy rainfall. The crawler-transporter, responsible for moving the massive Saturn V rocket, encountered communication issues, necessitating a two-hour halt that delayed arrival at the launch pad until after dark. Additionally, high winds prevented the mobile service structure from being moved into place for two days, further hindering progress.

The flight readiness test, a comprehensive assessment of the spacecraft's operational readiness, concluded on March 8, 1968. Three days later, a thorough review cleared Apollo 6 for launch, contingent upon resolving identified action items and successfully completing the remaining tests. Initially scheduled for launch on March 28, 1968, delays arose due to issues with guidance system equipment and fueling procedures, pushing the launch dates to April 1 and then April 3.

The countdown demonstration test, crucial for simulating the launch sequence, commenced on March 24 and was completed within a week. Despite this, another postponement occurred, setting the final countdown for April 3, with liftoff scheduled for the following day.

During the final countdown on April 3, any remaining issues were promptly addressed during built-in holds without further delaying the mission. This meticulous approach ensured that Apollo 6 was thoroughly vetted and prepared for its pivotal role in testing the Saturn V launch vehicle under near-operational conditions.

Apollo 6 lifted off from Launch Complex 39A at Kennedy Space Center on April 4, 1968, at 7:00 a.m. local time (1200 UT), initiating a crucial Saturn V launch vehicle test flight. The launch proceeded smoothly initially, with the massive rocket performing as expected during the first two minutes of flight.

However, as the S-IC first-stage engines continued to burn, the vehicle encountered unexpected pogo oscillations. These oscillations caused variations in thrust, resulting in the Saturn V experiencing accelerations of up to ±0.6 g (5.9 m/s^2)—more than double the maximum design specification of 0.25 g (2.5 m/s^2). Despite these intense vibrations, the Saturn V sustained no significant damage, though it did lose one of the panels from the Spacecraft-Lunar Module Adapter (SLA).

George Mueller, NASA's Associate Administrator for Manned Space Flight, provided a detailed explanation of the pogo oscillations experienced during Apollo 6 to a congressional hearing:

"Pogo arises fundamentally because you have thrust fluctuations in the engines. Those fluctuations are inherent to all engines due to variations in combustion. This variability causes the thrust of the first stage to fluctuate, which is a normal characteristic of engine operation."

"The engine is fed through a pipe that carries fuel from the tanks to the engine. This pipe acts like an organ pipe in terms of its length and has its resonance frequency. Just like an organ pipe, it can oscillate."

"The structure of the Saturn V rocket is similar to a tuning fork in its design. When you combine the thrust fluctuations with the resonant frequencies of the fuel feed lines and the structural elements of the rocket, it sets up an interaction between these frequencies. This interaction can cause longitudinal oscillations—like a tuning fork vibrating—throughout the vehicle."

Mueller explained how the complex interplay of engine thrust variations, fuel feed line resonances, and structural design characteristics contributed to the pogo oscillations observed during the Apollo 6 mission.

After successfully jettisoning the S-IC first stage, Apollo 6 encountered challenges with its S-II second stage, particularly with its J-2 engines. The issues began when engine number two started experiencing performance anomalies approximately 225 seconds after liftoff. These problems escalated abruptly at T+319 seconds. At T+412 seconds, the Instrument Unit shut down engine number two entirely due to its deteriorating condition. However, due to a wiring cross-connection issue, this command inadvertently shut down engine number three, which had been operating normally up to that point.

Despite this setback, the Instrument Unit compensated for the unexpected shutdowns. The remaining three engines of the S-II stage continued to burn for 58 seconds longer than originally planned. Subsequently, the S-IVB third stage had to extend its burn duration by 29 seconds to correct the trajectory deviations caused by the S-II stage issues. The S-IVB also experienced a slight performance degradation during this extended burn period.

Due to deviations from the planned mission profile, Apollo 6's Command and Service Module (CSM) and the S-IVB stage were placed into an elliptical parking orbit ranging from 173.14 kilometers (93.49 nautical miles) at its closest point to Earth to 360.10 kilometers (194.44 nautical miles) at its furthest—short of the intended 190-kilometer circular orbit. Despite this, the mission continued as planned.

During the initial orbit, adjustments were made to the S-IVB's orientation to test techniques for landmark tracking intended for future astronaut missions. Following two orbits to assess readiness for Trans-Lunar Injection (TLI), attempts to restart the S-IVB for the TLI maneuver were unsuccessful.

In response, flight director Clifford E. Charlesworth and Mission Control decided on a pre-planned alternate mission approach. The CSM's Service Propulsion System (SPS) engine was utilized to raise the spacecraft's orbit, achieving a high apogee of 22,204 kilometers (11,989 nautical miles) and a corresponding low perigee to ensure re-entry. This strategy, similar to Apollo 4, aimed to fulfill critical mission objectives.

The SPS engine burn lasted 442 seconds, achieving the desired orbit but leaving insufficient propellant for a second burn to accelerate atmospheric re-entry. Consequently, the spacecraft re-entered Earth's atmosphere at a speed of 10,000 meters per second (33,000 feet per second), slightly lower than the planned 11,000 meters per second (37,000 feet per second) intended to simulate a lunar return.

During its high-altitude phase, the Command Module (CM) gathered valuable data on the protection offered by its hull against the Van Allen Belts—a crucial consideration for future crewed missions. Approximately ten hours after launch, the CM splashed down 80 kilometers (43 nautical miles) north of Hawaii in the North Pacific Ocean, slightly off its intended landing point. The USS Okinawa swiftly recovered it.

Meanwhile, the Service Module (SM) was jettisoned before re-entry and burned up in the atmosphere. The S-IVB stage's orbit gradually decayed, leading to its controlled re-entry on April 26, 1968, marking the conclusion of Apollo 6's mission objectives despite the technical challenges encountered during the flight.

After Apollo 6's launch, Apollo Program Director Samuel C. Phillips acknowledged in a post-launch press conference that the mission was "less than a perfect mission," noting the challenges encountered. Despite losing two engines during the flight, Phillips highlighted reaching orbit as a significant unplanned accomplishment for the launch vehicle.

George Mueller, NASA's Associate Administrator for Manned Space Flight, also commented on Apollo 6. He initially described it as "a good job all around" and emphasized the launch's success and the mission's overall balance. However, he later revised his assessment, acknowledging that Apollo 6 would be regarded as a failure due to the issues faced during the mission.

The primary issue encountered, pogo oscillation, was a well-known phenomenon during Saturn V's first stage. NASA had previously believed that measures had been taken to "detune" the vehicle to prevent vibrations at natural frequencies. Following Apollo 6, NASA and its contractors extensively addressed these challenges for future missions. Approximately 1,000 engineers from both government and industry collaborated on solutions, including filling cavities in valves leading to the F-1 and J-2 engines with helium gas to dampen pressure oscillations effectively.

NASA engineers conducted thorough investigations following the issues encountered during Apollo 6 with the S-II and S-IVB stages, which were traced back to problems with the J-2 engines. They implemented corrective measures to prevent recurrence in future missions.

The problem identified was related to the propellant lines leading to the spark igniters in the J-2 engines. During ground testing, it was observed that the lines, which were equipped with metal bellows to accommodate thermal expansion, performed adequately due to the protective layer of frost and liquid air formed by cold propellants passing through them. However, this protective effect was absent in the vacuum of space encountered during Apollo missions. As a result, the bellows vibrated excessively and failed under peak flow conditions, leading to burn-through of the propellant lines.

Apollo 6 command module on display at the Fernbank Science Center in Atlanta, Georgia

To address this issue, NASA engineers replaced the bellows with rigid bends and strengthened the propellant lines to ensure they could withstand the operational conditions of space. This modification aimed to prevent similar failures in future flights of the Saturn V rocket.

In the aftermath of Apollo 6, discussions among NASA engineers included considerations for configuring the spacecraft's emergency detection system to abort missions during excessive pogo oscillations automatically. However, this proposal faced opposition, particularly from Director of Flight Crew Operations Deke Slayton. Instead, the focus shifted to developing a "pogo abort sensor" to allow the flight crew to decide whether to abort based on real-time data. By August 1968, it became evident through further analysis and testing that pogo issues could be effectively managed without such a sensor, leading to the project's abandonment.

The issues with the Spacecraft-Lunar Module Adapter (SLA) encountered during Apollo 6 were attributed to its honeycomb structure. As the Saturn V rocket accelerated through the atmosphere, air and water trapped within the honeycomb cells expanded, causing the adapter's surface to break free in places.

To mitigate this problem, engineers implemented several corrective measures. They drilled small holes in the adapter's surface to allow trapped gases to escape, preventing pressure buildup that could lead to surface failures. Additionally, a thin layer of cork was applied to the adapter to help absorb any moisture that could exacerbate the problem.

NASA's swift response and effective implementation of these fixes satisfied the Senate Committee on Aeronautical and Space Sciences. By late April, the committee reported that NASA had promptly analyzed and diagnosed the issues arising from Apollo 6 and had taken appropriate corrective actions.

Following detailed assessments of the Saturn V's performance and the efficacy of the implemented fixes, engineers at the Marshall Space Flight Center concluded that another uncrewed test flight of the Saturn V was unnecessary. Therefore, the next Saturn V mission, Apollo 8, would proceed with a crew aboard. Apollo 7, the first crewed Apollo mission, would be launched using the smaller Saturn IB rocket.

After completing its mission, Command Module CM-020 from Apollo 6 was transferred to the Smithsonian Institution for preservation and historical display. Currently, the Apollo 6 command module is on public display at the Fernbank Science Center in Atlanta, Georgia, allowing visitors to appreciate its role in the Apollo program's early space exploration efforts.

During the Apollo 6 mission, several cameras were integrated into the Saturn V rocket and the Command Module (CM) to capture critical moments and provide valuable data. However, not all of them operated as intended, leading to mixed results regarding footage retrieval.

Still from footage of Apollo 6's interstage falling away (NASA)

Four cameras were installed on the S-IC first stage of the Saturn V to eject them for recovery after specific events. Unfortunately, three of these four cameras failed to eject, destroying them. Only one camera was successfully recovered, intended to film the separation between the S-IC and S-II stages. The failure to eject the cameras was traced to insufficient nitrogen pressure in the ejection systems.

Similarly, on the S-II second stage, two cameras were installed, but only one was recovered. These cameras were designed to capture footage of the liquid oxygen tank and the stage separation events. The recovered camera successfully filmed the S-IC/S-II separation, providing valuable visual data despite the overall loss of most of the cameras.

Inside the Command Module, a motion picture camera was mounted to document the launch and re-entry phases of the mission. However, due to the mission's extended duration (about ten minutes longer than planned), the camera did not capture the re-entry events as initially intended.

Additionally, a 70 mm still camera operated within the Command Module during parts of the mission, focusing on Earth through the hatch window. This camera used haze-penetrating film and a specialized filter combination, producing high-resolution images with superior color balance compared to previous missions. The captured images included views of the United States, the Atlantic, Africa, and the western Pacific Ocean, proving valuable for cartographic, topographic, and geographic studies.

Despite the setbacks with some cameras, the images and footage successfully retrieved from Apollo 6 contributed to scientific research and provided essential data for future missions in the Apollo program.

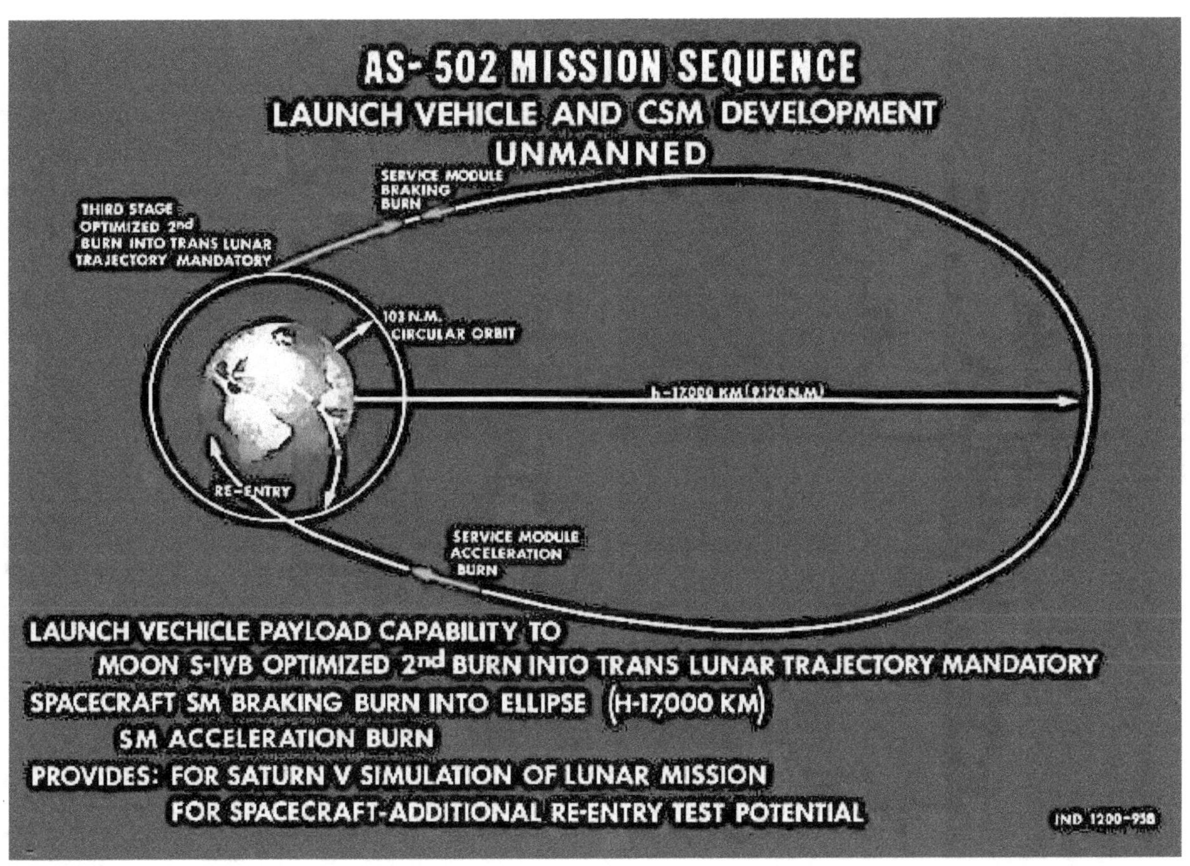
AS-502 MISSION SEQUENCE
LAUNCH VEHICLE AND CSM DEVELOPMENT
UNMANNED
SERVICE MODULE BRAKING BURN
THIRD STAGE OPTIMIZED 2nd BURN INTO TRANS LUNAR TRAJECTORY MANDATORY
103 N.M. CIRCULAR ORBIT
h-17000 KM (9120 N.M.)
RE-ENTRY
SERVICE MODULE ACCELERATION BURN
LAUNCH VECHICLE PAYLOAD CAPABILITY TO
MOON S-IVB OPTIMIZED 2nd BURN INTO TRANS LUNAR TRAJECTORY MANDATORY
SPACECRAFT SM BRAKING BURN INTO ELLIPSE (H-17,000 KM)
SM ACCELERATION BURN
PROVIDES: FOR SATURN V SIMULATION OF LUNAR MISSION
FOR SPACECRAFT-ADDITIONAL RE-ENTRY TEST POTENTIAL
IND 1200-958

Apollo 7

The First Crewed Mission

In October 1968, NASA launched Apollo 7, a pivotal crewed mission orbiting Earth aboard the Apollo Command and Service Module (CSM-101). Commanded by Walter M. Schirra and accompanied by crew members Donn F. Eisele and R. Walter Cunningham, the mission marked a significant step forward in the Apollo program.

The spacecraft, built by North American Rockwell, weighed 36,419 pounds at launch and orbited Earth for ten days, 20 hours, 9 minutes, and 3 seconds. For 163 orbits, Apollo 7 operated in a low Earth orbit with a perigee altitude of 227 kilometers and an apogee altitude of 301 kilometers, circling the Earth every 89.55 minutes.

Apollo 7 was launched from Cape Kennedy LC-34 atop a Saturn IB rocket, designated SA-205. The mission began on October 11, 1968, at 15:02:45 UTC, with the spacecraft taking on the callsign "Apollo 7". Its return to Earth concluded on October 22, 1968, when the USS Essex recovered the capsule from the North Atlantic Ocean at coordinates 27°32′N 64°04′W, having traveled 579,606.9 miles.

The Apollo program, renowned for its pioneering missions, distinguished each venture with unique call signs and distinctive mission insignias, embodying the essence and goals of human space exploration. These symbols, integral to the spirit of each mission, celebrated milestones, honored crew members, and ignited a generation's passion for the cosmos.

The Apollo 7 Saturn IB space vehicle is launched from the Kennedy Space Center's Launch Complex 34 at 11:03 a.m. October 11, 1968. A tracking antenna is on the left and a pad service structure on the right.

Apollo 7, the inaugural manned mission of the program, bore no call sign, marking its pivotal role in testing the Apollo spacecraft in Earth's orbit. Its mission insignia featured an elliptical trail of orange flame, symbolizing the spacecraft's orbital path around Earth against a deep navy backdrop representing the boundless expanse of space. The crew, commanded by Walter Schirra, included Major Donn Eisele and Walter Cunningham. Eisele, an accomplished astronaut with a background in astronautics from the U.S.

Naval Academy and the U.S. Air Force Institute of Technology, made his inaugural spaceflight. With a diverse U.S. Navy and Marine Corps career and a physics degree from UCLA, Cunningham also embarked on his first journey beyond Earth.

Initially slated for Apollo 1, Eisele joined Schirra's crew after a shoulder injury, replacing Roger Chaffee alongside Gus Grissom and Ed White. Formally designated as the Apollo crew on September 29, 1966, their mission initially scheduled as Apollo 2 faced potential cancellation amid internal NASA deliberations post-Apollo 1. As the Apollo program surged forward, meticulous preparation characterized every phase of their journey.

The backup crew for Apollo 7, destined for Apollo 10, included Thomas Stafford as commander, John W. Young as command module pilot, and Eugene A. Cernan as lunar module pilot. Their team originally featured Edward G. Givens until his tragic death in 1967, with William R. Pogue stepping in.

At Kennedy Space Center, Ronald E. Evans oversaw spacecraft readiness, while John L. 'Jack' Swigert served as launch capsule communicator (CAPCOM), coordinating operational details from Earth. William R. Pogue contributed to refining procedures, ensuring comprehensive support for both primary and backup crews.

Within Mission Control, astronaut CAPCOMs, including Stafford, Swigert, Young, Cernan, Evans, and Pogue, maintained crucial communications with the crew in space. Flight directors Glynn Lunney, Gene Kranz, and Gerry Griffin provided pivotal leadership, guiding Apollo 7 and subsequent missions toward NASA's lunar exploration goals with unwavering precision and safety.

Backed by his Navy background and experience, Commander John Watts Young was a backup for Walter Schirra, ready to step into forefront missions. Eugene Andrew "Gene" Cernan, later to command Apollo 17 and be the last human to walk on the Moon, served as the backup for Donn Eisele, highlighting the program's depth of talent and commitment.

The CAPCOM team exemplified NASA's rigorous selection process, ensuring seamless communication between Earth and the Apollo 7 crew throughout their mission. Meanwhile, support crew members Swigert, Evans, and Pogue upheld readiness and continuity, ensuring operational success.

In Mission Control, flight directors Lunney, Kranz, and Griffin provided steady guidance, navigating shifts and challenges to orchestrate each mission's intricate dance between Earth and space. They meticulously executed objectives with precision and dedication.

Apollo 7, scheduled to use the upgraded Block II spacecraft tailored for lunar missions, represented a pivotal step forward from the earlier Block I configuration used in Apollo 1. Unlike its predecessor, Block II was equipped for lunar module docking, a critical capability for future moon landings. The Command Module (CM) and astronauts' spacesuits underwent extensive redesigns to mitigate the risks that led to the Apollo 1 tragedy.

Schirra's crew, designated for this "open-ended" mission, aimed to thoroughly test life support systems, propulsion, guidance, and control capabilities. Initially planned for a 14-day duration, the mission was shortened to 11 days, reflecting adjustments made in response to the Apollo 1 findings.

Tragically, the challenges of space exploration were underscored by the loss of astronaut Donn Eisele in 1987, a poignant reminder of the risks borne by those who venture into the cosmos. Meanwhile, Major William Reid Pogue's pivotal contributions after replacing Edward Galen Givens Jr., who passed away in 1967, highlighted the resilience and determination of those committed to advancing humanity's reach into the stars.

Preparation for Apollo 7 was marked by a significant shift in perspective among its crew members. Initially, Wally Schirra had shown little enthusiasm for another spaceflight, focusing instead on his future beyond NASA. However, Schirra's outlook changed dramatically following the tragic Apollo 1 fire. As Cunningham recalls, "Wally Schirra was now seen as the man tasked with revitalizing the manned space program. That responsibility piqued Wally's interest."

The stakes were clear for Donn Eisele and the rest of the crew. Eisele reflected, "Coming right after the fire, we understood that the fate of the entire manned space program—and indeed our own lives—hinged on the success or failure of Apollo 7."The pressure was immense as they prepared to undertake a mission critical for their careers and the future of American space exploration.

In the aftermath of the Apollo 1 fire, the crew's confidence in North American Aviation's Downey, California facility, responsible for building the Apollo command modules, was initially shaken. Determined to ensure the safety of their mission, they insisted on overseeing every step of the craft's construction and testing. This commitment to vigilance, however, posed challenges to their training schedule. At that time, simulators for the Command Module (CM) were not yet operational at the Manned Spacecraft Center in Houston or Kennedy Space Center (KSC) in Florida, prolonging their stay in Downey.

Once simulators became available, the crew faced a new hurdle—juggling intensive training sessions alongside the demands of construction oversight. Despite the support of backup and support crews, the workload often extended their days to 12 or 14 hours. The Command Module was transported to KSC upon completion, shifting the crew's training focus to Florida. Despite periodic returns to Houston for planning and technical meetings, weekends at home became a rarity as they remained at KSC for continuous training and spacecraft testing.

Schirra's crew in training for Apollo 2, 1966

Former astronaut Tom Jones, reflecting on these events in 2018, noted that Wally Schirra, armed with a heightened awareness of the mission's risks, effectively leveraged his influence in negotiations with NASA and North American Aviation. Whether in conference rooms or on the assembly line, Schirra's insistence on rigorous safety measures left an indelible mark on the preparations for Apollo 7.

The preparations for Apollo 7 were thorough and rigorous, reflecting the crew's dedication to safety and readiness. The crew devoted five hours to training for every hour they anticipated to spend aboard the spacecraft during its potential eleven-day mission. This included technical briefings, pilots' meetings, and extensive personal study.

Their training regimen encompassed a range of critical scenarios. They underwent launch pad evacuation drills, practiced water egress procedures for exiting the spacecraft after splashdown, and received training in firefighting techniques. At MIT, they diligently trained on the Apollo Guidance Computer, essential for navigating in space. Each crew member logged 160 hours in

Command Module (CM) simulations, some of which involved live participation from Mission Control in Houston.

One pivotal test, the "plugs out" test, was conducted with the prime crew inside the spacecraft, albeit with the hatch open—a stark contrast to the tragic Apollo 1 test, in which crew members were unable to open the inward-opening hatch before the fire engulfed the cabin. This critical design flaw was rectified for Apollo 7, exemplifying the rigorous safety improvements implemented after the Apollo 1 incident.

In preparation for Apollo 7, command modules similar to the one used in the mission underwent extensive testing to ensure their reliability and safety. At the Manned Spaceflight Center in Houston, a three-astronaut crew consisting of Joseph P. Kerwin, Vance D. Brand, and Joe H. Engle spent eight days inside a command module placed in a vacuum chamber in June 1968. This test evaluated the spacecraft's systems under simulated space conditions.

Another test involved astronauts James Lovell, Stuart Roosa, and Charles M. Duke spending 48 hours at sea aboard a command module lowered into the Gulf of Mexico from a naval vessel in April 1968. This test assessed how the spacecraft's systems would respond to seawater exposure, a critical consideration for post-splashdown recovery operations.

Additional tests conducted in Houston included controlled fires aboard boilerplate command modules using various atmospheric compositions and pressures. These tests yielded crucial data that informed the decision to use a mixture of 60 percent oxygen and 40 percent nitrogen within the command module at launch. Within four hours after launch, this mixture would be replaced by a lower-pressure pure oxygen environment, enhancing fire safety measures.

Further validation tests involved dropping boilerplate spacecraft to simulate parachute deployments and evaluate potential damage if a command module landed on solid ground. All tests yielded satisfactory results, affirming the spacecraft's design and readiness for manned missions.

In the lead-up to Apollo 7, the Soviet Union launched unmanned probes, Zond 4 and Zond 5 (the latter famously carrying two tortoises), on circumlunar trajectories around the Moon, hinting at potential crewed lunar missions to come. Meanwhile, NASA faced delays with its Lunar Module (LM) development.

Amidst these developments, George Low, Apollo Program Spacecraft Manager, proposed a bold plan: if Apollo 7 proved successful, Apollo 8 would proceed to lunar orbit without the Lunar Module. This proposal, accepted by NASA, heightened the stakes for Apollo 7, making its success pivotal for advancing the lunar exploration timeline.

According to Thomas Stafford, backup crew commander Wally Schirra keenly felt the immense pressure of the program's expectations. This pressure manifested in Schirra's increasingly critical and sarcastic demeanor as he navigated the challenges leading up to the mission.

During the Mercury and Gemini programs, Guenter Wendt was pivotal as the leader of McDonnell Aircraft's spacecraft launch pad teams. His responsibility extended to ensuring the spacecraft's readiness for launch, which earned him deep respect and admiration from astronauts like Wally Schirra. Wendt's meticulous oversight and attention to detail became synonymous with mission success during these early manned space missions.

However, with the transition to the Apollo program, the spacecraft contractor shifted from McDonnell to North American. Consequently, Wendt was not initially assigned as the pad leader for Apollo 1. Undeterred, Schirra, known for his staunch confidence in Wendt's capabilities, actively campaigned for his return. Schirra's insistence led Deke Slayton, NASA's Director of Flight Crew Operations, to intervene and convince North American's management to recruit Wendt from McDonnell. Additionally, Schirra advocated for Wendt's shift change from midnight to the day shift, ensuring he could serve as pad leader for Apollo 7.

Wendt's role as pad leader throughout the Apollo program underscored his critical contribution to astronaut safety and mission readiness. His departure from the spacecraft area just before launch, a routine part of the evacuation protocol, was noted by astronauts Cunningham and Eisele, reflecting the profound trust placed in Wendt's expertise and leadership.

Guenter Wendt's tenure as pad leader symbolized the dedication and precision required for the monumental achievements of Project Apollo, where meticulous preparation on the launch pad was paramount to the success of lunar exploration missions.

The prime crew of the first manned Apollo space mission from left to right are: Command Module pilot, Don F. Eisele, Commander, Walter M. Schirra Jr. and Lunar Module pilot, Walter Cunningham. The photograph was taken inside the White Room which is attached to the crew access arm. From here astronauts ingress and egress the spacecraft. Commander Wally Schirra Jr. is seen inside the opening of the Command Module's main

Apollo 7 marked a significant milestone with the deployment of Command and Service Module 101 (CSM-101), the inaugural flight of the Block II variant. This upgraded spacecraft boasted enhanced capabilities, notably the ability to dock with a Lunar Module (LM), although none accompanied Apollo 7 on its mission.

The spacecraft configuration included crucial components such as the launch escape system and a spacecraft-lunar module adapter (SLA), designated as SLA-5. Despite its designation, SLA-5 did not house an LM; instead, it served as a structural interface between the Service Module (SM) and the Instrument Unit of the S-IVB stage. Without an LM, a structural stiffener fulfilled the adapter's role.

During the launch sequence, the launch escape system fulfilled its vital role and was discarded following the ignition of the S-IVB stage. Meanwhile, upon orbital separation from the spent S-IVB, the SLA remained attached to it, part of the spacecraft's separation protocol.

5 Aug. 1968) --- Apollo Spacecraft 101 Command/Service Modules being moved into position for mating with Spacecraft Lunar Module Adapter (SLA)-5 in the Kennedy Space Center's Manned Spacecraft Operations Building. Apollo Spacecraft 101 will be flown on the first manned Apollo space mission, Apollo 7 (Spacecraft

Following the tragic Apollo 1 fire, which led to a comprehensive reevaluation of spacecraft safety, the Block II Command and Service Module (CSM) underwent extensive redesign. Over 1,800 modifications were recommended, with approximately 1,300 explicitly implemented for the Apollo 7 mission. These changes aimed to enhance crew safety and spacecraft reliability significantly.

A pivotal improvement was introducing a new hatch design made from aluminum and fiberglass. This hatch could be opened by the crew from inside the spacecraft in just seven seconds and by ground personnel in ten seconds from outside—a crucial upgrade for emergency evacuation scenarios. The high-pressure oxygen system also saw a critical upgrade, replacing aluminum tubing with stainless steel to mitigate fire risks. Materials prone to combustion were substituted with non-flammable alternatives, including replacing plastic switches with metal ones. Furthermore, an emergency oxygen system was installed to shield astronauts from toxic fumes in case of a fire, complemented by onboard firefighting equipment.

In a cultural shift at NASA following the Gemini 3 mission, where Gus Grissom humorously named his spacecraft "Molly Brown," NASA enforced a policy against spacecraft naming. Despite this directive, Wally Schirra sought to name his craft "Phoenix" for Apollo 7. However, NASA denied his request, maintaining its stance against named spacecraft.

The necessity for distinct call signs became crucial with Apollo 9, which carried a Command Module (CM) and a Lunar Module (LM) that would separate and rock during its mission. This operational requirement mandated unique call signs for each vehicle, departing from the previous convention of using only the mission designation for identification.

Apollo 7 marked a pivotal moment in the Apollo program, unfolding against the backdrop of recovery and progress following the tragic Apollo 1 fire twenty-one months earlier. It heralded the beginning of the manned phase, aimed at demonstrating the Command and Service Module (CSM) capabilities and affirming the readiness of crew and ground support facilities. One of its key objectives was to showcase the CSM's rendezvous capability, a critical step toward lunar exploration.

The Apollo 7 mission, propelled by the Saturn IB launch vehicle designated SA-205, represented a pivotal milestone in NASA's Apollo program. The Saturn IB was an upgraded version of the Saturn rocket, specifically tailored to meet the demands of manned spaceflight missions.

Designated as Eastern Test Range #66, Apollo 7 aimed to validate crucial capabilities of the Command and Service Module (CSM). Notably, CSM-101 made its maiden flight in the Block II configuration, primed to support future lunar missions by accommodating the Lunar Module (LM) and integrating advanced systems.

Unlike the Saturn V, which powered subsequent lunar missions, the Saturn IB was chosen for Apollo 7 due to its capability to launch into low Earth orbit without the need for the larger Saturn V. SA-205, the fifth Saturn IB to fly, featured enhanced propellant lines for the J-2 engines' augmented spark igniters. This improvement prevented issues like the premature shutdown experienced during the unmanned Apollo 6 flight, where post-flight analysis revealed propellant leaks in the J-2 engine lines shared with the Saturn V.

The Saturn IB, a two-stage rocket, utilized an S-IVB second stage akin to the third stage of the Saturn V. This configuration proved versatile beyond the Apollo program's lunar missions, later deploying crews to Skylab and facilitating international cooperation through the Apollo-Soyuz Test Project.

Apollo 7's Saturn IB, SA-205, at Launch Complex 34

Apollo 7 holds a unique place in history as the sole crewed Apollo mission to launch from Cape Kennedy Air Force Station's Launch Complex 34. This site, significant for its role in early manned spaceflight, saw the deployment of Apollo 7 before being retired in 1969. Subsequent missions, including those of Skylab and the Apollo-Soyuz Test Project, departed from Launch Complex 39 at nearby Kennedy Space Center.

Launch preparations for Apollo 7 commenced with a meticulously planned countdown process, initiating at 19:00 GMT on October 6, 1968. The countdown sequence included strategic holds at critical junctures to address potential spacecraft issues: a six-hour hold at T-72 hours, a three-hour hold at T-33 hours, and a final hold at T-6 hours to ensure the launch crew had essential rest before the final push toward liftoff. These holds exemplified NASA's commitment to thorough preparation and safety in executing manned space missions during the Apollo era.

On October 11, 1968, at 09:00 GMT, the final countdown resumed smoothly until T-10 minutes, when preparations for the S-IVB stage of the launch vehicle encountered a minor setback. The initiation of the thrust chamber jacket chill down took longer than anticipated, risking a delay in the automatic countdown sequence. To ensure all temperature requirements were met, a hold was called at T-6 minutes 15 seconds, lasting 2 minutes 45 seconds. Subsequent analysis confirmed that a chill down would have naturally occurred without the hold, but the precautionary measure was prudent given the real-time conditions.

Apollo 7 marked a pivotal moment in American space exploration, breaking a 22-month hiatus in crewed missions since the tragic Apollo 1 incident. Launched from Launch Complex 34 at Cape Kennedy, Florida, precisely at 11:02:45 am EDT on October 11, 1968, the mission faced significant challenges from the outset.

Leading up to liftoff, concerns over weather conditions loomed large. An extensive high-pressure system centered over Nova Scotia brought powerful easterly surface winds to the launch site. This violated safety protocols, as these conditions posed a risk of blowing the Command Module (CM) back over land in case of an emergency abort instead of the planned water landing. Despite objections from Commander Walter Schirra, who expressed reservations about launching under such circumstances due to safety concerns related to the CM's old-style crew couches, management decided to proceed, overriding the safety rule.

On launch day, the conditions were meticulously recorded: scattered clouds and cumulonimbus clouds covering about 30 percent of the sky at 2,100 feet altitude, with clear visibility up to 10 statute miles. The temperature stood at 82.9°F with a relative humidity of 65 percent, while an anemometer measured winds blowing at 19.8 knots from a true north direction.

However, Apollo 7's ascent phase proceeded flawlessly, showcasing the precision and resilience of NASA's mission planning and execution. Liftoff occurred precisely within the planned launch window, ushering in a new space exploration chapter amidst the Apollo program's challenges and triumphs.

Moments after liftoff, the Saturn IB vehicle smoothly transitioned from a launch pad azimuth of 100° to a flight azimuth of 72° east of north. The first stage provided continuous thrust until the center engine cutoff at 000:02:20.65. The outboard engines shut down 3.67 seconds later, achieving an Earth-fixed velocity of 6,479.1 ft/sec, closely aligning with predicted cutoff conditions.

Following separating the S-IB stage from the upper stage at 000:02:25.59, the S-IVB engine ignited at 000:02:26.97. The stage's cutoff occurred at 00:10:16.76, exhibiting minimal deviations from the planned trajectory—just 2.3 ft/sec in velocity and 0.054 nautical miles in altitude. The S-IVB burn time of 469.79 seconds closely matched the predicted duration, demonstrating precise performance.

Throughout ascent, all structural load limits remained well within design tolerances, affirming the robustness of the Saturn IB's construction. Maximum wind conditions during this phase peaked at 81 knots at 172,000 feet. Wind shear in the high dynamic pressure region reached 0.0113 sec^{-1} in the pitch plane at 48,100 feet, while the highest wind speed observed in this region was 30.3 knots from 309° at 44,500 feet.

After separating from the Apollo 7 spacecraft, the spent S-IB stage followed a predicted trajectory until its impact in the Atlantic Ocean. The impact coordinates were determined to be at latitude 29.76° north and longitude 75.72° west, approximately 265.01 nautical miles from the launch site at Cape Kennedy.

After a flawless launch sequence, the Apollo 7 spacecraft gracefully entered Earth's orbit at 000:10:26.76, calculated from the moment of S-IVB cutoff with a 10-second allowance for engine tailoff and related effects. In this pivotal moment, the spacecraft's orbital conditions were meticulously measured: it reached an apogee, the highest point in its orbit, at 153.7 nautical miles, and a perigee, the lowest point, at 123.3 nautical miles. Its orbital inclination, the angle of its path relative to Earth's equator, was set at 31.58 degrees. With an orbital period of 89.70 minutes and a velocity of 25,538.6 feet per second, Apollo 7 settled into a stable orbit, marking a crucial milestone that paved the way for the mission's planned objectives and maneuvers in space.

Upon achieving orbit, the Apollo 7 spacecraft was internationally designated as 1968-089A, while the S-IVB stage was designated as 1968-089B. In the weightless space environment, the crew quickly adapted without experiencing disorientation issues related to movement inside the Command Module (CM) or observing Earth through the windows.

An interesting experiment during the early stages of the mission involved the lunar module pilot attempting to induce vertigo or motion sickness by moving their head rapidly in all directions. However, these efforts yielded negative results, indicating the crew's robust adaptation to microgravity conditions.

Apollo 7 S-IVB rocket stage in orbit

Despite this adaptability, the crew initially reported soreness in their back muscles around the kidney area. This discomfort was effectively alleviated through regular exercise and hyperextension of the back, demonstrating the importance of physical activity in mitigating the physiological effects of prolonged weightlessness.

Before separating from the S-IVB stage, the Apollo 7 crew executed a critical 2-minute 56-second manual takeover of attitude control from the launch vehicle stage at 002:30:48 mission time. This manual intervention tested and verified the closed-loop spacecraft/launch vehicle control system's capability for orbital attitude control.

In the context of international space operations, the Committee on Space Research (COSPAR) assigns designations to space objects based on their year of launch and the sequential number of launches during that year. Typically, COSPAR designates instrumented spacecraft with the letter A, rockets with B, and fragments with subsequent letters like C, D, E, etc.

During this manual takeover, the crew of Apollo 7 conducted pitch, roll, and yaw maneuvers manually to assess the responsiveness and accuracy of the spacecraft's control system. The system demonstrated proper functionality, affirming its readiness to maintain precise orientation and execute maneuvers essential for the mission's objectives.

After completing the manual takeover test of attitude control, the Apollo 7 crew switched back to the automatic launch vehicle system, which resumed its normal attitude timeline. The Command and Service Module (CSM) separated from the S-IVB stage at 002:55:02 mission time as the mission progressed. During this phase, venting of the S-IVB propellants contributed to raising the orbit to dimensions of 167.0 nautical miles at apogee and 125.3 nautical miles at perigee.

A primary objective of Apollo 7 was to perform "safing" procedures on the S-IVB stage. This involved reducing the pressure in the propellant tanks and high-pressure bottles to safe levels, essential for subsequent simulated rendezvous and docking maneuvers. The safing process was scheduled to occur in multiple stages. Initially, three pre-programmed venting procedures were intended to achieve this. However, the specific orbital conditions required four additional venting beyond the pre-programmed sequences. These extra procedures were crucial to ensure that the LH2 (liquid hydrogen) tanks were adequately safed under the dynamic orbital flight conditions.

During the Apollo 7 mission, the safing procedures for the S-IVB stage involved several critical steps spread over different phases of the mission. The first of these procedures began with venting operations that commenced at 000:10:17 and concluded at 005:11:15, totaling 3,274.1 seconds of venting time. Subsequently, a liquid oxygen dump was initiated at 001:34:28 and lasted for 721.00 seconds. Following this, a cold helium dump was executed twice: first at 001:42:28 for 2,868.00 seconds and again at 004:30:16 for 1,199.99 seconds. Additionally, a stage control sphere helium dump was initiated at 003:17:33, lasting 2,967 seconds before being terminated by the ground command to preserve the remaining helium for the LH2 tank vent-and-relief valve.

Despite these procedural adjustments, the safing process was completed, ensuring the stage's readiness for subsequent operational requirements, including rendezvous and docking simulations.

During the second revolution of the mission, the crew observed an anomaly with one of the spacecraft/IM (Instrument Unit) adapter panels on the S-IVB stage. Typically deployed at a 45° angle, this particular panel only opened to 25° due to a malfunction of the retention cable. Although designed to prevent the panel from fully closing, the cable had become stuck, causing the panel to partially close. This issue, however, did not pose a critical problem as the panels were intended for jettisoning in future missions. By the 19th revolution, the panel had corrected itself and moved to its full open position, resolving the initial observation made by the crew earlier in the mission.

Attached to the Saturn IV-B stage, the Lunar Module Adapter's four panels are retracted to the fully open position. This is where the Lunar Module (LM) is stored during launch. On missions requiring the use of a LM, the four panels would be retracted and jettisoned before rendezvous and docking. This photo was taken during the Apollo 7 mission, when no Lunar Module was carried. The SIV-B stage flew as the second stage on a Saturn IB rocket. It is also used as the third stage on the Saturn V. The Apollo 7 mission was designed to test the Apollo Command and Service Module spacecraft systems specifically. Apollo 9 was the first mission to fly the Lunar Module.

To prepare for rendezvous with the S-IVB stage during the Apollo 7 mission, a critical phasing maneuver was conducted at 003:20:09 using the service module reaction control system. This 16.3-second maneuver aimed to adjust the spacecraft's orbit to 165.2 nautical miles at apogee and 124.8 nautical miles at perigee, positioning the spacecraft approximately 76.5 nautical miles ahead of the S-IVB stage.

However, the S-IVB's orbit decayed faster than expected over the following six revolutions. To compensate for this orbital decay and achieve the necessary conditions for rendezvous, an additional phasing maneuver of 17.6 seconds duration was performed at 015:52:00. This maneuver successfully adjusted the spacecraft's orbit to 170.21 nautical miles at apogee and 123.01 nautical miles at perigee, ensuring alignment with the desired rendezvous parameters.

During the Apollo 7 mission, at 014:46, it was reported that the commander, Walter Schirra, had developed a severe head cold, which began approximately one hour after liftoff. He took two aspirins to alleviate symptoms. The following day, the other crew members, Donn Eisele and Walter Cunningham, began experiencing head cold symptoms. Throughout the mission, these colds persisted, causing significant discomfort due to the challenges of clearing their ears, noses, and sinuses in the weightless environment of space. Despite taking medication, the crew continued to suffer from these symptoms.

At 023:33, Commander Schirra canceled the first scheduled television transmission, which was supposed to begin in 20 minutes. He expressed frustration, noting that mission control had added additional tasks—two burns and a urine dump—to their workload while they were already dealing with the challenges of testing a new spacecraft and coping with their cold symptoms. Still affected by his cold, Schirra asserted that the television transmission would be delayed without further discussion.

During the Apollo 7 mission, achieving rendezvous with the S-IVB stage required precise maneuvers using the service propulsion system. The first firing, a 9.26-second corrective maneuver performed at 026:24:55 mission time, was crucial to adjust the spacecraft's phase by 1.32 degrees and achieve an altitude offset of 8.0 nautical miles. This positioning was essential to prepare for the second firing, which aimed to establish a coelliptic orbit with the S-IVB.

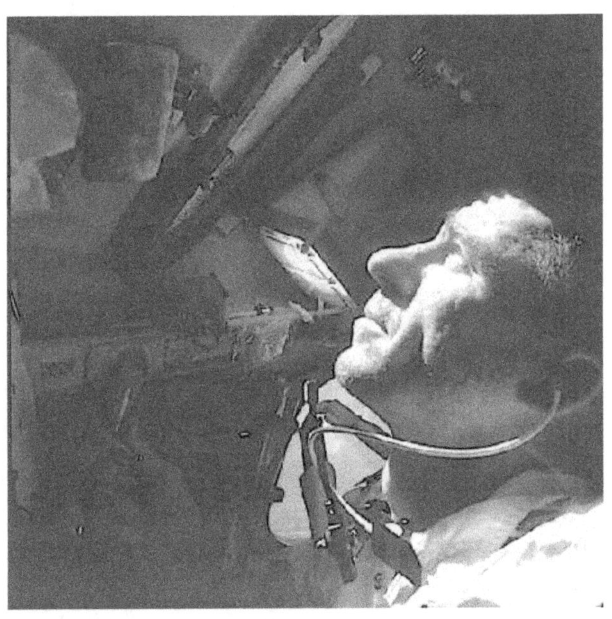

Cunningham during the mission

After the first firing, the spacecraft achieved an orbit of 194.1 nautical miles at apogee and 123.0 nautical miles at perigee. During this phase, the crew utilized the sextant to track the S-IVB, which was visible due to sunlight reflecting off its surface.

The second firing, lasting 7.76 seconds and executed at 028:00:56, occurred when the spacecraft was positioned 80 nautical miles behind and 7.8 nautical miles below the S-IVB. This maneuver successfully circularized the orbit, resulting in orbital parameters of 153.6 nautical miles at apogee and 113.9 nautical miles at perigee.

During the Apollo mission, meticulous planning and precise execution were paramount to achieving successful orbital maneuvers. Following two critical firings, the spacecraft successfully initiated the 46-second rendezvous terminal phase at 029:16:33, slightly ahead of schedule due to a minor orbital adjustment. This early execution necessitated a small midcourse correction at 029:37:48, ensuring the spacecraft maintained its trajectory toward the tumbling S-IVB stage.

Launching - October 11, 1968, at 11:03 am EST, Complex 34. The Saturn Launch Vehicle, carrying the Apollo 7 spacecraft, is photographed at more than 35,000 feet above the Atlantic Ocean from an C-135 aircraft. Aboard the spacecraft were Astronauts Walter Schirra, Donn Eisele, and Walter Cunningham. The Airborne Lightweight Optical Tracking System is an airborne precision photographic system that can provide information on launch vehicles during the early launch stage, separation and reentry phases of the flight.

At 029:43:55, a 708-second braking maneuver was meticulously performed, gradually bringing the spacecraft to within 70 feet of the S-IVB. This delicate proximity required 25 minutes of stationkeeping beginning at 029:55:43, as the spacecraft orbited at 161.0 by 122.1 nautical miles. A brief 5.4-second maneuver using the service module's reaction control system was executed to ensure precise positioning, gently moving the Command and Service Module (CSM) away from the S-IVB stage.

The crew of the Apollo mission executed a maneuver to position the Command and Service Module (CSM) around the S-IVB stage, undertaking crucial inspections and capturing photographs. This maneuver was pivotal as it simulated the spacecraft's capability to rendezvous with the Lunar Module (LM), symbolized by the S-IVB, in the event of an emergency where the LM's ascent stage malfunctioned after lunar departure.

During this maneuver, the crew faced challenges manually controlling the braking procedure. They encountered frustration due to the lack of reliable backup ranging information, a critical component typically available during actual LM rendezvous scenarios. This highlighted the rigorous demands and real-time problem-solving skills required of the crew in navigating and maneuvering the spacecraft.

Over the subsequent 24-hour period, the crew conducted tests and calibrations to ensure the spacecraft's systems were functioning optimally. These included a sextant calibration test at 041:00 and attitude control tests at 049:00 and 050:40. Additionally, they performed primary evaporator tests at 049:50 and 050:40 to verify essential life support systems.

Further demonstrating their proficiency, the crew performed a rendezvous navigation test using a sextant to track the S-IVB visually. This test extended to distances of 160 nautical miles at 044:40 and 320 nautical miles at 053:20, underscoring their ability to navigate and maintain positional awareness in deep space conditions accurately.

Later in the mission, the crew of Apollo 7 reported sighting the S-IVB at an impressive range of nearly 1,000 nautical miles. This observation underscored their ability to track distant objects in space and demonstrated the advanced capabilities of the spacecraft's visual systems.

The mission strategy aimed for maximum efficiency, completing as many primary and secondary objectives as possible early in the flight. By the end of the second day, over 90 percent of these objectives had been successfully achieved, showcasing the crew's productivity and the spacecraft's reliability.

A critical mission component involved testing the rendezvous radar transponder, essential for future docking operations between the Lunar Module (LM) ascent stage and the Command Module (CM) after liftoff from the lunar surface. The crew conducted three tests of this system: the first two at 061:00 and 071:40, followed by a pivotal third test during revolution 48 at 076:27. During this test, ground radar at White Sands Missile Range in New Mexico successfully acquired and locked onto the spacecraft's transponder at a range of 390 nautical miles, tracking it accurately to 415 nautical miles.

Another milestone during the mission occurred at 071:43 when the crew initiated the first of seven television transmissions, each lasting seven minutes. These transmissions marked the first live television broadcasts from a piloted American spacecraft, providing audiences on Earth with unprecedented views of space and the crew's activities.

During their historic telecast from space, the Apollo 7 crew began with a playful sign that read, "From the lovely Apollo room high atop everything." They then directed the camera out the window, capturing breathtaking views as the spacecraft flew over New Orleans and the Florida peninsula. The telecast vividly portrayed the spacecraft's orbital motion, offering viewers on Earth a firsthand glimpse of our planet from space.

Following this milestone, the mission continued with critical maneuvers and tests. The spacecraft's propulsion system was engaged six additional times, enhancing its operational readiness. Notably, the third firing at 075:48:00, performed ahead of schedule, was a 9.10-second adjustment by the stabilization and control system. This maneuver aimed to increase the reliability of the service module's reaction control system, altering the spacecraft's orbit to 159.7 by 91 nautical miles.

The service module's control system was cold-soaked for three hours to ensure optimal performance under extreme conditions. This involved exposing one side of the spacecraft to the frigid temperatures of space, simulating the challenges of prolonged exposure to the cosmic environment. The tests confirmed that the system's characteristics aligned with expectations, paving the way for future lunar missions.

At 097:00, an assessment was made to determine if the spacecraft's environmental control system and its thermal coating had withstood the rigors of space. Results indicated that the solar radiator panel tested was within operational parameters, validating its reliability for lunar flight.

The mission's educational broadcast at 095:25 included a comprehensive tour of the spacecraft's interior, showcasing various controls and systems. Additionally, efforts were made to demonstrate water condensation processes, a crucial aspect given the challenges of managing humidity in the spacecraft's confined environment—a significant concern for extended space missions.

This problem was anticipated in the cabin because cold coolant lines from the radiator to Apollo? 0, the environment control unit, and from the environment control unit to the inertial measurement unit were not insulated. The crew vacuumed the water overboard each time excessive condensation was noted on the coolant lines or in a puddle on the aft bulkhead after service propulsion system maneuvers. Experiment S005 (Synoptic Terrain Photography) began at 098:40, using a hand-held modified 70 mm Hasselblad 500C camera. The photographs were used to study the origin of the Carolina Bays in the United States, wind erosion in desert regions, coastal morphology, and the origin of the African rift valley.

Near-vertical, high sun-angle photographs of Baja California, other parts of Mexico, and the Middle East were useful for geologic studies. Photographs of New Orleans and Houston were generally better for geographic urban studies than those from previous programs. Mission commander Wally Schirra

During the Apollo 7 mission, significant milestones were achieved in Earth observation, particularly in oceanographic and geographical studies. For the first time, islands in the Pacific Ocean and extensive regions of northern Chile, Australia, and other areas were comprehensively photographed. Out of the 500 photographs taken to capture both land and oceanic features, approximately 200 were deemed usable, showcasing excellent color rendition and exposure quality.

The mission's photography efforts faced challenges typical of capturing precise moments in space. Crew members had to swiftly change film magazines, adjust filters, and fine-tune exposure settings as targets appeared. Maintaining stability while operating the camera was also crucial to avoid improper exposures, affecting some frames' outcomes.

View of the Sinai Peninsula from Apollo 7

Experiment S006, Synoptic Weather Photography, commenced at 099:10 using the same handheld modified 70 mm Hasselblad 500C camera used in Experiment S005. This experiment aimed to photograph a comprehensive range of 27 basic categories of weather phenomena. By systematically capturing these images, the mission contributed valuable data to meteorological studies, enhancing our understanding of global weather patterns from a vantage point in space.

During the Apollo 7 mission, the crew captured a wealth of data through their photography efforts, yielding approximately 500 photographs, of which around 300 focused on meteorological phenomena such as clouds and weather systems. Additionally, about 80 photographs highlighted significant features in oceanography, marking a substantial advancement in our visual understanding of Earth's natural environments.

The mission's photography encompassed a wide array of categories deemed scientifically valuable. These included detailed studies of weather systems, the effects of winds on cloud formations, observations of ocean surfaces, and even underwater views of Australian reefs, Pacific atolls, and regions like the Bahamas and Cuba. The photographs also documented landform effects, climate patterns, and hydrological features, providing comprehensive insights into Earth's dynamic systems.

The images of Hurricane Gladys on October 17 and Typhoon Gloria on October 20, 1968, were of particular note. These photographs represented the most detailed and clear views of tropical storms up to that time, offering unprecedented clarity and detail that advanced meteorological research significantly.

The quality of the photographs varied from fair to excellent, reflecting the challenges of maintaining camera stability in the weightless space environment. Despite these difficulties, the crew captured fine details, such as ocean swells from altitudes nearing 100 nautical miles, highlighting the remarkable capability of Apollo missions in remote sensing and Earth observation.

On the Apollo 7 mission, the third television transmission commenced at 119:08 and lasted approximately ten minutes. This broadcast showcased a food preparation demonstration in space, featuring the reconstitution of dried fruit juice with water. Viewers also witnessed the crew vacuuming water accumulated on cold glycol lines, highlighting the daily maintenance routines aboard the spacecraft. The telecast further provided insights into various controls at the commander's workstation, offering a glimpse into the operational aspects of managing spacecraft systems in orbit.

Shortly after that, at 120:43:00.44, the crew conducted the fourth service propulsion system firing. This maneuver was designed to assess the minimum-impulse capability of the service propulsion engine. Despite its brief duration of only 0.48 seconds, the firing successfully altered the spacecraft's orbit to 156.7 by 89.1 nautical miles, demonstrating precise orbital control capabilities.

Later, at 141:11, the fourth television transmission began with a tour of the Command Module (CM). The crew focused their camera on deposits observed on window 1 and the optical site markings used for measuring the pitch angle on window 2. Panning around the interior, viewers were treated to views of sleep stations, stowage areas, helmet bags, and pressure suit hoses, providing a comprehensive look at space, living, and working conditions.

During the latter phases of the Apollo 7 mission, the S-IVB stage continued its orbit around Earth until impacting the Indian Ocean at 09:30 GMT on October 18. The estimated impact coordinates were latitude 8.9° south and longitude 81.6° east.

A critical maneuver, the fifth service propulsion system firing, was executed at 165:00:00.42 to position the spacecraft optimally for the upcoming deorbit maneuver at the end of its planned orbital phase. This maneuver allowed for a potential additional orbit if needed, ensuring at least two minutes of tracking by the Hawaii ground station. The firing duration was extended beyond the original plan to validate the propellant gauging system. This firing produced the largest velocity change of the mission, accelerating the spacecraft by 1,691.3 feet per second. Notably, a manual thrust-takeover was incorporated midway through the maneuver. Post maneuver, the spacecraft achieved an orbit of 244.2 by 89.1 nautical miles.

Passive thermal control, essential for maintaining even external temperatures on future missions, was tested at 167:00 and 212:00. This involved adjusting spacecraft orientation to maximize solar exposure or shade as needed.

During the eighth day of the mission, the crew performed the second minimum-impulse maneuver at 193:00:00.45. This brief firing, lasting 0.50 seconds, was conducted to maintain orbital stability without significant changes.

The seventh television transmission began at 213:10, showcasing views of the Florida peninsula and demonstrating the crew's beards grown during the mission. This broadcast also included an analysis of solar flare data, confirming that the mission's orbital parameters would not be affected. The exercise served as a valuable test of contingency procedures for handling such events during future missions.

Following this, the seventh firing occurred at 224:00:07.70, strategically timed at the spacecraft's perigee. Critical for reentry and recovery planning, this maneuver adjusted the orbit further, lowering it to prepare for Earth reentry.

In the final television transmission at 236:18, lasting about 10 minutes, the crew again displayed their beards and shared observations of jet contrails below them over the Gulf Coast. They also described the vivid bands of color produced by Earth's day airglow, offering viewers a glimpse of the mesmerizing atmospheric phenomena observed from space.

During the latter stages of the Apollo mission, intricate maneuvers and critical decisions shaped the course of events. The midcourse navigation program, pivotal for ensuring precise trajectory adjustments, encountered unexpected challenges. Attempting to utilize the Earth's horizon and star sightings for orientation proved futile due to the indistinct and variable nature of the Earth's horizon as viewed through the spacecraft's sextant. The expansive air glow spanning about three degrees lacked definitive boundaries, complicating navigational fixes.

These issues, stemming partly from the spacecraft's low Earth orbit, contrasted sharply with the ease experienced when using lunar landmarks in conjunction with star sightings. Lunar features presented themselves with clarity comparable to terrestrial landmarks, facilitating accurate course corrections. Stars, visible at angles as shallow as 10° to 15° from the Moon's surface, provided reliable celestial reference points crucial for navigation throughout the mission.

As the mission neared its conclusion, preparations intensified for the crucial deorbit maneuver. Executed during the 163rd orbit, the eighth Service Propulsion System (SPS) firing, lasting 11.79 seconds, precisely positioned the spacecraft for reentry over Hawaii at 259:39:16 mission elapsed time. This final orbit parameters included an apogee of 225.3 nautical miles, a perigee of 88.2 nautical miles, a period of 90.39 minutes, and an inclination of 29.88°.

Controversy surrounding Apollo 7 extended into the preparations for reentry, exacerbated by the crew's battle with cold symptoms. A pivotal debate ensued over whether the crew should wear helmets and gloves during the critical reentry phase, balancing protection with operational effectiveness. Concerns focused on potential impediments caused by helmets, which could hinder crew members from clearing their throats and equalizing pressure in their ears as gravity drew mucus downward from areas affected by zero-gravity conditions.

Astronauts Walter M. Schirra Jr. (on right), mission commander; and Donn F. Eisele, command module pilot; are seen in the first live television transmission from space. Schirra is holding a sign which reads, "Keep those cards and letters coming in, folks!" Out of view at left is astronaut Walter Cunningham, lunar module pilot.

In a decision made 48 hours before reentry and driven by the crew's insistence, NASA opted to forgo helmets and gloves. This choice, though not without risks, aimed to enhance crew comfort and readiness during the descent and splashdown phases. The service module was jettisoned at 259 hours, 43 minutes, and 33 seconds into the mission, marking a critical transition toward the final reentry sequence.

Guided both automatically and manually, the command module (CM) initiated its descent into Earth's atmosphere at 259 hours, 53 minutes, and 26 seconds. By the time it reached an altitude of 400,000 feet, the CM had achieved a velocity of 25,846 feet per second. Real-time trajectory calculations and tracking indicated the service module's reentry into the Atlantic Ocean at 260 hours, 3 minutes, and 0 seconds, near latitude 29° North and longitude 72° West.

During the intense reentry, visual sightings were maintained for the CM, service module, and a 12-foot insulation disk positioned between them. The parachutes deployed flawlessly, guiding the CM to a gentle splashdown in the Atlantic Ocean southeast of Bermuda at 11:11:48 GMT (07:11:48 a.m. EDT) on October 22, 1968. This marked the mission's conclusion after 260 hours, 9 minutes, and 3 seconds.

However, challenges persisted throughout the mission. The crew's sleep schedule, designed for one member to remain awake while the others slept, proved difficult to maintain. Walter Cunningham later recounted instances where crew members struggled to work without disturbing their sleeping colleagues, highlighting the strain caused by sleep disruption.

Moreover, underlying tensions between the astronauts and Mission Control, compounded by Schirra's cold and sleep deprivation, occasionally surfaced during the flight. Schirra's frustration with NASA's decision to proceed with the launch despite adverse weather conditions underscored ongoing conflicts over mission management and risk assessment.

Apollo 7, while successful in its objectives, underscored the complex interplay between operational demands, crew health, and decision-making under pressure in the early days of the Apollo program.

Testing the television system aboard Apollo 7 sparked a notable disagreement between the crew and Mission Control in Houston. Commander Walter Schirra voiced his frustration, bluntly stating, "You've added two burns to this flight schedule, and you've added a urine water dump, and we have a new vehicle up here, and I can tell you at this point, TV will be delayed without any further discussion until after the rendezvous." His stance reflected the crew's steadfast commitment to prioritizing mission-critical objectives over what they perceived as secondary activities.

In hindsight, Schirra reflected on their stance, writing, "We'd resist anything that interfered with our main mission objectives. On this particular Saturday morning, a TV program clearly interfered." His sentiments were echoed by Donn Eisele, who noted in his memoirs, "We were preoccupied with preparations for that critical exercise and didn't want to divert our attention with what seemed to be trivialities at the time. ... Evidently, the earth people felt differently; there was a real stink about the hotheaded, recalcitrant Apollo 7 crew who wouldn't take orders."

Historians and analysts, such as French and Burgess, observed that Schirra's insistence was rational given the mission's priorities. They wrote, "When this point is considered objectively—that in a front-loaded mission, the rendezvous, alignment, and engine tests should be done before television shows—it is hard to argue with him." Despite the crew's resistance, Deke Slayton eventually acquiesced to Schirra's position, though the commander's assertiveness left flight controllers surprised by the crew's unwavering stance.

On Day 8 of the Apollo 7 mission, tensions between the crew and Mission Control escalated following a computer freeze caused by a new procedure sent from the ground. Frustrated by the failure of the procedure, Donn Eisele radioed back, "We didn't get the results that you were after. We didn't get a damn thing, in fact ... you bet your ass ... as far as we're concerned, somebody down there screwed up royally when he laid that one on us." Walter Schirra later acknowledged that this incident marked a significant moment of frustration for Eisele with Mission Control.

The following day, conflicts continued to arise. After multiple firings of the Reaction Control System (RCS) to stabilize the spacecraft during a test, Schirra vented his frustration to Mission Control, saying, "I wish you would find out the name of the idiot who thought up this test. I want to find out and talk to him personally when I get back down." Eisele chimed in, adding, "While you are at it, find out who dreamed up the 'P22 horizon test'; that is a beauty also."

During the final stages of Apollo 7, another significant point of contention arose between Mission Control and the crew, particularly Commander Walter Schirra, regarding the wearing of helmets during reentry. Schirra was concerned that the sinus pressure from their colds could potentially cause their eardrums to burst during reentry. He argued vehemently that they needed the freedom to pinch their noses and blow to equalize the pressure, a maneuver impossible with helmets on.

Over several days, Schirra repeatedly refused the advice from Mission Control to wear their helmets, asserting his authority as commander to make this decision. Deke Slayton, NASA's Director of Flight Crew Operations, warned Schirra that he must answer for this decision after the flight. Reflecting on this later, Schirra explained, "I had a cold, and I'd had enough discussion with the ground, and I didn't have much more time to discuss whether we would put the helmet on or off. I said, essentially, I'm on board, I'm commanding."

As they prepared for reentry, Schirra made the final call that no helmets would be worn. This decision did not go unnoticed by NASA's leadership. Christopher C. Kraft, Director of Flight Operations, demanded an explanation from the Capsule Communicator (CAPCOM), Stafford, regarding what he perceived as Schirra's insubordination. Kraft later acknowledged, "Schirra was exercising his commander's right to have the last word, and that was that."

Despite the controversy surrounding the helmet issue, Apollo 7 concluded successfully. The spacecraft splashed down at 11:11:48 UTC on October 22, 1968, approximately 200 nautical miles south-southwest of Bermuda and just 7 nautical miles north of the recovery ship USS Essex. The mission lasted 10 days, 20 hours, 9 minutes, and 3 seconds, achieving its objectives despite the internal tensions and challenges the crew and Mission Control faced.

Post-splashdown, the CM initially assumed an apex-down flotation attitude, which the inflatable bag uprighting system swiftly rectified within 13 minutes to its normal flotation position. However, intermittent recovery beacon visibility and voice communication interruptions posed additional challenges during this period.

Swift retrieval operations ensued, with the crew successfully retrieved by helicopter and aboard the recovery ship within 56 minutes of splashdown. The CM itself was recovered just 55 minutes later, with its estimated weight at splashdown recorded at 11,409 pounds. The mission's total journey spanned an estimated distance of 3,953,842 nautical miles, encapsulating a historic voyage of exploration and accomplishment in the annals of space exploration history.

Upon retrieval of the Command Module (CM), conditions aboard the USS Essex were precisely detailed: light rain showers under a 600-foot ceiling, visibility limited to 2 nautical miles, and a 16-knot wind blowing from 260° true north. The air and water temperatures were 74°F and 81°F, respectively, with waves reaching up to 3 feet from the same direction.

Following its recovery, the CM was offloaded from the Essex on October 24 at Norfolk Naval Air Station, Virginia. Here, the Landing Safing Team promptly commenced evaluation and deactivation procedures at 14:00 GMT. By 01:30 GMT on October 27, 1968, deactivation was completed. Subsequently, the CM embarked on a journey to Long Beach, California, where it was transported by truck to the North American Rockwell Space Division facility in Downey. This site became the hub for comprehensive postflight analysis.

The Apollo 7 mission culminated as a resounding success on multiple fronts. All spacecraft systems functioned flawlessly throughout the mission, with nearly all detailed test objectives achieved. As an engineering test flight, Apollo 7 provided crucial insights into the performance of various systems, including the orbital safing experiment and the reliability of attitude control in both manual and automatic modes. Notably, the mission marked the inaugural use of a mixed cabin atmosphere comprising 65% oxygen and 35% nitrogen, a significant departure from the previously used 100% oxygen environment implemented in response to the Apollo 1 fire investigation recommendations.

Additionally, Apollo 7 pioneered the provision of hot and cold drinking water through the service module's fuel cells, a vital advancement crucial for future lunar missions. Consumable usage remained within safe parameters, enabling the inclusion of supplementary flight activities as the mission progressed. Thus, Apollo 7 validated the operational readiness of spacecraft systems and laid foundational groundwork for subsequent lunar exploration missions, affirming its pivotal role in the broader Apollo program.

Apollo 7 and 8 astronauts at the White House with President Lyndon and First Lady Lady Bird Johnson, Vice President Humphrey, NASA Administrator James E. Webb and Charles

One of the most significant aerodynamic phenomena encountered during the Apollo 7 mission was the unexpected "perigee torquing," observed as a rotation of the Command and Service Module (CSM), particularly noticeable when the spacecraft's perigee was at 90 nautical miles. Analysis of post-mission data yielded several important conclusions:

The Apollo 7 mission and findings from previous missions and ground tests confirmed that the CSM was fully qualified for operations within Earth's orbital environment and was poised for further testing in cislunar and lunar orbital settings.

The functionality and operational concepts governing crew interactions with spacecraft interfaces—including procedures, accommodations, and display controls—proved satisfactory.

The prime crew of the first manned Apollo mission (Spacecraft 101/Saturn 205) participates in water egress training in the Gulf of Mexico. Left to right, are astronauts Walter M. Schirra Jr. (stepping into life raft), Donn F. Eisele, and Walter Cunningham. They have just egressed Apollo Command Module Boilerplate 1102 and are awaiting helicopter pickup. Inflated bags were used to upright the boilerplate. MSC swimmers assisted in the training exercise.

Thermal management across the active and passive spacecraft exceeded predictions for the near-Earth environment, ensuring optimal system performance.

The mission demonstrated the endurance required for sustained systems operation during lunar missions, a critical factor for extended space missions.

While Apollo 7 successfully showcased rendezvous capabilities using onboard optical and data systems alone, adding ranging information was deemed highly advantageous for enhancing precision during the final approach phase.

In the context of Project Apollo, validating navigation techniques was crucial for ensuring the success and safety of lunar missions. Here's an overview of specific findings and achievements related to navigation techniques:

The feasibility of using landmark tracking techniques in Earth orbit was successfully validated. This method allowed astronauts to navigate by observing and tracking identifiable landmarks on Earth's surface, proving essential for orientation and course correction during orbital operations.

Apollo 7 astronauts (l-r) Walter Cunningham, Donn F. Eisele, and Walter M. Schirra, Jr., compare notes at a mission debriefing conducted today at the Kennedy SpaceCenter. October 23, 1968. Scan by Ed Hengeveld. Caption by Kipp Teague.

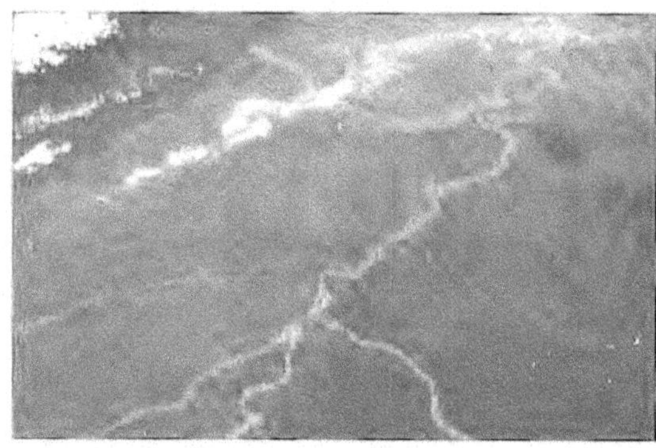

Example of Synoptic Terrain Photography: India, Nepal, Tibet, and Himalayas from 126 n mi altitude (NASA AS07-11-1980).

Observations of the Earth limb in low Earth orbit posed challenges due to optical design considerations and operational techniques. Addressing these challenges was critical, as precise Earth limb observations are fundamental for determining spacecraft orientation and relative positioning.

During venting operations, frozen liquid particles temporarily block star visibility. These particles were anticipated to dissipate quickly without significantly contaminating optical surfaces in Earth orbit, highlighting the need to plan spacecraft venting procedures to minimize operational disruptions carefully.

Example of Synoptic Weather Photography: a view of Hurricane Gladys over the Pacific Ocean at an altitude of 99 n mi (NASA AS07-07-1877).

In Cislunar space, under proper spacecraft orientation shielding optics from solar and Earth/Moonlight, the scanning telescope effectively provided clear star visibility data. This capability enabled reliable recognition of constellations essential for maintaining the inertial platform's orientation, critical for precise navigation beyond Earth's orbit.

The sextant proved effective for star visibility even during daylight, utilizing designated Apollo navigation stars positioned as close as 30° from the Sun's line of sight. This capability allowed for accurate realignment of the spacecraft's guidance systems without reliance solely on nighttime observations.

Apollo 7, within its detailed objectives, underscored several key goals for its launch vehicle:

A member of the Apollo 7 crew is hoisted up to a U.S. Navy Sikorsky SH-3A Sea King recovery helicopter (BuNo 149918) from Helicopter Anti-Submarine Squadron 5 (HS-5) during recovery operations. The Apollo 7 spacecraft splashed down at 07:11 hrs, 22 October 1968, approximately 200 nautical miles south-southwest of Bermuda. Recovery ship was the aircraft carrier USS Essex (CVS-9).

Demonstrating the adequacy of the launch vehicle's attitude control system for seamless orbital operations is a milestone crucial for achieving and maintaining precise trajectories in space.Apollo 7 marked a pivotal milestone in manned space exploration by meticulously validating a myriad of critical capabilities essential for successful missions beyond Earth's atmosphere:

The mission effectively demonstrated the S-IVB stage's orbital staging capability, a significant achievement that underscored NASA's mastery in managing and utilizing propulsion stages in the vacuum of space. This capability was crucial for subsequent deep space maneuvers and lunar operations.

On 22 October 1968, the Apollo 7 crew is welcomed aboard the U.S. Navy aircraft carrier USS Essex (CVS-9), the prime recovery ship for the mission. This was the first Apollo splashdown and, therefore, the first three person 'landing' for NASA. Left to right, are astronauts Walter M. Schirra Jr., commander; Donn F. Eisele, command module pilot; and, Walter Cunningham, lunar module pilot. In left background is Dr. Donald E. Stullken, NASA Recovery Team Leader from the Manned Spacecraft Center's Landing and Recovery Division.

Furthermore, evaluations of the S-IVB J-2 engine's augmented spark igniter line modifications were completed successfully. These enhancements significantly bolstered ignition reliability, critical for propulsion during deep space missions, where reliable engine start-ups are paramount.

The S-IVB/instrument unit orbital coast lifetime capability assessment proved adequate among the secondary objectives. This capability ensured the sustained functionality of essential systems during extended mission phases, crucial for the longevity and success of manned space missions. The integration of command and service module piloted launch vehicle orbital attitude control systems was successfully demonstrated. This highlighted the spacecraft's capability to maintain precise orientations essential for executing complex mission maneuvers and ensuring crew safety.

Primary spacecraft objectives focused on validating the performance of the command and service module and the crew under operational conditions. These objectives were achieved, affirming the reliability and functionality of the spacecraft and its support systems in the unforgiving space environment.

Detailed test objectives further emphasized precision and operational capability:

The inertial measurement unit was successfully aligned using the sextant, ensuring accurate navigation relative to celestial bodies crucial for course corrections.

Orientation determination of the internal measurement unit and daylight visibility checks of star patterns were accomplished, verifying essential navigational aids and the spacecraft's autonomy.

Optical tracking of a target vehicle using the sextant during rendezvous operations showcased precise navigational techniques essential for docking maneuvers.

The guidance navigation control system demonstrated both automatic and manual attitude control reaction control system maneuvers, validating essential control capabilities despite partial rate checks.

Various maneuvers using the guidance navigation control system controlled the service propulsion and reaction control system velocities at different mission phases, highlighting operational flexibility and precision.

Successful demonstration of the guidance navigation control system's ability to guide Earth orbit entry ensured safe and controlled reentry into Earth's atmosphere.

Attempts to establish an Earth horizon model through star and Earth horizon sightings were unsuccessful due to visibility challenges, underscoring the complexities of navigating in space environments.

Performance data for the inertial measurement unit and accelerometer bias tests were successfully obtained, contributing to ongoing spacecraft navigation improvements.

The entry monitoring system was monitored during critical propulsion burns and reentry phases, providing crucial data for mission safety and success.

The stabilization control system executed automatic and manual attitude control reaction control system maneuvers, excluding high and auto rate modes, demonstrating robust operational control capabilities.

Apollo 7's achievements extended beyond basic spacecraft operations, encompassing terrain and weather photography experiments, studies on bone demineralization and cardiovascular health in space, and evaluations of spacecraft communication systems and environmental controls. These comprehensive tests and validations laid a solid foundation for subsequent Apollo missions, pushing the boundaries of human exploration and paving the way for future space endeavors.

After Apollo 7 was successfully completed, NASA honored Walter Schirra, Donn Eisele, and Walter Cunningham with its Exceptional Service Medal in recognition of their achievements. This accolade underscored their pivotal roles in demonstrating the viability and safety of crewed spaceflight following the Apollo 1 tragedy.

On November 2, 1968, President Lyndon Johnson hosted a ceremony at the LBJ Ranch in Johnson City, Texas, where he personally presented the astronauts with their medals. During the same event, Johnson also bestowed NASA's highest honor, the Distinguished Service Medal, upon James E. Webb, who had recently retired as NASA

Administrator. Webb was recognized for his outstanding leadership throughout America's space program, particularly during the critical early phases of the Apollo missions.

Additionally, Johnson invited the Apollo 7 crew to visit the White House, which they accepted in December 1968. This gesture highlighted the national pride and recognition bestowed upon the astronauts for their successful mission, solidifying their place in the annals of space exploration history.

Despite the challenges and tensions between the Apollo 7 crew and Mission Control, the mission successfully achieved its primary objectives. It rigorously tested the Apollo command and service module, affirming its flightworthiness under various conditions. This crucial validation paved the way for the next ambitious step in the Apollo program: Apollo 8's historic mission to orbit the Moon, scheduled just two months later.

John T. McQuiston reflected in The New York Times after Donn Eisele's passing in 1987, noting that Apollo 7's success significantly bolstered confidence in NASA's space program. The mission's accomplishments demonstrated NASA's ability to overcome setbacks and proceed with complex missions.

According to author Andrew Chaikin, "Three weeks after the Apollo 7 crew returned, NASA administrator Thomas Paine authorized Apollo 8 to launch in late December and orbit the Moon. Apollo 7 had delivered NASA from its trial by fire—it was the first small step down a path that would lead another crew, nine months later, to the Sea of Tranquility." This pivotal decision marked a turning point, propelling NASA forward in its quest to achieve President Kennedy's goal of landing humans on the Moon before the end of the decade.

Reflecting on Apollo 7, General Sam Phillips, the Apollo Program Manager, declared, "Apollo 7 goes into my book as a perfect mission. We accomplished 101 percent of our objectives." This statement underscored the mission's resounding success in meeting and exceeding its goals.

Christopher C. Kraft, Director of Flight Operations, praised Walter Schirra and his crew for their exceptional performance, stating, "Schirra and his crew did it all—or at least all of it that counted ... They proved to everyone's satisfaction that the SPS engine was one of the most reliable we'd ever sent into space. They operated the Command and Service Modules with true professionalism."

Reflecting on the dynamics during the mission, Donn Eisele acknowledged, "We were insolent, high-handed, and Machiavellian at times. Call it paranoia, call it smart—it got the job done. We had a great flight." His candid assessment highlighted the challenges and determination required to succeed.

Gene Kranz, Flight Director during Apollo 7, provided perspective years later, noting, "We all look back now with a longer perspective. Schirra wasn't on us as bad as it seemed at the time. The bottom line was that we did the job as a team even with a grumpy commander." Kranz's reflection emphasized the teamwork and resilience that defined NASA's approach to overcoming obstacles and achieving milestones in space exploration.

Following Apollo 7, none of its crew members flew in space again. Jim Lovell states, "Apollo 7 was a very successful flight—they did an excellent job—but it was a very contentious flight. They all teed off the ground people quite considerably, and I think that put a stop on future flights [for them]." The mission's internal tensions and strained relations with ground control contributed to this outcome.

Walter Schirra had already announced his retirement from both NASA and the Navy before the flight, effective July 1, 1969. This decision marked the end of his distinguished career in space exploration. The other two crew members, Donn Eisele and Walter Cunningham saw their spaceflight opportunities diminish following Apollo 7. There were reports that Chris Kraft, Director of Flight Operations, expressed reluctance to work with any crew member again, impacting their future assignments within NASA.

Cunningham was assigned to lead the Astronaut Office's Skylab division but later declined an informal offer to command the first Skylab crew. Instead, this role went to Pete Conrad, commander of Apollo 12, with Cunningham offered the backup commander position. Frustrated with this decision, Cunningham resigned from NASA in 1971.

Donn Eisele's career trajectory was also affected. He became the first active astronaut to divorce, followed by a quick remarriage, which drew attention and speculation within NASA. His performance as the backup Command Module Pilot for Apollo 10 was considered indifferent, and he resigned from the Astronaut Office in 1970. He remained with NASA at the Langley Research Center in Virginia until his retirement in 1972.

In October 2008, NASA administrator Michael D. Griffin decided to posthumously award the Distinguished Service Medal to Walter Schirra, Donn Eisele, and Walter Cunningham, the crew of Apollo 7. This recognition came years after their mission, which had not initially resulted in the award. The medals were awarded "[f]or exemplary performance in meeting all the Apollo 7 mission objectives and more on the first crewed Apollo mission, paving the way for the first flight to the Moon on Apollo 8 and the first crewed lunar landing on Apollo 11."

By this time, only Cunningham was still alive, as Eisele had passed away in 1987 and Schirra in 2007. Eisele's widow accepted his medal on his behalf, and Bill Anders, a member of the Apollo 8 crew, accepted Schirra's medal. Other Apollo astronauts, including Neil Armstrong, Buzz Aldrin, and Alan Bean, attended the award ceremony.

The Apollo 7 command module on display

critical events that ensured the spacecraft's readiness and reliability. Beginning on March 18, 1968, with individual and combined tests of the Command Module (CM) and Service Module (SM) systems at the factory, the mission's path was marked by systematic milestones:

By March 28, the Saturn IB stage arrived at Kennedy Space Center (KSC), followed closely by the Saturn IV-B stage on April 7 and the Saturn IB instrument unit on April 11. These deliveries laid the groundwork for subsequent assembly and integration phases.

On April 29, the factory completed an integrated test of the CM and SM systems, validating their operational synergy. By May 29, CM #101 and SM #101 arrived at KSC, and by May 30, they were successfully mated together, forming the complete Command and Service Module (CSM) for Apollo 7.

The rigorous testing continued through June and July, with combined systems tests, altitude tests, and the relocation of the space vehicle to Cape Kennedy Launch Complex 34 by July 29. August saw the completion of integrated systems tests and the electrical mating of CSM #101 to its launch vehicle by August 27.

As September approached, final preparations intensified: an overall space vehicle test by August 30, a successful countdown demonstration test by September 4, and the completion of the flight readiness test by September 17. Each milestone represented a critical step forward, demonstrating the meticulous planning and technical precision required for a crewed mission into Earth orbit.

Apollo 8

In December 1968, NASA launched Apollo 8, a monumental mission in space exploration history, intended as a crewed lunar orbital flight. Commanded by Frank F. Borman II and accompanied by crew members James A. Lovell Jr. and William A. Anders, Apollo 8 was a daring venture into uncharted territory.

The spacecraft, consisting of Apollo CSM-103 and the Lunar Module Test Article (LTA-B), was manufactured by North American Rockwell. Its launch mass breakdown was as follows: the Command Module (CM) at 5,621 kilograms, the Service Module (SM) at 23,250 kilograms, and additional components such as the Spacecraft/Lunar Module Adapter and the LTA fixed to the rocket. At liftoff, the total mass was 28,870 kilograms.

Apollo 8's mission lasted six days, 3 hours, and 42 seconds, during which it orbited the Moon and returned safely to Earth. The journey began on December 21, 1968, at 12:51:00 UTC, launched atop a powerful Saturn V rocket (SA-503) from Kennedy Space Center's Launch Complex 39A.

The spacecraft achieved lunar orbit insertion on December 24, 1968, at 9:59:20 UTC, marking a historic first as humans circled the Moon. Apollo 8 conducted critical observations throughout ten lunar orbits and provided invaluable data before departing lunar orbit on December 25, 1968, at 6:10:17 UTC.

Launch of Apollo 6 (identifiable by its white-painted service module) as seen from the top of the launch tower

Apollo 8's return to Earth concluded on December 27, 1968, at 15:51:42 UTC, with recovery by the USS Yorktown in the North Pacific Ocean. The mission's success demonstrated NASA's capabilities in lunar exploration and set the stage for subsequent Apollo missions that would ultimately land astronauts on the lunar surface.

Apollo 8 stands as a transformative mission within NASA's Apollo program, denoted as a pivotal Type "C prime" mission that departed significantly from earlier endeavors. Initially conceived as an unmanned Earth orbital mission known as

SA-503, plans swiftly evolved following the success of Apollo 6 in April 1968. NASA promptly reclassified SA-503 as a crewed lunar mission, marking a dramatic shift in scope and ambition.

Equipped with both the Command and Service Module (CSM) and Lunar Module (LM), Apollo 8, launched on December 21, 1968, became the first crewed spacecraft to depart Earth's gravitational sphere of influence, setting course for the Moon. It was also the third flight and the maiden crewed launch of the powerful Saturn V rocket, lifting off from Kennedy Space Center in Florida, adjacent to Cape Kennedy Air Force Station.

The Apollo 8 crew: Frank Borman, Bill Anders and Jim Lovell.

Originally intended by astronaut Jim McDivitt's crew as a test of the lunar module in low Earth orbit, Apollo 8's mission profile was redefined in August 1968 to a more audacious endeavor: a command-module-only lunar orbital flight scheduled for December.

Apollo 8 had originally been planned as an Earth orbit mission to check out the spacecraft. Flight Director Chris Kraft felt the CSM was in the best shape of any spacecraft ever. It was passing all its tough tests with flying colors. It was the LM development that was lagging, with a number of problems still to be resolved. It didn't help to receive CIA reports inferring the Russians might be working on a lunar flight with a new Soyuz spacecraft.

This adjustment was necessitated by the LM's developmental timeline, which delayed its readiness for an initial flight. As a result, Command Module Pilot Frank Borman and his crew—James Lovell and William Anders—shifted their focus to mastering translunar navigation rather than the LM-specific training originally planned.

This transition necessitated rigorous testing and modifications. The Saturn V rocket's S-II stage, crucial for lunar missions, underwent meticulous scrutiny at the Mississippi Test Facility to ensure it met stringent safety standards for human spaceflight, a process known as "manrating." Following successful tests, the stage was cleared for use and transported back to Kennedy Space Center in June 1968.

In early August 1968, George Low, the Apollo Program manager, had this crazy idea of just going to the Moon with no LM on the first manned flight of the mighty Saturn V. The Russians 'spectacular fireballs had shown what it was like when things went wrong during a launch of these big rockets. At this point Low only saw a circumlunar flight. He bounced the idea off Chris Kraft and Bob Gilruth. "His idea was a shocker," said Kraft, "but if we could pull it off it would be absolutely pivotal to landing men on the Moon," and proposed they go into lunar orbit as well.

After consulting Deke Slayton and von Braun and getting the go ahead from a surprised NASA hierarchy and President Johnson, NASA decided to officially go for orbits around the Moon with a CSM flight only, subject to a successful Apollo 7 manned flight. It was decided to make the first attempt on 21 December with a liftoff time of 1251 GMT (0751 USEST or 2251 AEST). This day and time would allow the crew to observe the first planned landing site at the ideal Sun elevation of 6.7°. At least it would make sure of beating the Russians and confirm that a manned lunar flight in orbit was possible.

The sudden change in flight plans caused by the decision to move Apollo 8 to a lunar orbit mission disrupted Deke Slayton's meticulously crafted crew schedules. Originally, James McDivitt, Dave Scott, and Rusty Schweickart were slated for an Earth orbit mission to test the Lunar Module. Despite this shift, Slayton opted to keep McDivitt's crew on track for their LM mission and swapped them with the crew of Apollo 9: Frank Borman, Jim Lovell, and Bill Anders, with Neil Armstrong, Buzz Aldrin, and Fred Haise as their backups.

Slayton, confident in Borman's crew's readiness, remarked, "I thought that this crew could be ready in four months, no problem." On August 10, 1968, he summoned Jim McDivitt to his office and informed him of the change, explaining that NASA didn't want to discard their training. McDivitt understood the reasoning, noting his and Schweickart's intimate familiarity with the specific lunar module being tested, and agreed to the swap.

Two days later, Slayton extended the offer to Frank Borman, who immediately accepted the opportunity to command Apollo 8.

The new crew assignment was publicly announced on August 19, 1968, although the mission's details remained confidential until after the safe return of Apollo 7. In mid-September, NASA issued a carefully worded news release hinting at plans beyond Apollo 7, mentioning possibilities like a long-duration high orbit flight or a circumlunar/lunar orbit mission. Surprisingly, the media did not discern the hint that the impending lunar orbit mission would become one of NASA's most pivotal and historic achievements in space exploration.

However, challenges persisted. By August 1968, delays in LM development prompted NASA to revise its plans. A non-flightworthy LM test article, weighing approximately 19,900 pounds, was substituted into the spacecraft for balance and mass-loading purposes. This decision, made on August 19, coincided with the official designation of the mission as "Apollo 8." Crew training promptly shifted focus to lunar operations, signifying a decisive step toward NASA's ambitious goal of reaching the Moon.

The prospect of a lunar mission first emerged during discussions with the crew on August 10th, against the backdrop of Apollo 7's pending October launch. The outcome of Apollo 7, NASA's first manned mission in the Apollo program, would dictate the nature of the subsequent mission—whether it would orbit the Moon, fly around it, or remain in Earth orbit. Immediately, training efforts pivoted to prepare for the most challenging scenario: a lunar orbital mission. Ground support teams accelerated their preparations in tandem.

The first crucial simulation exercise took place on September 9th, underscoring the intensity and focus of the preparations. By October 9th, the space vehicle was relocated to the launch site, marking a critical milestone in the mission timeline. Apollo 7's successful completion on October 22nd provided essential data on spacecraft performance during ten days in Earth orbit. This pivotal mission paved the way for a thorough evaluation, culminating in the official decision on November 12th to proceed with a lunar orbit mission—just five weeks before the scheduled launch.

Central to this decision were assessments of the risks inherent in a lunar mission. These included the critical dependence on the service propulsion engine for maneuvering within lunar orbit and the extended return journey of three days compared to the rapid return windows of Earth orbit missions, typically lasting between 30 minutes to three hours. Equally weighed was the mission's strategic value in advancing NASA's overarching goal of landing a human on the Moon before the close of 1969.

The objectives of a lunar mission encompassed crucial advancements in deep space navigation, communication systems, and spacecraft thermal management. These missions were pivotal in preparing for future lunar landings. Apollo 8 marked a historic milestone as the first manned mission launched atop the powerful three-stage Saturn V rocket. Before Apollo 8, Saturn V had undertaken two unmanned test flights.

The spacecraft utilized for Apollo 8 was a Block II Command and Service Module (CSM). Notably, its spacecraft/launch vehicle adapter introduced a novel mechanism: panels designed to cover the Lunar Module (LM) could be jettisoned, laying the groundwork for subsequent missions.

The primary goals of Apollo 8 were twofold: to validate the operational Apollo 8 aimed to test the mettle of both crew and spacecraft in the unforgiving realm of deep space aboard the formidable Saturn V rocket. This mission was pivotal in NASA's Apollo program, setting the stage for future lunar landings.

The crew of Apollo 8, led by Colonel Frank Frederick Borman II of the United States Air Force, Commander of the mission, brought together a wealth of experience and expertise in spaceflight. Born on March 14, 1928, in Gary, Indiana, Borman was 40 years old during Apollo 8. He held a B.S. from the U.S. Military Academy and an M.S. in Aeronautical Engineering from the California Institute of Technology. Borman had previously commanded Gemini 7, demonstrating his proficiency and leadership in earlier missions.

Captain James Arthur Lovell, Jr., of the United States Navy, served as a Command Module Pilot. Also selected in the 1962 astronaut group, Lovell was an experienced pilot with prior missions on Gemini 7 and as commander of Gemini 12. Born in Cleveland, Ohio on March 25, 1928, Lovell was 40 during Apollo 8 and held a B.S. from the U.S. Naval Academy. His extensive background and piloting skills made him a key asset to the crew.

Major William Alison Anders joined as a Lunar Module Pilot, marking his debut in spaceflight. Born in Hong Kong on October 17, 1933, Anders was 35 years old during Apollo 8. He held a B.S. in Electrical Engineering from the U.S. Naval Academy and an M.S. in Nuclear Engineering from the U.S. Air Force Institute of Technology. Anders was selected as an astronaut in 1963, bringing technical expertise and a fresh perspective to the mission.

Apollo 8 crew is photographed posing on a Kennedy Space Center (KSC) simulator in their space suits. From left to right are: James A. Lovell Jr., William A. Anders, and Frank Borman.

The original crew lineup for Apollo 8 included Michael Collins as Command Module Pilot, but due to health reasons, he was replaced by Jim Lovell in July 1968. This adjustment in crew composition underscored the adaptability and readiness of the astronauts selected for the mission, paving the way for a successful lunar orbit mission that would become a landmark achievement in space exploration.

During the preparations for Apollo 8, a support crew was crucial in ensuring mission readiness and success alongside the prime and backup crews. The backup crew, initially comprised of Neil A. Armstrong as Commander, Edwin E. Aldrin Jr. as Command Module Pilot (CMP), and Fred W. Haise Jr. as Lunar Module Pilot (LMP), mirrored the structure of the prime crew. This backup team stood ready to step in if needed, exemplifying the redundancy built into NASA's mission planning.

When James A. Lovell Jr. was reassigned to the prime crew as CMP due to health reasons, Buzz Aldrin shifted roles to CMP, and Fred Haise was brought in as the new backup LMP. This adjustment highlighted the flexibility and adaptability of NASA's astronaut corps, ensuring that crews were optimally configured for each mission.

The Capsule Communicators (CAPCOMs) for Apollo 8 were pivotal figures who relayed critical information between Mission Control and the astronauts in space. Notable among them were Michael Collins, Thomas Kenneth "Ken" Mattingly II, Gerald Paul Carr, Neil Armstrong, Buzz Aldrin, Vance DeVoe Brand, and Fred Haise. Their expertise and direct communication with the crew were essential for operational clarity and decision-making during the mission.

Supporting the prime and backup crews, the support crew for Apollo 8 included Ken Mattingly, Vance Brand, and Gerald Carr. They played a vital role in maintaining the flight plan, checklists, and ground rules, ensuring consistency and readiness across all aspects of the mission. Additionally, the support crew developed and refined simulator procedures, particularly focusing on emergency scenarios, to prepare the astronauts comprehensively for any contingencies they might encounter in space.

In Mission Control at Houston, Texas, the mission was overseen by three shifts of flight directors: Clifford E. Charlesworth (Green team), Glynn S. Lunney (Black team), and Milton L. Windler (Maroon team). Each team of flight directors guided the mission through their respective shifts, coordinating with the CAPCOMs and support crew to ensure the smooth execution of Apollo 8's historic journey to the Moon and back.

The Apollo program faced significant challenges when the Lunar Module (LM) production fell behind schedule, jeopardizing NASA's ambitious timeline for lunar exploration. By June 1968, LM-3, designated for Apollo 8, arrived at the Kennedy Space Center with over a hundred critical defects. This situation led Bob Gilruth, director of the Manned Spacecraft Center (MSC), and other officials to conclude that LM-3 would not be flight-ready in 1968, possibly delaying its availability to February or March 1969.

This delay threatened to disrupt NASA's meticulously planned seven-step mission sequence, designed to culminate in a lunar landing by the end of 1969. George Low, Manager of the Apollo Spacecraft Program Office, proposed a creative solution in August 1968 to mitigate the impact of the LM delay. He suggested advancing the launch of CSM-103, the next Command and Service Module, which would be ready three months earlier than LM-3. Instead of a repeat of Apollo 7's Earth-orbit mission, CSM-103 could be utilized for a more ambitious mission: a journey to the Moon, potentially entering lunar orbit before returning safely to Earth.

This revised mission plan maintained the overall timeline and accelerated critical lunar landing procedures originally planned for Apollo 10. Consequently, the medium Earth orbit "E" mission was eliminated from the schedule. The only mission directly affected by the delay was the "D" mission, which would now be postponed. However, this adjustment ensured that NASA remained on track for a mid-1969 lunar landing, aligning with the program's ultimate goal.

George Low's strategic decision exemplified NASA's adaptability and resolve in overcoming setbacks, demonstrating the agency's commitment to achieving one of humanity's most audacious goals—landing humans on the Moon and returning them safely to Earth within the decade.

On August 9, 1968, George Low presented his proposal to key NASA figures, including Bob Gilruth, Chris Kraft, and Donald Slayton, at the Marshall Space Flight Center (MSFC) in Huntsville, Alabama. They were joined by Kurt Debus, Samuel C. Phillips, Rocco Petrone, and Wernher von Braun. Jerry Wittenstein from flight mechanics detailed potential trajectories for the new mission.

Chris Kraft, the Flight Director, found the proposal technically feasible for flight control. Kurt Debus and Rocco Petrone confirmed that the Saturn V rocket designated AS-503 could be prepared by December 1, meeting the timeline requirements. Wernher von Braun expressed confidence that the pogo oscillation issues, which had troubled Apollo 6, had been resolved. Nearly all senior NASA managers supported the new mission enthusiastically, highlighting confidence in both the spacecraft and the personnel involved. They recognized the mission's potential to boost morale significantly by attempting a circumlunar flight.

The primary hurdle remained convincing James E. Webb, the NASA Administrator, of the mission's viability. With strong backing from within NASA, Webb ultimately authorized the mission. Consequently, Apollo 8 was officially reclassified from a "D" mission to a "C-Prime" lunar-orbit mission, marking a pivotal shift in NASA's Apollo program strategy. This decision underscored NASA's ability to adapt swiftly to challenges and capitalize on opportunities, maintaining momentum toward the ambitious goal of landing humans on the Moon.

The first stage of AS-503 being erected in the Vehicle Assembly Building (VAB) on February 1, 1968

Following the change in mission plans for Apollo 8, Deke Slayton approached Jim McDivitt to offer him the opportunity to command the mission. However, McDivitt declined, expressing his team's strong commitment to their extensive preparations for testing the Lunar Module (LM). Slayton then decided to swap the prime and backup crews of the originally planned D and E missions. This swap also necessitated a change in spacecraft: Frank Borman's crew would now use CSM-103, while McDivitt's crew would switch to CSM-104 due to readiness constraints.

Erection and mating of spacecraft 103 to Launch Vehicle AS-503 in the VAB for the Apollo 8

David Scott, who had closely overseen the testing of CM-103, was disappointed about giving up this spacecraft for CM-104, even though the two were nearly identical. William Anders also had reservations about being the Lunar Module Pilot (LMP) on a mission that no longer included a Lunar Module. Instead, Apollo 8 would carry the LM test article, a boilerplate model designed to simulate the weight and balance characteristics of the actual LM-3.

The urgency to meet the 1969 lunar landing goal was heightened by developments in the Soviet Union's space program. The Zond 5 mission, launched on September 15, 1968, carried living creatures, including Russian tortoises, on a circumlunar trajectory around the Moon and returned them safely to Earth on September 21. This mission raised concerns within NASA and the media that the Soviets might attempt a similar manned circumlunar mission before the end of 1968.

Compounding these concerns, American reconnaissance satellites detected preparations for a Soviet N1 rocket, a rival to the American Saturn V, at the Baikonur Cosmodrome.

The night before the launch of Apollo 8, the crew, staying in the crew quarters at Kennedy Space Center, had a memorable visit from Charles Lindbergh and his wife, Anne Morrow Lindbergh. Lindbergh recounted his historic 1927 solo flight across the Atlantic Ocean during their visit. He shared how, before embarking on his journey, he used a piece of string to measure the distance from New York City to Paris on a globe, which helped him calculate the fuel needed for the flight. Remarkably, his total fuel was just a tenth of what the Saturn V rocket would burn every second during Apollo 8's launch.

Apollo 8 atop Saturn V being rolled out to Pad 39A atop the crawler-transporter

The following day, the Lindberghs had the opportunity to witness firsthand the launch of Apollo 8 from a nearby dune. This encounter bridged the gap between two iconic moments in aviation and space exploration history, connecting Lindbergh's pioneering transatlantic flight with NASA's ambitious mission to reach and orbit the Moon.

The Saturn V rocket designated AS-503 for the Apollo 8 mission was initially planned for an uncrewed Earth-orbit test flight with a boilerplate command and service module. However, due to issues encountered during Apollo 6's flight in April 1968, including severe pogo oscillation and engine failures, NASA needed to address these problems before committing to a crewed mission.

Apollo 6's problems, particularly the pogo oscillation during the first stage and engine failures in subsequent stages, posed significant risks to crew safety. Pogo oscillation occurred when the engines vibrated at frequencies that resonated with the spacecraft's own natural vibrations, causing potentially dangerous oscillations. A Marshall Space Flight Center (MSFC) team worked on solutions to mitigate this issue, including installing a helium gas system to dampen the vibrations.

The failure of three engines during Apollo 6's flight was a critical issue that needed to be addressed to ensure the reliability of the Saturn V rocket for future missions, including Apollo 8. Researchers determined that the root cause of the engine failures was primarily related to a leaking hydrogen fuel line.

During Apollo 6, a leaking hydrogen fuel line ruptured when exposed to vacuum conditions, causing a loss of fuel pressure in engine two. As a safety measure, an automatic shutoff system attempted to close the liquid hydrogen valve for engine two to prevent further issues. However, due to a miswired connection, this action accidentally shut down engine three's liquid oxygen supply instead of engine two's hydrogen valve. Consequently, engine three failed within one second of engine two's shutdown.

Similar issues were found with the third-stage engine, where a faulty igniter line contributed to its failure. To address these problems, the engineering team modified the igniter lines and fuel conduits. These modifications aimed to prevent similar issues on future Saturn V launches, including Apollo 8.

In August 1968, teams at the Marshall Space Flight Center (MSFC) conducted rigorous testing to validate their solutions to the critical issues encountered during Apollo 6's flight. One Saturn stage was equipped with shock-absorbing devices to mitigate pogo oscillation, while another stage had modified fuel lines tested to resist leaks and ruptures under vacuum conditions. These tests aimed to demonstrate the effectiveness of the engineering solutions developed by NASA.

Once NASA administrators were satisfied that these problems had been successfully addressed, they approved using AS-503 for a crewed mission. On September 21, the Apollo 8 spacecraft was mated with the Saturn V rocket. On October 9, the entire assembly was transported to the launch pad, 3 miles (4.8 km) away, atop one of NASA's crawler-transporters.

Testing continued throughout December to ensure readiness for the mission. This included various levels of readiness tests from December 5 to 11. Final checks and modifications to address issues such as pogo oscillation, fuel line integrity, and igniter lines were conducted on December 18, just three days before the scheduled launch date.

Apollo 8, designated Eastern Test Range #170, featured the Command and Service Module (CSM-103) and the lunar module test article LTA-B. This mission was meticulously planned due to its lunar objectives, necessitating precise launch timing within daily and monthly windows. These constraints aimed to align the spacecraft's trajectory with optimal lighting conditions over targeted lunar sites, crucial for future landing missions.

The initial launch opportunity was set for December 1968, with a daily window spanning December 20-27. January 1969 served as a backup. To maximize daylight for optimal visibility during critical phases, December 21 was chosen for the first attempt. This ensured the spacecraft could pass over a future lunar landing site at coordinates 2.63° latitude and 34.03° longitude, with the Sun at an elevation angle of 6.74°.

The launch window for December 21 stretched from 12:50:22 to 17:31:40 GMT, with liftoff scheduled precisely at 12:51:00 GMT. Factors such as lunar orbit inclination, free return trajectory inclination, and spacecraft propellant reserves were meticulously calculated and factored into the mission's planning, highlighting NASA's rigorous approach to achieving its lunar exploration goals.

The terminal countdown sequence for Apollo 8 commenced at 01:51 GMT on December 20, marking T-28 hours before scheduled liftoff. Early in the countdown, a critical issue surfaced: the spacecraft's liquid oxygen supply, vital for environmental control and fuel cell systems, was found contaminated with nitrogen. Swift measures were taken to replace the liquid oxygen, necessitating rescheduled tank pressurization by T-10 hours.

Despite these setbacks, meticulous coordination during the planned six-hour hold at T-9 hours allowed teams to realign virtually all delayed countdown tasks. By the time operations resumed at T-9 hours, the spacecraft was effectively back on track. At T-8 hours, liquid oxygen loading operations (S-NB) commenced. These cryogenic loading procedures concluded precisely at 08:29 GMT on December 21, aligning with the one-hour scheduled hold and positioning Apollo 8 for its historic launch.

At 0235 USEST on December 21, 1968, Deke Slayton woke the Apollo 8 crew for their customary steak and scrambled eggs breakfast. By 0458 USEST (1958 AEST), the three astronauts had settled into their cramped spacecraft, which would be their home for the next six days.

Commander Frank Borman, positioned on the left, was the first to enter. His responsibilities included monitoring the Saturn V rocket's performance. Jim Lovell, the most experienced of the crew, crawled in last and sat in the middle seat. Lovell managed the Command Module's computer systems and navigational tasks. Bill Anders, the team's rookie, occupied the right seat, overseeing the spacecraft's electrical systems and communications.

Borman and Lovell had previously flown together during Gemini VII, spending 14 days in space, and Lovell had further experience from his flight with Buzz Aldrin on the final Gemini mission. This made Lovell the most seasoned astronaut among the trio.

The hatch sealed shut at 0534 USEST, marking the crew's departure from the outside world. The countdown proceeded smoothly as they prepared for their historic mission.

At T-3 hours 30 minutes, the countdown resumed at 09:21 GMT, marking a pivotal moment as preparations intensified for the historic Apollo 8 mission. The crew, composed of Colonel Frank Borman, Captain James Lovell, and Major William Anders, entered the spacecraft at T-2 hours 53 minutes, readying themselves for the journey ahead.

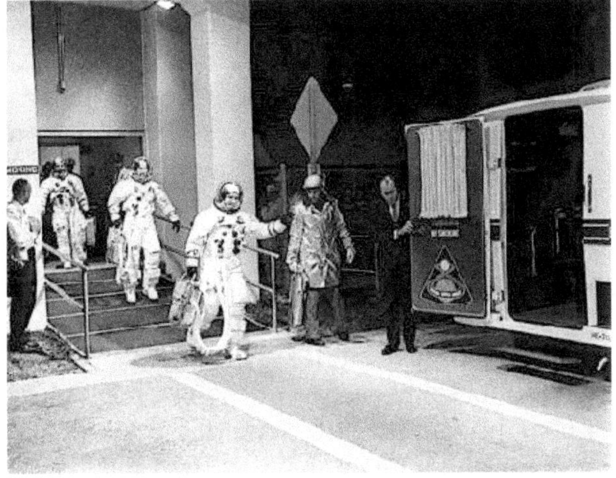

The Apollo 8 astronauts (from left: Anders, Lovell, Borman) walk out to the crew transfer van for the drive to the pad.

Leading up to the launch, weather conditions played a crucial role. A stationary front, originating as a cold front the previous afternoon, positioned itself over the Miami area. At launch time, surface winds initially came from the north, shifting to westerly at an altitude of 4,900 feet and prevailing from the west above that altitude. The sky was adorned with cirrus clouds covering 40 percent of the area, although the cloud base was not recorded. Visibility stood at a clear ten statute miles, with a temperature of 59.0°F, relative humidity at 88 percent, and a dew point of 56 percent. The barometric pressure measured 14.804 lb/in². Winds were measured at 18.7 ft/sec at 348° true north, recorded by the anemometer atop a light pole 60.0 feet above the launch site ground level.

Apollo 8 lifted off from Launch Complex 39, Pad A, at the Kennedy Space Center in Florida, precisely at 12:51:00 GMT (07:51:00 a.m. EST) on December 21, 1968, within its designated launch window. The ascent proceeded smoothly and according to plan.

Shortly after liftoff, the vehicle executed a roll maneuver, transitioning from a launch pad azimuth of 90° to a flight azimuth of 72.124° east of north. The first stage (S-IC) engine shut down at 000:02:33.82 into the flight, separated by the S-IC stage and ignition of the second stage (S-II) engine. The S-II engine operated until 000:08:44.04, after which the stage separated, initiating the ignition of the third stage (S-NB).

The S-NB stage performed as expected, with its first engine cutoff occurring at 000:11:24.98 into the mission. Deviations from the planned trajectory were minimal, with velocity showing a slight increase of only +1.44 ft/sec and altitude variance negligible at -0.01 nautical miles.

Apollo 8 Launch

Following a successful ascent, the stages of Apollo 8 impacted the Atlantic Ocean as planned. The S-IC stage impacted at 000:09:00.410, located at latitude 30.2040° north and longitude 74.1090° west, approximately 353.462 nautical miles from the launch site. Shortly after, the S-II stage impacted at 000:19:25.106, at latitude 31.8338° north and longitude 37.2774° west, approximately 2,245.913 nautical miles from the launch site.

Four recoverable film camera capsules were carried aboard the S-IC stage during ascent. Two capsules were positioned in the forward interstage, providing views of the S-IC/S-II separation and S-II engine ignition. The remaining capsules, equipped with pulse cameras, were mounted atop the S-IC stage's LOX tank, offering views aft into the tank via fiber optics bundles. One LOX tank capsule was successfully recovered by helicopter at 00:19:30, located at latitude 30.22° north and longitude 73.97° west. Despite damage from seawater and dye marker intrusion, the film contained valuable data. The status of the other three capsules remains unknown.

Additionally, two television cameras mounted on the S-IC stage provided high-quality propulsion and control system components data throughout the ascent phase.

Maximum wind conditions encountered during ascent reached 114.1 ft/sec at 284° true north at 49,900 feet, within a high dynamic pressure region. Wind shears affecting components were generally of low magnitude, with the largest recorded as a pitch plane shear of 0.0103 seconds at 52,500 feet.

At 000:11:34.98, Apollo 8 achieved Earth orbit, defined as the S-IVB cutoff plus 10 seconds to account for engine tail-off and transient effects. At orbital insertion, the spacecraft's conditions were as follows: apogee and perigee distances were 99.99 nautical miles and 99.57 nautical miles, respectively, with an inclination of 32.509°, a period of 89.19 minutes, and a velocity of 25,567.06 feet per second. These orbital parameters were calculated based on a spherical Earth with a radius of 3,443.934 nautical miles.

Upon achieving Earth orbit, the spacecraft was designated as 1968-118A, while the S-IVB stage was designated as 1968-118B. The initial phases of the mission proceeded as planned, marking critical milestones in Apollo 8's journey.

After reaching Earth orbit, both the crew and Houston flight controllers embarked on a meticulous 2-hour and 38-minute process to ensure the spacecraft's systems were fully operational and ready for the critical Trans-Lunar Injection (TLI). This thorough check was essential, given the history of the S-IVB third stage's previous uncrewed test failure to reignite for this crucial burn.

As the first CAPCOM on duty, Michael Collins played a pivotal role in the communication chain. At 2 hours, 27 minutes, and 22 seconds after launch, he radioed the landmark message, "Apollo 8.

You are Go for TLI." This official clearance from Mission Control marked the pivotal moment when Apollo 8 received authorization to begin its journey to the Moon.

The subsequent ignition of the S-IVB engine proceeded flawlessly at the designated time, initiating the TLI burn with precision. Over the next five minutes, the spacecraft's velocity surged from 7,600 to 10,800 meters per second (25,000 to 35,000 ft/s), propelling Apollo 8 on its historic trajectory toward lunar orbit.

After the S-IVB stage had successfully positioned Apollo 8 on its trajectory toward the Moon, the command and service modules (CSM) separated from the spent rocket stage. With the stage now drifting away, the crew maneuvered the spacecraft to capture photographs of the discarded stage and practiced flying in formation with it.

During this maneuver, as the crew rotated the spacecraft, they were treated to their first panoramic views of Earth from a distance—the first time in history that humans beheld our planet in its entirety at once. This awe-inspiring sight marked a profound moment for the crew and for the world watching back on Earth.

However, Commander Frank Borman expressed concern that the S-IVB stage was lingering too close to the CSM. He communicated this worry to Mission Control and proposed a separation maneuver to increase the distance between the two. Initially, Mission Control suggested a subtle adjustment using the small reaction control system (RCS) thrusters on the service module (SM), which would add 1.1 ft/s (0.34 m/s) to their velocity away from Earth. Borman hesitated, reluctant to lose visual contact with the drifting stage.

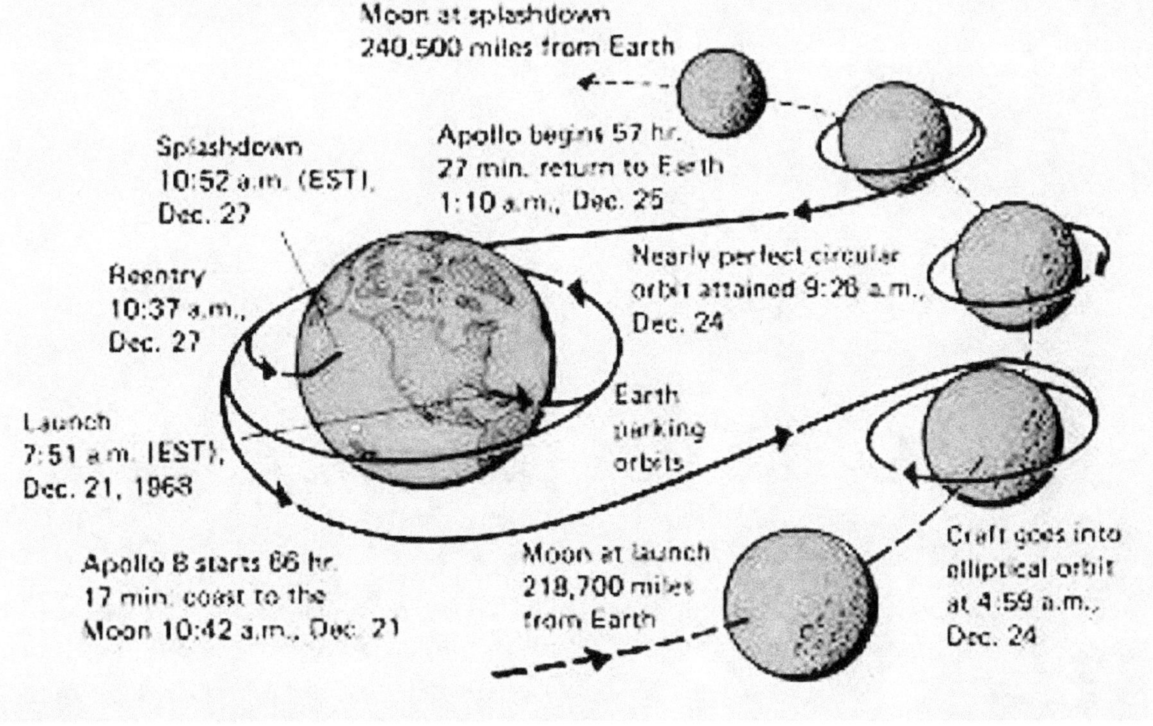

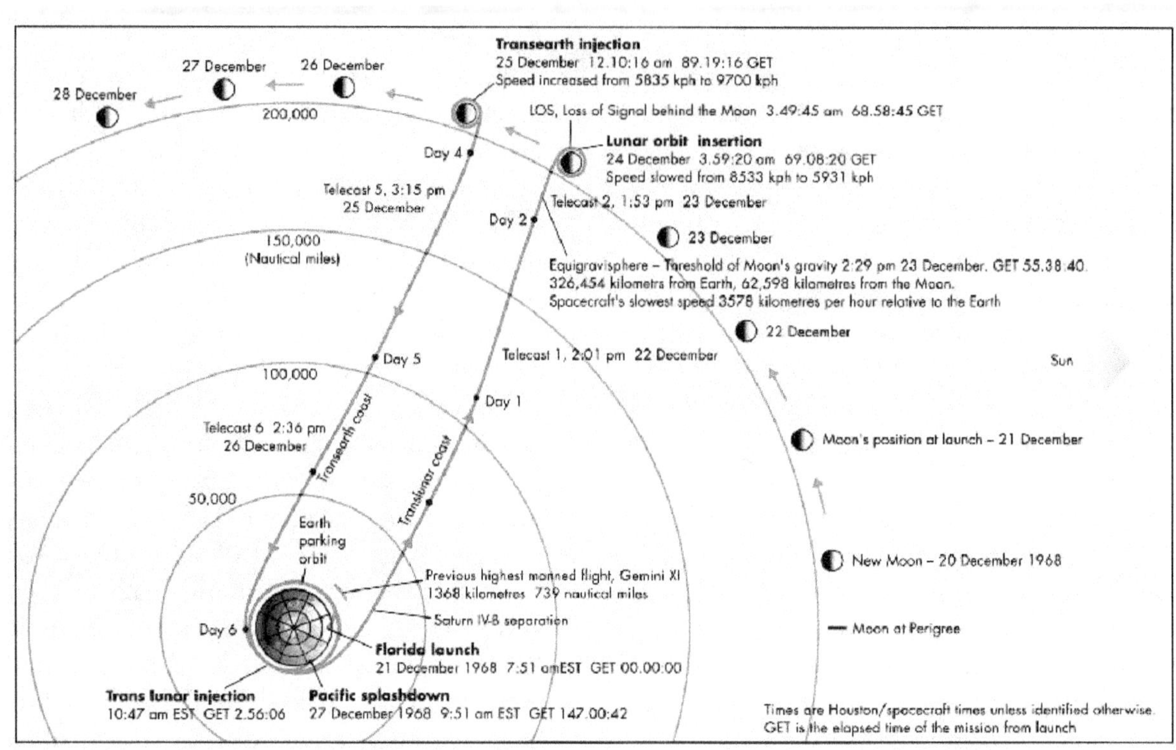
Transearth injection
25 December 12.10:16 am 89.19:16 GET
Speed increased from 5835 kph to 9700 kph
27 December
26 December
28 December
200,000
LOS, Loss of Signal behind the Moon 3.49:45 am 68.58:45 GET
Lunar orbit insertion
24 December 3.59:20 am 69.08:20 GET
Speed slowed from 8533 kph to 5931 kph
Day 4
Telecast 5, 3:15 pm
25 December
Telecast 2, 1:53 pm 23 December
Day 2
23 December
150,000
(Nautical miles)
Equigravisphere – Threshold of Moon's gravity 2:29 pm 23 December. GET 55.38:40.
326,454 kilometres from Earth, 62,598 kilometres from the Moon.
Spacecraft's slowest speed 3578 kilometres per hour relative to the Earth
22 December
Day 5
Telecast 1, 2:01 pm 22 December
Sun
100,000
Day 1
Telecast 6 2:36 pm
26 December
Transearth coast
Translunar coast
Moon's position at launch – 21 December
50,000
Earth
parking
orbit
New Moon – 20 December 1968
Previous highest manned flight, Gemini XI
1368 kilometres 739 nautical miles
Saturn IV-B separation
Day 6
Moon at Perigree
Florida launch
21 December 1968 7:51 am EST GET 00.00:00
Trans lunar injection
10:47 am EST GET 2.56:06
Pacific splashdown
27 December 1968 9:51 am EST GET 147.00:42
Times are Houston/spacecraft times unless identified otherwise.
GET is the elapsed time of the mission from launch

After deliberation between the crew and Mission Control, they agreed on a compromise. They executed a burn in the direction of Earth, accelerating the spacecraft by 7.7 ft/s (2.3 m/s) to widen the gap between Apollo 8 and the S-IVB stage. However, the unplanned maneuver and the time required to prepare for it caused the crew to fall an hour behind schedule with their onboard tasks.

Five hours after launch, Mission Control issued a command to the S-IVB stage to vent its remaining fuel, altering its trajectory to ensure it posed no further risk to Apollo 8. With the lunar module test article still attached, the stage continued on a path that carried it beyond the Moon's orbit and into a solar orbit ranging from 0.99 to 0.92 astronomical units (148 to 138 million kilometers) from the Sun. This orbit, inclined at 23.47° relative to Earth's equatorial plane, gave the S-IVB a lengthy orbital period of 340.80 days. Designated a derelict object, it will persist in orbiting the Sun for years unless intercepted or retrieved.

A pivotal milestone for the Apollo 8 crew was their passage through the Van Allen radiation belts, an intense region of charged particles extending up to 15,000 miles (24,000 km) from Earth's surface. Prior scientific calculations had indicated that traversing these belts swiftly at the spacecraft's high velocity would subject the crew to a radiation dose roughly equivalent to a standard chest X-ray, amounting to about one milligray (mGy). To accurately measure and monitor radiation exposure, each crew member wore a Personal Radiation Dosimeter that transmitted real-time data to Earth. Additionally, three passive film dosimeters recorded cumulative radiation levels throughout the mission.

By the mission's conclusion, the average radiation dose received by each crew member totaled 1.6 mGy. This careful monitoring provided valuable data for future missions and reassured scientists and mission planners about the feasibility of crewed space travel through the Van Allen belts.

At 000:42:05 into the mission, the optics cover was jettisoned, and the crew conducted star checks over the Carnarvon, Australia, tracking station to verify the alignment of the spacecraft platform. By 001:56:00, all spacecraft systems were confirmed operational and approved for translunar injection. The mission was structured with three commit points: launch, Earth parking orbit, and translunar coast. These points ensured that alternate missions could be implemented to prioritize crew safety and scientific goals if any issues arose. A "free-return" trajectory plan was also in place, allowing the spacecraft to loop behind the Moon and return directly to Earth if necessary.

During inflight systems checks, it was discovered that liquid oxygen venting through the J-2 engine had increased the apogee by 6.4 nautical miles, slightly exceeding predictions by 0.7 nautical miles. Nevertheless, this minor deviation did not significantly impact the mission trajectory.

The critical translunar injection maneuver, a 317.72-second firing of the S-IVB engine, occurred at 002:50:37.79. The engine shutdown followed at 002:55:55.51, with translunar injection achieved ten seconds later, propelling the spacecraft to a velocity of 35,504.41 feet per second. This maneuver spanned 1.5 Earth revolutions, lasting 2 hours, 44 minutes, and 30.53 seconds.

Upon reaching the translunar phase, the spacecraft separated from the S-IVB stage at 003:20:59.3 through a small maneuver executed by the service module reaction control system. Simultaneously, the high-gain antenna was deployed, a pivotal step for communications. Following a spacecraft turnaround, the crew observed and photographed the departing S-IVB stage, practicing station-keeping procedures.

At 003:40:01, a maneuver using the service module reaction control system adjusted the spacecraft's distance from the S-IVB by 1.1 feet per second. However, the desired separation rate was not achieved, prompting a second maneuver at 004:45:01, which increased the distance by 7.7 feet per second.

One of the objectives of the Apollo 8 mission was to send the S-IVB stage into a solar orbit, a maneuver often referred to as a "slingshot" to maximize the mission's scientific returns. This complex maneuver involved several critical steps to alter the trajectory of the S-IVB stage.

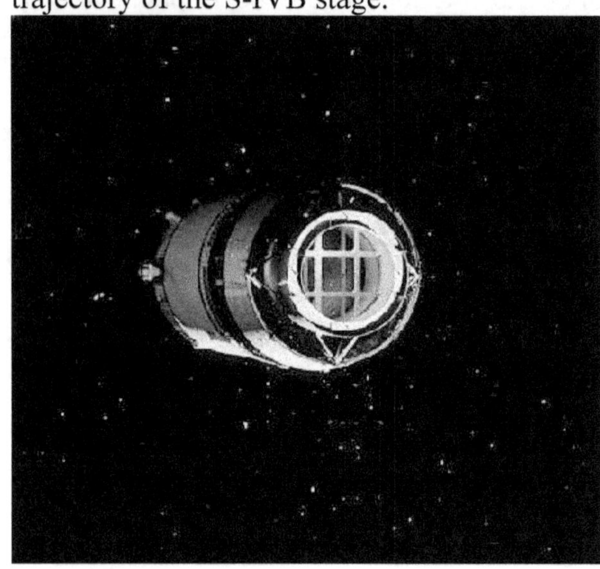

Apollo 8 S-IVB rocket stage shortly after separation. The LM test article, a circular boilerplate model of the LM, is visible with four triangular legs connecting it to the stage.

At 004:55:56.02, the LH2 (liquid hydrogen) vent valve was opened, initiating the process. This was followed by a dump of remaining liquid oxygen and a burn of the auxiliary propulsion system propellant in the S-IVB. The expulsion of liquid oxygen through the J-2 engine began at 005:07:55.82 and continued for five minutes.

Subsequently, the auxiliary propulsion motors were fired from 005:25:55.85 until depletion at 005:38:34.00. This firing provided the necessary velocity increment to propel the S-IVB stage past the trailing edge of the Moon. The closest approach of the S-IVB to the Moon occurred at 682 nautical miles at 069:58:55.2, located at latitude 19.2 degrees north by longitude 88.0 degrees east.

After passing through the lunar sphere of influence, the trajectory of the S-IVB entered a solar orbit with specific orbital parameters: a semi-major axis of 77.130 million nautical miles, an aphelion (farthest point from the Sun) and perihelion (closest point to the Sun) of 79.770 million by 74.490 million nautical miles, an inclination of 23.47 degrees, and a period of 340.8 days.

The accuracy of the translunar injection maneuver was remarkable, requiring only a small midcourse correction to achieve the desired lunar orbit insertion altitude of 65 nautical miles. However, due to the trajectory adjustment caused by the second maneuver that separated the spacecraft from the S-IVB, a 2.4-second midcourse correction burn of 20.4 feet per second was necessary at 010:59:59.2 to precisely align with the intended trajectory.

During the translunar coast phase, which followed the midcourse corrections, the crew conducted systems checks, performed navigation sightings, and tested the spacecraft's high-gain antenna. The four-dish unified S-band type antenna swung out from the service module after separation from the S-IVB, allowing for improved communication with Earth.

Apollo 8 marked a significant milestone as the first piloted U.S. mission where crew members experienced symptoms akin to mild motion sickness, similar to incipient seasickness. Shortly after leaving their couches, all three astronauts reported nausea attributed to rapid body movements within the spacecraft. The duration of these symptoms varied among the crew members, lasting between 2 and 24 hours, but they did not impair operational effectiveness.

At 016:00:00, after waking from a restless period of sleep, the mission commander experienced additional symptoms, including a headache, continued nausea, vomiting, and diarrhea. These effects were managed onboard with the available medical supplies and support systems.

During the post-mission medical debriefing, the commander noted that the symptoms experienced inflight, initially thought to be viral gastroenteritis, might have been exacerbated by a sleeping tablet he took at 011:00:00. This tablet, identified as Seconal™, had previously caused similar symptoms during pre-mission drug testing. The association suggested a potential side effect influencing the commander's physical condition during the mission.

Despite these health challenges, the crew continued operational duties, including conducting two live television transmissions during the translunar flight phase. The first transmission lasted 23 minutes and 37 seconds, starting at 031:10:36. It featured wide-angle views inside the spacecraft, showcasing Lovell as he prepared a meal. However, issues with the telephoto lens affected Earth images of poor quality due to excessive light intake. To address this, a procedure was devised to apply filters from the still camera onto the television camera, improving the quality of subsequent transmissions.

The second live transmission occurred at 055:02:45 and lasted 25 minutes and 38 seconds. It provided viewers on Earth with scenes from the western hemisphere, offering a unique perspective of our planet during the historic Apollo 8 mission.

At 055:38:40, a historic moment unfolded aboard Apollo 8: the crew received confirmation that they had ventured beyond Earth's gravitational dominance, now feeling the faint tug of the Moon. Positioned 176,250 nautical miles from Earth and 33,800 nautical miles from their lunar destination, their spacecraft's velocity had decelerated to 3,261 feet per second. As Apollo 8 traversed deeper into the Moon's gravitational influence, their speed gradually increased.

The critical moment arrived at 069:08:20.4 when the service propulsion system roared to life, executing a precise lunar orbit insertion burn. Positioned 76.6 nautical miles above the Moon's surface, this 246.9-second burn altered their trajectory, establishing an elliptical orbit spanning 168.5 by 60.0 nautical miles at 5,458 feet per second. This milestone concluded a translunar coast lasting 66 hours, 16 minutes, and 21.79 seconds.

Apollo 8 marked a historic milestone in space exploration as the first crewed spacecraft to venture beyond Earth's orbit and orbit another celestial body. Its mission profile was meticulously planned, beginning with a standard 100-nautical-mile (185.2 km) circular Earth parking orbit. Upon launch, Apollo 8 entered an initial orbit with specific parameters: an apogee of 99.99 nautical miles (185.18 km), a perigee of 99.57 nautical miles (184.40 km), and an inclination of 32.51° relative to the Equator. This orbit had a precise orbital period of 88.19 minutes, crucial for mission planning and trajectory calculations.

The next critical phase was the trans-lunar injection (TLI) maneuver. Lasting 318 seconds, this burn of the S-IVB third stage propelled the combined 63,650 lb (28,870 kg) command and service module, along with a 19,900 lb (9,000 kg) lunar module test article, from an initial orbital velocity of 25,567 feet per second (7,793 m/s) to an injection velocity of 35,505 ft/s (10,822 m/s). This acceleration achieved a record-breaking speed for human travel relative to Earth, approaching but slightly below Earth's escape velocity of 36,747 feet per second (11,200 m/s).

Apollo 8's trajectory placed it into an elongated elliptical orbit around Earth, strategically positioned to intersect with the Moon's gravitational influence. This daring maneuver allowed Apollo 8 to journey beyond Earth's immediate vicinity, setting the stage for subsequent Apollo lunar missions culminating in humanity's first steps on the lunar surface.

The standard lunar orbit established for Apollo missions was designed as a precise 60-nautical-mile (110 km) circular path above the Moon's surface. Upon initial lunar orbit insertion, the trajectory formed an elliptical shape with a perilune (closest point to the Moon) of 60.0 nautical miles (111.1 km) and an apolune (farthest point from the Moon) of 168.5 nautical miles (312.1 km). This orbit had an inclination of 12° relative to the lunar equator and initially spanned a range from 60.7 to 59.7 nautical miles (112.4 to 110.6 km) once circularized, completing a full orbit every 128.7 minutes.

During its orbital phase around the Moon, Apollo 8 encountered significant gravitational anomalies known as lunar mass concentrations, or "mascons," which exerted greater influence than anticipated. These mascons caused the lunar orbit to deviate, fluctuating between 63.6 and 58.6 nautical miles (117.8 and 108.5 km) throughout ten orbits, spanning approximately twenty hours.

At its farthest point from Earth, Apollo 8 achieved an impressive distance of 203,752 nautical miles (234,474 statute miles or 377,349 kilometers). This remarkable journey demonstrated the spacecraft's capability to navigate vast distances in space and laid the groundwork for future Apollo missions that would ultimately land astronauts on the lunar surface.

Entering lunar orbit, the crew of Apollo 8 encountered a sight no human eye had witnessed before the far side of the Moon. As communication momentarily faded behind the lunar horizon, they ventured into uncharted visual territory. In the subsequent hours, meticulous navigation checks and ground-based calculations refined their orbital parameters, paving the way for a pivotal maneuver. At 073:35:06.6, a brief 9.6-second circularization burn refined their orbit to 60.7 by 59.7 nautical miles.

As Command Module Pilot, Jim Lovell's primary responsibility was navigation, a crucial role, especially in scenarios where direct communication with Mission Control might be interrupted. While Mission Control typically handled navigation calculations, having a crew member skilled in celestial navigation was essential for contingency planning. Lovell utilized a sextant installed on the spacecraft to determine their position relative to stars, measuring angles between stars and the horizon of either Earth or the Moon. This method, however, was challenged by a significant debris cloud surrounding the spacecraft, which obscured star sightings and added complexity to navigation efforts.

Approximately seven hours into the mission, the crew ran about 1 hour and 40 minutes behind schedule due to difficulties encountered during their separation from the S-IVB stage and Lovell's obscured star sightings. To manage thermal conditions aboard the spacecraft, they initiated Passive Thermal Control (PTC), commonly known as the "barbecue roll." This maneuver involved rotating the spacecraft about its long axis once per hour to ensure even heat distribution across its surface. In the direct sunlight of space, parts of the spacecraft's exterior could reach temperatures exceeding 200 °C (392 °F), while shaded areas plummeted to around -100 °C (-148 °F). These extreme temperature differentials posed risks such as potential heat shield cracking and propellant lines bursting.

As Command Module Pilot, Jim Lovell's primary duty centered on navigation, a pivotal role crucial for ensuring the mission's success, particularly in scenarios where direct communication with Mission Control could be disrupted. While Mission Control typically handled navigation calculations, having a crew member adept in celestial navigation was essential for contingency planning. Lovell relied on a sextant installed aboard the spacecraft to determine their precise position relative to the stars, measuring angles between stars and either Earth's or the Moon's horizon. However, this method was significantly challenged by a substantial debris cloud enveloping the spacecraft, which obscured star sightings and complicated navigation efforts.

Approximately seven hours into the mission, the crew found themselves approximately 1 hour and 40 minutes behind schedule. This delay stemmed from challenges encountered during the separation from the S-IVB stage and difficulties with Lovell's obscured star sightings. To effectively manage the spacecraft's thermal conditions, they initiated Passive Thermal Control (PTC), also known as the "barbecue roll." This maneuver involved rotating the spacecraft about its long axis once per hour to ensure even heat distribution across its surface. In the unfiltered sunlight of space, sections of the spacecraft's exterior could reach scorching temperatures exceeding 200 °C (392 °F), while shaded areas plummeted to frigid lows around -100 °C (-148 °F). Such extreme thermal differentials posed risks such as potential heat shield cracks and propellant line ruptures.

Maintaining the PTC maneuver presented challenges due to the inability to achieve a perfectly stable rotation. As the spacecraft rotated, it traced out a cone-shaped pattern, necessitating minor adjustments every thirty minutes to counteract the widening sweep of the cone. This meticulous thermal management was crucial for safeguarding the spacecraft's integrity and ensuring the crew's safety throughout the demanding journey to and from the Moon.

About an hour into his scheduled sleep shift, Frank Borman sought permission from ground control to take a Seconal sleeping pill, hoping to rest during the mission. However, the medication had minimal effect, and Borman struggled to fall asleep. When he finally did, he woke feeling extremely unwell. He experienced vomiting twice and had a bout of diarrhea, resulting in the spacecraft being littered with small globules of vomit and feces.

Despite initially preferring to keep his medical issues private, Borman's condition prompted Jim Lovell and Bill Anders to advocate for informing Mission Control. They decided to use the Data Storage Equipment (DSE), which can record voice transmissions and telemetry, to document Borman's illness. They then requested Mission Control to review the recording, specifically asking for an evaluation of their voice comments regarding Borman's health condition. This step was crucial for ensuring that Mission Control was fully informed and could provide appropriate guidance and support to the crew amidst the challenging circumstances of the mission.

Following Frank Borman's illness aboard Apollo 8, the crew and medical personnel at Mission Control convened a conference in an unused second-floor control room (Houston had two identical control rooms, typically using only one during missions). During this meeting, it was concluded that there was minimal cause for concern. Borman's symptoms were attributed either to the 24-hour flu, as initially thought by Borman himself, or potentially to a reaction from the Seconal sleeping pill he had taken.

However, modern research suggests that Borman may have been experiencing space adaptation syndrome, affecting approximately a third of astronauts during their initial exposure to weightlessness. Unlike earlier spacecraft like Mercury and Gemini, where astronauts had limited mobility within cramped cabins, the Apollo command module offered significantly more space. This increased freedom of movement allowed astronauts to move about more freely but also contributed to symptoms of space sickness due to the vestibular system adapting to weightlessness.

The onset of space adaptation syndrome during Apollo 8 marked a significant shift in understanding the physiological challenges of space travel, highlighting the importance of acclimatization and adaptation strategies for future missions. Subsequent missions, such as Apollo 9 with astronaut Rusty Schweickart, also encountered similar challenges as astronauts adjusted to the unique environment of space.

Over the next twelve hours, the crew's activities centered on scientific exploration and documentation. They meticulously photographed both the near and far sides of the Moon, focusing on potential landing sites and geological features of interest. Utilizing the spacecraft's sextant, they captured images crucial for mapping and future mission planning. These photographs aimed to document lunar geography and uncover the elevation and precise geographic coordinates of lunar features, enhancing our understanding of the Moon's enigmatic far side.

The inclusion of sextant photography aboard Apollo 8 served dual purposes: first, it provided crucial image comparisons for evaluating lunar landmarks and for navigation training. Second, it aimed to photograph one of the certified Apollo landing sites, marking a significant milestone in lunar exploration.

The first image taken by humans of the whole Earth disk, probably photographed by William

Apollo 8's photographic efforts went beyond mere documentation; they provided unprecedented insights into lunar surface illumination. Unlike Earth-based telescopic photography affected by atmospheric distortion or satellite imagery prone to electronic processing losses, Apollo 8's photographs captured the Moon in its unfiltered glory. This allowed for precise analysis of lunar light's intensity and spectral distribution, offering scientists a clearer view than ever before.

Throughout their mission, the crew meticulously executed their photographic tasks. They obtained over 800 still photographs using 70 mm film, 600 of which were high-quality reproductions showcasing intricate lunar surface features. The remaining photographs documented key moments such as the separation and venting of the S-IVB stage and striking long-distance views of both Earth and the Moon.

Apollo 8 also utilized over 700 feet of 16 mm film to capture pivotal moments and activities during its mission. This included footage of the S-IVB stage separation, detailed lunar landmark photography using the sextant, sequences documenting the lunar surface, and activities inside the spacecraft.

The crew's still photography proved invaluable in expanding our understanding of the lunar environment. Beyond visual documentation, their observations provided critical qualitative insights. During the initial lunar orbit phase, crew members described the lunar surface as appearing "black-and-white," without discernible color—often likened to "whitish gray, like dirty beach sand." This observation highlighted the stark contrast between the Moon's monochromatic landscape and the vivid colors seen on Earth.

Despite these descriptive observations, the crew noted challenges in capturing the full detail of lunar surface features through photography. Features within shadowed areas and intensely bright regions were identifiable to the crew's eyes but proved more challenging to delineate clearly in photographic form. This discrepancy underscored the complexities of lunar surface imaging and the ongoing need for visual and photographic data to study and map the Moon's terrain comprehensively.

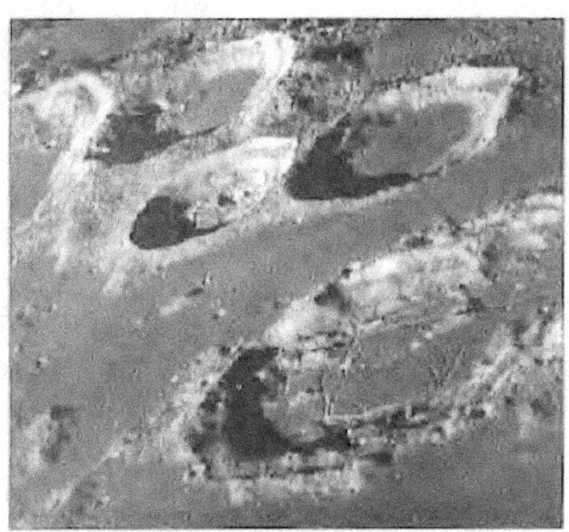

Craters Goclenius (foreground), Columbo A, and Maegelhans (background) (NASA AS08-13-2224).

The insights from Apollo 8's observations and photographic documentation revolutionized our understanding of lunar surface conditions and significantly influenced the planning for future lunar landing missions. By combining crew observations with detailed photographic data, researchers could reinterpret lunar surface features and phenomena in new ways.

One critical outcome was the expansion of lunar-surface lighting parameters deemed suitable for safe landing operations. Before Apollo 8, it was believed that a minimum sun angle of 6° was necessary for adequate surface visibility during landings. However, Apollo 8's crew observed and documented surface details even at sun angles as low as 2° or 3°. They concluded that these lower angles did not significantly hinder safe lunar landings except in areas deeply shadowed for prolonged periods.

Conversely, they established that sun angles above 20° were less optimal for manual landing maneuvers due to reduced surface contrast and visibility. Between 16° and 20°, lighting conditions were deemed acceptable for critical phases of the descent to the lunar surface, offering good definition of surface features.

The crew's observations also highlighted challenges related to color perception on the Moon. They noted an absence of sharp color boundaries, especially from altitudes around 60 nautical miles. This reduced the reliability of color as a distinguishing factor for geological units during close lunar operations.

Just before sunrise on a pivotal early orbit around the Moon, the command module pilot peered through the telescope and observed what appeared to be zodiacal light and the solar corona. This rare observation provided a unique perspective on solar phenomena from the Moon's orbit. Meanwhile, during two successive revolutions, the lunar module pilot noted a conspicuous cloud or bright area in the lunar sky. If confirmed, this sighting would mark the observation of one of the Magellanic clouds, distant galaxies rarely visible from Earth.

The crew also engaged in long-distance photography of Earth, capturing compelling images of global weather patterns and terrain features. However, constraints on the spacecraft's attitude prevented lunar photography during the translunar coast phase. Nevertheless, they successfully conducted high-quality lunar photography during the return journey toward Earth.

Initially adhering to the meticulously planned lunar orbit mission, the crew diligently executed all scheduled tasks. However, recognizing the accumulating effects of fatigue, the commander made a crucial decision at 084:30 to prioritize crew rest during the final four hours in lunar orbit. Consequently, all scheduled activities were canceled except for essential platform alignments and preparations for the journey back to Earth.

During the relatively uneventful cruise phase of the Apollo 8 mission, the crew focused on routine checks to ensure the spacecraft's systems were functioning correctly and that it remained on its intended course toward the Moon. NASA had scheduled a significant event during this phase—a television broadcast planned for 31 hours after launch.

For this broadcast, the Apollo 8 crew utilized a 2-kilogram (4.4 lb) camera equipped with a Vidicon tube, capable of broadcasting in black-and-white. The camera featured two lenses: a wide-angle lens with a 160° field of view and a telephoto lens with a narrower 9° field of view.

During their first broadcast, the crew intended to provide a tour of the spacecraft and showcase the Earth as seen from space. However, they encountered challenges with aiming the narrow-angle lens accurately, exacerbated by the lack of a monitor to aid in framing their shots. This difficulty made it impossible to capture and display clear Earth images effectively. Furthermore, without appropriate filters, any bright light sources overexposed the Earth's image, resulting in a featureless, bright blob in the broadcast.

The broadcast lasted for 17 minutes until the spacecraft's rotation moved the high-gain antenna out of alignment with Earth's receiving stations, prompting the crew to conclude the transmission. Before signing off, Jim Lovell wished his mother a happy birthday, adding a personal touch to this historic television moment from deep space.

The Apollo 8 crew had abandoned their planned sleep shifts by this point in the mission. Jim Lovell decided to go to sleep 32 and a half hours into the flight, which was three and a half hours earlier than scheduled. Shortly after, Bill Anders also took a sleeping pill and followed suit by going to sleep.

Astronaut William A. Anders, lunar module pilot, is shown during intravehicular activity (IVA) on the Apollo 8 lunar orbit mission.

During the outward cruise toward the Moon, the crew faced significant challenges in observing the lunar surface. Most notably, three of the spacecraft's five windows fogged up due to out-gassed oils from the silicone sealant. Additionally, the spacecraft's attitude needed for passive thermal control further hindered their visibility of the Moon. As a result, when the spacecraft had maneuvered behind the Moon, the crew finally had their first visual encounter with it.

At 55 hours into the flight, Apollo 8 conducted a second television broadcast. For this broadcast, the crew improvised by using filters intended for their still cameras to capture images of the Earth through the telephoto lens of their onboard camera. Despite the difficulty in aiming, requiring them to adjust the entire spacecraft, the crew successfully transmitted the first television pictures of our planet to Earth as seen from deep space. During the transmission, they described their observations of Earth, highlighting the visible features and colors they could discern. This historic broadcast lasted 23 minutes and provided humanity with a stunning and unprecedented view of Earth from the Moon's vicinity.

At approximately 55 hours and 40 minutes into their journey, a significant milestone marked Apollo 8's historic mission.

Positioned 38,759 miles (62,377 km) from the Moon, the spacecraft entered the Moon's gravitational sphere of influence. This meant the Moon's gravitational pull on Apollo 8 became stronger than Earth's—a first for human space exploration. Despite the monumental nature of this event, the crew's primary focus remained on precise trajectory calculations relative to their launch pad at Kennedy Space Center.

This photograph of the Moon was taken from Apollo 8 at a point above 70 degrees east longitude.

They diligently continued these calculations until their final mid-course correction, transitioning to a reference frame crucial for their upcoming maneuvers in lunar orbit. This meticulous approach underscored the crew's dedication to navigating the complexities of space travel with precision and accuracy, laying the groundwork for their imminent orbital insertion around the Moon.

The crew executed a second mid-course correction to refine their trajectory as Apollo 8 approached its crucial Lunar Orbit Insertion (LOI) maneuver. This retrograde burn, slowing the spacecraft by 2.0 ft/s (0.61 m/s), adjusted their path to pass closer to the Moon. At precisely 61 hours into the mission, approximately 24,200 miles (38,900 km) from the lunar surface, the crew initiated an 11-second RCS burn. This maneuver ensured that Apollo 8 skimmed just 71.7 miles (115.4 km) above the lunar terrain during its upcoming approach.

A portion of the lunar far side as seen from

By the 64-hour mark, preparations for LOI-1 were underway. This critical maneuver had to be flawlessly executed on the far side of the Moon, where direct communication with Earth was impossible. After receiving a final "go/no go" decision from Mission Control at 68 hours, the crew received the affirmative and were assured they were "riding the best bird we can find." In response, Command Module Pilot Jim Lovell calmly remarked, "We'll see you on the other side."

This marked a historic moment as Apollo 8 became the first human mission to venture behind the Moon, entering a radio silence known as a blackout. Frances "Poppy" Northcutt, a key figure in NASA's mission control, vividly described the tension during this phase: "That was a very nerve-racking period... You've got this big mystery on the backside of the Moon. You do not know what's happening and there's not a darn thing anybody here can do about it until we hear from them."

As Apollo 8 approached the critical Lunar Orbit Insertion 1 (LOI-1), the crew meticulously verified every spacecraft system, ensuring all switches were set correctly with just ten minutes remaining. It was during this final preparation phase that they caught their first glimpses of the Moon. Flying over the unlit side, Command Module Pilot Jim Lovell was the first to notice shafts of sunlight casting an oblique glow on the lunar surface.

However, their fleeting moment of observation was cut short as the LOI burn loomed just two minutes away. Despite the limited time to appreciate the lunar vista, the crew's brief encounter with Earth's celestial companion added a profound dimension to their historic journey, underscoring the awe-inspiring nature of their mission.

At 69 hours, 8 minutes, and 16 seconds after launch, the Apollo 8 mission reached a pivotal moment: its Service Propulsion System (SPS) ignition. This engine burned for 4 minutes and 7 seconds, successfully placing the spacecraft into orbit around the Moon. This burn was a nerve-wracking experience for the crew onboard, famously described as the longest four minutes of their lives. The precision of this burn was critical; any deviation could have resulted in a dangerously elliptical orbit around the Moon or even a trajectory leading away from it. Conversely, the spacecraft risked impacting the lunar surface if the burn had been too long.

Once assured that the spacecraft was functioning as expected, the crew finally had a moment to gaze at the Moon, which would be their companion for the next 20 hours as they orbited this celestial body.

Meanwhile, Mission Control awaited confirmation of Confirma's success on Earth. If the engine had not fired or if the burn duration had been incorrect, the spacecraft would have emerged from behind the Moon sooner than expected. Precisely on schedule, the signal arrived confirming Apollo 8's placement into a planned orbit measuring 193.3 by 69.5 miles (311.1 by 111.8 km) around the Moon. This achievement marked a monumental step in human space exploration, demonstrating the meticulous planning and execution required for such unprecedented missions.

After confirming the spacecraft's status, Jim Lovell provided the first detailed description of the lunar surface from Apollo 8:

"The Moon is essentially grey, no color; it looks like plaster of Paris or a kind of greyish beach sand. We can see quite a bit of detail. The Sea of Fertility doesn't stand out as well here as it does back on Earth. There's not as much contrast between it and the surrounding craters. The craters appear rounded off, with many of them showing signs of recent impacts—some look like they've been struck by meteorites or other projectiles. One notable feature is Langrenus, a vast crater with a central cone. Its walls are terraced, with about six or seven distinct terraces descending into the crater."

Lovell's observations provided Earth with its first close-up glimpse of the lunar landscape, highlighting the stark, monochromatic nature of the Moon's surface and revealing the geological intricacies of its craters and formations. This detailed account marked a significant milestone in human exploration as Apollo 8 continued to orbit the Moon, capturing imagery and data that would deepen our understanding of Earth's nearest celestial neighbor.

As Apollo 8 continued its orbit around the Moon, Jim Lovell and the crew diligently documented the terrain passing beneath them. A crucial objective of their mission was reconnaissance of potential future landing sites on the lunar surface, with special attention given to Mare Tranquillitatis, which was designated as the planned landing site for Apollo 11. The launch timing of Apollo 8 had been carefully chosen to ensure optimal lighting conditions for observing and photographing these sites.

Inside the spacecraft, a film camera had been positioned at a window, capturing one frame per second of the lunar landscape below.

Bill Anders, one of the crew members, devoted much of the next 20 hours to photographing targets of interest with meticulous detail. By the end of their mission, the crew had amassed a remarkable collection: over eight hundred 70 mm still photographs and 700 feet (210 m) of 16 mm movie film, providing invaluable visual data that would inform future lunar exploration efforts and deepen humanity's understanding of Earth's satellite.

During Apollo 8's fourth pass across the front of the Moon, a historic moment unfolded as the crew witnessed an "Earthrise" for the first time in human history. This breathtaking phenomenon occurs when the Earth emerges from behind the lunar horizon, visible from the Moon's surface or orbit.

The significance of this event was not lost on astronaut Bill Anders, who excitedly alerted his fellow crew members as he captured the moment with a black-and-white photograph. Inspired by the sight, Anders quickly requested color film from Jim Lovell and took another photograph, now famously known as "Earthrise." This iconic image later gained recognition as one of Life magazine's hundred photos of the century.

The rarity of Earthrise stems from the Moon's synchronous rotation, which keeps one hemisphere facing Earth at all times. This means that from any single point on the lunar surface, Earth generally remains fixed in the sky, either above or below the horizon. Therefore, Earthrise can only be observed during lunar orbit or from specific locations near the Moon's limb, where libration—a slight rocking motion of the Moon—occasionally brings Earth into view.

As the intense mission of Apollo 8 continued, Bill Anders remained focused on photographing the lunar surface. At the same time, Jim Lovell took over spacecraft control to allow Commander Frank Borman some much-needed rest. However, sleeping in the confined and noisy spacecraft proved challenging for Borman, who managed to rest for two orbits. Throughout this period, Borman would wake intermittently to check on their status, only fully awakening when he noticed his crewmates starting to make errors and struggling to comprehend questions without needing them to be repeated. It became evident to Borman that the crew's fatigue had reached a critical point, having gone without a proper night's sleep for over three days.

Realizing the importance of alertness, especially with a planned second television broadcast expected to draw significant viewership, Borman made a decisive decision. He ordered Anders and Lovell to get some rest and canceled the remaining part of the flight plan focused on observing the Moon. Initially reluctant, Anders eventually agreed that Borman set up the camera to continue taking automatic pictures of the Moon. Additionally, Borman ensured preparations were in place for the upcoming broadcast to maintain high crew readiness.

For the next two orbits, while Anders and Lovell rested, Borman took charge at the spacecraft controls, overseeing operations and ensuring readiness for the next phases of their mission. His leadership and concern for crew well-being during this critical phase of Apollo 8 exemplified the teamwork and dedication necessary for the success of pioneering space missions.

Taken by Apollo 8 crewmember Bill Anders on December 24, 1968, at mission time 075:49:07 [8] (16:40 UTC), while in orbit around the Moon, showing the Earth rising for the third time above the lunar horizon.

The astronauts initiated their second television transmission as Apollo 8 completed its ninth orbit around the Moon. Commander Frank Borman began by introducing the crew, each member then sharing their impressions of the lunar surface and the profound experience of orbiting the Moon. Borman notably described it as "a vast, lonely, forbidding expanse of nothing," capturing their surroundings' stark beauty and isolation.

After discussing the lunar terrain they were traversing, Bill Anders conveyed a message from the crew to all those watching back on Earth. Each astronaut took turns reading a section from the Biblical creation story found in the Book of Genesis. This deeply symbolic gesture reflected their contemplative mood and the awe-inspiring nature of their mission.

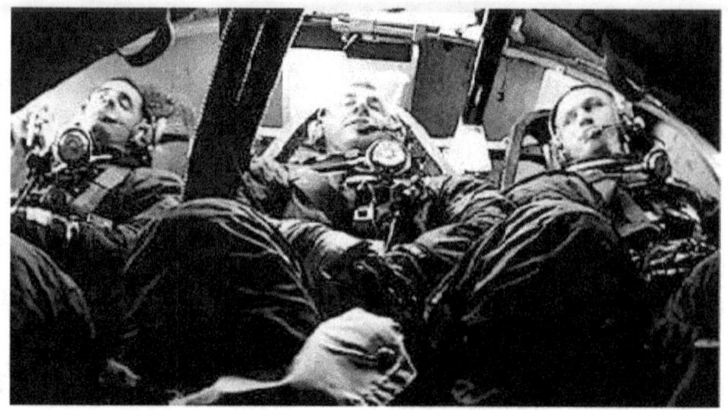

Apollo 8: the Christmas mission around the Moon

Concluding the broadcast, Borman extended heartfelt wishes, encapsulating the sentiments shared by all three crew members from their unique vantage point in lunar orbit. He expressed, "And from the crew of Apollo 8, we close with good night, good luck, a Merry Christmas, and God bless all of you—all of you on the good Earth." This message resonated globally, uniting people in a moment of reflection and goodwill during the holiday season, reaffirming the profound impact of Apollo 8's historic journey beyond Earth's bounds.

With the completion of their second television transmission, the crew of Apollo 8 focused on their final critical task: the trans-Earth injection (TEI), scheduled just two and a half hours after the broadcast ended. This maneuver was pivotal, marking the most crucial burn of their mission. Failure to ignite the Service Propulsion System (SPS) would have left the crew stranded in lunar orbit, with little possibility of rescue.

Like previous burns, TEI had to be executed above the Moon's far side, out of direct communication with Earth. The crew executed the burn precisely on schedule, ensuring the spacecraft maintained its trajectory back toward Earth. Telemetry signals were reestablished as Apollo 8 emerged from behind the Moon at exactly 89 hours, 28 minutes, and 39 seconds, confirming their calculations' accuracy and the maneuver's success.

Upon regaining voice contact, Commander Jim Lovell humorously announced, "Please be informed, there is a Santa Claus," to which Capcom Ken Mattingly replied, "That's affirmative, you are the best ones to know." This light-hearted exchange highlighted the crew's spirits as they began their journey back to Earth on December 25th, Christmas Day—a fitting conclusion to a mission that had captured the world's imagination and marked a historic milestone in space exploration.

Later in the mission, Jim Lovell utilized some free time to conduct navigational sightings, adjusting the module to observe various stars using the computer keyboard. However, he inadvertently erased some of the computer's memory in an unintended mishap. This action caused the Inertial Measurement Unit (IMU) to register erroneous data, indicating that the module's orientation was unchanged from its pre-launch position. Consequently, the IMU fired the thrusters to correct what it perceived as a deviation in attitude.

Upon realizing the cause of the computer's response, the crew understood the need to re-input accurate orientation data into the system. Lovell spent about ten minutes recalculating the correct numbers, maneuvering the spacecraft using the thrusters to align with stars like Rigel and Sirius. It then took 15 minutes to update the computer with the corrected data.

This incident highlighted the crew's adaptability and problem-solving skills in managing unexpected challenges during the mission. It also foreshadowed a critical moment during the Apollo 13 mission, sixteen months later, when Lovell would face a similar task under far more perilous circumstances after the module's IMU was disabled to conserve power.

During their mission on Apollo 8, the crew maintained constant communication with Earth, with Commander Frank Borman particularly focused on monitoring the performance of the spacecraft's Service Propulsion System (SPS). He repeatedly inquired about the status of the SPS, ensuring it was operational and ready for contingencies such as an early return to Earth if needed. Before each pass behind the Moon, Borman insisted on receiving a "go/no go" decision, highlighting his commitment to safety and mission readiness.

On their second approach to the Moon, the crew set up equipment to broadcast a live view of the lunar surface. Astronaut Bill Anders described the craters they were flying over, offering viewers on Earth a firsthand glimpse of the rugged lunar terrain. After this orbit, an 11-second burn of the SPS, known as LOI-2, was executed to circularize their orbit at approximately 70.0 by 71.3 miles (112.7 by 114.7 km).

The crew continued their meticulous checks of the spacecraft's systems throughout subsequent orbits while dedicating time to observing and photographing the Moon. During the third pass over the lunar surface, Commander Borman, a devout Christian, fulfilled a poignant request. Originally scheduled to participate in a service at St. Christopher's Episcopal Church near Seabrook, Texas, Borman could not attend due to the Apollo 8 mission. At the suggestion of Rod Rose, an engineer and fellow parishioner at Mission Control, Borman read a small prayer. This recording was later transmitted to Earth and played during the church service, symbolizing the crew's connection with their faith and community despite their extraordinary journey in space.

Crew observations of the lunar surface revealed that the "washout" effect, where surface details are obscured by backscatter, was significantly less severe than anticipated. Furthermore, smaller lunar surface features were visible in shadowed areas under low sun angles, indicating that lighting conditions for future lunar landings would likely meet photometric standards.

The shift of Apollo 8 from an Earth orbital mission to a lunar mission necessitated rapid adjustments in pre-mission planning, crew training, and ground support configurations. Despite the challenging circumstances, these adaptations were completed within a notably condensed timeframe, highlighting NASA's capability to respond effectively to unforeseen mission changes.

To evaluate the effectiveness of the passive thermal control system during a lunar orbit mission. Achieved, confirming its ability to maintain spacecraft temperatures within acceptable ranges.

These objectives underscore the successful execution of critical mission activities and tests during Apollo 8. The spacecraft's command and service module, supported by robust mission facilities, demonstrated exceptional performance throughout the piloted Saturn V mission. Key achievements included precise Saturn targeting for translunar injection, successful execution of long-duration service propulsion burns and midcourse corrections, and flawless implementation of pre-translunar injection procedures. The command and service module's orbital navigation capabilities were also effectively validated.

Detailed test objectives were also met, such as controlled entry guidance and navigation during lunar return, star-lunar and star-Earth horizon sightings despite optical challenges, and effective acquisition, tracking, and communication using the high-gain S-band antenna with the Manned Space Flight Network throughout the lunar mission. Data on the passive thermal control system further affirmed its reliability under the rigors of lunar orbit conditions.

In December 1968, Apollo 8 embarked on a pioneering mission that captured the world's imagination. As the spacecraft approached the Moon, anticipation grew for a historic television broadcast. On Christmas Eve, a transmission began, watched live by an estimated one billion people in 64 countries, with delayed broadcasts reaching 30 more nations. Against the backdrop of space, the crew delivered a message that transcended mere scientific achievement.

"Ladies and gentlemen, from the crew of Apollo 8, we have a message that we would like to send to you. In the beginning..." With these profound words, the astronauts began reading from the Book of Genesis, sharing the timeless verses of creation and light. Command Module Pilot Jim Lovell's voice filled the void of space, "And God called the light 'Day'..." This broadcast, capturing the first-ever live images of a lunar sunrise, became an iconic historical moment, blending scientific exploration with spiritual reflection.

Commander Frank Borman solemnly concluded, "And from the crew of Apollo 8, we close with good night, good luck, a Merry Christmas, and God bless all of you, all of you on the good Earth." Their heartfelt message resonated across the globe, uniting people of all nations in awe and wonder at humanity's audacious journey into space. It was a poignant reminder of our shared humanity and the boundless possibilities of exploration set against the stark beauty of the lunar landscape.

Following this broadcast, Apollo 8 executed critical maneuvers to prepare for its return journey to Earth. Transearth injection and the Service Propulsion System's precise firing ensured the spacecraft's trajectory back home. Amidst the technical precision, communication glitches briefly interrupted their connection with Mission Control, a reminder of the challenges inherent in lunar missions.

Nevertheless, Apollo 8's crew persevered. Soon, two-way communication was restored, and the crew exchanged words with Houston again. Command Module Pilot Jim Lovell humorously remarked, "Roger. Please be informed there IS a Santa Claus," prompting a lighthearted exchange with Mission Control, affirming the crew's unique perspective from space.

During the cruise back to Earth, the crew of Apollo 8 enjoyed a period of relative relaxation while attentively monitoring the spacecraft's systems. Their confidence in the trajectory calculations meant that barring unforeseen circumstances, the spacecraft was on course to reenter Earth's atmosphere and splash down in the Pacific Ocean two and a half days after the Trans-Earth Injection (TEI) maneuver.

On Christmas afternoon, the crew engaged in their fifth television broadcast from space. This broadcast was unique as it provided viewers with a tour of the spacecraft, offering insights into daily life for an astronaut in space. After concluding the broadcast, the crew made a delightful discovery: a small present from Deke Slayton in the food locker. It was a genuine turkey dinner with stuffing, packaged like those given to troops serving in Vietnam.

This unexpected gesture added a festive touch to their Christmas in space, highlighting the camaraderie and thoughtfulness that characterized their mission. It was a moment of comfort and connection amid the vastness of space, reflecting both the mission's technical achievements and the human spirit that united them with Earth during the holiday season.

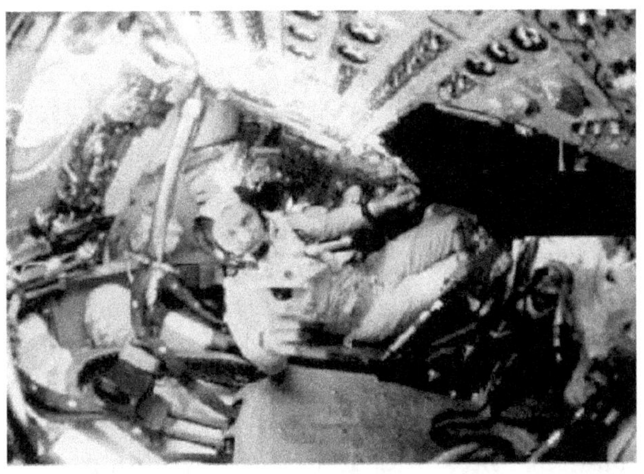

Still from film of the crew taken while they were in orbit around the Moon. Frank Borman is in the center.

Another thoughtful surprise from Deke Slayton came in the form of three miniature brandy bottles intended as gifts for the crew. Commander Borman, mindful of the mission's safety protocols, instructed the crew to refrain from opening them until they had safely landed. True to his directive, the bottles remained unopened, symbolically remembering their journey even years later.

In addition to Slayton's gifts, the crew received small presents from their wives, further enhancing their connection with their loved ones during this historic mission.

As they approached the end of their journey, at about 124 hours into the mission, Apollo 8 conducted its sixth and final television transmission. This broadcast featured the mission's best video images of Earth, providing a stunning view of their home planet during the four-minute transmission.

With two uneventful days behind them, the crew began preparations for reentry. The onboard computer would manage the critical reentry process, requiring the crew only to ensure the spacecraft was properly oriented with its blunt end facing forward. Despite the computer's primary role, Commander Borman stood ready to take manual control in the unlikely event of a computer failure, ensuring a safe return to Earth.

After discovering previously unknown mass concentrations, or "mascons," which perturbed Apollo 8's orbit, adjustments were made to the final lunar orbit parameters. This resulted in an elliptical orbit with an apogee (farthest point from the Moon) of 63.6 nautical miles and a perigee (nearest point to the Moon) of 58.6 nautical miles. The critical transearth injection maneuver was executed at 089:19:16.6 using the service propulsion system, positioned at an altitude of 60.2 nautical miles after ten lunar revolutions and 20 hours and 10 minutes in lunar orbit. The spacecraft achieved a velocity of 8,842 feet per second during transearth injection, marking the transition toward the return journey to Earth. Throughout the mission, Apollo 8 reached a maximum distance from Earth of 203,752.37 nautical miles.

During the transearth phase, Apollo 8 encountered its most significant communication challenge. The spacecraft initially struggled to establish two-way voice contact and telemetry synchronization emerging from lunar occlusion post-transearth injection. Despite establishing a two-way phase lock at 089:28:47, it took until 089:33:28 and 089:43:00 to achieve full voice contact and telemetry synchronization, respectively. Analysis indicated that difficulties may have arisen from attempts to acquire the high-gain antenna signal within the service module reflection region, potentially causing the antenna to track on a side lobe instead of the main signal. Additionally, an accidental high-bit-rate transmission configuration further complicated communications, necessitating a corrective command at 089:29:29 to normalize voice transmission and playback of stored data.

During the transearth phase, essential navigational activities continued, including star and horizon observations to maintain course alignment between the Moon and Earth. A steady roll rate supported these operations to stabilize onboard systems. At 104:00:00, a minor midcourse correction using the service module propulsion system was performed, adjusting velocity by 4.8 feet per second to refine the trajectory.

Unfortunately, at 106:00:26, the onboard state vector computations encountered an error, resulting in their loss until recovery at 106:45. Despite these challenges, crucial transmissions continued, including a 9-minute 31-second television broadcast from within the spacecraft at 104:24:04, providing a glimpse into Apollo 8's interior during its historic journey back to Earth.

At 146:28:48, the Apollo 8 command module jettisoned its sixth stage, marking the transition to an automatically guided entry profile. Throughout the reentry phase, data from the service module provided critical telemetry, complementing photographic coverage that aligned closely with predicted trajectory parameters regarding altitude and latitude.

Apollo 8's command module reentered Earth's atmosphere at 146:46:12.8, descending from an altitude of 400,000 feet at a velocity of 36,221.1 feet per second. The intense ionization during entry bathed the interior of the command module in a vivid cold-blue light, resembling daylight brightness. As anticipated, atmospheric forces caused the command module to experience a lift, temporarily raising its altitude to 210,000 feet before resuming its descent toward splashdown. On their return journey, Apollo 8 encountered a series of critical events as they returned to Earth. Following midcourse corrections and a series of television transmissions, an unexpected error corrupted the onboard state vector and platform alignment data, which the crew swiftly corrected. Testing the high-gain antenna for automatic acquisition, they continued their transmissions, fostering communication with ground stations even as they neared Earth.

During the intense re-entry phase, the Command Module (CM) separated from the

Service Module (SM) as planned. Entering the Earth's atmosphere, they faced a communication blackout due to the intense heat and ionization. The crew experienced peak gravitational forces, reaching a maximum of 6.84 g during descent.

As Apollo 8 descended, recovery efforts began swiftly. Aircraft and ships tracked the CM's descent, establishing radar and visual contact as it descended through the atmosphere. At various points, radar contact was established from distances as far as 270 nautical miles down to a close proximity of 60 nautical miles.

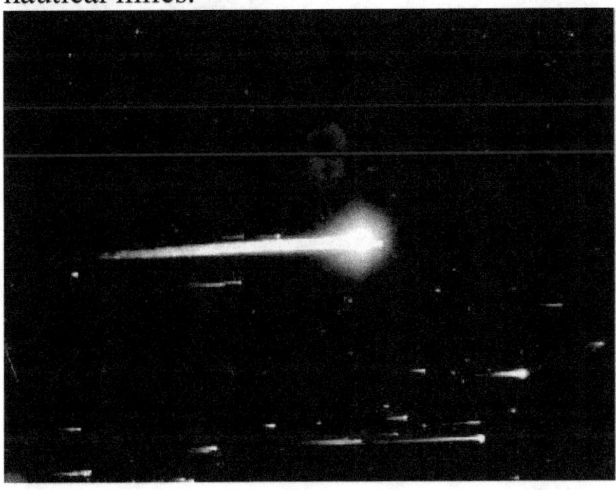

Reentry, December 27, 1968, photographed from a KC-135 Stratotanker at 40,000 feet

The spacecraft deployed its drogue and main parachutes as it descended further, ensuring a safe landing trajectory. Voice and recovery beacon signals were established, allowing recovery helicopters and aircraft to pinpoint the CM's location and prepare for retrieval.

After separating from the service module, the command module of Apollo 8 was prepared for reentry into Earth's atmosphere. This involved exposing the heat shield and shedding the unneeded mass of the service module, which would burn up as planned upon reentry.

As Apollo 8 approached Earth, about six minutes before hitting the top of the atmosphere, the crew witnessed a spectacular sight: the Moon rising above the Earth's horizon, exactly as predicted by trajectory specialists. This serene moment contrasted with the impending reentry, where the spacecraft would encounter the intense heat of atmospheric entry.

Entering the thin outer atmosphere, the crew observed the outside becoming hazy as glowing plasma enveloped the spacecraft. The onboard computer took control, adjusting the spacecraft's attitude to manage its descent. As Apollo 8 decelerated, the forces peaked at 6 standard gravities (59 m/s^2), a testament to the intense reentry conditions.

During the descent, Apollo 8 briefly skipped like a stone across the atmosphere before settling into a trajectory toward the ocean. The drogue parachute deployed at 30,000 feet (9.1 km), stabilizing the spacecraft. The three main parachutes deployed at 10,000 feet (3.0 km), further slowing the descent.

The parachute system deployed successfully, guiding the command module to a splashdown in the Pacific Ocean precisely at 10:51:42 GMT (05:51:42 a.m. EST) on December 27th. The mission's total duration was 147 hours, 0 minutes, and 42.0 seconds, culminating in an impact point just 1.4 nautical miles from the target and 2.6 nautical miles from the recovery ship, U.S.S. Yorktown. The splashdown coordinates were estimated at approximately latitude 8.10° north and longitude 165.00° west.

Apollo 8 splashed down safely in the North Pacific Ocean southwest of Hawaii at 15:51:42 UTC on December 27, 1968. The precise splashdown position was recorded as 8°8′N 165°1′W, marking the successful conclusion of the mission and the safe return of its crew to Earth.

After Apollo 8 splashed down in the North Pacific Ocean, the spacecraft landed upside down, termed Stable 2 position, with the parachutes dragging it over due to the swell. The crew, experiencing the buffeting of a 10-foot (3.0 m) swell, faced additional discomfort, with Borman even vomiting during this period.

Upon splashdown, the command module initially assumed an apex-down flotation posture due to impact forces. However, within 6 minutes and 3 seconds, the inflatable bag uprighting system successfully righted it to its normal flotation position. As planned, helicopters and aircraft maintained vigilance over the spacecraft until local sunrise, 43 minutes after splashdown. At dawn, pararescue personnel deployed and swiftly retrieved the crew aboard the recovery ship 88 minutes after splashdown. The command module itself was recovered an hour later.

Approximately six minutes after splashdown, the spacecraft was successfully righted into a normal apex-up (Stable 1) orientation by its inflatable bag uprighting system. This allowed for a more stable position as the crew awaited recovery.

The first frogman from the aircraft carrier USS Yorktown arrived 43 minutes after splashdown to assist with the recovery operations. Within 45 minutes of the frogman's arrival, the crew was safely aboard the Yorktown's flight deck, marking the end of the mission and the safe return of the Apollo 8 astronauts.

The entire mission spanned an estimated 504,006 nautical miles, concluding with the safe return of the Apollo 8 command module and its crew. This was a pivotal achievement in space exploration and a testament to human ingenuity and bravery.

During the Apollo 8 mission, as recovery swimmers were dispatched, the weather conditions aboard the USS Yorktown were meticulously documented: scattered clouds at 2,000 feet with an overcast layer at 9,000 feet, visibility extending up to ten miles, and a northward wind blowing at 19 knots. The water temperature was a balmy 82°F, with waves reaching up to six feet from a direction of 110° true north.

Apollo 8 CM is hoisted aboard the recovery ship (NASA S68-56304).

Command module on the deck of

Following its retrieval, the Command Module (CM) was carefully offloaded from Yorktown on December 29th at Ford Island, Hawaii. Once on land, the Landing Safing Team commenced their evaluation and deactivation protocols starting at 21:00 GMT, completing them by January 1, 1969. The CM was transported to Long Beach, California, and trucked to the North American Rockwell

Space Division facility in Downey, California, for detailed postflight analysis. Its arrival at Downey occurred precisely at 21:00 GMT on January 2, 1969.

The mission's outcomes were marked by success with only minor issues. All systems aboard the Apollo 8 spacecraft performed flawlessly, achieving all primary mission objectives. The crew's performance was exemplary throughout, despite challenges such as fogging three spacecraft windows due to exposure of window sealant to the space environment. Crew activities were curtailed early due to fatigue, yet approximately 90 percent of planned photographic objectives were met, including 60 percent of additional lunar photographs requested as "targets of opportunity."

The photographic documentation during the Apollo 8 mission unveiled numerous previously unknown lunar features, particularly concentrated on the far side of the Moon. These discoveries represented areas previously only captured by automated spacecraft from much greater distances. The heat shield system also performed exceptionally well, showing no signs of degradation from exposure to cislunar space or the lunar environment, meeting expected performance standards.

Conclusions drawn from the comprehensive post-mission analysis included:

The Command and Service Module (CSM) systems demonstrated full operational capability for future crewed lunar missions.

Throughout the cislunar and lunar orbit phases, all system parameters and consumable quantities were meticulously maintained within their designated operational limits.

Passive thermal control, achieved through a slow rolling maneuver perpendicular to the Sun line, effectively regulated spacecraft temperatures, ensuring they remained within optimal ranges.

The navigation techniques developed for translunar and lunar orbit flights proved highly effective in maintaining precise lunar orbit insertion and transearth injection guidance accuracies.

Apollo 8 crew safely aboard the recovery ship U.S.S

Crew of Apollo 8 addressing the crew of USS Yorktown after successful splashdown and recovery

Non-simultaneous sleep schedules disrupted crew circadian rhythms, adversely affecting their normal sleep cycles and providing inadequate rest during the lunar orbit coast phase.

Communication and tracking capabilities at lunar distances were outstanding across all modes. The newly implemented high-gain antenna performed exceptionally well, enduring dynamic structural loads and vibrations that surpassed expected operational levels.

Apollo 8, arriving at the end of 1968, marked a pivotal moment amidst a tumultuous year globally, which included political assassinations, social unrest, and international upheavals like the Prague Spring. Despite these challenges, Time magazine named the crew of Apollo 8 as its Men of the Year for 1968, recognizing their historic achievement and influence on global events.

Apollo 8 was significant because it was the first mission to travel beyond Earth's gravitational pull and orbit another celestial body, the Moon. The mission was daring and perilous, with even the crew acknowledging its uncertain odds of success. However, they completed their mission, orbiting the Moon and capturing iconic images such as the famous Earthrise photograph during their fourth orbit.

The Earthrise image, showing Earth rising above the lunar horizon, became an iconic symbol of the beauty and fragility of our planet seen from space. It profoundly impacted global consciousness about environmental issues and inspired the first Earth Day in 1970. Life magazine later included it in their list of 100 Photographs That Changed the World, cementing its cultural and historical significance.

In retrospect, Apollo 8's successful mission and its imagery's profound impact helped restore hope and optimism during a challenging year, earning gratitude from people worldwide who saw their achievement as a bright spot in 1968's troubled history.Apollo 8 marked a historic turning point in space exploration, achieving numerous groundbreaking milestones:

It was the first time humans left Earth, ventured into space, and saw our planet from afar.

Astronauts experienced continuous daylight without sunsets or sunrises.

They were exposed to solar radiation beyond Earth's protective magnetic field.

Apollo 8 demonstrated the full power of the Saturn V rocket, with its unprecedented 3.4 million kilograms of thrust.

Humans entered another gravitational field for the first time and orbited the Moon, witnessing its far side and the Earthrise phenomenon.

The mission also marked the first lunar orbit and the return journey from the Moon through Earth's atmosphere, traveling farther and faster than ever before.

Additionally, Apollo 8 symbolized America's leap ahead in the space race with the Soviet Union. While the Soviets initially led, focusing on Earth orbital platforms, the Americans quickly advanced with the lunar orbit rendezvous concept. This strategic shift significantly accelerated the Apollo program, allowing them to close the gap and achieve these historic milestones.

Meanwhile, the Soviets faced setbacks in their space efforts during late 1968. Issues with docking techniques in Soyuz missions and critical failures in unmanned Zond flights underscored their challenges. Despite planning a manned Zond 7 mission in December, technical difficulties forced its cancellation, leaving Soviet cosmonauts to watch as Apollo 8 embarked on its historic lunar journey.

Apollo 8's historic significance resonated deeply with both contemporaries and historians alike. Michael Collins, an astronaut from Apollo 11, emphasized its momentous importance in space exploration history. Space historian Robert K. Poole went further, considering Apollo 8 as the most historically significant of all the Apollo missions.

The mission garnered unprecedented media coverage, reminiscent of the attention received by John Glenn's Mercury-Atlas 6 flight in 1962. Approximately 1,200 journalists covered Apollo 8, and the BBC broadcast the mission in 54 countries and 15 languages. Even the Soviet Union, America's Cold War rival, acknowledged the mission's achievement. Pravda featured praise from Boris Nikolaevich Petrov, Chairman of the Soviet Interkosmos program, highlighting

Apollo 8 as an outstanding American space science and technology accomplishment.

Apollo 8's impact reached a vast global audience. It's estimated that a quarter of the world's population at the time watched either the live or delayed Christmas Eve transmission during Apollo 8's ninth orbit around the Moon. This broadcast was particularly memorable as the crew read from the Book of Genesis, emphasizing the mission's significance against the backdrop of human history and exploration.

Apollo 8's broadcasts from space also earned recognition beyond the scientific community. They received an Emmy Award, the highest honor from the Academy of Television Arts & Sciences, further solidifying the mission's cultural and historical impact. The images and messages transmitted from Apollo 8 captivated the world and inspired generations, setting a profound precedent for future space exploration endeavors.

Madalyn Murray O'Hair, an atheist activist, sparked controversy by filing a lawsuit against NASA following the reading of the Book of Genesis by the Apollo 8 crew during their lunar orbit broadcast. O'Hair argued that American astronauts, as government employees, should not engage in public prayer in space. However, the United States Supreme Court rejected the case, citing a lack of jurisdiction over activities in outer space.

Despite the lawsuit's dismissal, the incident prompted NASA to adopt a cautious approach regarding religious expressions throughout the rest of the Apollo program. This sensitivity continued into subsequent missions, such as Apollo 11, where Buzz Aldrin privately observed Presbyterian Communion on the Moon's surface after landing. Aldrin refrained from publicly discussing this act for several years and only alluded to it indirectly at the time.

In commemorating the historic Apollo 8 mission, the United States Post Office Department issued a postage stamp in 1969 (Scott catalog #1371). This stamp featured a detail from Anders' famous photograph of the Earthrise over the Moon, taken on Christmas Eve, accompanied by the words "In the beginning God...", the opening words of the book of Genesis. This stamp served as a poignant reminder of the mission's significance and the cultural impact of its crew's actions.

Additionally, shortly after their return to Earth in January 1969, the Apollo 8 crew made a public appearance during the Super Bowl III pre-game show. They participated by reciting the Pledge of Allegiance, contributing to their role in American cultural and historical memory beyond the realm of space exploration.

Apollo 9

The Third Crewed Mission

On March 3 - 13, 1969, NASA launched Apollo 9, a crewed Earth orbital mission crucial for testing the Lunar Module (LM) in space. Apollo 9 was pivotal in NASA's quest to land humans on the Moon. This mission was designed to thoroughly test the capabilities of the Apollo spacecraft and its crew in Earth orbit, paving the way for lunar exploration.

Commanded by James A. McDivitt and joined by David R. Scott and Russell L. Schweickart, Apollo 9 aimed to validate the performance and capabilities of the Apollo Command and Service Module (CSM-104) and the Apollo Lunar Module (LM-3).

North American Rockwell (CSM) and Grumman (LM) manufactured the spacecraft with a launch mass of 95,231 pounds. It was propelled into space on March 3, 1969, atop the powerful Saturn V rocket (SA-504) from Kennedy Space Center's Launch Complex 39A.

Apollo 9's mission duration spanned 10 days, 1 hour, and 54 seconds, during which it completed 151 orbits of Earth. The CSM was aptly nicknamed "Gumdrop," while the LM was referred to as "Spider."

The mission achieved several critical milestones, including the first docking of the CSM with the LM on March 3, 1969, and subsequent undocking and redocking maneuvers. On March 7, 1969, Schweickart conducted an Extra-Vehicular Activity (EVA) lasting 77 minutes and validating the LM's ability to support astronaut activity outside the spacecraft.

Apollo 9's return to Earth concluded on March 13, 1969, with the recovery by the USS Guadalcanal in the North Atlantic Ocean, near coordinates 23°15′N 67°56′W, having traveled 4,214,543 miles and conducted 151 orbits.

Apollo 9, launched from Kennedy Space Center on March 3, 1969, marked a pivotal moment in NASA's Apollo program. This mission, classified as a Type D flight, was designed to test the lunar module (LM) in Earth orbit, crucially validating its capabilities ahead of the upcoming lunar missions. Preceded by an unmanned test flight of the LM on Apollo 5, Apollo 9 represented the first piloted endeavor with the "lunar ferry."

Over ten days, Apollo 9 orbited the Earth, showcasing the complete Apollo spacecraft ensemble for the first time. At its core were two essential components: the command and service module (CSM) and the LM. This mission was the second crewed Apollo flight launched atop the towering Saturn V rocket, underscoring its role in preparing for the imminent lunar exploration.

Central to Apollo 9's objectives was the rigorous assessment of the LM's functionalities, pivotal for subsequent lunar landings. The mission demonstrated the LM's ability to navigate independently in space, validating its descent and ascent propulsion systems critical for lunar operations. The successful rendezvous and docking of the LM with the CSM were equally significant, essential maneuvers for lunar missions that ensured safe return and lunar orbit operations.

Moreover, Apollo 9 conducted crucial tests that would prove invaluable in future

Apollo missions. These included firing the LM's descent engine to maneuver the spacecraft stack—a capability later served as a vital backup during the troubled Apollo 13 mission. Additionally, the mission tested the portable life support system, essential for extravehicular activities, affirming its reliability in the space environment.

Apollo 9 undertook rigorous tests that pushed the limits of lunar module (LM) operations, exceeding conditions anticipated for an actual lunar landing. This strategic approach was driven by NASA's commitment to ensure that key mission objectives could still be achieved in the event of an early termination. Compared to previous missions like Apollo 7 and Apollo 8, Apollo 9's schedule for its initial phase was notably packed, designed to maximize crew engagement and spacecraft utilization.

Launched on March 3, 1969, Apollo 9 aimed to validate the performance of the entire Apollo spacecraft configuration—comprising the command and service modules (CSM) and the LM—in the challenging environment of a piloted Saturn V mission. Central to its objectives was demonstrating the LM's capability to function precisely, mirroring conditions anticipated during lunar operations. This included testing the LM's crew handling characteristics and evaluating its operational readiness for nominal and backup lunar orbit rendezvous scenarios.

The meticulous evaluation of the LM across three distinct piloting periods was key to achieving these goals. Each phase demanded repeated activation and deactivation of onboard systems, a demanding sequence unique to Apollo 9. This comprehensive testing regimen was critical for assessing the LM's reliability and performance under varying operational scenarios.

Additionally, Apollo 9's mission scope extended to evaluating the consumables aboard the CSM and LM, essential for sustaining crew operations throughout extended space missions. This assessment ensured that the spacecraft could adequately support future lunar missions and confirmed the viability of planned consumable management strategies.

In preparation for these intensive tests, the LM underwent rigorous altitude chamber testing, ensuring its readiness to withstand the harsh conditions of space and lunar descent. This meticulous preparation underscored NASA's commitment to meticulous planning and thorough testing, essential for the success of subsequent Apollo missions, including the historic Apollo 11 lunar landing later that year.

Apollo 9, therefore, not only validated the technical prowess of the LM and its associated systems but demonstrated NASA's capability to execute complex space missions with precision and reliability—a testament to the agency's commitment to lunar exploration and pioneering achievements in human spaceflight.

In April 1966, James McDivitt, David Scott, and Russell Schweickart were chosen by Deke Slayton, Director of Flight Crew Operations, as the second crew for NASA's ambitious Apollo program. Selected initially as backups to the crew of Gus Grissom, Ed White, and Roger Chaffee for the inaugural crewed Earth orbital test flight of the block I command and service module (AS-204), they were slated to fly AS-204 in 1967. However, setbacks in the development of the block I CSM delayed their mission until the following year.

The crew of Apollo 9 - James A. McDivitt (CMDR), David R. Scott (CMP) and Russell L. Schweickart (LMP)

James Alton McDivitt, born on June 10, 1929, in Chicago, Illinois, and a Colonel in the United States Air Force (USAF), was appointed the mission commander. McDivitt had already distinguished himself as the command pilot for Gemini 4 and held a Bachelor of Science in aeronautical engineering from the University of Michigan. During Apollo 9, at the age of 39, he played a pivotal role in guiding the mission toward achieving its objectives.

David Randolph Scott, born on June 6, 1932, in San Antonio, Texas, was also a Colonel in the USAF and served as the command module pilot for Apollo 9. Scott's prior experience included piloting Gemini 8 alongside Neil Armstrong, where they achieved the first successful spacecraft docking in history. He earned a Bachelor of Science from the U.S. Military Academy and a Master of Science in aeronautics and astronautics from MIT. Selected as an astronaut in 1963, Scott, aged 36 during Apollo 9, provided critical support for mission operations.

Russell Louis "Rusty" Schweickart, born on October 25, 1935, in Neptune, New Jersey, joined Apollo 9 as the lunar module pilot. A civilian astronaut with a background in the United States Air Force and Massachusetts Air National Guard, Schweickart brought expertise in aeronautical engineering, holding both a Bachelor's and Master's Degree from MIT. At 33 years old during Apollo 9, Schweickart's role was instrumental in testing the lunar module's operational capabilities.

The backup crew for Apollo 9 included Commander Charles "Pete" Conrad, Jr., Commander Richard Francis "Dick" Gordon, Jr., and Commander Alan LaVern Bean, all from the United States Navy, ready to step in if needed. Several capsule communicators (CAPCOMs) assisted the mission, including Major Stuart Allen Roosa, Lt. Commander Ronald Ellwin Evans, Major Alfred Merrill Worden, Conrad, Gordon, and Bean themselves. Under the leadership of Eugene F. Kranz, Gerald D. Griffin, and M.R. "Pete" Frank, the support crew and flight directors ensured the smooth execution of operations and communication during this critical phase of NASA's lunar exploration program.

The revised plan positioned the McDivitt crew for the second crewed CSM mission, which was to rendezvous in Earth orbit with an unmanned lunar module (LM) launched separately. This mission was crucial for testing the combined capabilities of the CSM and LM in preparation for lunar missions. Meanwhile, the third crewed mission, under the command of Frank Borman, was planned to be the maiden launch of the Saturn V rocket carrying a crew.

Tragically, on January 27, 1967, during a launch-pad test for their scheduled February 21 mission, Apollo 1, a fire erupted in the cabin, claiming the lives of Grissom, White, and Chaffee. This devastating event prompted a comprehensive safety review of the Apollo program, leading to significant spacecraft design and safety protocol revisions.

Amidst these changes, Apollo 5 was launched as an unmanned mission to test the first lunar module (LM-1), marking a critical step forward in validating the LM's performance in space.

Following the tragic loss of the Apollo 1 crew in 1967, NASA restructured its Apollo program schedule to ensure safety and mission readiness. Under this revised plan, Apollo 7 was slated as the first crewed mission, scheduled for October 1968. This pivotal mission aimed to test the block II command module without incorporating a lunar module, focusing instead on validating the spacecraft's performance in Earth orbit.

NASA adopted a systematic approach with lettered missions, each building upon the successes of its predecessor toward the ultimate goal of a crewed lunar landing—referred to as the "G mission." Apollo 7 was designated as the "C mission" in this sequence. However, the subsequent "D mission" required essential testing of the crewed lunar module, which was experiencing delays, posing a risk to President John F. Kennedy's ambitious goal of landing Americans on the Moon and safely returning them to Earth by the decade's end.

In November 1967, NASA announced Colonel James McDivitt's crew as the prime crew for the critical "D mission." This mission, Apollo 9, was tasked with extensive testing of both the command module and the lunar module in Earth orbit. The thorough evaluation of these spacecraft components was essential for validating their readiness for lunar missions, ensuring that subsequent missions could proceed safely toward achieving NASA's lunar exploration objectives. McDivitt's crew, comprised of David Scott, Russell Schweickart, and himself, undertook rigorous training and preparation to execute the demanding objectives of Apollo 9. Their mission would prove instrumental in advancing lunar exploration capabilities and play a crucial role in reinvigorating public confidence in the Apollo program after the setbacks of Apollo 1.

McDivitt, Scott, and Schweickart train for the AS-205/208 mission in the first Block II spacecraft and space suits, which still had most of the fire hazards the Apollo 1 spacecraft had.

In August 1968, with the successful completion of Apollo 7 in October on the horizon, George M. Low, the Apollo Program Manager, proposed an ambitious plan to keep President Kennedy's lunar landing goal on track. He suggested that if Apollo 7 proceeded smoothly, Apollo 8 should be modified to go directly to lunar orbit without a lunar module (LM). Originally slated as the "D mission," Apollo 8 would now advance its schedule, positioning it as a pivotal mission in NASA's lunar exploration timeline.

Initially, Apollo 9 was designated as the "E mission," tasked with comprehensively testing spacecraft systems in medium Earth orbit. However, after NASA approved the plan to send Apollo 8 to lunar orbit, Apollo 9 was reassigned as the "D mission," responsible for testing the LM in Earth orbit. This decision posed a critical choice for James McDivitt, commander of the Apollo 9 mission. Deke Slayton, Director of Flight Crew Operations, offered McDivitt the opportunity for his crew to switch to Apollo 8 and thus become the first humans to orbit the Moon. However, McDivitt, representing his crew's consensus, opted to stay with the original plan for Apollo

9, prioritizing the thorough testing of the LM's capabilities in Earth orbit.

Following the successful mission of Apollo 7, the crews of Apollo 8 and Apollo 9 were swapped as per the revised schedule. This crew swap also affected the lineup for future lunar landing attempts. According to NASA's protocol, backup crews typically became the prime crew three missions later. As a result, Neil Armstrong's crew, originally the backup for Frank Borman's Apollo 8 mission, now found themselves in a position to make the historic first attempt to land on the Moon with Apollo 11. This switch placed Pete Conrad's crew, initially slated for Apollo 8, in line for the second lunar landing attempt on Apollo 12.

Apollo 9 was tasked with qualifying the Lunar Module (LM) for crewed lunar missions, proving its capability to perform crucial maneuvers in space necessary for a successful lunar landing, including docking with the Command and Service Module (CSM). In their comprehensive book on the Apollo Program, Colin Burgess and Francis French highlighted McDivitt's crew as exceptionally well-trained. They began their intensive preparation in January 1966, initially serving as backups for the ill-fated Apollo 1 mission and always designated to be the first to pilot the LM in space.

Flight Director Gene Kranz lauded the Apollo 9 crew as the best prepared for their mission, particularly noting the expertise of David Scott as Command Module Pilot (CMP). The crew underwent rigorous mission-specific training totaling approximately 1,800 hours, equating to about seven hours of training for every hour they were expected to spend in flight. Their training commenced even before the tragic Apollo 1 fire, starting with the first Block II spacecraft intended for their mission. They participated in exhaustive vehicle checkouts for the CSM at North American Rockwell's facility in Downey, California, and for the LM at Grumman's plant in Bethpage, New York. Additionally, they actively tested the modules at the launch site to ensure everything was meticulously prepared for their mission.

The Apollo 9 crew's training regimen encompassed a diverse array of simulations and preparations crucial for their mission's success. Among these, they engaged in simulated zero-gravity environments underwater and aboard the famed "Vomit Comet." These exercises were pivotal for practicing extravehicular activities (EVAs) planned during their mission, allowing the crew to familiarize themselves with the challenges of working in microgravity.

Traveling to Cambridge, Massachusetts, the crew underwent intensive training on the intricacies of the Apollo Guidance Computer (AGC) at the Massachusetts Institute of Technology (MIT). Mastery of the AGC was essential for navigating and controlling their spacecraft, particularly during critical maneuvers such as lunar module operations and rendezvous procedures.

Schweickart with the life support backpack

The crew visited the Morehead Planetarium and Griffith Planetarium to deepen their understanding of celestial navigation. Here, they focused on studying the 37 stars that the AGC utilized for navigation purposes, ensuring they were adept at identifying these stars and using them for precise guidance.

Most of their training involved spending over 300 hours each in Command Module (CM) and Lunar Module (LM) simulators. These simulators, located at Kennedy Space Center (KSC) and Houston, provided realistic scenarios that simulated various aspects of their mission. Many of these sessions involved live participation from Mission Control, allowing for thorough integration and coordination between the crew and ground support.

Beyond KSC and Houston, the crew also trained in simulators located in other facilities, further honing their skills and readiness for the challenges ahead.

Apollo 9 marked a significant milestone as the first mission to utilize the Command and Service Module (CSM), Lunar Module (LM), and Saturn V rocket together. At Kennedy Space Center (KSC), preparations for this ambitious mission provided crucial opportunities for the launch team to simulate procedures essential for future lunar landing missions.

The Lunar Module, delivered by Grumman in June 1968, underwent rigorous testing, including sessions in the altitude chamber to replicate the harsh conditions of space. Concurrently, technicians at KSC worked on assembling the towering Saturn V rocket inside the iconic Vehicle Assembly Building (VAB). However, integrating the Command Module (CM) and Service Module (SM) with the Saturn V proved challenging, even for the experienced team from North American Aviation.

Upon completing its altitude chamber tests, the LM was prepared to install vital equipment such as rendezvous radar and antennas. This meticulous process ensured the LM was fully equipped for its crucial role in orbital maneuvers and docking operations during the mission.

Without significant delays, on January 3, 1969, the fully assembled launch vehicle emerged from the VAB and began its slow journey to Launch Complex 39A, transported by the massive crawler-transporter. In the ensuing weeks, comprehensive flight readiness reviews for the CM, LM, and Saturn V confirmed that all systems were prepared and functioning correctly for the upcoming mission.

The Apollo 9 launch vehicle was a Saturn V, designated SA-504. The mission also carried the designation Eastern Test Range #9025. The CSM was designated CSM-104 and had the call sign "Gumdrop," derived from the appearance of the command module when it was transported on Earth. During shipment, it was covered in blue wrappings, making it look like a wrapped gumdrop. The lunar module was designated LM-3 and had the call sign "Spider," derived from its arachnid-like configuration.

In preparation for Apollo 9, NASA deployed the Saturn V rocket designated AS-504, marking a pivotal step forward in lunar exploration. This Saturn V was the fourth to take flight and, notably, the second to carry astronauts into space. It was also the first to integrate a lunar module, underscoring its role in advancing the Apollo program's lunar landing ambitions.

LM-3 arrives at KSC, June 1968

Compared to its predecessor in Apollo 8, AS-504 underwent significant modifications to enhance efficiency and reduce weight. Notably, engineers eliminated the inner core of the F-1 engine chamber in the S-IC first stage, achieving a slight increase in specific impulse—an indicator of propulsion efficiency. Weight savings were realized by adopting lighter skins for the liquid oxygen tanks and redesigning other components for decreased mass.

Apollo 9 launches from Kennedy Space Center, March 3, 1969

Improvements extended into the S-II second stage, where uprated J-2 engines were employed alongside a closed-loop propellant utilization system. This replaced the open-loop system used in Apollo 8, enhancing overall fuel efficiency and mission reliability. A substantial reduction of approximately 3,250 pounds (1,470 kg) in the second stage's weight was achieved, with about half attributable to a 16 percent reduction in the thickness of the tank side walls. Apollo 9 marked a pivotal moment in NASA's lunar exploration strategy. It utilized the Command and Service Module (CSM-104), the third Block II CSM to carry astronauts. This mission introduced crucial advancements in spacecraft capabilities that were instrumental for future lunar landings.

The launch vehicle for Apollo 9 being taken to Pad 39A

Unlike Apollo 8, which lacked docking equipment and a lunar module, Apollo 9 featured significant upgrades. The spacecraft was equipped with a probe-and-drogue assembly near the forward hatch of the Command Module (CM), facilitating rigid docking between the CSM and the Lunar Module (LM). This innovation enabled seamless internal crew and equipment transfers between the two spacecraft—a capability vital for lunar missions.

Interestingly, had the mission order not been adjusted, Apollo 8 would have flown with CSM-103, the spacecraft originally intended for Earth-orbit missions. This switch underscored NASA's adaptive approach to mission planning, ensuring that each flight maximized technological advancements and operational readiness for future lunar endeavors.

The preparation for the Apollo missions involved meticulous scrutiny and engineering refinement of the Lunar Modules (LMs), essential for lunar landings. Initially slated for the Earth-orbit mission, LM-2 was the first flight-ready model from Grumman's production line. However, thorough testing by the crew revealed multiple flaws, prompting the decision to delay its use. This delay allowed for LM-3, a significantly improved version, to become available.

Both LM-2 and LM-3 faced challenges regarding weight; they were deemed too heavy for lunar surface missions. Grumman initiated a comprehensive weight reduction program, which saw significant success with LM-5, slated for Apollo 11—a mission that successfully landed humans on the Moon.

Despite its improvements, LM-3 encountered structural issues such as small cracks in its aluminum alloy framework, attributed to stresses like rivet insertion. Grumman engineers diligently worked to rectify these issues until the module's integration with the Saturn V rocket in December 1968. It was encapsulated within the Spacecraft-Lunar Module Adapter (SLA-11A), essential for its journey to space.

LM-2, on the other hand, never made it to space and remains on display at the National Air and Space Museum,

NASA introduced innovative changes during the Apollo 9 mission to enhance crew comfort and operational clarity. Notably, astronauts were equipped with early versions of the Sony Walkman, portable cassette recorders intended to aid in mission observations. This mission also marked a unique allowance: each astronaut could bring a mixtape for personal enjoyment during the flight.

Commander James McDivitt and Pilot David Scott favored easy listening and country music, while Lunar Module Pilot Rusty Schweickart had chosen classical music. Interestingly, Schweickart's cassette initially went missing but was serendipitously rediscovered on the ninth day of the ten-day mission, courteously presented to him by Scott.

In a departure from NASA's earlier policies on spacecraft naming, which forbade the practice after Gemini 3's "Molly Brown," the Apollo 9 crew faced a practical challenge. The Command and Service Module (CSM) and Lunar Module (LM) required distinct call signs when they separated during missions.

During simulations, the crew informally dubbed the CSM as "Gumdrop," inspired by its appearance in the blue protective wrapping from the manufacturer. The LM was named "Spider," reflecting its likeness with deployed landing legs.

Initially perceived as too informal by NASA public relations, these call signs—Gumdrop for the CSM and Spider for the LM—ultimately received official sanction for Apollo 9. However, for subsequent missions, starting with Apollo 11, NASA opted for more formal call signs, reflecting the agency's evolving protocols and increasing the visibility of lunar missions to the public.

During the Apollo 9 mission, significant advancements in spacewalking equipment and procedures were tested, laying crucial groundwork for future lunar exploration. The Extravehicular Mobility Unit (EMU) backpack made its inaugural flight, deployed specifically for Rusty Schweickart's Extravehicular Activity (EVA).

Schweickart's EMU included the Portable Life Support System (PLSS), a pivotal component providing oxygen for breathing and water circulation through the Liquid Cooling Garment (LCG). This garment played a crucial role in regulating body temperature and preventing overheating during periods of extravehicular activity.

Apollo 9 backup crew training in Gumdrop

The Oxygen Purge System (OPS), a safety feature housed atop the backpack, was Accompanying the PLSS. This emergency system could supply oxygen for approximately one hour during PLSS failure, ensuring astronaut safety during critical moments.

The EMU used on Apollo 9 served as a precursor to the more advanced versions employed during the historic lunar landing of Apollo 11. These iterations incorporated further refinements and enhancements, reflecting ongoing technological developments and lessons learned from earlier missions.

During Apollo 9, David Scott conducted a stand-up EVA without wearing a PLSS in a separate operational context. Instead, he remained tethered to the Command Module's (CM) life support systems via an umbilical connection. This setup utilized a Pressure Control Valve (PCV) developed in 1967. The PCV enabled stand-up EVAs from the Lunar Module (LM) hatches and CM or brief excursions outside these spacecraft. Scott later employed this system during his stand-up EVA on the lunar surface during Apollo 15 and by command module pilots in the final Apollo missions exploring deep space.

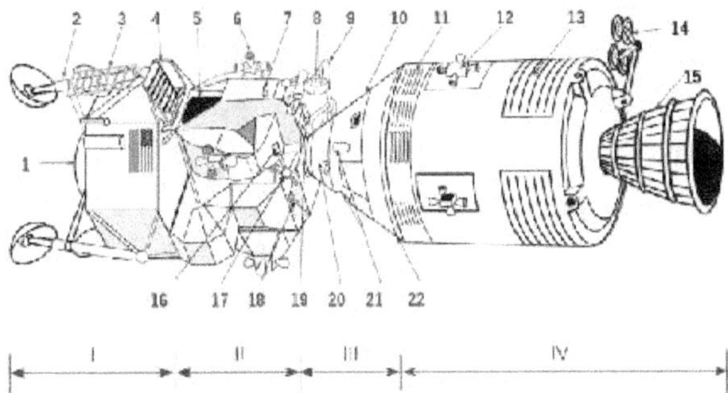

Apollo spacecraft configuration with CSM (right) and LM docked I - Lunar module descent stage; II - Lunar module ascent stage; III - Command module; IV - Service module. 1 LM descent engine skirt; 2 LM landing gear; 3 LM ladder; 4 Egress platform; 5 Forward hatch; 6 LM reaction control system quad; 7 S-band inflight antenna (2); 8 Rendezvous radar antenna; 9 S-band steerable antenna; 10 Command Module crew compartment; 11 Electrical power system radiators; 12 SM reaction control system quad; 13 Environmental control system radiator; 14 S-band steerable

The launch of Apollo 9, initially slated for February 28, 1969, faced an unexpected delay when all three astronauts fell ill with colds. NASA postponed the liftoff to ensure the mission's success and the crew's well-being. During this period, intensive around-the-clock efforts were undertaken to maintain spacecraft readiness, albeit at a cost of approximately $500,000 due to the delay.

Finally, on March 3, 1969, at 11:00:00 EST (16:00:00 GMT), Apollo 9 successfully launched from Kennedy Space Center (KSC). The launch occurred comfortably within the allocated launch window, which would have remained open for another three and a quarter hours.

Notably, the firing control room had a distinguished guest on this occasion: Vice President Spiro Agnew, representing the newly inaugurated Nixon administration, was present to witness this significant milestone in space exploration.

Apollo 9 Command/Service Modules (CSM) nicknamed "Gumdrop" and Lunar Module (LM), nicknamed "Spider" are shown docked together as Command Module pilot David R. Scott stands in the open hatch. Astronaut Russell L. Schweickart, Lunar Module pilot, took this photograph of Scott during his EVA as he stood on the porch outside the Lunar Module.

Launch preparations for Apollo 9 were meticulously planned but not without challenges. Originally scheduled for 28 February 1969, the terminal countdown commenced at T-28 hours on 27 February. However, a setback occurred when, thirty minutes into the scheduled hold at T-16 hours, the countdown was recycled to T-42 hours. This delay was necessary to allow the crew to recover from a mild viral respiratory illness. The countdown resumed at 17:30:00 GMT on 1 March.

Weather conditions posed another significant hurdle. A low-pressure disturbance southwest of Cape Kennedy caused overcast skies. At launch, the weather report indicated stratocumulus clouds covering 70 percent of the sky at a base of 3,500 feet, with altostratus clouds covering 100 percent at 9,000 feet. The temperature was 67.3°F, the relative humidity was 61 percent, and the barometric pressure measured 14.642 lb/in^2. Winds, measured at

13.4 knots from 160° true north at the launch site's anemometer 60 feet above ground, added to the atmospheric challenges.

Astronaut Russell L. Schweickart, lunar module pilot, operates a 70mm Hasselblad camera during his extravehicular activity on the fourth day of the Apollo 9 earth-orbital mission. The Command/Service Module and the Lunar Module 3 "Spider" are docked. This view was taken from the Command Module "Gumdrop". Schweickart, wearing an Extravehicular Mobility Unit (EMU), is standing in "golden slippers" on the Lunar Module porch. On his back, partially visible, are a Portable Life Support System (PLSS) and an Oxygen Purge System (OPS).

The ascent phase of Apollo 9 began on 3 March 1969 at 16:00:00 GMT (11:00:00 a.m. EST) from Kennedy Space Center Launch Complex 39, Pad A. The launch window was extended until 19:15:00 GMT to accommodate precise timing. During ascent, the vehicle's trajectory involved a roll from a launch pad azimuth of 90° to a flight azimuth of 72° between 000:00:13.3 and 000:00:33.0. Key events included the shutdown of the S-IC engine at 000:02:42.76, followed by separation from the S-II stage and ignition of its engine. The S-II engine later shut down at 000:08:56.22, followed by separation from the S-IVB stage, which ignited at 000:09:00.82. The first cutoff of the S-IVB engine occurred at 000:11:04.66, with slight deviations in velocity (+2.86 ft/sec) and altitude (-0.17 n mi) from the planned trajectory.

During the launch of Apollo 9, Commander James McDivitt and his crew encountered a journey marked by both expected and unexpected sensations. Despite some noticeable vibrations, McDivitt reported a generally smooth ride. One notable moment occurred when Saturn V's first stage shut down, momentarily causing a forward push against their seats before the second stage ignited, pressing them back into their couches.

Although both the first and second stages of the rocket slightly underperformed relative to expected metrics, the S-IVB third stage effectively compensated for any deficiencies. This crucial component completed its burn at 00:11:04.7 into the mission, precisely placing Apollo 9 into a stable parking orbit around Earth. The spacecraft now orbited at an altitude ranging from 102.3 to 103.9 miles (164.6 to 167.2 km), achieving a critical milestone toward fulfilling the mission's objectives.

Apollo 9 marked a historic milestone as the first piloted mission to extensively test the Lunar Module (LM) in Earth orbit. Lift-off occurred from Kennedy Space Center Pad 39A, encapsulating NASA's Apollo program's aspirations to prepare for lunar exploration.

During ascent, the stages of the Saturn V rocket performed as planned but encountered challenging wind conditions. Maximum winds of 148.1 knots were recorded at 38,480 feet, with a maximum wind shear of 0.0254 sec^{-1} observed at 48,160 feet altitude.

Following a successful launch, the S-IC stage impacted the Atlantic Ocean at 000:08:56.44 GMT, approximately 346.64 nautical miles from the launch site, while the S-II stage impacted at 000:20:25.35 GMT, 2,413.2 nautical miles downrange.

After the S-IVB third stage completed its burn and reached parking orbit at 000:11:14.65 GMT, Apollo 9 settled into an orbit with an apogee and perigee of 100.74 by 99.68 nautical miles, an inclination of 32.552°, and a period of 88.20 minutes. The spacecraft achieved a velocity of 25,569.78 feet per second relative to Earth.

Designated internationally as 1969-018A, the Command and Service Module (CSM) continued its mission in orbit, while the S-IVB stage was designated 1969-018B. Upon their respective operations, the Lunar Module would be designated 1969-018C for the ascent stage and 1969-018D for the descent stage in future mission phases.

After achieving a stable parking orbit, the crew of Apollo 9 embarked on critical orbital maneuvers that would validate essential procedures for future lunar missions. At 02:41:16 into the mission, they separated the Command and Service Module (CSM) from the S-IVB stage. This pivotal step marked the beginning of their maneuver to rendezvous and dock with the Lunar Module (LM), which was nestled at the end of the S-IVB stage. The success of this docking operation was crucial, as it would determine the feasibility of subsequent lunar landings.

Commander David Scott, responsible for piloting the CSM, skillfully maneuvered the spacecraft and completed the docking using the probe-and-drogue docking assembly. Following this achievement, astronauts Jim McDivitt and Rusty Schweickart meticulously inspected the tunnel connecting the Command Module (CM) and the Lunar Module (LM). Satisfied with their inspection, the crew separated the combined spacecraft from the now-redundant S-IVB stage.

The next significant task involved demonstrating the capability to maneuver two docked spacecraft using a single engine. At 05:59:01.1 into the mission, a brief five-second burn of the Service Module's Service Propulsion System (SPS) engine was executed. Commander Scott enthusiastically confirmed the successful completion of this maneuver, ensuring that the LM remained securely attached and operational.

Following this critical maneuver, the S-IVB stage underwent another firing sequence, propelling it into a solar orbit. This action was essential to prevent any potential orbital debris from harming the ongoing Apollo 9 mission.

After separating from the S-IVB stage, the adapter that protected the Command Module (CM) during launch was jettisoned. The CM was then maneuvered so that its apex, housing the docking probe, faced toward the Lunar Module (LM). The docking procedure between the CM and the LM was completed at 003:01:59.3 into the mission.

After completing the docking procedure between the Command Module (CM) and the Lunar Module (LM), Commander James McDivitt and Lunar Module Pilot Rusty Schweickart began preparing for entry into the LM. They pressurized the tunnel connecting the two spacecraft and, with assistance from Command Module Pilot David Scott, removed the CM hatch. They meticulously checked the latches on the docking ring to ensure a secure seal. Then, they connected the electrical umbilical lines, establishing a power supply to the LM while it remained docked to the CM. Once these preparations were complete, the hatch was securely replaced.

At 004:08:06 into the mission, a newly implemented ejection mechanism ejected the docked spacecraft from the S-IVB stage. Following this separation maneuver, the S-IVB was restarted at 004:45:55.54 GMT and burned for 62.06 seconds. Ten seconds later, the S-IVB entered an intermediate coasting orbit measuring 1,671.58 by 105.75 nautical miles, designed to allow the engine to sufficiently cool down before its next restart within one orbital revolution. This orbit had a period of 119.22 minutes, an inclination of 32.302°, and the insertion velocity was 27,753.61 feet per second.

At 005:59:01.07 GMT, the crew performed the first of eight service propulsion firings, executing a 5.23-second maneuver that raised the orbit of the combined Command and Service Module (CSM) with the LM to 127.6 by 113.4 nautical miles.

Lunar module inside S-IVB stage following separation (NASA AS09-19-2919).

The third and final ignition of the S-IVB stage occurred at 006:07:19.26 GMT, involving a 242.06-second burn aimed at demonstrating the engine's restart capability after an 80-minute coasting period. This maneuver also tested the engine's performance under "out-of-specification" conditions and ensured better ground tracking and lighting conditions for the upcoming rendezvous maneuvers.

After achieving escape orbit 10 seconds after the S-IVB engine cutoff, the spacecraft reached a velocity of 31,619.85 feet per second. However, the performance of the S-IVB stage did not meet expectations due to several anomalies encountered during the mission. One significant issue was the failure of an LH2 (liquid hydrogen) and WX (waste expulsion) dump. The LH2 dump through the engine was unsuccessful due to a loss of pneumatic control over the engine valves. Similarly, the WX dump was not executed because of pneumatic control loss during the third burn. The WX tank was safely managed by utilizing the LOX (liquid oxygen) non-propulsive venting system to mitigate risks.

The third ignition of the S-IVB stage also placed it into a solar orbit with specific orbital parameters: an aphelion and perihelion of 80,280,052 by 69,417,732 nautical miles, an inclination of 24.390°, an eccentricity of 0.07256, and a period of 325.8 days. These precise orbital calculations ensured the stage's safe trajectory away from Earth.

On the second day of the mission, crew activities focused on systems checks, maneuvers to adjust pitch, roll, and yaw, and executing the service propulsion system's second, third, and fourth burns while docked to the Lunar Module (LM). The second burn, conducted at 022:12:04.07 GMT and lasting 110.29 seconds, raised the combined spacecraft's orbit to 192.5 by 110.76 nautical miles. The third burn, performed at 025:17:39.27 GMT and lasting 279.88 seconds, further elevated the orbit to 274.9 by 112.4 nautical miles. This maneuver also reduced the spacecraft's mass, making it more maneuverable using the reaction control system engines for subsequent mission activities and ensuring optimal positioning for rendezvous operations.

During these two burns, tests were made to measure the oscillatory response of a docked spacecraft to provide data to improve the autopilot response for this configuration. The fourth burn, at 028:24:41.37, was a 27.87-second phasing maneuver to shift the node east and put the spacecraft in a better position later for lighting, braking, and docking.

During the Apollo 9 mission, the scheduled sleep period was extended from 09:00:00 to 19:30:00, allowing the astronauts a necessary rest period. Despite sleeping well, they were occasionally disturbed by non-English transmissions, which David Scott speculated possibly originated from Chinese sources.

The second day in orbit, March 4, featured a significant series of maneuvers, with the highlight being three burns of the Service Propulsion System (SPS). The initial burn commenced at 22:12:04.1 and lasted for 110 seconds. This burn included testing the engine's ability to swivel or "gimbal" to dampen induced oscillations, a capability successfully demonstrated within five seconds by the autopilot system. Two additional SPS burns followed, effectively reducing the propellant load in the Service Module (SM).

Throughout these maneuvers, both the spacecraft and its engine performed admirably, often exceeding expected performance metrics. Notably, the stability of the Command and Service Module (CSM) during engine gimbal tests on Apollo 9 would prove instrumental in future missions. In 1972, Jim McDivitt, then the Apollo Spacecraft Program manager, recalled this capability when approving the continuation of Apollo 16 despite encountering similar unstable gimbal issues post-separation from its Lunar Module (LM) in lunar orbit.

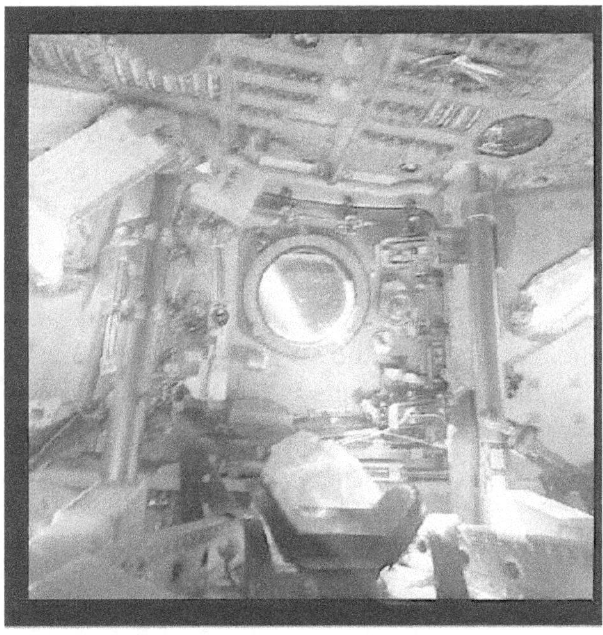

The interior of Gumdrop

On the third day of the Apollo 9 mission, the flight plan called for the commander and lunar module pilot to enter the Lunar Module (LM) to conduct system checks and utilize its descent engine to maneuver the combined spacecraft. This descent engine, serving as a backup to the Service Propulsion System (SPS), would later prove crucial during the Apollo 13 mission.

However, the planned activities faced an unexpected challenge when Rusty Schweickart, experiencing space adaptation sickness, vomited during preparations. Commander James McDivitt also felt queasy due to the contortion-like movements required to don their space suits for the LM checkout. Despite these discomforts, they pressed on with the day's agenda.

View of the Apollo 9 Lunar Module "Spider" in a lunar landing configuration photographed by Command Module pilot David Scott inside the Command/Service Module "Gumdrop" on the fifth day of the Apollo 9 earth-orbital mission. The landing gear on "Spider" has been deployed. lunar surface probes (sensors) extend out from the landing gear foot pads. Inside the "Spider" were astronauts James A. McDivitt, Apollo 9 Commander; and Russell L. Schweickart, Lunar Module pilot.

The astronauts successfully entered the LM, marking the first intra-vehicle transfer in the US space program without needing a spacewalk, contrasting with earlier Soviet cosmonaut methods. Although the hatches between the Command Module (CM) and LM were closed, the modules remained docked, demonstrating the LM's independent communications and life support systems.

At 043:15, the lunar module pilot activated and checked the LM systems, followed by the commander at 044:05. The LM's landing gear was deployed at 045:00, and despite Schweickart experiencing illness, they proceeded with the mission objectives. Due to the crew's health and schedule delays, plans for extravehicular activity were restricted to a single daylight pass, focusing solely on opening the hatches between the CM and LM. Schweickart remained connected to the environmental control system hoses to manage his condition.

Following communication checks and a brief television transmission from inside the LM at 046:25, demonstrating the interior and crew activities, the descent engine was fired for 371.51 seconds at 49:41:34.46 while the vehicles were still docked. This burn showcased attitude control with the digital autopilot and manual throttle adjustments, vital for lunar descent maneuvers.

The transfer back to the CM began at 050:15, and the LM was deactivated at 051:00. To prepare for the LM's active rendezvous, a fifth Service Propulsion System firing of 43.26 seconds duration occurred at 054:26:12.27, circularizing the orbit to facilitate the rendezvous sequence. Although the resulting orbit was slightly elliptical, it was deemed acceptable for mission objectives.

In the Lunar Module (LM), Rusty Schweickart experienced another bout of vomiting, prompting Commander James McDivitt to request a private communication channel with doctors in Houston. The first incident had gone unreported due to its brief nature, but once details emerged to the media, it led to negative press coverage.

Despite these challenges, the crew completed the LM checkout, which included a crucial test: the successful firing of the descent engine. This burn, lasting 367 seconds, simulated the throttle pattern essential for the Moon landing. Subsequently, they returned to the Command Module (CM), piloted by David Scott in Gumdrop.

Following their return, a fifth firing of the Service Propulsion System (SPS) was executed to circularize Apollo 9's orbit, a necessary step for the upcoming rendezvous maneuvers. This maneuver occurred at 54:26:12.3 GMT, raising the spacecraft's orbit to 142 by 149 miles (229 by 240 km), positioning it optimally for the next mission phase.

On the fourth day of the mission, March 6, the planned activity for Rusty Schweickart involved exiting the Lunar Module (LM) and traversing to the Command Module's (CM) hatch. This maneuver aimed to demonstrate the feasibility of such an emergency procedure. Schweickart wore the Portable Life Support System (PLSS), the backpack designed for lunar surface extravehicular activities (EVAs), marking the first test of the PLSS in space before any lunar landings.

Commander James McDivitt initially considered canceling the EVA due to Schweickart's earlier health concerns. However, as Schweickart's condition improved, McDivitt authorized him to exit the LM. Schweickart successfully moved along the exterior of the LM using handholds, demonstrating mobility and agility outside the spacecraft. Meanwhile, David Scott stood ready in the CM's hatch, assisting as needed and documenting the EVA with photographs.

Both astronauts performed various experiments and tasks during this spacewalk, confirming Schweickart's capability to maneuver effectively in space. Schweickart adopted the call sign "Red Rover", a playful reference to the color of his hair, as he undertook this critical test of extravehicular mobility and operations.

On the fourth day of the mission, extravehicular operations were scheduled to showcase the astronauts 'capabilities outside their spacecraft. The original plan outlined a detailed procedure lasting 2 hours and 15 minutes, in which the lunar module pilot would exit the Lunar Module (LM), transfer to the open hatch of the Command Module (CM), and then return.

However, due to unforeseen circumstances, including bouts of nausea experienced by the lunar module pilot the previous day and the extensive preparations required for rendezvous, the operation was shortened significantly to just 39 minutes.

At 072:45, the LM was depressurized, followed by the opening of the forward hatch at 072:53. With precise timing, the lunar module pilot initiated his egress to the forward platform at 072:59:02, maneuvering feet first and face up. Equipped with the extravehicular mobility unit backpack, essential for communication, oxygen supply, and cooling water circulation, he relied on a 25-foot nylon rope as his lifeline to prevent drifting into the void of space.

To stabilize himself outside the LM, he securely anchored his feet into the "golden slippers," distinctive gold-painted restraints attached to the exterior surface near the hatch, affectionately dubbed the "front porch" by the astronauts. This positioning allowed him to conduct operations and experiments while ensuring his safety and stability outside the spacecraft's weightless environment.

Simultaneously, the command module pilot, relying on life support systems within the Command/Service Module (CSM), proceeded with his tasks. At 073:02:00, he depressurized the CM and opened its side hatch. Stepping partially outside, he observed the surrounding environment, captured detailed photographs, and attempted to collect thermal samples from the CM's exterior. However, upon finding these samples missing, he retrieved thermal samples from the Service Module at 073:26.

Meanwhile, at 073:39, the lunar module pilot retrieved thermal samples from the Lunar Module (LM). Shortly after, he initiated an abbreviated evaluation of translation and body-attitude-control capabilities using specially designed extravehicular transfer handrails. The planned hand-over-hand journey from the LM to the CM did not occur during this operational phase.

Throughout these activities, the lunar module pilot meticulously documented the mission using 16 mm and 70 mm photography, capturing the command module pilot's maneuvers and detailed views of both spacecraft exteriors.

The lunar module pilot began his ingress at 073:45 and completed it at 073:46:03. By 073:53, the forward hatch was closed and locked, and the LM was repressurized. The CM hatch was then closed and locked at 073:49. The CM was repressurized by 074:02. The second television transmission was made at 074:55. The commander returned to the CM at 075:15, followed by the lunar module pilot at 076:55.After the lunar module pilot came back inside, both spacecraft were repressurized. A second and final 10-minute television broadcast was telecast from inside the LM. The voice and pictures were good, improving the previous day's transmission.

On March 7, the fifth day of the mission, Apollo 9 marked a pivotal event: the separation and rendezvous of the lunar module (LM) and the command module (CM). This maneuver was critical as it demonstrated the capability to dock two spacecraft in orbit, a vital precursor to lunar landings.

The lunar module designated Spider, could not return astronauts to Earth, emphasizing its role solely for lunar descent and ascent operations. McDivitt and Schweickart entered Spider early, without their helmets and gloves, to facilitate setting up the LM for its operational tasks.

When David Scott in Gumdrop initiated the release sequence, Spider initially hung on the latches of the docking probe. After a second attempt, Spider was successfully released. Initially orbiting slightly below Gumdrop, Spider maneuvered into a somewhat higher orbit over time, setting the stage for a controlled separation where Gumdrop would lead.

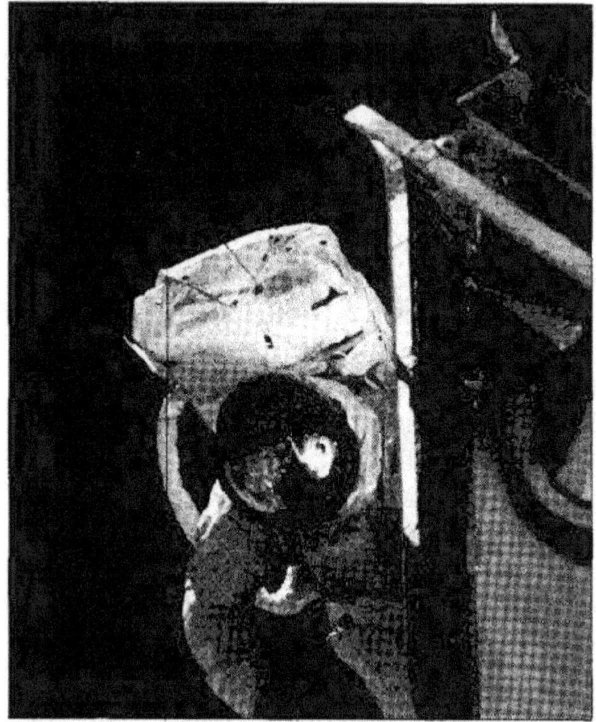

Schweickart on "porch" of LM during EVA activities (NASA AS09-19-2994).

Throughout the day, McDivitt conducted multiple engine burns using the LM's descent engine at various throttle settings. These tests thoroughly evaluated the LM's flight capabilities in space.

Later, when Spider was about 115 miles (185 km) away from Gumdrop, another engine burn was performed to lower its orbit, initiating a gradual approach toward Gumdrop. This intricate process took over two hours to complete. Additionally, Spider's descent stage was jettisoned during this mission phase.

During the approach and rendezvous on March 7, Apollo 9 followed procedures designed to simulate future lunar missions as closely as possible. Spider, the lunar module, took an active role in the maneuver to demonstrate that either spacecraft could initiate rendezvou

Image of the Moon taken from Apollo 9

Jim McDivitt piloted Spider close to Gumdrop, the command module, and maneuvered the LM to allow David Scott to inspect its exterior for any signs of damage. This visual inspection was crucial to ensure the safety and integrity of the spacecraft. McDivitt then proceeded to dock Spider with Gumdrop, a task that posed some challenges due to glare from the Sun. Scott guided from inside Gumdrop to assist in the docking process, a procedure that would typically fall to the command module pilot in later lunar missions.

After successfully docking, McDivitt and Schweickart returned to Gumdrop, marking the end of Spider's operational use. Spider was jettisoned further to test its systems, including the ascent engine, while its engine was fired remotely from Mission Control until its fuel was depleted. This simulation mimicked the ascent stage's climb from the lunar surface, raising Spider to an orbit with an apogee exceeding 3,700 nautical miles (6,900 km; 4,300 mi).

While most of the lunar module's systems were thoroughly tested during the mission, the landing radar remained untested in Earth orbit, as its functionality was specifically designed for lunar landings. This highlighted one of the mission's limitations in fully simulating lunar descent operations.

On the fifth day, the lunar module pilot transferred to the LM at 088:05, followed by the commander at 088:55, to prepare for the first LM free flight and active rendezvous. Apollo 9 The CSM was maneuvered to the inertial undocking attitude at 092:22. Undocking was attempted at 092:38:00, but the capture latches did not release immediately. Undocking occurred at 092:39:36, and the LM was rolled on its axis so the CMP could visually inspect it. A small separation maneuver at 093:02:54, using the service module reaction control system, placed the LM 2.0 n mi behind the CSM 45 minutes later. The maximum range between the LM and CSM was 98 n mi, achieved about halfway between the coelliptical sequence initiation and constant differential height maneuver.

The LM engine ran smoothly during this maneuver until throttled to 20 percent, chugging noisily. The commander stopped throttling and waited. Within seconds, the chugging stopped. He accelerated to 40 percent before shutting down and had no more problems. The LM crew then checked their systems and fired the descent engine again to 10 percent. It ran evenly.

The first rendezvous phasing maneuver was carried out precisely at 093:47:35.4 using the descent propulsion system under abort guidance control. This maneuver strategically positioned the Lunar Module (LM) into a nearly equiperiod orbit, with apogee and perigee altitudes approximately 12.2 nautical miles above and below the Command/Service Module (CSM).

Scott opens the CM hatch during EVA activities (NASA AS09-20-3064).

The second maneuver, while planned, remained unused during this phase as it was reserved for contingencies requiring an LM abort. Calculated at 094:57:53, it stood ready as a precautionary measure.

At 095:39:08.06, the third critical rendezvous maneuver was executed, refining the LM's orbit to approximately 138.9 by 133.9 nautical miles. Following this, the coelliptic sequence initiation took place at 096:16:06.54. Shortly after, the descent stage was promptly jettisoned, coinciding with the commencement of thrust from the reaction control system.

The strategic jettison left the LM trailing the CSM by 10 nautical miles and positioned 82 nautical miles below. Having fulfilled its purpose, the descent stage remained in Earth orbit until its controlled reentry, impacting the Indian Ocean off the eastern coast of Africa at 03:45 GMT on March 23.

Schweickart during EVA (NASA AS09-19-

LM in first free flight following separation from CSM (NASA AS09-21-3199).

Post-jettison, the ascent stage assumed an orbit measuring 116 by 111 nautical miles. Despite initial challenges, including the failure of the ascent stage tracking light, rendezvous radar tracking was reestablished through coordinated efforts with the CSM using reaction control systems.

At 096:58:15.0, a critical maneuver known as the constant differential height maneuver was executed, marking the first use of the ascent stage engine. This maneuver was pivotal in achieving precise orbital alignment and preparing for the terminal phase initiation. At 097:57:59, the onboard solution for terminal phase initiation was successfully carried out, resulting in an ascent stage orbit of approximately 126 by 113 nautical miles.

Two small midcourse corrections were performed 10 and 22 minutes after the terminal phase's initiation. Terminal phase braking began at 098:30:03, followed by stationkeeping, formation flying, photography, and docking at 099:02:26. The ascent stage had been separated from the CSM for 6 hours 22 minutes 50 seconds.

.After docking, the crew transferred back to the CSM by 101:00. The ascent stage was jettisoned at 101:22:45.0, and the ascent engine fired for 362.4 seconds at 101:53:15.4 until oxidizer depletion. The final orbit for the ascent stage was 3,760.9 by 126.6 n mi, with an expected orbital lifetime of five years; however, entry occurred on 23 October 1981. The sixth service propulsion bum, a 1.43-second maneuver at 123:25:06.97, had been postponed for one revolution because the reaction control translation required before ignition for propellant settling was improperly programmed. Initially scheduled for 121:48:00, the orbit-shaping retrograde maneuver was intended to lower the perigee so that the reaction control system deorbit capability would be enhanced in the event of a contingency.

During the later days of Apollo 9's mission (March 8–13), the focus shifted to assessing the long-duration performance of the Command and Service Module (CSM) and conducting additional scientific tasks at a more relaxed pace.

With the mission's primary objectives achieved earlier, including crucial rendezvous and docking maneuvers, the crew had more time for secondary activities. One significant task during this period was conducting specialized Earth photography. The crew captured detailed images of Earth's surface using a unique setup of four Hasselblad cameras coupled together, each equipped with film sensitive to different parts of the electromagnetic spectrum. This innovative approach allowed them to track phenomena such as water pollution entering seas from river mouths and to highlight agricultural areas using infrared photography.

LM ascent stage following separation from descent stage, preparing to redock with CM (NASA AS09-21-3236)

The camera system used on Apollo 9 was experimental and served as a prototype for future Earth observation missions. It laid the groundwork for the Earth Resources Technology Satellite (ERTS), which later evolved into the Landsat series of satellites that revolutionized remote sensing of the Earth's resources and environment.

Due to the extended duration of their orbital mission, the crew had the luxury of waiting for optimal conditions to capture these images, allowing them to avoid cloud cover and ensure the highest quality data. The insights gained from this photographic survey would prove valuable for future missions, including planning for the Skylab space station.

During the later stages of Apollo 9's mission, the crew engaged in various activities to further test equipment and refine their operational skills.

Scott utilized a sextant to track Earth landmarks and practice celestial navigation techniques, which would be essential for later lunar missions. This included observing Jupiter and practicing techniques for navigating by the stars.

In addition to their Earth observations, the crew also tracked other objects in space, including the Pegasus 3 satellite launched in 1965 and the ascent stage of the Lunar Module Spider, demonstrating their ability to monitor and interact with objects in different orbits.

On the sixth day of the mission, the crew performed the sixth burn of the Service Propulsion System (SPS) engine. This burn was initially delayed by one orbit due to issues with the reaction control system (RCS) thruster burn, which needed to settle the reactants in their tanks properly. The SPS burn was crucial as it lowered the perigee (the closest point to Earth) of Apollo 9's orbit. This adjustment enhanced the RCS thruster deorbit capability, providing a backup to the SPS for controlled reentry into Earth's atmosphere.

Apollo 9 concluded its mission with significant testing of the Command and Service Module (CSM) systems and preparations for reentry and recovery.

During the latter stages of the mission, Scott primarily focused on testing the CSM systems, while McDivitt and Schweickart had the opportunity to observe and document Earth from space. They alerted Scott to any notable sights, allowing him brief breaks from his tasks and appreciating the view.

On the eighth day of the mission, March 10, the seventh burn of the Service Propulsion System (SPS) was conducted. This burn aimed to enhance the RCS (Reaction Control System) deorbit capability and extend the orbital lifetime of Gumdrop. It shifted the apogee of Apollo 9's orbit to the Southern Hemisphere, providing a longer free-fall time toward Earth's atmosphere during reentry. The burn also included tests of the propellant gauging system, which had shown anomalies in earlier SPS burns. With these tests completed, Apollo 9's RCS thrusters were confirmed capable of guiding it back to Earth and landing in the primary recovery zone even if the SPS engine had failed.

The eighth and final SPS burn to initiate the return trajectory to Earth occurred on March 13, shortly after the mission's ten-day mark. Following this burn, the Service Module was jettisoned. However, unfavorable weather conditions in the primary landing zone necessitated a one-orbit delay in the landing process. Ultimately, Apollo 9 splashed down safely approximately 160 nautical miles east of the Bahamas, close to the recovery carrier USS Guadalcanal. The mission duration totaled 10 days, 1 hour, and 54 seconds.

During the final four days in orbit, the crew conducted Earth resources and multispectral terrain photography experiments over the southern United States, Mexico, Brazil, and Africa. One objective, designated experiment S065, was to determine how multiband photography in the visible and near-infrared regions from orbit may be effectively applied to the Earth resources disciplines.

Another crucial mission objective was to capture simultaneous photographs using four different film/filter combinations from orbit to advance future multispectral photographic systems. The outcomes surpassed expectations, setting a new standard in orbital imaging quality and subject matter diversity unmatched by any previous mission. These achievements promised invaluable insights for future program planning.

Several factors contributed to the outstanding results. The mission's four days allowed the crew flexibility to wait out cloud cover, ensuring optimal photographic conditions. The spacecraft's orbital inclination of 33.6° provided unparalleled vertical and near-vertical coverage of previously unphotographed areas. Ample reserves of reaction control propellants enabled precise spacecraft orientation as needed, essential for capturing targeted imagery.

Additionally, the spacecraft's windows remained uncontaminated, preserving pristine visibility crucial for high-resolution photography. Continuous support and evaluation from the science support room at the Manned Spacecraft Center further enhanced mission effectiveness.

Beyond photographic accomplishments, the crew conducted an inertial measurement unit alignment using Jupiter sightings, a pioneering use of planetary reference. They also performed various celestial observations, including daylight star sightings, landmark alignments, and star sextant sightings, underscoring their scientific engagement and the mission's multifaceted objectives.

During two consecutive orbits, at 192:43 and 194:13, the crew successfully tracked the Pegasus III satellite from 1,000 nautical miles. Launched on July 30, 1965, Pegasus III's observation provided valuable data during the mission.

Thunderhead over South America as seen in nearly vertical view from Apollo 9 (NASA AS09-

While passing over Hawaii, the crew sighted the ascent stage from 222:38:40 to

222:45:40, marking a significant visual contact during the mission.

Earlier in the mission, at 169:30:00.36, the service propulsion system fired for the seventh time, executing a 24.90-second maneuver. This burn raised the apogee to 253.2 nautical miles and established optimal conditions for the nominal deorbit phase. Contingency plans were also in place, with the reaction control system capable of executing a deorbit maneuver from this apogee if necessary, ensuring a safe landing near the primary recovery area.

The actual deorbit maneuver took place after 151 orbits, delayed by one revolution to 240:31:14.84 due to adverse weather conditions at the planned recovery site. This critical maneuver, the eighth service propulsion firing lasting 11.74 seconds, ensured a controlled descent and successful crew recovery, concluding the mission's operational phase.

Following the jettisoning of the service module at 240:36:03.8, the Command Module (CM) began its descent back to Earth following a primary guidance system profile. Reentry into Earth's atmosphere occurred at 240:44:10.2 at an altitude of 400,000 feet, with the CM entering at a velocity of 25,894 feet per second.

Although the service module did not survive reentry intact, radar tracking data accurately predicted its impact point in the Atlantic Ocean near latitude 22.0° north and longitude 65.3° west. The estimated impact location was approximately 175 nautical miles downrange from the CM.

The CM deployed its parachute system, leading to a successful splashdown in the Atlantic Ocean at 17:00:54 GMT (12:00:54 p.m. EST) on March 13. The entire mission lasted 241 hours and 54 seconds.

The CM landed about 2.7 nautical miles from the targeted point and approximately 3 nautical miles from the recovery ship, U.S.S. Guadalcanal. This concluded the mission with a successful return and recovery of the crew, achieving the mission's objectives and demonstrating significant advancements in space exploration capabilities.

Following splashdown, the estimated site was latitude 23.22° north and longitude 67.98° west. Upon landing, the Command Module (CM) assumed an apex-up flotation attitude. The crew was swiftly retrieved by helicopter and safely aboard the recovery ship just 49 minutes after splashdown. The CM itself was recovered 83 minutes later.

At splashdown, the CM's estimated weight was 11,094 pounds. It traveled an impressive total distance of 3,664,820 nautical miles throughout the mission.

Weather conditions recorded onboard the Guadalcanal during retrieval indicated scattered clouds at 2,000 feet, broken clouds at 9,000 feet, visibility of 10 nautical miles, and a wind speed of 9 knots from 200° true north. Air temperatures were recorded at 79°F, while water temperatures measured 76°F, and waves reached up to seven feet.

The crew departed from the Guadalcanal via helicopter at 15:00 GMT on March 14, arriving at Eleuthera, Bahamas, by 16:30 GMT. They were flown to Houston for further debriefing and medical evaluations.

On March 16, the CM was offloaded from the Guadalcanal at Norfolk Naval Air Station, Virginia. Immediately, the Landing Safing Team commenced evaluation and deactivation procedures at 16:00 GMT, completing the process by March 19.

The CM was then transported to Long Beach, California, and subsequently trucked to the North American Rockwell Space Division facility in Downey, California, arriving on March 21 for thorough postflight analysis. This meticulous evaluation marked the conclusion of a successful mission, contributing significant data and advancements to the field of space exploration.

Upon analyzing the extensive data collected post-mission during Project Apollo, a series of pivotal conclusions emerged, offering insights into the mission's achievements and areas requiring attention.

The precision of onboard rendezvous equipment and procedures within both spacecraft proved crucial for executing intricate maneuvers essential for lunar landings. Command Module pilots adeptly managed computations and mirror-image maneuvers on schedule, underscoring the meticulous planning and technical expertise pivotal to mission success.

The Extravehicular Mobility Unit (EMU) demonstrated outstanding performance throughout extravehicular activities, affirming its reliability in supporting astronauts during lunar surface operations. These successes and rigorous qualification tests incorporating minor design adjustments bolstered confidence in the EMU's operational integrity.

The mission also underscored the feasibility of extravehicular crew transfers in contingency scenarios, with demonstrated proficiency in cabin depressurization and repressurization procedures across both spacecraft—a pivotal capability for ensuring astronaut safety and mission flexibility.

Apollo 9 CM on parachute system just before splashdown (NASA S69-20364)

Despite overall operational readiness, evaluations identified areas for improvement in lunar module systems. Notably, the non-operational steerable antenna and challenges assessing the landing radar in Earth orbit highlighted technological nuances yet to be fully addressed.

Upon thorough review, it became evident that none of the identified anomalies significantly threatened the mission's success. The crew and spacecraft interfaces, encompassing procedures, provisioning, restraints, displays, and controls, functioned satisfactorily for piloted lunar module operations. This included seamless interaction between the two spacecraft, whether docked or undocked.

Consumable usage within the lunar module remained well within predicted limits, demonstrating ample margins to sustain lunar mission requirements effectively.

An issue arose with gas in the Command Module's potable water supply, affecting food rehydration and thereby impacting food taste and palatability. However, water from the lunar module was deemed acceptable, mitigating potential concerns.

The Apollo 9 Command Module (CM), with flotation collar still attached, is hoisted aboard the prime recovery ship, USS Guadalcanal, during recovery operations. The Apollo 9 crew, astronauts James A. McDivitt, David R. Scott, and Russell L. Schweickart, had already been picked up earlier by helicopter and flown to the dock of the carrier. Splashdown occurred at 12:00:53 p.m. (EST), March 13, 1969, only 4.5 nautical miles from the aircraft carrier, to conclude a successful 10-day Earth-orbital space mission.

Mission support, facilitated by the Manned Space Flight Network, effectively coordinated simultaneous ground control of two piloted spacecraft, ensuring robust operational oversight throughout the mission's duration. These combined assessments underscored the meticulous planning and execution essential to the success of Project Apollo, advancing humanity's exploration capabilities beyond Earth's confines.

Apollo 9 crew aboard recovery ship U.S.S. Guadalcanal (1. tor.: Schweickart, Scott, McDivitt) (NASA S69-27921).

The Command Module's orbital navigation, utilizing yaw-control techniques for landmark tracking, was successfully demonstrated and deemed sufficient. Nevertheless, establishing the stir visibility threshold for the Command Module's scanning telescope in a docked configuration remained inconclusive. This uncertainty suggested potential reliance on solar, lunar, and planetary references if inertial orientation were inadvertently compromised during translunar flight.

The success of Apollo 9 cemented the path forward to place a man on the Moon

Apollo 10

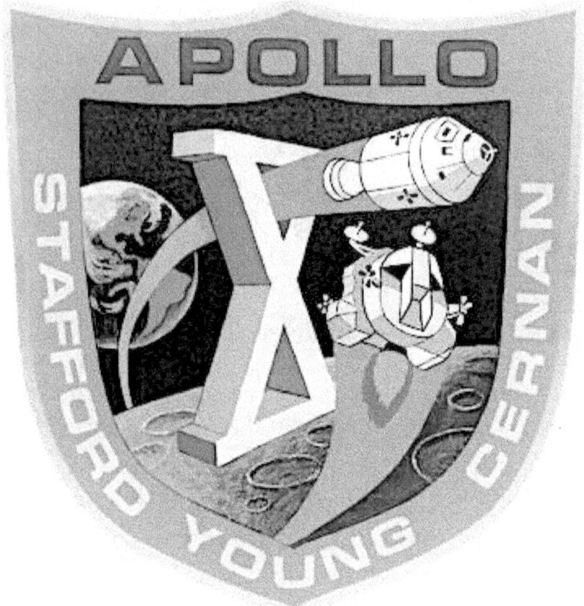

In May 1969, NASA launched Apollo 10, a critical precursor to the historic Apollo 11 moon landing mission. Commanded by Thomas P. Stafford and accompanied by crew members John W. Young and Eugene A. Cernan, Apollo 10 was designed as a crewed lunar orbital flight, thoroughly testing all aspects of lunar landing procedures without actually landing on the Moon.

The spacecraft comprised Apollo CSM-106, manufactured by North American Rockwell, and LM-4, built by Grumman. Apollo 10 had a launch mass of approximately 42,775 kg and was launched on May 18, 1969, at 16:49:00 UTC, atop the Saturn V SA-505 rocket from Kennedy Space Center's Launch Complex 39B.

During its mission, the CSM was nicknamed "Charlie Brown," while the LM was called "Snoopy." The crew conducted a series of crucial maneuvers and tests, including the first lunar orbit by a crewed spacecraft. They achieved orbital insertion around the Moon on May 21, 1969, and performed several orbits before undocking the LM for independent operations.

Apollo 10's LM descended to within 14.4 kilometers of the lunar surface, providing critical data and testing the lunar landing radar and other systems essential for Apollo 11's subsequent landing. After completing their tasks, the LM ascent stage redocked with the CSM on May 23, 1969, before the crew began their journey back to Earth.

The mission concluded successfully on May 26, 1969, with recovery by the USS Princeton in the Pacific Ocean at coordinates 15°2′S 164°39′W, having flown 829,437.5 miles and conducting 31 orbits.

Apollo 10's achievements paved the way for the historic Apollo 11 mission, ensuring that all systems and procedures were in place for humanity's first steps on the Moon later that year.

Apollo 10 marked a pivotal milestone in NASA's ambitious Project Apollo, meticulously planned as a Type F mission. Positioned as a crucial lunar module piloted flight demonstration in lunar orbit, its overarching purpose was to serve as the final dress rehearsal before attempting mankind's first historic landing on the Moon. This mission also held the distinction of being the first where every member of its three-person crew had previous spaceflight experience, a testament to NASA's evolving expertise and confidence in its astronaut corps.

The primary objectives of Apollo 10 were manifold. Foremost among them was to rigorously test and validate the performance of the crew, the spacecraft itself, and the extensive mission support infrastructure crucial for a piloted lunar endeavor. This comprehensive evaluation spanned both the command and service modules and, critically, the lunar module (LM), which would play a pivotal role in the upcoming lunar landing missions.

Throughout the meticulously orchestrated mission, which faithfully simulated the sequence of events planned for an actual lunar landing, Apollo 10 undertook a series of vital tasks. These included intricate visual observations and the execution of stereoscopic strip photography aimed at Apollo Landing Site 2, designated as the primary site for the imminent lunar landing.

By 1967, NASA had meticulously categorized its Apollo missions into various types, denoted by letters, each designed to build toward the ultimate goal of a lunar landing incrementally. The uncrewed test flights, denoted as "A" or "B" missions, laid the groundwork. Apollo 7 subsequently validated the Command and Service Module (CSM) crewed flight capabilities as the "C" mission. At the same time, Apollo 9 achieved a significant milestone with the first crewed orbital test of the Lunar Module (LM), classified as the "D" mission. Apollo 8, although initially slated as a "C-prime" mission due to its circumlunar trajectory without an LM, ultimately provided NASA with the confidence to bypass the "E" mission, originally intended to test the full Apollo spacecraft in Earth orbit. Thus, Apollo 10, meticulously planned and executed, solidified its place as the "F" mission—a critical dress rehearsal that set the stage for the triumphant lunar landing of Apollo 11.

During the planning stages of Project Apollo, there was significant debate within NASA about the possibility of skipping the "F" mission, Apollo 10, and attempting the first lunar landing earlier. Some within the agency argued passionately for this approach, questioning the rationale of bringing astronauts so close to the lunar surface only to retreat without attempting a landing. The proposal gained traction especially because the lunar module slated for Apollo 10 was deemed too heavy to carry out a lunar landing mission effectively. Proponents suggested that delaying Apollo 10 by a month from its planned May 1969 launch could allow the use of the lighter lunar module intended for Apollo 11, thus enabling a landing attempt sooner.

George Mueller, a prominent NASA official known for his assertive leadership in advancing the Apollo program, advocated strongly for attempting the lunar landing on Apollo 10. His proactive stance reflected his belief in accelerating progress toward landing astronauts on the Moon.

Launch of Apollo 10 on May 18, 1969

However, not everyone at NASA supported this bold proposal. Christopher C. Kraft, the Director of Flight Operations, and others voiced concerns about the feasibility and risks involved. They pointed out that attempting a landing on Apollo 10 would require developing new procedures for a lunar orbit rendezvous, a complex maneuver that had yet to be fully tested. Moreover, uncertainties persisted regarding the Moon's gravitational variations (mass concentrations), potentially affecting the spacecraft's trajectory and safety.

Amid these deliberations, Lieutenant General Sam Phillips, the Apollo Program Manager, carefully weighed the arguments from both sides. Ultimately, he sided with those advocating for prudence and the necessity of a comprehensive dress rehearsal. This decision underscored the critical importance of Apollo 10 as the final preparatory mission before embarking on the historic lunar landing with Apollo 11.

The crew of Apollo 10, consisting of Colonel Thomas Patten Stafford (USAF) as commander, Commander John Watts Young (USN) as command module pilot, and Commander Eugene Andrew "Gene" Cernan (USN) as lunar module pilot, represented a pinnacle of experience and skill in NASA's astronaut corps.

Colonel Thomas Stafford, born on September 17, 1930, in Weatherford, Oklahoma, brought a wealth of experience to Apollo 10, marking his third spaceflight. A graduate of the U.S. Naval Academy in 1952, Stafford had previously piloted Gemini 6-A and commanded Gemini 9-A. His backup for Apollo 10 was Colonel Leroy Gordon Cooper, Jr. (USAF).

Commander John Young, born on September 24, 1930, in San Francisco, California, joined Stafford in being 38 years old during the Apollo 10 mission. He earned a B.S. in aeronautical engineering from the Georgia Institute of Technology in 1952 and was selected as an astronaut in 1962. Young had already flown on Gemini 3 and commanded Gemini 10, establishing himself as a seasoned spacefarer. His backup was Lt. Colonel Donn Fulton Eisele (USAF).

Commander Eugene Cernan, born on March 14, 1934, in Chicago, Illinois, was 35 years old at the time of Apollo 10. He held a B.S. in electrical engineering from Purdue University and an M.S. in aeronautical engineering from the U.S. Naval Postgraduate School. Selected as an astronaut in 1963, Cernan had previously piloted Gemini 9-A. His backup for Apollo 10 was Commander Edgar Dean Mitchell (USN).

The prime crew of the Apollo 10 lunar orbit mission at the Kennedy Space Center. They are from left to right: Lunar Module pilot, Eugene A. Cernan, Commander, Thomas P. Stafford, and Command Module pilot John W. Young.

Announced on November 13, 1968, the crew of Apollo 10 was notable not only for their individual achievements but also for their collective experience. With a combined total of five previous spaceflights among them, they were the most seasoned crew to have flown in space up to that point, a distinction that held until the Space Shuttle era. Moreover, they were the first American space mission crew composed entirely of astronauts who had already flown in space, underscoring their expertise and readiness for the challenges of the Apollo 10 mission.

During the Apollo 10 mission, a dedicated team of professionals meticulously filled the support and operational roles, ensuring every aspect of the mission was meticulously managed.

The capsule communicators (CAPCOMs) were crucial in maintaining communication between mission control and the Apollo 10 crew. Major Charles Moss Duke, Jr. (USAF), Major Joe Henry Engle (USAF), Major Jack Robert Lousma (USMC), and Lt. Commander Bruce McCandless, II (USN) served in this capacity, providing real-time updates, instructions, and support to the astronauts during their mission.

A team of experienced astronauts known as the support crew supported the mission. This group included Major Joe Engle (USAF), Lt. Col. James Benson Irwin (USAF), and Major Charles Duke (USAF). Their role was to assist the primary crew as directed by the mission commander, ensuring that mission rules, flight plans, and checklists were meticulously prepared and updated.

The flight directors, led by Glynn S. Lunney and Gerald D. Griffin on the first shift, Milton L. Windler on the second shift, and M. P. "Pete" Frank on the third shift, provided overall guidance and decision-making authority from mission control. Their responsibility encompassed making critical decisions for crew safety and mission success, exemplifying their pivotal roles in the Apollo 10 mission's execution.

Looking ahead, the backup crew for Apollo 10 comprised L. Gordon Cooper Jr. as commander, Donn F. Eisele as command module pilot, and Edgar D. Mitchell as lunar module pilot. Originally slated for future missions, Cooper and Eisele were later replaced by Alan Shepard and Stuart Roosa, respectively, due to operational considerations and the need for additional training time.

During the Apollo missions, every detail, from the spacecraft names to their insignias, was significant for the crew and the public. For Apollo 10, the command module was aptly named "Charlie Brown," evoking the beloved character from the Peanuts comic strip. Its lunar module counterpart, "Snoopy," embodied the adventurous spirit associated with its comic namesake, Snoopy the dog. These names, chosen by the astronauts themselves and approved by Peanuts creator Charles Schulz, initially raised eyebrows within NASA for their informality. However, they quickly became emblematic of the missions.

Apollo 10, known as the "F" mission, served as a crucial dress rehearsal for the impending lunar landing of Apollo 11. Its primary objectives were manifold: to validate the performance of crew, spacecraft, and mission support systems during a crewed mission to lunar orbit, and specifically to evaluate the lunar module's capabilities in the lunar environment. Additionally, Apollo 10 aimed to capture detailed photography of Apollo Landing Site 2 (ALS-2) in the Sea of Tranquillity, the prospective landing site for Apollo 11.

"Snoopy," tasked with reconnaissance around the lunar landing site, resonated deeply with the public, reflecting a nod to NASA's diligent workers. Those who excelled were awarded silver "Snoopy pins," a symbol of excellence within the space program. At NASA facilities, Snoopy posters adorned the walls, depicting the cartoon dog with his trademark aviator helmet, now replaced with a space helmet.

Ultimately, the names "Charlie Brown" and "Snoopy" for Apollo 10 became a public relations triumph, captivating imaginations worldwide. They highlighted the mission's playful yet profound spirit, demonstrating how humanity found ways to connect through familiar and endearing characters even in the vastness of space.

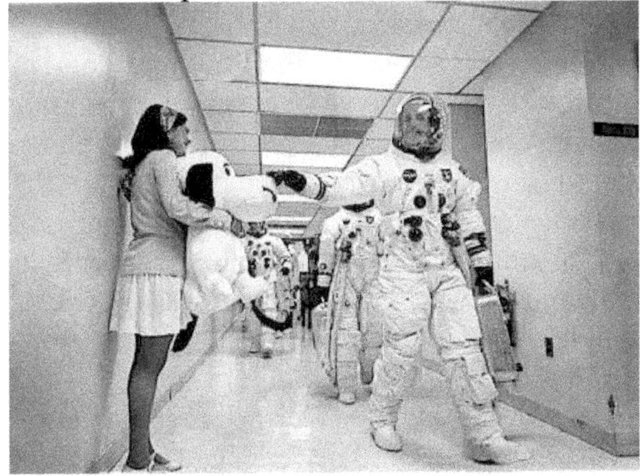

En route to launch, Stafford touches a "Snoopy" doll

As Commander Stafford outlined, their mission involved piloting the lunar module to within ten miles above the lunar surface, conducting radar and photographic surveys to meticulously map out the first potential landing site, executing the first lunar rendezvous, and assessing future landing sites before returning home.

The mission closely adhered to the trajectory and timeline for Apollo 11, aligning even the solar angles at ALS-2 for accurate photography. However, unlike Apollo 11, Apollo 10 was not slated to attempt a lunar landing. ALS-1, which had already been extensively documented by the Apollo 8 mission, was located further east in the Sea of Tranquillity. At the suggestion of scientist-astronaut Harrison Schmitt, the launch date was adjusted to May 18, 1969, to ensure optimal conditions for photographing ALS-2.

ALS-2, chosen for its relatively smooth terrain and scientific interest, offered a suitable alternative to ALS-1, deemed too far east for practical landing operations. One notable deviation from Apollo 11's plan was Apollo 10's extended stay of an additional day in lunar orbit following the rendezvous of the Command and Lunar Modules. This extra time allowed for comprehensive testing of the lunar module's systems and extensive photography of potential future landing sites, enhancing mission readiness and scientific exploration.

In preparation for their pivotal role in Apollo 10, the astronauts underwent rigorous training that far exceeded the mission's eight-day duration. For every hour spent in the mission itself, they devoted five hours to formal training. This comprehensive regimen included a spectrum of activities beyond standard mission preparations.

Technical briefings, pilot meetings, and intensive study sessions formed the foundation of their preparation. They engaged in hands-on testing of the Command and Service Module (CSM) at North American Rockwell's facility in Downey, California, ensuring they were intimately familiar with the spacecraft's operations. Similarly, they meticulously tested and honed their skills with the Lunar Module (LM) at Grumman's Bethpage, New York facility.

The astronauts journeyed to Cambridge, Massachusetts, receiving detailed briefings on the Apollo Guidance Computer at the Massachusetts Institute of Technology's Instrumentation Laboratory. This training was crucial, as the guidance computer would be pivotal in navigating the spacecraft during critical mission phases.

Much of their preparation involved simulator training at the Manned Spacecraft Center (MSC) in Houston and Kennedy Space Center (KSC) in Florida. Here, they logged over 300 hours each, meticulously rehearsing maneuvers and procedures in simulated Command Module and Lunar Module environments. These simulations were designed to replicate the conditions and challenges they would face in space, ensuring they were prepared for any scenario.

Stafford (right) and Cernan in the lunar module simulator, April 1969

The astronauts endured sessions in MSC's centrifuge to acclimate to the high-acceleration conditions of reentry into Earth's atmosphere. This training was vital for preparing them physically and mentally for the demanding return journey.

During Apollo 10, the lunar module (LM) played a pivotal role in simulating the procedures and conditions of a lunar landing, but it was not equipped for actual touchdown and return. The LM used on Apollo 10, designated LM-4, differed significantly from the one used for Apollo 11, LM-5, which successfully completed the first lunar landing.

The LM-4 carried a total mission-loaded weight of 13,941 kilograms (30,735 lb), while the Apollo 11 LM-5 weighed 15,095 kilograms (33,278 lb). This weight difference was critical because the ascent stage of the LM-4 was intentionally under-fueled. This precaution ensured that if the crew attempted a landing, they would not have sufficient fuel to lift off from the lunar surface and rendezvous with the command module (CSM). This decision was a deliberate safety measure to prevent any possibility of an unauthorized landing attempt by the crew.

According to Craig Nelson's book "Rocket Men," there was cautious speculation within NASA about the crew's intentions, with concerns that they might attempt an unauthorized landing. Eugene Cernan, one of the astronauts on Apollo 10, acknowledged the precautions taken, stating that the ascent module of their LM was intentionally short-fueled, making it impossible to conduct a lunar liftoff.

NASA's Associate Administrator for Manned Space Flight, Mueller, underscored that Apollo 10's LM was specifically a test module designed for the dress rehearsal of a lunar landing mission. Due to its weight and fuel limitations, it was not configured to support a full lunar landing and return. This approach ensured that the mission adhered strictly to its testing procedures and systems objectives without the risk of an unintended lunar landing attempt.

The preparation and assembly of the Apollo 10 spacecraft at Kennedy Space Center (KSC) in Florida was a meticulously orchestrated process, ensuring every component was ready for its crucial mission.

The descent stage of the Lunar Module (LM) arrived at KSC on October 11, 1968, followed five days later by the arrival of the ascent stage. These components were carefully mated together on November 2. Subsequently, the Service Module (SM) and Command Module (CM) reached KSC on November 24, with their mating completed just two days later.

The Saturn V launch vehicle, essential for propelling the spacecraft toward the Moon, saw portions arrive throughout November and December 1968. By December 30, the entire launch vehicle was assembled in the iconic Vehicle Assembly Building (VAB). Prior to its integration with the spacecraft, the Command and Service Module (CSM) underwent rigorous testing in an altitude chamber, ensuring its readiness for the mission's demands. It was finally mounted atop the launch vehicle on February 6, 1969.

On March 11, 1969, the fully assembled Apollo 10 spacecraft made its monumental journey to Launch Complex 39B. This event marked the first use of High Bay 2 within the VAB for spacecraft assembly, requiring the crawler-transporter to navigate a unique path: exiting the VAB through its rear, looping around the building, and joining the main crawler way toward the launch pad. This intricate maneuver was executed flawlessly using Mobile Launch Platform-3 (MLP-3), just eight days after the launch of Apollo 9, which was still in orbit then.

The Saturn V rocket, designated AS-505, served as the launch vehicle for Apollo 10, marking the fifth flight-ready Saturn V to be launched and the third to carry astronauts into orbit. This formidable rocket differed from its predecessor used on Apollo 9 in several significant aspects.

Firstly, the Saturn V for Apollo 10 boasted a lower dry weight in its first two stages, excluding propellant, contributing to enhanced efficiency and performance. The interstage connecting these stages was also notably reduced in weight. Despite the S-IVB third stage being slightly heavier, all three stages could accommodate a greater payload of propellant. Moreover, the S-II second stage of the Apollo 10 Saturn V generated increased thrust compared to its counterpart on Apollo 9, enhancing the rocket's overall propulsion capabilities.

The Apollo spacecraft used for the Apollo 10 mission integrated several crucial components essential for its successful operation:

Command Module 106 (CM-106): This module housed the crew during the mission and served as the primary living and working quarters. It contained controls and instruments necessary for spacecraft navigation, communication, and life support.

Service Module 106 (SM-106): The service module supported the command module with propulsion, electrical power, and other essential systems required for the mission. It also housed the main engine used for mid-course corrections and lunar orbit insertion.

Lunar Module 4 (LM-4): The lunar module was designed for lunar landing missions. For Apollo 10, LM-4 served as a test vehicle for the lunar descent and ascent stages, although it did not land on the Moon.

Spacecraft-Lunar Module Adapter (SLA), designated as SLA-13A: The SLA-13A provided structural support and the interface between the Saturn V rocket's Instrument Unit and the CSM. It also housed the lunar module during launch and transit.

Launch Escape System (LES): The LES included rockets mounted on the command module to propel it away from the Saturn V rocket in case of an emergency during launch. It ensured the crew's safety by providing a means to escape if the launch had to be aborted.

Weighing approximately 76.99 metric tons at liftoff, Apollo 10 set a new record as the heaviest spacecraft ever launched into orbit up to that point. This impressive mass underscored the complexity and scale of the mission, highlighting the advancements made in spaceflight technology and engineering by NASA and its partners.

During the countdown to Apollo 10's launch, any issues that arose were swiftly managed within the scheduled holds, ensuring that they did not disrupt the mission timeline. However, a notable incident occurred the day before liftoff involving Gene Cernan, one of the mission's astronauts. On that day, Cernan had been returning from a final visit with his wife and child when he was stopped for speeding. Complicating matters, he lacked proper identification and was under strict orders not to disclose his identity.

In his autobiography, Cernan recalled feeling genuine concern about being arrested. Fortunately, Gunther Wendt, the launch pad leader who happened to be nearby and recognized Cernan, intervened. Wendt explained the situation to the skeptical police officer, emphasizing Cernan's status as an astronaut preparing for a historic space mission. Thanks to Wendt's intervention, Cernan was released without further incident, allowing him to proceed with final preparations for the momentous launch of Apollo 10.

Apollo 10 embarked on its historic journey from Kennedy Space Center (KSC) on May 18, 1969, precisely at 12:49:00 EDT (16:49:00 UT), beginning a carefully planned 4.5-hour launch window. This timing was meticulously chosen to ensure optimal lighting conditions over Apollo Landing Site 2 during the Lunar Module's upcoming maneuvers near the lunar surface. Unlike its predecessors, Apollo 10 departed from Launch Complex 39B, as preparations for Apollo 11 were already underway at Pad 39A, making it the only Apollo mission to lift off from 39B and to be managed from Firing Room 3.

The countdown had commenced smoothly on May 16 at 21:00:00 EDT (01:00:00 UT on May 17), proceeding without any significant delays until a minor setback occurred when the primary LOX replenish pump failed to start due to a blown fuse in its motor starter circuit. However, prompt troubleshooting and fuse replacement efforts ensured that LOX loading, delayed by only 50 minutes, caught up with the schedule during a planned hold at T-3 hours 30 minutes.

Mission Control in Houston during an Apollo 10 telecast

Weather conditions on launch day posed challenges, with overcast skies caused by a high-pressure cell off the New England coast. Approximately 40% of the sky was covered by cumulus clouds at 2,200 feet, while altocumulus clouds covered 20% at 11,000 feet, and cirrus clouds obscured 100% of the sky at higher altitudes. Despite these conditions, Apollo 10 lifted off under the watchful eye of favorable ground and flight teams. The launch temperature was 80.1°F, with a relative humidity of 75% and a barometric pressure of 14.779 lb/in². Winds measured 19.0 knots from 142° true north, influencing the launch dynamics.

The Saturn V rocket roared to life at 16:49:00 GMT (11:49:00 p.m. EDT), marking the first manned launch from Launch Complex 39B, Pad B. The launch window was extended until 21:09 GMT to optimize visibility and operational conditions on the lunar surface, with a sun elevation angle of 11°.

During the ascent phase of the Apollo 10 mission, the Saturn V rocket executed a series of critical maneuvers:

Apollo 10 becomes the first piloted mission to lift off from Kennedy Space Center Pad 39B (NASA S69-34145).

Initial Azimuth Rotation: The rocket initially rotated from a launch pad azimuth of 90° to a flight azimuth of 72.028°. This maneuver occurred between 000:00:13.05 and 000:00:32.3 after liftoff, aligning the rocket's trajectory with its intended flight path.

S-IC Stage Shutdown and Separation: At 000:02:41.63 into the flight, the S-IC first-stage engine shut down. Shortly after shutdown, the S-IC stage separated from the rocket, followed by the ignition of the S-II second-stage engine.

S-II Stage Operation and Separation: The S-II engine operated until 000:09:12.64, providing thrust to propel the spacecraft further into space. After the engine shut down, the S-II stage separated from the rocket, making way for the S-IVB third stage.

S-IVB Stage Ignition: At 000:09:16.9, the S-IVB third stage ignited its engine after separation from the S-II stage. This stage was crucial for inserting the spacecraft into its planned trajectory towards the Moon.

First S-IVB Cutoff (SECO-1): The first cutoff of the S-IVB engine occurred at 000:11:43.76 into the flight. This marked the end of powered flight for the S-IVB stage, achieving the initial velocity and altitude targets with minor deviations: the velocity was slightly lower than planned at -0.23 ft/sec, and the altitude was slightly off by -0.08 nautical miles.

Following their missions, the spent stages of the Saturn V descended into the Atlantic Ocean:

The S-IC stage impacted at 000:08:59.12, approximately 348.80 nautical miles from the launch site.

The S-II stage followed at 000:20:17.89, impacting approximately 2,389.29 nautical miles from the launch site.

During Apollo's ascent phase, the spacecraft encountered challenging weather conditions. Maximum wind speeds reached 82.6 knots at 46,520 feet altitude, blowing from the northwest. At 50,200 feet, a significant wind shear of 0.0203 sec-1 was observed, indicating turbulent atmospheric conditions.

Upon reaching parking orbit at insertion, specific parameters were meticulously recorded: the spacecraft achieved an elliptical orbit around Earth with an apogee of 100.32 nautical miles and a perigee of 99.71 nautical miles. The orbital inclination was set at 32.546 degrees, completing a full orbit every 88.20 minutes at a velocity of 25,567.88 feet per second. These figures were calculated relative to a spherical Earth model with a radius of 3,443.934 nautical miles.

Each phase of the Apollo mission was meticulously cataloged with international designations: the Command and Service Module (CSM) upon reaching Earth orbit was designated 1969-043 A, while the S-IVB stage was labeled 1969-043B. Following undocking at the Moon, the Lunar Module (LM) received the designation 1969-018C.

Moving into the translunar phase, after initial systems checks in Earth orbit, a critical 343.08-second translunar injection maneuver was executed at 002:33:27.5. This second firing of the S-IVB engine concluded at 2:39:10.58, propelling the spacecraft into a trajectory that would lead it toward the Moon. The translunar injection occurred precisely ten seconds later, initiating a journey lasting one and a half Earth orbits, totaling 2 hours, 27 minutes, and 16.82 seconds, at a velocity of 35,585.83 feet per second.

At 003:02:42.4, the CSM was separated from the S-IVB stage, followed by a careful transposition and docking with the LM at 003:17:36.0. The docked spacecraft were then ejected at 003:56:25.7, initiating a separation maneuver at 004:39:09.8. These crucial maneuvers were broadcasted live to Earth, beginning at 003:06:00 for 22 minutes and again from 003:56:00 for 13 minutes and 25 seconds. Additional televised broadcasts during the translunar coast included various views of Earth and the spacecraft interior at specified durations.

A ground command was later executed to vent residual propellants from the S-IVB stage, ensuring it continued its trajectory past the Moon. This marked a pivotal step in the Apollo mission's complex journey toward lunar exploration.

Crew boarding the command module before launch

During its ascent to orbit, Apollo 10 encountered a challenging ride caused by pogo oscillations, creating a rough journey for the crew. Approximately 12 minutes after liftoff, the spacecraft successfully entered a low Earth orbit, with its trajectory reaching a high point of 185.79 kilometers and a low point of 184.66 kilometers.

Following thorough systems review in Earth orbit, the crew proceeded with the critical trans-lunar injection (TLI) maneuver. During this crucial burn of the S-IVB third stage to propel them toward the Moon, the vehicle experienced another bout of shaking. Gene Cernan, one of the astronauts aboard, expressed concern that they might need to abort the maneuver. However, despite the unsettling vibrations, the TLI burn was executed successfully without further incident.

Once on its way to the Moon, the crew performed the transposition, docking, and extraction maneuvers. This involved separating the Command and Service Module (CSM) from the S-IVB stage, maneuvering around, and docking the CSM's nose to the top of the lunar module (LM). This pivotal moment was televised back to mission controllers in Houston using the mission's new color television camera, marking the first time such footage was broadcast inside a spacecraft. The live broadcast also captivated a worldwide audience with vibrant color views of Earth.

Apollo 10 during rollout

However, not all aspects of the mission went perfectly. An issue arose when the mylar cover of the Command Module's hatch came loose, causing fiberglass insulation to spill into the tunnel connecting the CSM and LM and into both spacecraft.

After completing its tasks, ground command fired the S-IVB stage one last time, propelling it into a solar orbit for 344.88 days, effectively concluding its role in the Apollo 10 mission.

During Apollo 10's mission, the S-IVB stage made its closest approach to the Moon at 1,680 nautical miles, occurring precisely at 078:51:03.6 on May 21 at 23:40 GMT. Following this, as the spacecraft passed beyond the lunar sphere of influence, it entered a solar orbit with specific orbital parameters. The trajectory led to a solar orbit with an aphelion (farthest point from the Sun) of 82.160 million nautical miles and a perihelion (closest point to the Sun) of 73.330 million nautical miles. The orbital inclination was 23.46 degrees, and the orbit completed a full cycle in approximately 344.88 days.

A critical midcourse correction maneuver planned and executed with precision, took place at 026:32:56.8 into the mission. This 7.1-second burn adjusted the spacecraft's trajectory by 49.2 feet per second, aligning it perfectly with the trajectory intended for a lunar landing in July. Remarkably accurate, this maneuver rendered two additional planned midcourse corrections unnecessary.

The crew employed a passive thermal control technique throughout the translunar coast phase to regulate spacecraft temperatures. This method maintained optimal thermal conditions except when specific spacecraft orientations were required for operational purposes, ensuring the crew's comfort and equipment performance during their journey toward the Moon.

Upon reaching a critical phase of its mission, Apollo 10 executed maneuvers to enter lunar orbit. At 75:55:54 into the mission, the Command and Service Module's (CSM) service propulsion system (SPS) engine fired for 356.1 seconds. This burn, conducted 176.1 kilometers (95.1 nautical miles) above the far side of the Moon, successfully slowed the spacecraft into an initial lunar orbit measuring 314.8 by 111.5 kilometers (170.0 by 60.2 nautical miles).

Following two orbits around the Moon, a 13.9-second SPS firing further refined the orbit, circularizing it to 113.0 by 109.6 kilometers (61.0 by 59.2 nautical miles) at 80:25:08.1. These precise orbital adjustments ensured optimal conditions for the crew to begin their lunar observations and operations. Within hours after the initial orbit insertion and circularization burns, the crew commenced tracking planned landmarks on the lunar surface, capturing observations and photographs. They documented notable features such as the craters Coriolis, King, and Papaleksi, among others on both the near and far sides of the Moon.

During the lunar orbit phase, after two revolutions of tracking and ground updates, a 13.9-second maneuver was performed at 080:25:08.1 to circularize the orbit at 61.0 by 59.2 n mi.

About an hour after the second burn that circularized their lunar orbit, Apollo 10's Lunar Module (LM) crew, Stafford and Cernan, entered the LM to conduct system checks. However, they encountered an unexpected issue: a flurry of fiberglass particles from a previous problem filled the interior. Using a vacuum cleaner, they diligently cleaned up the particles as best they could. Stafford assisted Cernan in removing smaller bits from his hair and eyebrows, likening his appearance afterward to someone who had just emerged from a chicken coop.

Earth as seen from 100,000 n mi

Despite their efforts, the fiberglass particles proved persistent. They caused discomfort, making the crew itch, and some even made their way into the spacecraft's air conditioning system. Throughout the mission, they had to clear the particles from the filter screens continuously.

Earthrise as seen from Apollo 10 (NASA AS10-27-3890)

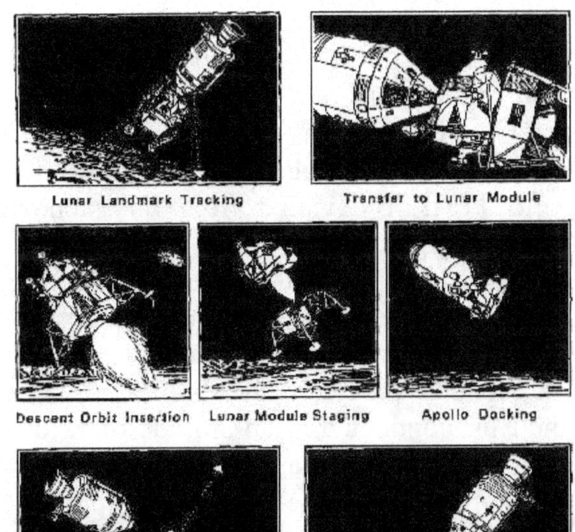

Another concern was whether these particles had affected the docking ring that connected the LM to the Command Module (CM), potentially causing a slight misalignment. Mission Control evaluated the situation and determined that the misalignment was minor and still within safe operational limits, reassuring the crew about the integrity of their spacecraft. This incident, while an annoyance, did not compromise the mission's objectives but underscored the challenges and meticulous maintenance required during lunar exploration.

As the crew of Apollo 10 embarked on their journey to the Moon, they settled into a routine aboard the spacecraft. With a relatively light workload, much of their time was spent studying the flight plan or resting. They utilized their downtime wisely, occasionally making television broadcasts back to Earth, which captivated a global audience of over a billion viewers. In June 1969, in recognition of their pioneering broadcasts, the crew later accepted a special Emmy Award on behalf of the first four Apollo missions.

During the mission, a minor course correction was required to align Apollo 10 precisely with the trajectory planned for Apollo 11. This adjustment, lasting 7.1 seconds and occurring at 26:32:56.8 into the mission, ensured that subsequent lunar missions followed a consistent path toward the lunar surface.

Despite their preparations, the crew faced challenges typical of space travel. One notable issue arose from the taste of their food, affected by an accidental overdosing of chlorine in their drinking water by Stafford. This surplus chlorine had to be mixed with their dehydrated food, affecting its taste and adding a unique flavor to their lunar journey experience.

During Apollo 10's mission, a significant television broadcast of the lunar surface lasted 29 minutes and 9 seconds, beginning precisely at 080:45:00. The crew provided detailed descriptions of the lunar features visible beneath them. The picture quality of the lunar scenes was reported to be excellent, offering viewers on Earth a vivid glimpse of the Moon.

Later, at 081:55, the lunar module pilot entered the Lunar Module (LM) for a two-hour session of housekeeping activities and communication tests specific to the LM. Although some tests had to be cut short due to time constraints, initial results were excellent, with plans to resume the remaining tests later in the mission.

At 095:02, the commander and lunar module pilot returned to the LM to activate its systems, discovering that it had shifted 3.5 degrees out of alignment with the Command Module (CM). Concerns arose about potentially shearing off latching pins if the spacecraft were separated, jeopardizing redocking prospects. However, mission control assured them that there would be no issue if the misalignment remained under six degrees.

Undocking occurred at 098:11:57 and was captured on television for 20 minutes and 10 seconds, starting at 098:13:00. During this period, the LM's landing gear was deployed, and all LM systems underwent thorough checks.

Following undocking, an 8.3-second maneuver using the Command and Service Module's (CSM) reaction control system separated it to a distance of about 30 feet from the LM. At the time, the CSM was in orbit at 62.9 by 57.7 nautical miles. Station keeping maneuvers were then initiated as the command module pilot visually inspected the LM.

To finalize the separation, the CSM's reaction control system executed a maneuver directed radially downward toward the Moon's center, achieving a separation at descent orbit insertion of approximately two nautical miles from the LM. These maneuvers marked critical stages in Apollo 10's mission as it prepared for its subsequent lunar orbital operations and eventual return to Earth.

After Stafford and Cernan completed their checks on Snoopy, they returned to Charlie Brown for a brief rest before re-entering Snoopy and initiating its undocking from the command module (CSM) at 98:29:20. During this pivotal moment, astronaut Young, remaining aboard the CSM, achieved the historic feat of becoming the first person to solo-fly in lunar orbit. Following undocking, Stafford and Cernan deployed the lunar module's landing gear and meticulously inspected its systems to ensure operational readiness for upcoming maneuvers and observations.

Meanwhile, aboard the command module (CSM), astronaut Young maintained vigilant oversight of the LM's location and status, prepared to intervene and assist the LM crew if needed. Stafford and Cernan, navigating Snoopy, approached within 15.6 kilometers (8.4 nautical miles) of the lunar surface at ALS-2's eastern point, then executed a precise phasing burn at 100:58:25.93. This maneuver, lasting just under 40 seconds, allowed for a secondary pass over ALS-2, reducing the distance to the Moon to 14.4 kilometers (7.8 nautical miles), their closest approach.

Stafford's observations during these low passes provided critical insights into ALS-2's characteristics, noting its unexpectedly smooth terrain reminiscent of the desert near Blythe, California. Despite this assessment, concerns about potential rougher terrains awaiting Apollo 11 if it deviated from its intended trajectory were raised.

Following the initial maneuvers, the lunar module (LM) crew executed a critical descent orbit insertion by firing the descent engine for 27.4 seconds precisely at 99:46:01.6. This maneuver, conducted at an altitude of 15,000 meters (50,000 feet), was crucial as it mirrored the conditions where the subsequent Apollo 11 mission would commence its powered descent to the lunar surface. Notably, this phase also marked the first real-world test of the LM's landing radar, previously only evaluated under terrestrial conditions.

CM after separation from LM (NASA AS10-27-3873).

NASA mission planners, buoyed by Apollo 10's detailed observations from these low-altitude maneuvers, confidently designated ALS-2 as the primary landing site for Apollo 11.

The next critical task was to prepare for separating the lunar module (LM) ascent stage from the descent stage, followed by the jettisoning of the descent stage. This maneuver would enable the ascent propulsion system to fire, propelling the ascent stage back toward the command and service module (CSM).

However, as Stafford and Cernan began these preparations, an unexpected complication arose: the LM started to gyrate uncontrollably, causing concern among the crew. Reacting to the situation, Cernan exclaimed in frustration, "Son of a bitch!" This exclamation, captured on a live mic broadcast back to Earth, along with other candid language used by the crew during the mission, sparked some controversy and complaints from the public and officials monitoring the mission.

CSM Charlie Brown

Apollo Lunar Module Snoopy about to dock with the command module

Shortly after the tumbling began, Stafford swiftly discarded the descent stage of the lunar module (LM), approximately five seconds into the incident. He then engaged in a manual struggle to regain control, suspecting a malfunction such as an "open thruster" causing the gyration. This critical action was essential to realign the spacecraft for rendezvous with Charlie Brown.

The root cause of the issue was traced back to a switch governing the abort guidance system mode. Both crew members inadvertently operated the switch, inadvertently returning it to its original position instead of moving it as intended. This mistake had potentially grave consequences; firing Snoopy in the wrong direction could have led to missing the rendezvous with Charlie Brown or, worse, a collision with the lunar surface.

Stafford stabilized the LM ascent stage after approximately eight seconds of intense effort. The crew then fired the ascent engine at the LM's lowest orbital point, mirroring the crucial orbital insertion maneuvers planned for future lunar landings. Snoopy followed this trajectory for about an hour before initiating another engine burn to refine its approach toward Charlie Brown further.

Snoopy successfully rendezvoused and re-docked with Charlie Brown at 106:22:02, concluding the mission just under eight hours after its initial undocking. The historic docking was broadcast in color from the command and service module (CSM), marking a significant moment in history of space exploration.

Once Cernan and Stafford returned to Charlie Brown, Snoopy was sealed off and separated from the command module. The remaining fuel in the lunar module's ascent-stage engine was then expended to propel Snoopy on a trajectory that took it past the Moon and into a heliocentric orbit around the Sun.

This fate was unique to Snoopy among the Apollo lunar modules. For subsequent missions, such as Apollo 11, the ascent stages were intentionally left in lunar orbit to crash into the Moon's surface eventually. Post-Apollo 11 missions used ascent stages to gather data from seismometers placed on the lunar surface, except for Apollo 13's ascent stage, which was repurposed as a lifeboat for the crew's safe return to Earth before being released to burn up in the Earth's atmosphere. Unfortunately, Apollo 16's ascent stage was lost after NASA lost control of it following its jettison.

Snoopy's journey into a heliocentric orbit marked the end of its operational life, but it contributed uniquely to our understanding of lunar exploration and space travel during the Apollo era.

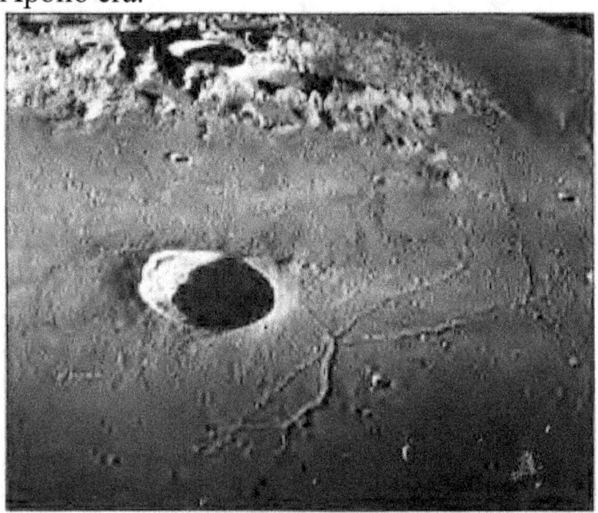

Northwest view of Triesnecher crater with associated

Following stationkeeping, a crucial 27.4-second burn of the lunar module's descent propulsion system occurred at 099:46:01.6. This maneuver precisely positioned the LM into a descent orbit measuring 60.9 by 8.5 nautical miles. The orbit's lowest point was strategically placed approximately 15 degrees ahead of lunar landing site 2.

During this phase, the crew diligently captured numerous photographs of the lunar surface despite encountering some camera malfunctions and intermittent communications issues. Throughout, they maintained a continuous commentary on their observations, ensuring a detailed record of their findings.

Approximately an hour later, the LM conducted a low-level pass over Apollo landing site 2, marking a pivotal moment in the mission. This pass included a critical test of the landing radar system, visual assessments of lunar lighting conditions, stereoscopic strip photography, and executing a phasing maneuver using the descent engine. The LM's trajectory dipped to its lowest measured point, reaching 47,400 feet (7.8 nautical miles) above the lunar surface precisely at 100:41:43.

The second critical maneuver for the lunar module (LM) occurred at 100:58:25.93, involving a 39.9-second burn of the descent propulsion system for phasing. This burn was meticulously timed to establish a lead angle equivalent to that encountered during a lunar landing mission's powered ascent cutoff phase, positioning the LM into a specific orbit measuring 190.1 by 12.1 nautical miles.

Later, at 102:44:49, as preparations for rendezvous with the command and service module (CSM) were underway, the LM began a slow yaw motion, followed shortly by a rapid roll with minor pitch and yaw adjustments at 102:45:12. This unexpected movement prompted the separation of the ascent stage from the descent stage at 102:45:17, occurring at an altitude of 31.4 nautical miles. The motion was swiftly brought under control just eight seconds later.

Subsequently, a critical 15.55-second firing of the ascent engine at 102:55:02.13 propelled the ascent stage into a new orbit measuring 46.5 by 11.0 nautical miles. Meanwhile, the descent stage entered lunar orbit, marking the conclusion of its operational role in the mission.

The investigation into the lunar module's (LM) unexpected motion revealed that human error was to blame. It was discovered that the LM's abort guidance system's control mode had been left in AUTO instead of the required ATTITUDE HOLD mode during the staging maneuver. In AUTO mode, the abort guidance system autonomously directed the LM to acquire the command and service module (CSM), contrary to the planned attitude timeline. Recognizing the deviation, the commander swiftly assumed manual control to correct the LM's orientation.

Despite this initial setback, the critical insertion maneuver using the LM ascent propulsion system was executed precisely at the orbital low point as scheduled. This burn established an orbit typical of a standard LM insertion for a lunar landing mission, measuring 45 by 11.2 nautical miles. Following the maneuver, the LM coasted in this orbit for approximately one hour.

As the mission progressed into darkness, the terminal maneuver was initiated around the midpoint of this phase. The finalization of the terminal phase, including braking maneuvers, was conducted manually as per the planned procedures. These adjustments ensured the mission remained on track despite the earlier operational challenges, demonstrating the crew's ability to adapt and respond effectively to unexpected situations in the demanding environment of lunar exploration.

The rendezvous simulation replicated a scenario resembling a typical return from the lunar surface. It commenced with a precise 27.3-second LM coelliptic sequence initiation maneuver at 103:45:55.3. This maneuver adjusted the spacecraft's trajectory, placing it into an orbit measuring 48.7 by 40.7 nautical miles.

Docking between the lunar module (LM) and the command and service module (CSM) was completed at 106:22:02, occurring at 54.7 nautical miles after 8 hours, 10 minutes, and 5 seconds of lunar flight. Upon docking, the LM crew transferred the exposed film packets from the LM to the CM, ensuring the safe preservation of mission data.

Subsequently, the LM ascent stage was jettisoned at 108:24:36. To adjust its orbit, a 6.3-second separation maneuver was executed at 108:43:23.3, raising the orbit parameters to 64.0 by 56.3 nautical miles. Following this adjustment, a significant 249.0-second firing of the ascent engine was initiated remotely, lasting until depletion at 108:52:05.5.

Photograph of ALS-2 taken by Apollo 10

Following this initial adjustment, a brief 1.65-second constant differential height maneuver was performed at 104:43:53.29. This maneuver specifically raised the orbit's perigee to 42.1 nautical miles, further refining the spacecraft's path.

The final phase of the rendezvous simulation involved a critical 16.50-second terminal phase initiation maneuver at 105:22:55.28. This maneuver effectively raised the orbit's parameters to 58.3 by 46.8 nautical miles, ensuring the spacecraft was optimally positioned for the subsequent rendezvous and docking maneuvers with the command and service module (CSM).

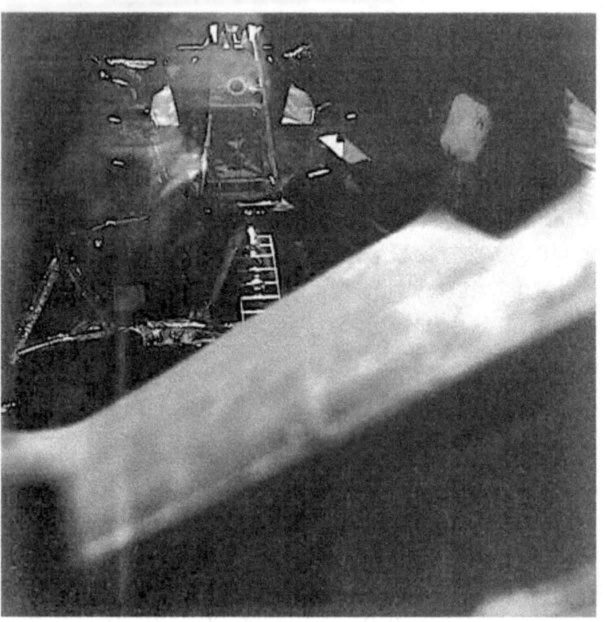

LM Snoopy containing Stafford and Cernan, as inspected by Young after separation from Charlie Brown

Approximately one orbit after docking, the planned depletion burn of the LM ascent propulsion system commenced as scheduled, utilizing the LM ascent engine arming assembly. This burn was specifically designed to place the LM into a solar orbit, away from lunar influence.

Communications with the LM ascent stage was maintained until its batteries were depleted around 120:00, approximately 12 hours after its jettison. This final phase marked the conclusion of the LM's operational role in the mission.

After jettisoning the LM ascent stage, the crew of Apollo 10 proceeded with planned activities, including rest periods, photography sessions, and lunar surface observations from orbit. Despite crew fatigue, they successfully located and photographed 18 landmarks on the lunar surface, capturing various features for detailed analysis.

Due to the mission's and crew's well-being demands, two scheduled television broadcasts had to be canceled. Subsequently, the main Service Propulsion System (SPS) engine of the command and service module (CSM) was reignited for approximately 2.5 minutes at 137:39:13.7. This maneuver was crucial in setting Apollo 10 on a trajectory back toward Earth, effectively initiating its return journey.

Before the crucial maneuver known as transearth injection, the Apollo spacecraft provided Earth with a breathtaking view of the lunar surface and its interior. For 24 minutes and 12 seconds, starting precisely at 132 hours, 7 minutes, and 12 seconds into the mission, these transmissions conveyed the stark, rugged landscape of the Moon, juxtaposed against the controlled environment within the spacecraft.

Following this transmission, the astronauts rested well, preparing for the meticulous tasks ahead. Their next objective was landmark tracking and photography, crucial exercises that demanded precision and focus. Throughout their remaining time in lunar orbit, they meticulously recorded 18 distinct landmark sightings, capturing both stereo and oblique photographs that would later serve as invaluable data for lunar exploration and research.

Despite their rigorous schedule, which included extended periods of intense focus and physical exertion, the crew faced challenges. Two scheduled television broadcasts had to be canceled due to the onset of crew fatigue, highlighting the demanding nature of their mission and the resilience required to navigate the complexities of space travel.

Transearth injection, a pivotal moment in the mission, was successfully executed at 137 hours, 39 minutes, and 13.7 seconds into the journey. Achieved at an impressive velocity of 8,987.2 feet per second, this critical maneuver commenced with a precisely timed 164.8-second engine firing. Executed at 56.0 nautical miles, it marked the conclusion of their lunar orbital phase, which spanned 31 orbits over 61 hours, 37 minutes, and 23.6 seconds.

During their journey homeward bound, the Apollo crew engaged in several observational tasks critical to their safe return. They meticulously conducted star-Earth horizon sightings, using these celestial landmarks to navigate their spacecraft through the vast expanse of space.

Among their scheduled activities was a pivotal test assessing the effectiveness of the Command Module's high-gain antenna, crucial for maintaining communication with mission control on Earth. This test was complemented by a series of six television broadcasts, varying in duration, which provided captivating views from within the spacecraft and stunning perspectives of both Earth and the Moon as seen from the astronauts' unique vantage point.

In a notable moment of personal care and innovation, Eugene Cernan and his fellow crewmates achieved a historic first: they successfully became the inaugural astronauts to shave in space. Utilizing a safety razor and thick shaving gel—a departure from earlier mission protocols that prohibited such items due to safety concerns—they demonstrated adaptability and resourcefulness in the face of the unknown challenges of space travel.

As their journey progressed, the crew's responsibilities included a critical mid-course correction burn of the Command Module's engine. This precise maneuver, executed at 188 hours, 49 minutes, and 58 seconds into the mission, lasted approximately 6.7 seconds. It was the sole adjustment needed to fine-tune their trajectory before the later separation of the Command Module from the Service Module, marking another milestone in their voyage back to Earth.

As the Apollo 10 spacecraft hurtled toward Earth on its final approach, the crew achieved record-breaking speeds unmatched by any humans before or since, relative to our planet: a staggering 39,897 kilometers per hour (11.08 kilometers per second or 24,791 miles per hour). This velocity was necessitated by a streamlined return trajectory designed to shorten the journey to just 42 hours, significantly quicker than the usual 56-hour return for previous missions.

Not only did the Apollo 10 crew travel faster than any other humans relative to Earth, but they also journeyed farther from their homes in Houston than any astronauts before, reaching 408,950 kilometers (220,820 nautical miles). Although surpassed by the Apollo 13 crew by a mere 200 kilometers overall, this distance remains unparalleled in terms of the distance traveled away from Earth.

During their return journey, the crew continued to execute critical activities essential for their safe return. They conducted star-Earth horizon navigation sightings and tested the reflectivity of the Command Module's S-band high-gain antenna, ensuring clear communication with mission control. They also employed passive thermal control techniques and navigation procedures similar to those used during their outbound journey toward the Moon.

In a testament to their precision and the meticulous planning of mission controllers, the crew required only one midcourse correction burn during their return trip. This brief maneuver, lasting just 6.7 seconds and adjusting their velocity by 2.2 feet per second, occurred approximately three hours before the planned separation of the Command Module from the Service Module.

Necho crater on the far side of the Moon

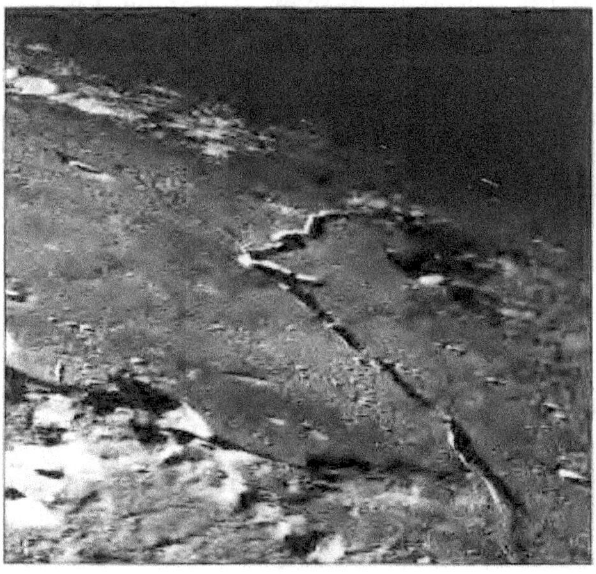

Photograph of the lunar nearside; crater Hyginus, near Central Bay, seen from the CM (NASA ASI0-31-4650).

As they rounded the far side of the Moon during their lunar orbit, they reached their farthest point from Earth just as Houston rotated nearly a full diameter away due to the Earth's own rotation—an intricate dance of orbital mechanics and precise timing that characterized Apollo 10's remarkable achievements.

Apollo landing site #3. Crater Bruce is seen at the bottom right (NASA AS10-27-3907)

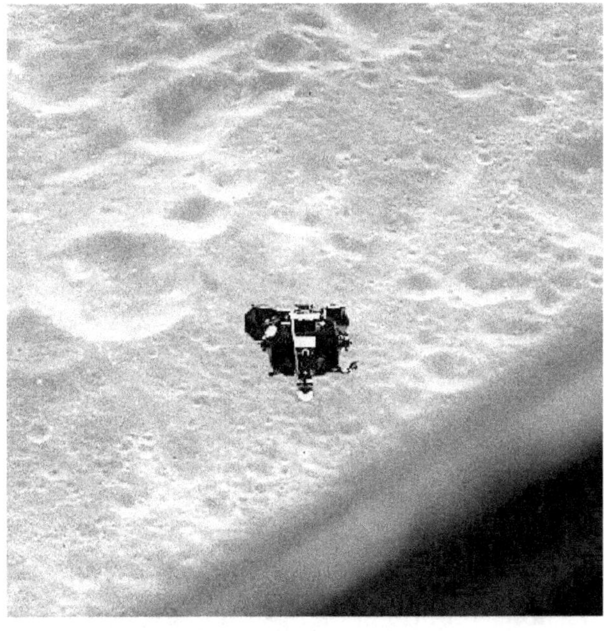

LM ascent stage prior to docking with the CM

At 191 hours, 33 minutes, and 26 seconds into their historic journey, the Command Module (CM) of Apollo 10, containing the intrepid crew, safely separated from the Service Module (SM) in preparation for their reentry into Earth's atmosphere. Approximately 15 minutes later, precisely at 191 hours, 48 minutes, and 54.5 seconds, reentry occurred, marking the beginning of their descent back to the planet's surface.

Splashdown of the Command Module took place about 15 minutes after reentry, precisely at 16:52:23 UTC on May 26, 1969. The splashdown occurred in the Pacific Ocean approximately 740 kilometers (400 nautical miles) east of American Samoa, concluding their mission at 192 hours, 3 minutes, and 23 seconds. Their safe return was facilitated by the USS Princeton, which swiftly recovered the astronauts from the sea.

Onboard the USS Princeton, the crew spent approximately four hours receiving a congratulatory phone call from President Richard Nixon—an honor recognizing their remarkable achievement and the culmination of their journey. Unlike the crews of later Apollo missions that touched down on the lunar surface, the Apollo 10 astronauts were not required to undergo quarantine upon their return.

Following their recovery, the astronauts were flown to Pago Pago International Airport in Tafuna for a warm reception. From there, they boarded a C-141 cargo plane bound for Ellington Air Force Base near Houston, where they were welcomed home as heroes.

Throughout their return journey, the crew shared their experiences with the world through six television transmissions, showcasing breathtaking views of both the Moon and Earth from their unique perspective in space. These broadcasts captured pivotal moments such as views after transearth injection, receding images of the Moon, and serene glimpses of Earth against the backdrop of their spacecraft's interior.

The Service Module was jettisoned at 191 hours, 33 minutes, and 26 seconds, followed by a controlled entry profile into Earth's atmosphere. Reentry occurred at an altitude of 400,000 feet, with the Command Module reaching speeds of 36,314 feet per second. This followed a tranquil transearth coast lasting 54 hours, 3 minutes, and 40.9 seconds—an awe-inspiring conclusion to

Apollo 10's daring voyage into the unknown depths of space.

At 16:52:23 GMT (11:52:23 p.m. EDT) on May 26, the Command Module (CM) of Apollo 10 gracefully splashed down in the Pacific Ocean, marking the successful conclusion of its epic journey. The mission had spanned a total duration of 192 hours, 3 minutes, and 23 seconds.

Apollo 10 CM on parachutes prior to splashdown

The CM touched down approximately 1.3 nautical miles from its intended target point and 2.9 nautical miles from its recovery ship, the U.S.S. Princeton. Upon landing, the CM assumed a stable apex-up flotation attitude designed to facilitate safe recovery operations.

Just 39 minutes after splashdown, the crew was swiftly retrieved by helicopter and brought aboard the waiting recovery ship. The Command Module itself was recovered only 57 minutes later, ensuring the safe retrieval of all mission data and equipment.

At the moment of splashdown, the Command Module weighed an estimated 10,901 pounds. Throughout its journey, Apollo 10 traveled an impressive estimated distance of 721,250 nautical miles, a testament to the crew's endurance and the precision of mission planning. Following the successful retrieval of the Command Module (CM) by the USS Princeton, the weather conditions recorded onboard showed a relatively clear sky with 10% cloud cover at 2,000 feet and 20% at 7,000 feet. Visibility was excellent at ten nautical miles, with a gentle breeze of five knots from the north-northeast. The air temperature was not specified, but the water temperature was a warm 85°F, and waves were recorded at up to three feet high.

On May 31, the CM was offloaded from the USS Princeton at Ford Island, Hawaii. Shortly after that, at 18:00 GMT, the Landing Safing Team commenced evaluation and deactivation procedures. These meticulous tasks were completed with precision by 05:56 GMT on June 3.

The Command Module was transported to Long Beach, California, at 10:15 GMT on June 4. From there, it was promptly trucked to the North American Rockwell Space Division facility in Downey, California, where postflight analysis and detailed examination of all systems were conducted.

Despite encountering some minor issues during the mission, none of these posed significant constraints on achieving the mission's objectives. Notably, valuable data on lunar gravitation were gathered during the 61 hours spent in lunar orbit, contributing significantly to our understanding of celestial mechanics.

It is noteworthy that Apollo 10 holds a place in the Guinness Book of World Records for the fastest speed ever attained by humans, recorded at 24,791 statute miles per hour at an altitude of 400,000 feet during entry on May 26, 1969. However, the official Apollo 10 mission report specifies an even higher maximum speed at entry: 36,397 feet per second, equivalent to approximately 24,816 statute miles per hour. This distinction underscores the extraordinary achievements and technological advancements made during the Apollo program, setting the stage for future space explorations.

Apollo 10 CM Astronaut recovery

Conclusions drawn from the analysis of post-mission data from Apollo 10 underscored several key achievements and insights pivotal to subsequent lunar landing missions:

The systems aboard the Command and Service Modules (CSM) and the Lunar Module (LM) were fully operational and capable of supporting a piloted lunar landing mission. This validation was crucial as it confirmed the spacecraft's technological readiness for the demanding lunar surface operations.

The crew activity timeline, particularly in tasks aligning with the lunar landing profile, demonstrated that critical activities such as LM checkout, initial descent, and rendezvous maneuvers were feasible. Importantly, these tasks were achievable without imposing unreasonable workload burdens on the astronauts, highlighting the effectiveness of mission planning and crew training.

Post-splashdown crew recovery operations in the Pacific Ocean

Despite encountering minor challenges, the Apollo 10 mission successfully achieved its detailed test objectives. Notably, the primary exception was the LM steerable antenna and relay modes for voice and telemetry communications, which remained unresolved but did not impede the overall mission success.

Further conclusions drawn from the Apollo 10 mission data highlighted critical advancements and capabilities that paved the way for subsequent lunar missions:

The S-band communications capability of the Lunar Module (LM), utilizing both the steerable and omnidirectional antennas, was deemed satisfactory even at lunar distances. This capability was crucial for maintaining reliable communications between the LM and mission control during critical phases of lunar operations.

The landing radar's performance in the lunar environment, particularly during descent propulsion firings, was effective and reliable for altitudes experienced during lunar descent. This radar played a vital role in providing essential altitude data to the LM crew during the crucial final stages of landing.

The Lunar Module's rendezvous radar successfully demonstrated its range capability in the lunar environment, yielding excellent results. Additionally, the innovative use of VHF ranging information from the Command Module (CM) provided consistent correlation with radar range and range-rate data, enhancing precision and reliability during rendezvous maneuvers.

The lunar module's abort guidance system exhibited robust capability in controlling ascent propulsion maneuvers and guiding spacecraft during rendezvous operations. This system's reliability was crucial for ensuring crew safety and mission success in contingency scenarios.

The Mission Control Center and the Manned Space Flight Network effectively controlled and monitored both CM and LM vehicles at lunar distances during descent and rendezvous operations. This capability was crucial for managing complex mission dynamics and ensuring real-time support for astronauts navigating the lunar environment.

Significant advancements were made in refining the lunar potential model compared to previous missions like Apollo 8. Orbit determination and prediction procedures were notably more precise for both spacecraft in lunar orbit. The combined analysis of trajectory reconstructions from Apollo 8 and 10 enhanced the lunar potential model, providing confidence in its adequacy to support future lunar descent and ascent operations.

Apollo 10's mission objectives were meticulously designed to advance the United States' lunar exploration capabilities, building crucial steps toward the ultimate goal of landing astronauts on the Moon—the mission aimed to achieve a series of critical milestones across multiple spacecraft and launch vehicle systems.

The primary objectives for the spacecraft included demonstrating the performance of crew, space vehicle, and mission support facilities during a piloted lunar mission. This encompassed the command and service module's interaction with the lunar module, evaluating their functionality in both lunar and cislunar environments. Key tasks included executing undocked descent orbit insertions, conducting high-thrust maneuvers, and testing the lunar module's systems such as the landing radar and guidance controls.

Throughout the mission, secondary objectives were equally vital, covering a wide range of systems and procedures. These included tests of the spacecraft's navigation and guidance systems, communications capabilities using different antennas, and assessments of thermal control and propulsion systems under various operational conditions. Notably, the mission also sought to gather data on crew procedures during lunar orbit operations and to simulate critical maneuvers like rendezvous and lunar landing scenarios.

The launch vehicle objectives were equally ambitious, focusing on verifying the capabilities of the S-IVB stage for trajectory injections, maintaining specified attitudes for docking maneuvers, and safely disposing of propellants after use. Engine performance, including modifications to enhance operational reliability, was rigorously tested across stages to ensure they met the demanding requirements of lunar missions.

Apollo 11

The Fifth Crewed Mission

On July 16, 1969, NASA launched Apollo 11, a historic crewed lunar landing mission that captivated the world.

Apollo 11 was a Type G mission, a piloted lunar landing demonstration. The primary objective of the Apollo program was to perform a piloted lunar landing and return safely to Earth. It was only the second time an all-experienced crew had flown an American mission, and it would be the last until Space Shuttle mission STS-26 nearly two decades later.

Commanded by Neil A. Armstrong, with Michael Collins as Command Module Pilot and Edwin E. Aldrin Jr. as Lunar Module Pilot, Apollo 11 aimed to achieve humanity's first landing on the Moon.

The spacecraft consisted of Apollo CSM-107, manufactured by North American Rockwell, and Apollo LM-5, built by Grumman. With a launch mass of 109,646 pounds, Apollo 11 blasted off from Kennedy Space Center's Launch Complex 39A atop the mighty Saturn V SA-506 rocket.

The CSM was nicknamed "Columbia," while the LM was famously called "Eagle." After a journey of approximately 240,000 miles, Apollo 11 entered lunar orbit on July 19, 1969. Armstrong and Aldrin then boarded the LM "Eagle" to descend to the lunar surface while Collins remained in orbit aboard the CSM "Columbia."

On July 20, 1969, at 20:17:40 UTC, Neil Armstrong became the first human to set foot on the Moon, uttering the iconic words, "That's one small step for [a] man, one giant leap for mankind." Buzz Aldrin shortly followed him, and together they conducted a historic Extra-Vehicular Activity (EVA) lasting 2 hours, 31 minutes, and 40 seconds.

The landing site, Tranquility Base, was located in the Mare Tranquillitatis (Sea of Tranquility) region on the lunar surface at coordinates 0.67416°N 23.47314°E. They collected 21.55 kilograms of lunar samples before returning to the LM ascent stage.

After spending approximately 21 hours on the lunar surface, the LM ascent stage successfully redocked with the CSM on July 21, 1969. The crew then began their journey back to Earth, splashing down in the North Pacific Ocean on July 24, 1969, and being recovered by the USS Hornet, having flown 952,354 miles.

Apollo 11's successful mission marked a monumental achievement in human exploration, fulfilling President John F. Kennedy's vision and setting the stage for future lunar missions and space exploration endeavors.

In November 1967, the crew assignments for Apollo 9 were set, marking a pivotal moment in the Apollo program. Commander Neil Armstrong, alongside Command Module Pilot (CMP) Jim Lovell and Lunar Module Pilot (LMP) Buzz Aldrin, were named as the backup crew. This decision followed a series of adjustments driven by technical delays in the Lunar Module (LM), prompting Apollo 8 and Apollo 9 to swap their primary and backup crews. Consequently, Armstrong's crew transitioned to backup status for Apollo 8, setting the stage for Armstrong's anticipated command of Apollo 11.

Apollo 11 crew (1. to r.): Neil Armstrong, Mike Collins, Buzz Aldrin (NASA S69-31740).

A significant personnel shift occurred due to health concerns within the astronaut corps. Michael Collins, slated initially as CMP for Apollo 8, faced medical issues necessitating surgery. Lovell stepped in, temporarily joining the Apollo 8 crew while Collins recuperated. Upon Collins' recovery, he rejoined Armstrong's team as CMP. Meanwhile, Fred Haise filled in as backup LMP, with Aldrin assuming backup CMP duties for Apollo 8.

Apollo 11 marked the second American mission, in which all crew members boasted prior spaceflight experience, a milestone first achieved by Apollo 10. This continuity of experience underscored NASA's strategy to leverage seasoned astronauts for critical missions.

During crew assignment deliberations, Deke Slayton presented Armstrong with the option to replace Aldrin with Lovell, citing interpersonal dynamics. Despite considerations, Armstrong chose to retain Aldrin, emphasizing his capability and the importance of continuity within the team. This decision reflected Armstrong's belief in maintaining stable crew relationships, crucial for the success of high-stakes missions like Apollo 11.

In contrast to the close-knit camaraderie of some Apollo crews, the Apollo 11 prime crew cultivated a professional and amiable working relationship. Armstrong, known for his reserved demeanor, and Collins, who preferred independence, described their dynamic with Aldrin as that of "amiable strangers." This characterization, however, was not universally shared, as Armstrong affirmed the effective collaboration within all his crews.

In the annals of space exploration, few missions rival the historic significance of Apollo 11, commanded by Neil Alden Armstrong. Born in Wapakoneta, Ohio, Armstrong's journey to the moon exemplified both technological prowess and human bravery. Alongside him was Lieutenant Colonel Michael Collins, serving as Command Module Pilot, and Colonel Edwin Eugene "Buzz" Aldrin, Jr., as Lunar Module Pilot.

Armstrong, a decorated pilot selected as an astronaut in 1962, had already made history commanding Gemini 8, achieving the first-ever orbital docking. His academic background in aeronautical and aerospace engineering from Purdue University and USC highlighted his technical expertise. At 38 years old during Apollo 11, Armstrong stood as a stalwart leader in the realm of human space exploration.

Collins, born in Rome, Italy, brought military discipline and Gemini 10 experience to Apollo 11. A West Point graduate and seasoned astronaut, Collins commanded the Command Module Columbia, orbiting the moon solo while his colleagues descended to the lunar surface.

Aldrin, born in Montclair, New Jersey, distinguished himself as the first astronaut with a doctorate to fly in space. His background in engineering and astronautics, coupled with prior Gemini 12 experience, complemented Armstrong's command of the mission.

The mission support structure for Apollo 11 was extensive and meticulously organized, reflecting NASA's commitment to mission success. Serving as capsule communicators (CAPCOMs) were a distinguished group including Major Charles Moss Duke, Jr. (USAF), Lt. Commander Ronald Ellwin Evans (USN), Lt. Commander Bruce McCandless II (USN), and others, each bringing crucial expertise from their military and scientific backgrounds.

The support crew featured notable figures such as Major William Reid Pogue (USAF) and John Leonard "Jack" Swigert, Jr., ready to step in if needed. Flight directors, responsible for orchestrating mission operations, included Clifford E. Charlesworth, Gerald D. Griffin, Eugene F. Kranz, and Glynn S. Lunney, overseeing operations across shifts to ensure round-the-clock coverage.

In the event of contingencies, a backup crew was on standby, with Jim Lovell poised to command, joined by William Anders as CMP and Fred Haise as LMP. Having previously flown with Lovell on Apollo 8, Anders was initially slated for parallel training as backup CMP with Ken Mattingly. However, Anders' impending departure from NASA necessitated Mattingly's readiness as a backup, anticipating any delays beyond Apollo 11's planned July launch.

During the Apollo program, the crew rotations saw Lovell, Mattingly, and Haise initially scheduled for Apollo 14 but reassigned to Apollo 13 due to crew dynamics and training requirements. Mattingly's role as CMP was later assumed by Jack Swigert for Apollo 13 following George Mueller's decision to reject the initial crew selection—a first in the Apollo program's history.

During the Apollo missions, NASA introduced a third crew category, the support crew, a pivotal addition to mission operations. Unlike the prime and backup crews, the support crew, comprising astronauts like Ken Mattingly, Ronald Evans, and Bill Pogue for Apollo 11, played a crucial role in maintaining mission readiness. They meticulously managed flight plans, checklists, and mission protocols, ensuring the prime and backup crews were well-informed of any updates or changes. Their responsibilities extended to developing and refining emergency procedures, crucial for simulator training sessions where crews focused on honing their skills and familiarizing themselves with mission-critical tasks.

At the heart of mission communication was the capsule communicator (CAPCOM), an astronaut stationed at NASA's Mission Control Center in Houston, Texas. The CAPCOM was the direct link between ground control and the flight crew, relaying vital instructions, updates, and information during all mission phases. For Apollo 11, the CAPCOM team included a distinguished lineup of astronauts such as Charles Duke, Ronald Evans, Bruce McCandless II, James Lovell, William Anders, Ken Mattingly, Fred Haise, Don L. Lind, Owen K. Garriott, and Harrison Schmitt. Their role was instrumental in ensuring smooth communication and support for the historic lunar landing mission.

Apollo 11's success hinged on the expertise and dedication of a diverse group of key personnel, each contributing essential skills and knowledge to the mission.

Flight Directors: The flight directors for Apollo 11 were pivotal in guiding the mission through its critical phases:

Clifford E. Charlesworth (Green Team) oversaw the launch and extravehicular activity (EVA).

Gerald D. Griffin (Gold Team) served as the backup for the Green Team's shift.

Gene Kranz (White Team) took charge during the crucial lunar landing.

Glynn Lunney (Black Team) managed the lunar ascent phase.

Milton Windler (Maroon Team) focused on mission planning.

Other Key Personnel: Among the many crucial contributors were:

Farouk El-Baz, a geologist who played a vital role in studying the Moon's geology, identifying landing sites, and training pilots.

Kurt Debus, a rocket scientist oversaw the construction of launch pads and critical launch infrastructure.

Jamye Flowers provided administrative support as the astronauts' secretary.

Eleanor Foraker, a tailor responsible for designing and developing space suits tailored for the mission's needs.

Jack Garman, a computer engineer and technician who played a critical role in ensuring the onboard computer systems operated flawlessly.

Millicent Goldschmidt, a microbiologist who designed techniques for aseptic lunar material collection and trained astronauts in their implementation.

Eldon C. Hall designed crucial hardware for the Apollo Guidance Computer, which was essential for navigation and control during the mission.

Margaret Hamilton, a software engineer who developed and implemented software for the onboard flight computer, ensuring critical systems operated smoothly.

John Houbolt, a key figure in planning the trajectory and route to the Moon, ensuring precise navigation for the spacecraft.

Gene Shoemaker was a geologist who trained astronauts in field geology, preparing them for lunar surface activities.

Bill Tindall coordinated mission techniques and protocols, ensuring seamless integration of procedures throughout the mission.

Together, these individuals formed a multidisciplinary team that exemplified NASA's collaborative spirit and commitment to excellence, ultimately achieving one of humanity's greatest accomplishments—the successful landing of Apollo 11 on the Moon.

The emblem of the Apollo 11 mission, designed by Michael Collins, aimed to symbolize the peaceful lunar landing by the United States. Collins, with input from Jim Lovell and others, chose the bald eagle as the central motif, representing the national bird of the United States. To emphasize the mission's peaceful intent, Tom Wilson suggested adding an olive branch in the eagle's beak. Collins further enhanced the design by incorporating a lunar background with Earth in the distance, though the initial artwork had some directional issues with sunlight and shadows.

The team decided on natural colors for the eagle and moon, surrounded by a blue and gold border. Concerned about international understanding, Neil Armstrong proposed using "Apollo 11" instead of just "Eleven" on the emblem. They also opted not to include their names, aiming to symbolically represent all those who contributed to the lunar landing effort.

The artwork was crafted by an illustrator at the Manned Spacecraft Center (MSC) and submitted to NASA for approval. Initially, the design faced rejection due to concerns that the eagle's talons appeared too aggressive. Following discussions, the olive branch was moved to the eagle's talons, aligning with the mission's peaceful theme.

The emblem's enduring design influence extended beyond Apollo 11. It notably inspired the reverse side of the Eisenhower dollar coin released in 1971 and later influenced the design of the smaller Susan B. Anthony dollar introduced in 1979. Thus, the emblem symbolized a historic achievement in space exploration and became a lasting cultural icon, representing the spirit of discovery and cooperation.

The call signs and names chosen for the spacecraft of Apollo 11 reflected both practical considerations and historical inspirations, as well as a nod to literary and symbolic references.

Initially, during early mission planning, the Command Module (CM) was designated as "Snowcone" and the Lunar Module (LM) as "Haystack" for internal and external communications. However, following a suggestion from Julian Scheer, assistant manager for public affairs, these names were eventually replaced with more symbolic and meaningful choices.

The Lunar Module (LM) was named "Eagle," inspired by its prominent depiction on the mission insignia. This choice symbolized strength, grace, and the pioneering spirit associated with the mission's lunar landing.

The Command Module (CM) received the name "Columbia," proposed by Scheer. This name drew its inspiration from multiple sources, including the giant cannon "Columbiad" featured in Jules Verne's novel "From the Earth to the Moon," which launched a spacecraft from Florida—a clear connection to NASA's launch site. Additionally, "Columbia" carries historical significance as a poetic and historical name for the United States, often associated with exploration and discovery. Michael Collins later elaborated in his 1976 book, suggesting it was also a reference to Christopher Columbus, further highlighting the exploration theme.

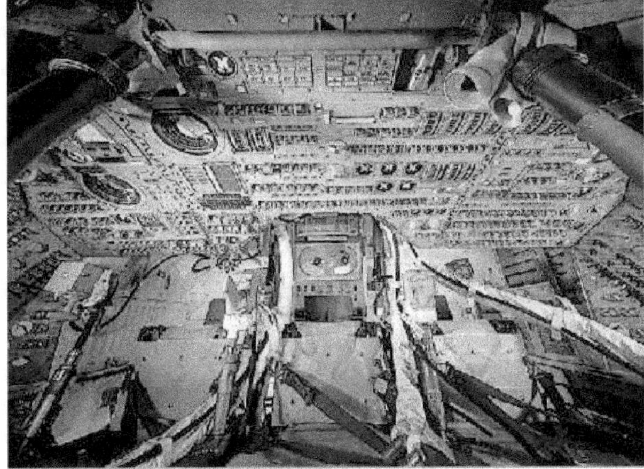

Original cockpit of the command module (CM) with three seats,

The site selection process for Apollo 11 involved careful evaluation and consideration of several criteria to ensure a safe and successful lunar landing. Initiated by NASA's Apollo Site Selection Board and informed by data from the Lunar Orbiter and Surveyor programs, the selection aimed to identify the most optimal landing site on the Moon.

Five potential landing sites were initially identified in February 1968 based on extensive photographic data from lunar orbiters and surface condition assessments by the Surveyor program. These sites were meticulously analyzed to meet stringent requirements: proximity to the lunar equator to minimize propellant use, no major obstacles to simplify landing maneuvers, and a relatively smooth terrain with minimal cratering.

Despite challenges like the need to relax the original requirement for a crater-free site, the selection criteria ultimately focused on ensuring operational safety and ease for the lunar module's landing. Factors considered included the absence of large hills, tall cliffs, or deep craters that could interfere with landing radar readings, the trajectory allowing for a free return to Earth in an emergency, and optimal lighting conditions during descent.

Map of Moon showing prospective sites for Apollo 11. Site 2 was chosen.

Site 2, located within the Sea of Tranquility (Mare Tranquillitatis), emerged as the primary choice due to its favorable terrain characteristics and acceptable conditions reported by Apollo 10's lunar module, which conducted a close flyby in May 1969. Sites 3 and 5 were designated as backup options should any delays affect the mission's launch schedule.

The decision regarding who would be the first to step onto the lunar surface during the Apollo 11 mission sparked considerable discussion and speculation in early 1969. Initially, during press conferences after the crew announcement, when asked about who would take the first step, Deke Slayton, director of flight crew operations, indicated that the decision had yet to be made. Neil Armstrong added that the choice was not based on individual preference.

In the early planning stages, the lunar module pilot (LMP) was slated to be the first to exit the spacecraft following the precedent set by Gemini missions. This approach aligned with previous missions where commanders had not conducted spacewalks. Reports circulating in early 1969 suggested that Buzz Aldrin, as the LMP for Apollo 11, would be the first to walk on the Moon. Even Associate Administrator George Mueller reportedly indicated Aldrin would have this honor.

However, tensions arose when Aldrin learned of rumors suggesting Armstrong, as the mission commander and a civilian astronaut, might be chosen instead. Feeling strongly about his role and qualifications, Aldrin attempted to rally support among other LMPs but encountered skepticism and perceived lobbying within the astronaut corps.

To resolve the issue and maintain mission cohesion, Deke Slayton ultimately clarified that Neil Armstrong would be the first to step onto the lunar surface. This decision was officially announced during a press conference on April 14, 1969, to quell any potential interdepartmental conflicts and reinforce the chain of command aboard Apollo 11.

The decision on who would be the first person to step onto the lunar surface during the Apollo 11 mission involved practical considerations related to the lunar module's hatch location and the astronauts' ability to maneuver in their bulky spacesuits. Buzz Aldrin believed for many years that these factors heavily influenced the final choice. During a simulation where Aldrin attempted to exit the spacecraft first, he encountered difficulties and inadvertently damaged the simulator. This incident highlighted the challenges of maneuvering in the confined space of the lunar module.

Despite this, the final decision was communicated to Aldrin and Neil Armstrong in late spring. After consulting with Armstrong, Deke Slayton informed him of the plan to have Armstrong exit the spacecraft first. Armstrong agreed with the decision, viewing it as the most practical approach.
However, the media and some insiders later accused Armstrong of using his authority as mission commander to assert his position as the first person on the Moon. In his autobiography, Chris Kraft revealed a deliberate meeting among senior NASA officials to ensure Armstrong would be the first to walk on the Moon. They argued that the first lunar walker should embody qualities similar to Charles Lindbergh—calm and composed—and decided to alter the flight plan accordingly.

Saturn V SA-506, the rocket carrying the Apollo 11 spacecraft, moves out of the Vehicle Assembly Building towards Launch Complex 39.

In the lead-up to launch day, the towering Saturn V SA-506 emerges from the immense

Vehicle Assembly Building, its colossal form slowly approaching Launch Complex 39 at Kennedy Space Center. This iconic rocket, bearing the weight of humanity's ambitions, symbolizes the culmination of years of meticulous planning and engineering prowess.

At Kennedy Space Center, the components of the Apollo 11 spacecraft began their journey well before launch day. The ascent stage of LM-5, famously named Eagle, arrived first on January 8, 1969. This was followed by the descent stage four days later, each meticulously checked and prepared for the lunar landing mission. Unlike its predecessor in Apollo 10, Eagle was equipped with advanced features: a VHF radio antenna crucial for direct communication with astronauts during their historic moonwalks, a lighter ascent engine for enhanced maneuverability, and increased thermal protection on the landing gear. Notably, it also carried a suite of scientific experiments known as the Early Apollo Scientific Experiments Package (EASEP), underscoring NASA's commitment to maximizing scientific returns from the mission.

CAPCOM Charles Duke (left), with backup crewmen Jim Lovell and Fred Haise listening in during Apollo 11's descent

Meanwhile, the command and service module (CSM-107), named Columbia, joined the assembly process on January 23. The spacecraft underwent meticulous checks and integration processes, ensuring every component functioned flawlessly under the extreme conditions of space. Modifications to Columbia were minimal, primarily involving adjustments such as removing insulation from the forward hatch to streamline operations during the mission.

On January 29, the CSM was successfully mated with the lunar module, marking a critical milestone in the assembly phase. The integrated spacecraft, now fully prepared for its journey to the moon, was transferred to the Vehicle Assembly Building on April 14, where final preparations and inspections would take place to ensure readiness for the historic launch.

The meticulous assembly of Saturn V AS-506, the rocket destined to carry Apollo 11 to the moon, unfolded with precision and careful planning. The journey began with the arrival of the S-IVB third stage on January 18, followed by the S-II second stage on February 6, the S-IC first stage on February 20, and finally, the Saturn V Instrument Unit on February 27. Each component, meticulously crafted and rigorously tested, represented a crucial piece of the technological marvel that would propel humanity toward its lunar destiny.

On May 20, at precisely 12:30 PM, the fully assembled 5,443-tonne (5,357-long-ton; 6,000-short-ton) rocket emerged from the imposing Vehicle Assembly Building, perched atop the crawler-transporter. Its slow, deliberate journey to Launch Pad 39A at Kennedy Space Center's Launch Complex 39 coincided with the ongoing Apollo 10 mission, underscoring NASA's relentless pace in advancing lunar exploration.

Countdown preparations intensified as a critical countdown test commenced on June 26 and culminated on July 2. The night of July 15 illuminated Launch Complex 39A as floodlights bathed the scene, marking the final preparations before launch. During the early hours, the fuel tanks of the S-II and S-IVB stages were meticulously filled with liquid hydrogen, a precise operation essential for optimal performance.

By dawn, three hours before the historic launch, the fueling process was completed flawlessly. Launch operations, guided by 43 meticulously written programs in the ATOLL programming language, showcased NASA's pioneering efforts in automation, ensuring every mission step was executed with unparalleled precision and reliability.

The final hours before meticulous preparations and a sense of anticipation marked the historic launch of Apollo 11. Deke Slayton, the veteran astronaut and crew operations director, woke the crew shortly after 04:00. After refreshing with showers and shaving, they gathered with Slayton and the backup crew for the traditional pre-flight meal of steak and eggs, a symbolic moment steeped in NASA tradition.

Fully suited in their iconic white spacesuits, the astronauts began breathing pure oxygen to purge nitrogen from their bodies, essential for adapting to the low-pressure environment of space. By 06:30, they embarked on their journey to Launch Complex 39, the culmination of years of training and preparation.

Inside the spacecraft, the sequence of final preparations unfolded with precision. Fred Haise, the backup crew member, assisted Neil Armstrong into the left-hand couch at 06:54, followed shortly by Michael Collins on the right-hand couch. Buzz Aldrin completed the trio, settling into the center couch, positioning themselves for the imminent journey beyond Earth's bounds.

As the clock ticked toward launch time, Haise departed around two hours and ten minutes before liftoff, leaving the prime crew sealed inside their spacecraft. The closeout crew meticulously sealed the hatch, ensuring the cabin was purged and pressurized to the exacting standards required for the mission's safety.

Approximately an hour before launch, the closeout crew left Launch Complex 39, leaving the astronauts to their final moments of solitude and focus. Inside the firing room, over 450 personnel monitored every aspect of the countdown, their expertise and diligence ensuring that every system and procedure was executed flawlessly.

With the countdown automated at three minutes and twenty seconds before liftoff, the culmination of decades of scientific advancement, technological innovation, and human courage stood poised for history.

The launch of Apollo 11 on July 16, 1969, captivated the world with its magnitude and historical significance. Around one million spectators gathered along highways and beaches near the Kennedy Space Center, their eyes fixed on Launch Complex 39A as humanity prepared to embark on its first journey to the moon.

Among the distinguished guests were General William Westmoreland, the Chief of Staff of the United States Army, four cabinet members, 19 state governors, 40 mayors, 60 ambassadors, and 200 congressmen. Vice President Spiro Agnew, who shared this momentous occasion with former President Lyndon B. Johnson and his wife, Lady Bird Johnson, also graced the scene.

Media presence was extensive, with approximately 3,500 representatives from around the globe, two-thirds from the United States and the remainder from 55 other countries. The launch proceedings were broadcast live in 33 countries, captivating an estimated 25 million viewers in the United States alone. Millions more tuned in via radio broadcasts worldwide, underscoring the global interest and significance of the mission.

President Richard Nixon, situated in the White House, watched the historic launch from his office alongside his NASA liaison officer, Apollo astronaut Frank Borman. From these global vantage points, the launch of Apollo 11 united humanity in awe and anticipation, marking a pivotal moment in the annals of space exploration and human achievement.

The launch of Apollo 11, propelled by the mighty Saturn V rocket designated SA-506, commenced on July 16, 1969, at precisely 13:32:00 UTC (9:32:00 EDT). This monumental mission, also known as Eastern Test Range #5307, bore the weight of human ambition and scientific endeavor.

Just 13.2 seconds into the flight, SA-506 began its controlled roll to achieve its flight azimuth of 72.058°, setting the trajectory toward the moon. The first-stage engines burned fiercely for approximately 2 minutes and 42 seconds before achieving full shutdown, marking the end of their crucial role in the launch sequence. Shortly after, the S-IC stage separated, making way for the ignition of the S-II engines.

As the mission progressed, the S-II stage powered the spacecraft higher into Earth's orbit, with its engines cutting off and separating around 9 minutes and 8 seconds after liftoff. This pivotal moment allowed the S-IVB stage to take over, igniting its engines moments later for the first time in this historic journey.

With each stage performing flawlessly, Apollo 11 and its crew, aboard the Command and Service Module CSM-107 named "Columbia" and the Lunar Module LM-5 named "Eagle," embarked on their trajectory toward the moon.

The Apollo 11 Saturn V space vehicle lifts off with astronauts Neil A. Armstrong, Michael Collins and Edwin E. Aldrin Jr. at 9:32 a.m. EDT July 16, 1969, from Kennedy Space Center's Launch Complex 39A.

NASA's Apollo Site Selection Board meticulously scrutinized potential landing sites for Apollo 11 over more than two years. Initially considering thirty locations, the board rigorously evaluated each based on a set of critical criteria to ensure a safe and successful mission to the Moon.

High-resolution photographs captured by the Lunar Orbiter satellite provided invaluable insights into the lunar surface, while close-up images and data from Surveyor spacecraft further informed the selection process. The final decision was narrowed down to three prime candidates that met stringent requirements:

Firstly, smoothness was paramount; chosen sites exhibited relatively few craters and boulders, crucial for safe spacecraft touchdown and lunar module stability.

Secondly, propellant requirements were carefully calculated. Sites were selected to minimize the spacecraft's fuel expenditure, crucial for both landing and return journeys.

Thirdly, the concept of a free return trajectory played a pivotal role. Selected sites were within reach of the spacecraft's trajectory, allowing a return to Earth without additional engine firings—a contingency crucial for mission safety.

Moreover, slope gradients played a decisive role. Landing sites were meticulously chosen with slopes of less than 2 degrees, ensuring a gentle approach and stable landing conditions for the lunar module.

Ultimately, the timing of the lunar landing was intricately tied to the specific location selected. Factors such as the Moon's position relative to Earth and the optimal lighting conditions for landing and subsequent operations governed the precise timing of Apollo 11's historic descent.

The original selection of lunar landing sites focused on regions within 45° east and west of the Moon's center and 5° north and south of its equator, all situated on the visible side of the Moon. These criteria ensured optimal visibility and communication conditions with Earth throughout the mission.

Several critical factors meticulously guided final site choices. One key consideration was ensuring a smooth approach path free of large hills, high cliffs, or deep craters that could potentially interfere with accurate altitude readings from the lunar module's landing radar. This precision was essential for a safe descent and landing.

Another vital factor was the capability for effective launch preparation recycling, crucial in case the countdown to launch was delayed. This recycling capability ensured that the mission could proceed smoothly even during unforeseen delays.

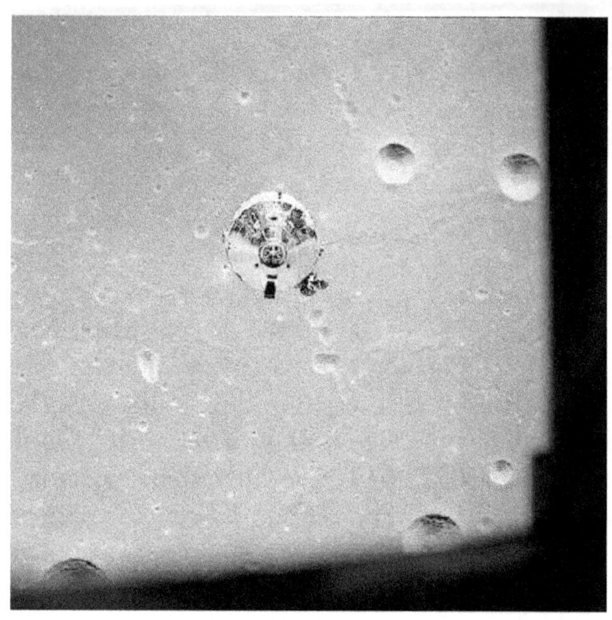

Columbia in lunar orbit, photographed from Eagle

Determining launch windows for lunar landing missions involved a complex interplay of factors. These included assessing illumination conditions both at launch and at the planned landing site on the Moon, the launch pad azimuth (the compass direction of the launch), the geometry of the translunar injection trajectory, and the angle of the sun's elevation relative to the lunar landing site. Additionally, considerations extended to illumination conditions at Earth splashdown points and the distribution of sun elevation angles across potential landing sites. These angles typically ranged from 5° to 14° and varied from east to west.

Under the conditions specified, lunar landing site selection also considered the intricate dynamics of sunlight and its impact on visibility and operational safety. Visible shadows cast by lunar craters were crucial for aiding astronauts in identifying topographical features during descent and landing.

The optimal sun angle for lunar landings was around 16°, a mean value that provided adequate shadowing for clear visibility without excessive glare. However, as the sun angle approached the descent angle, critical for landing, a phenomenon known as "washout" could occur. This effect, caused by high backward reflectance, could reduce visual contrast and make surface features harder to discern.

Sun angles significantly above the flight path were less favorable because shadows became less visible unless the sun was sufficiently outside the descent plane. Sun angles exceeding 18° were typically excluded from consideration unless the landing was scheduled a day earlier when the lighting conditions would still provide at least 5° of elevation angle.

Considering that lunar sunlight incidence changed by approximately 0.5° per hour, a strict sun elevation angle restriction limited landing attempts to specific 16-hour periods, occurring every 29.5 days due to lunar orbital dynamics. This constraint effectively dictated the number of feasible launch opportunities within a lunar month, correlating directly with the number of candidate landing sites under consideration.

The timing of each launch was meticulously calculated based on allowable variations in launch pad azimuth and the lunar position relative to the spacecraft's arrival trajectory. Launches were planned to place the spacecraft into an orbital plane intersecting both the Moon and its antipode at the time of arrival. A launch pad azimuth variation of 34° provided a launch window typically spanning 4 hours and 30 minutes, optimizing the alignment necessary for successful lunar landing missions.

During the Apollo missions, the "daily launch window" referred to the specific period when the launch direction was within the required range to intercept the Moon's orbital path from Earth. This window presented two distinct opportunities each day. One launch window typically occurred over the Pacific Ocean, facilitating a preferred daytime launch due to optimal lighting conditions. The second launch opportunity was situated over the Atlantic Ocean.

Launch preparations were meticulously orchestrated to capitalize on these windows of opportunity. Weather conditions played a critical role in determining launch feasibility. During one such launch scenario, a high-pressure system situated off the North Carolina coast and a weak low-pressure trough in the northeastern Gulf of Mexico resulted in light southerly surface winds and increased moisture levels around Cape Kennedy (now Cape Canaveral). These atmospheric conditions contributed to cloudy skies and isolated thunderstorms observed at launch.

Specific meteorological data recorded at the launch site included a cloud cover of 10% cumulus clouds at 2,400 feet altitude, 20% altocumulus clouds at 15,000 feet altitude, and 90% coverage of cirrostratus clouds (base altitude not recorded). The temperature at launch time was 84.9°F, with a relative humidity of 73% and a barometric pressure reading of 14.798 lb/in². Wind speed, as measured by an anemometer located 60.0 feet above ground level at the launch site, registered at 6.4 knots from a direction of 175° true north.

The terminal countdown for Apollo 11 commenced at T-28 hours, precisely at 21:00:00 GMT on July 14th. Scheduled holds were strategically placed in the countdown sequence: a pause of 11 hours at T-9 hours and another of 1 hour 32 minutes at T-3 hours 30 minutes were the only planned interruptions.

During the countdown, a delay of 25 minutes occurred in the loading of LH2 (Liquid Hydrogen) into the S-II stage due to a communications issue in the Pad Terminal Connection Room. However, this delay was successfully recuperated later during the scheduled hold at T-3 hours 30 minutes, ensuring the countdown remained on track.

Apollo 11 lifted off from Kennedy Space Center's Launch Complex 39, Pad A, precisely at 13:32:00 GMT (09:32:00 a.m. EDT) on July

16, 1969, marking the beginning of its ascent phase. The planned launch window extended until 17:54:00 GMT, carefully chosen to capitalize on optimal lighting conditions with a sun elevation angle of 10.8° on the lunar surface. This strategic timing was crucial for facilitating clear visibility and operational conditions during the mission's initial phases. After a flawless launch, Apollo 11 began its journey to the Moon with precision and careful execution of critical maneuvers. Between 000:00:13.2 and 000:00:31.1, the spacecraft smoothly transitioned from a launch pad azimuth of 90° to a flight azimuth of 72.058°. The initial S-IC stage powered the ascent until shutdown at 000:02:41.63, followed by separating the S-IC and S-II stages and igniting the S-II engine. This stage operated until shutdown at 000:09:08.22, marking another successful phase before the S-II stage separated from the S-IVB, which ignited at 000:09:12.2.

Throughout ascent, the mission encountered maximum wind conditions of 18.7 knots at 297° from true north at 37,400 feet, with a peak wind shear of 0.0077 seconds^-1 at 48,490 feet. Despite these challenges, deviations from the planned trajectory were minimal, with velocity varying by only -0.6 ft/sec and altitude by just -0.1 nautical miles.

Lunar landing site 2 in the Sea of Tranquility, compared to the size of Washington, DC (NASA).

The S-IC stage impacted the Atlantic Ocean at 000:09:03.70, positioned at latitude 30.212° north and longitude 74.038° west, approximately 357.1 nautical miles from the launch site. The S-II stage impacted further along at 000:20:13.7, touching down at latitude 31.535° north and longitude 34.844° west, a considerable distance of 2,371.8 nautical miles from the launch site.

Upon achieving a near-circular Earth orbit at 000:11:49.34, the spacecraft's parameters included an apogee and perigee of 100.4 by 98.9 nautical miles, an inclination of 32.521°, and a velocity of 25,567.9 ft/sec. This insertion into Earth orbit set the stage for subsequent maneuvers, culminating in the critical translunar injection (TLI) burn at 002:44:16.20. The S-IVB engine fired for 346.87 seconds, pushing Apollo 11 toward its trajectory to the Moon.

The second ignition of the S-IVB engine at 002:50:03.03 facilitated translunar injection, boosting the spacecraft's velocity to 35,567.3 ft/sec. This maneuver was executed after completing one and a half Earth orbits over 2 hours 38 minutes and 23.73 seconds. Approximately 30 minutes after TLI, a precise transposition, docking, and extraction maneuver occurred. This involved separating Columbia from the spent S-IVB stage, docking with Eagle, and extracting the Lunar Module (LM) for the journey to the Moon.

To avoid any potential collision risks, the spent S-IVB stage was sent on a trajectory past the Moon, utilizing a gravitational slingshot effect to enter a solar orbit. This strategic maneuver ensured the safety and trajectory stability of the combined spacecraft as it embarked on its historic mission to land on the lunar surface.

Eagle in lunar orbit photographed from Columbia

During the Translunar Phase of Apollo 11's mission, several critical maneuvers and events ensured the spacecraft's trajectory toward the Moon and prepared for the subsequent lunar landing. At 003:15:23.0, the Command and Service Module (CSM) separated from the S-IVB stage. Shortly after, at 003:24:03.7, the CSM transposed and docked with the Lunar Module (LM), forming a unified spacecraft for the lunar journey.

At 004:17:03.0, the docked spacecraft was ejected from the S-IVB stage, initiating the next phase of the mission. A precise separation maneuver followed at 004:40:01.72, lasting 2.93 seconds to ensure proper distancing between components. Ground commands were then issued to vent residual propellants from the S-IVB, redirecting it into a solar orbit after passing the Moon.

The closest approach of the S-IVB to the Moon occurred at 1,825 nautical miles at 20:14:00 GMT on July 19th, marking its trajectory beyond the lunar sphere of influence into a stable solar orbit. The resultant solar orbit parameters included an aphelion and perihelion of 82.000 million by 72.520 million nautical miles, an inclination of 0.3836°, and an orbital period of 342.00 days.

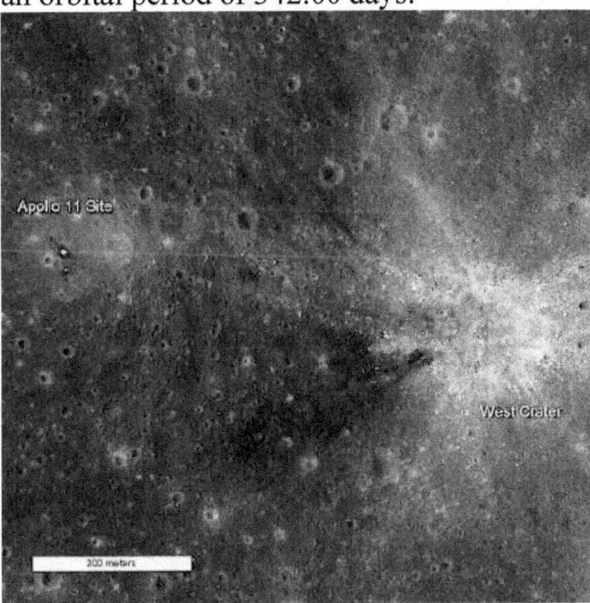

Landing site relative to West crater

After executing the critical maneuver behind the Moon on July 19th at 17:21:50 UTC, Apollo 11 proceeded with its mission, achieving nearly flawless trajectory parameters following the Translunar Injection firing. A minor midcourse correction of 3.13 seconds, adjusting the velocity by 20.9 feet per second, was conducted at 026:44:58.64 during the translunar phase to ensure precise alignment toward the Moon.

Throughout the subsequent phases of its journey, the spacecraft utilized passive thermal control, a methodical "barbecue" maneuver, to regulate internal temperatures effectively. This rotating maneuver helped maintain optimal spacecraft conditions amidst the varying thermal environments of space.

During the mission, several televised transmissions provided unprecedented views inside and outside the Command Module (CM) and Lunar Module (LM). An unscheduled 50-minute transmission occurred at 030:28, followed by a scheduled 36-minute broadcast starting at 033:59. The highlight was a 96-minute color television transmission that began at 055:08, showcasing high-resolution images of crew activities. These broadcasts included interior scenes of both spacecraft, Earth views, and detailed operations such as probe and drogue removal, spacecraft hatch openings, LM preparation, and equipment testing.

During the extended television transmission, at 055:30, the mission commander and lunar module pilot transitioned to the LM to conduct initial inspections and systems checks. They returned to the CM at 058:00 to finalize preparations for lunar orbit insertion.

At 075:49:50.37, with the spacecraft positioned 86.7 nautical miles above the lunar surface, the Service Propulsion Engine (SPE) fired for 357.53 seconds. This critical burn successfully inserted Apollo 11 into lunar orbit with an elliptical path measuring 169.7 by 60.0 nautical miles. The journey from Earth to lunar orbit had spanned 73 hours, 5 minutes, and 34.83 seconds, marking a crucial milestone in the mission's progression toward landing on the Moon.

During the Lunar Orbit and Lunar Surface Phase of Apollo 11, pivotal operations and checks were conducted to prepare for the historic lunar landing.

View of Earth at 98,000 n mi altitude following translunar injection (NASA AS11-36-5355).

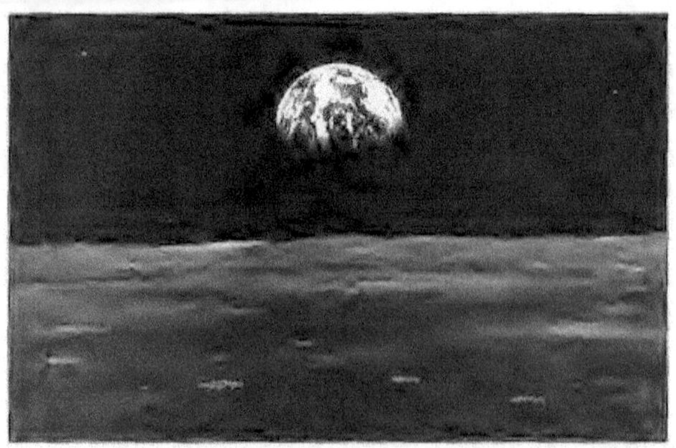

Earthrise over lunar surface following lunar orbit insertion

During the second lunar orbit, precisely at 078:20, a scheduled live color television transmission provided awe-inspiring views of the lunar surface, showcasing the approach path toward landing site 2 in the southwest region of the Sea of Tranquility. This broadcast offered a rare glimpse of the Moon's terrain as seen from the Lunar Module (LM) while still docked to the Command and Service Module (CSM).

Following two orbital revolutions and a navigation update, a critical maneuver known as the second Service Propulsion System (SPS) retrograde burn was executed at 080:11:36.75. This 16.88-second engine firing circularized the spacecraft's orbit, stabilizing it at a precise distance of 66.1 by 54.5 nautical miles around the Moon.

Subsequently, at approximately 080:11:36.75, the mission commander and lunar module pilot transitioned to the LM. Over the next two hours, they meticulously performed various housekeeping tasks, voice and telemetry tests, and a thorough check of the LM's oxygen purge system. Both LM systems and consumables were thoroughly inspected and found to be in optimal condition. Additionally, comprehensive checks were performed on both LM cameras, confirming their operational readiness.

Approach to lunar landing site 2 in southwest Sea of Tranquillity from the LM while still docked to the CSM

At 095:20, the crew returned to the LM once again to conduct a comprehensive systems check in preparation for the critical descent phase to the lunar surface. These meticulous preparations were essential to ensure the safety and success of the upcoming lunar landing, marking a pivotal moment in the Apollo 11 mission as humanity prepared to set foot on another celestial body for the first time in history.

At 12:52:00 UTC on July 20, 1969, Neil Armstrong and Buzz Aldrin embarked on the pivotal phase of their lunar mission as they entered the lunar module Eagle. They meticulously finalized their preparations for the descent to the Moon's surface inside the compact spacecraft. Meanwhile, Michael Collins continued his solitary vigil aboard the command module, Columbia, observing Eagle through the vast expanse of space.

LM Eagle seen from CM Columbia following undocking (NASA ASll-44-6574).

At 17:44:00 UTC, the historic moment arrived as Eagle gracefully detached from Columbia. While stationed in Columbia, Collins carefully inspected Eagle as it rotated before him, ensuring that all systems were functioning flawlessly and the landing gear was deployed as planned. Armstrong, as Eagle began its descent, remarked with poetic clarity, "The Eagle has wings!"

However, as they descended toward the lunar surface, Armstrong and Aldrin encountered unexpected challenges. They noted passing landmarks ahead of schedule, indicating they were "long" and would land several miles west of their intended target point. The lunar module, Eagle, was moving too swiftly, possibly influenced by mascons—areas of concentrated mass beneath the Moon's surface causing gravitational anomalies. Flight Director Gene Kranz and his team discussed potential causes, from air pressure fluctuations in the docking tunnel to the effects of Eagle's earlier maneuvers.

Critical maneuvers followed in rapid succession as Eagle continued its descent. At 100 hours, 12 minutes into the mission, Eagle undocked from Columbia at an altitude of 62.9 nautical miles. A precise separation maneuver, executed at 100 hours, 39 minutes and 52.9 seconds, adjusted Eagle's trajectory toward the lunar surface. The lunar module's descent orbit insertion maneuver, achieved by firing the descent propulsion system for 30 seconds at 101 hours, 36 minutes, and 14 seconds, further refined Eagle's path into a planned orbit of 58.5 by 7.8 nautical miles.

At 102 hours, 33 minutes, and 5.01 seconds, the critical powered descent engine burn commenced as scheduled. Despite the precise timing, Eagle initiated its descent about 4 nautical miles farther downrange than anticipated, necessitating adjustments to their landing coordinates. Shortly after, the first of several alarms sounded at 102 hours, 38 minutes, and 22 seconds, indicating a computer overload. Assured of the mission's safety, Armstrong and Aldrin pressed on.

As they approached the final stages of their descent, Armstrong manually maneuvered Eagle over rugged lunar terrain, navigating past boulders as large as five to ten feet in diameter. At 102 hours, 44 minutes, and 28 seconds, just 72 seconds before landing, a red-line low-level fuel warning light illuminated, adding to the tension of the final moments. Five minutes into the descent burn, with Eagle just 6,000 feet (1,800 meters) above the lunar surface, the tranquility of Armstrong and Aldrin's approach was interrupted by the LM guidance computer (LGC) sounding the first of several unexpected alarms: the 1201 and 1202 program alarms. These alarms, signaling "executive overflows", indicated that the guidance computer was struggling to keep up with its workload in real-time, necessitating delays in executing some tasks.

In Mission Control Center, computer engineer Jack Garman swiftly assessed the situation and relayed to Guidance Officer Steve Bales that it was safe for the crew to proceed with the descent. Bales promptly transmitted this crucial decision to Armstrong and Aldrin aboard Eagle, allowing them to focus on the task at hand.

Reflecting on these critical moments, Margaret Hamilton, Director of Apollo Flight Computer Programming at MIT's Charles Stark Draper Laboratory, later emphasized the vital role of the guidance computer's programming. She said, "To blame the computer for the Apollo 11 problems is like blaming the person who spots a fire and calls the fire department." The computer was programmed to do more than recognize error conditions. A complete set of recovery programs was incorporated into the software. The software's action, in this case, was to eliminate lower-priority tasks and re-establish the more important ones. The computer, rather than almost forcing an abort, prevented an abort. If the computer hadn't recognized this problem and taken recovery action, I doubt if Apollo 11 would have been the successful Moon landing it was.

Software engineer Don Eyles later investigated the root cause, attributing it to a hardware design flaw first observed during tests of the unmanned Lunar Module (LM) in Apollo 5. Normally, having the rendezvous radar on standby (to be ready in case of an emergency abort) shouldn't have affected the LGC. However, an electrical phasing mismatch within the radar system caused the stationary antenna to appear electrically as if it were oscillating between two positions. Depending on random power-up conditions, this discrepancy led the LGC to interpret erroneous data updates as legitimate signals, triggering the cycle-stealing that overwhelmed the computer and led to the alarms.

Eyles' findings, presented in a 2005 Guidance and Control Conference paper, highlighted the complexity and precision required in Apollo's hardware and software integration. Despite the hiccup, the mission's success was ultimately ensured by swift troubleshooting and decisive action from both Mission Control and the crew aboard Eagle. This incident underscored the rigorous testing and adaptive problem-solving that defined NASA's approach to pioneering space exploration during the Apollo era.

At 20:17:39 GMT (16:17:39 EDT) on July 20, 1969, the Lunar Module (LM), named Eagle, made history as it touched down on the Moon's surface during NASA's Apollo 11 mission. Engine shutdown followed just 1.5 seconds later, marking the culmination of an intense descent and a monumental achievement in human exploration.

The landing site was Mare Tranquilitatis (Sea of Tranquility), positioned at latitude 0.67408° north and longitude 23.47297° east. However, Eagle landed at a slight angle of 4.5 degrees and approximately 3.75 nautical miles southwest of the planned touchdown point. Despite this deviation, the LM landed safely with about 45 seconds of fuel remaining, underscoring the precision and efficiency of the mission's planning and execution.

Upon landing, Neil Armstrong immediately assessed their surroundings through the LM's window. He noticed the computer's intended landing target was in a rugged area north and east of a 300-foot-diameter crater later identified as West crater. Armstrong opted to take semi-automatic control of Eagle to ensure a safe touchdown, navigating it away from the hazardous boulder-strewn terrain.

Throughout the descent, Buzz Aldrin provided vital navigation data to Armstrong, who focused intensely on piloting the LM toward a suitable landing spot. At 107 feet (33 meters) above the lunar surface, with fuel running critically low, Armstrong identified a clear patch of ground and maneuvered Eagle toward it. However, as they descended to 250 feet (76 meters), he realized the chosen spot contained a crater. Undeterred, Armstrong expertly guided the LM over the crater and identified another level patch of ground.

As Eagle descended to just 100 feet (30 meters) from the surface, with a mere 90 seconds of fuel remaining, lunar dust kicked up by the LM's engine began to obscure Armstrong's visibility. Focusing on large rocks protruding from the dust cloud, he carefully gauged the spacecraft's speed and trajectory for a safe landing.

Moments before touchdown, a light indicator signaled that Eagle's footpads had made contact with the lunar surface. Aldrin immediately called out, "Contact light!" According to procedure, Armstrong was to shut down the descent engine promptly to prevent potential damage from exhaust pressure. However, in the heat of the moment, Armstrong momentarily forgot. Three seconds later, with precision timing, Eagle settled onto the lunar surface, and Armstrong swiftly shut down the engine.

Aldrin immediately confirmed, "Okay, engine stop. ACA—out of detent." Armstrong responded with equal clarity, "Out of detent. Auto." Aldrin continued to report on the status of various control systems, ensuring everything was secure after landing.

After confirming with Aldrin that the post-landing checklist had been completed, Neil Armstrong acknowledged the status by stating, "Engine arm is off." Then, with a calm and historic announcement to Mission Control, he transmitted, "Houston, Tranquility Base here. The Eagle has landed." This spontaneous change from referring to the LM as "Eagle" to "Tranquility Base" underscored the significance and success of their lunar landing.

In response, Charles Duke, the CAPCOM (Capsule Communicator) at Mission Control, expressed relief and joy, saying, "Roger, Tranquility, we copy you on the ground. You got a bunch of guys about to turn blue. We're breathing again. Thanks a lot." Duke's words reflected the collective tension and exhilaration felt by the entire Mission Control team, now assured of the crew's safe arrival on the Moon's surface.

Two and a half hours after the successful landing on the Moon, as preparations for the Extra-Vehicular Activity (EVA) were underway, Buzz Aldrin, the Lunar Module (LM) pilot, made a heartfelt radio transmission back to Earth:

"This is the LM pilot. I want to take this opportunity to ask every person listening in, whoever and wherever they may be, to pause for a moment and contemplate the events of the past few hours and to give thanks in his or her own way."

Following this message, Aldrin privately took communion. Given the ongoing legal challenges from Madalyn Murray O'Hair, an atheist activist who had previously filed lawsuits against NASA regarding religious activities in space, Aldrin opted not to mention taking communion on the lunar surface explicitly. As an elder at Webster Presbyterian Church, Aldrin had prepared for this moment with a communion kit provided by his pastor, Dean Woodruff. The chalice used for this historic communion is now kept at Webster Presbyterian Church, where the event is commemorated annually on the Sunday closest to July 20th.

Despite the mission's schedule allowing for a five-hour sleep period following landing, both Armstrong and Aldrin decided to forego sleep and instead began early preparations for their planned EVA. Their anticipation and eagerness to commence the historic moonwalk overrode any immediate need for rest, reflecting their dedication and commitment to the mission's success.

For the first two hours after landing on the lunar surface, the Apollo 11 crew diligently conducted system checkouts, adjusted controls for their lunar stay, and enjoyed their initial post-landing meal. Originally scheduled for rest before their Extravehicular Activity (EVA), the astronauts were too excited and prepared to venture out sooner than planned.

Neil Armstrong, the mission commander, equipped himself with the back-mounted Portable Life Support System (PLSS) and oxygen purge system before exiting the Lunar Module (LM). At 109:07:35, he opened the forward hatch, and by 109:19:16, he began his descent down the LM ladder. Armstrong deployed the Modular Equipment Stowage Assembly during this descent from the LM's descent stage. A live television camera on the module broadcasted his historic descent to millions of viewers on Earth.

Armstrong's left foot touched the lunar surface at 02:56:15 GMT on July 21 (22:56:15 EDT on July 20), marking humanity's first step onto another celestial body. His iconic words, "That's one small step for man, one giant leap for mankind," resonated across the globe.

According to the Apollo 11 Mission Report (MSC-00171), post-flight analysis clarified that approximately 45 seconds of fuel remained at the LM's touchdown, contradicting earlier reports of only 7 seconds. Armstrong proceeded to conduct a brief exterior check of the LM, noting minimal penetration of the footpads into the lunar soil and negligible collapse of the LM's footpad struts. He observed sinking about one-eighth inch into the fine lunar dust, which adhered lightly to his boots. There was no crater around the descent engine's landing site, and he remarked on the challenges of visibility in the LM's shadowed areas due to the stark darkness.

He then collected a contingency sample of lunar soil from the vicinity of the LM ladder. He reported that although loose material created a soft surface, he encountered very hard, cohesive material as he dug down six or eight inches. The commander then photographed the lunar module pilot as he exited at 109:39:00 and descended to the lunar surface at 109:43:15.

Preparations for Neil Armstrong and Buzz Aldrin's historic walk on the Moon commenced at 23:43 UTC, although they took longer than anticipated—three and a half hours instead of the planned two. On Earth, training scenarios had meticulously laid out everything required. Still, on the Moon, the Lunar Module (LM) cabin was cluttered with additional items such as checklists, food packets, and tools.

After six hours and thirty-nine minutes of meticulous preparation, Armstrong and Aldrin were finally prepared to exit Eagle, and the LM was depressurized accordingly. The hatch of Eagle swung open at 02:39:33 UTC. Armstrong, encumbered by his Portable Life Support System (PLSS), initially struggled to maneuver through the narrow hatch opening.

Armstrong's heart rate, among the highest recorded for Apollo astronauts, reflected the tension and physical exertion involved in exiting and re-entering the LM. At 02:51 UTC, Armstrong carefully descended the nine-rung ladder to the lunar surface. Due to the placement of the remote control unit on his chest, he had limited visibility of his feet during the descent. As he descended, Armstrong activated the TV camera mounted on the Modular Equipment Stowage Assembly (MESA), which had been folded against Eagle's side, ensuring that the historic moment was captured and broadcast live to audiences back on Earth.

Aldrin inside LM during first LM inspection (NASA AS11-36-5390).

During the Apollo 11 mission, the television images from the Moon were transmitted using slow-scan television (TV), which was incompatible with standard broadcast TV technology. This meant that the images were displayed on a special monitor, and another conventional TV camera captured these images from the monitor, essentially creating a broadcast of a broadcast. This process significantly reduced the quality of the picture.

The initial signal was received at NASA's Goldstone tracking station in the United States. However, the signal was quickly switched to the Honeysuckle Creek Tracking Station near Canberra, Australia for better fidelity. Shortly after that, the feed was further improved by switching to the more sensitive Parkes radio telescope, also in Australia.

Despite encountering technical and weather-related difficulties, black-and-white images of the first lunar Extra-Vehicular Activity (EVA) were successfully received and broadcast to an estimated audience of at least 600 million worldwide. These images became iconic, symbolizing humanity's first steps on another celestial body.

Honeysuckle Creek Tracking Station

While copies of the broadcast-format video were preserved and are widely accessible, recordings of the original slow-scan source transmission from the lunar surface were likely erased during NASA's routine practice of reusing magnetic tape. This has contributed to the historical significance and nostalgia surrounding the event, as the original, raw footage from the Moon is no longer available in its original form.

After describing the lunar surface dust as "very fine-grained" and "almost like a powder," Neil Armstrong took his historic step at 02:56:15 UTC, six and a half hours after Eagle had landed. Stepping off Eagle's landing pad, he famously declared: "That's one small step for [a] man, one giant leap for mankind."

Armstrong intended to say, "That's one small step for a man," emphasizing the individual significance amidst the broader achievement for humanity. However, due to the transmission quality and possibly his accent, the indefinite article "a" was not audible in the live broadcast and was initially missed by most observers. When later asked about his quote, Armstrong maintained that he believed he had said "for a man." Subsequent printed versions of the quote included the "a" in square brackets to reflect his intention.

Several explanations have been proposed for omitting the "a". Some suggest that Armstrong's pronunciation or the intermittent nature of the audio and video links to Earth, possibly affected by storms near the Parkes Observatory, obscured the article. Digital analyses of the original tape have since indicated that the "a" may have been spoken but was obscured by static.

However, critics have argued that claims of static or slurring as explanations are attempts to save face, suggesting instead that Armstrong may have misspoken or simplified his intended statement under the pressure of the moment. Despite these debates, Armstrong's words remain etched in history as a profound symbol of humanity's first steps beyond Earth.

About seven minutes after Neil Armstrong first set foot on the Moon's surface, he collected a contingency soil sample using a sample bag attached to a stick. This precaution ensured that there would be lunar soil to bring back in case an emergency required the astronauts to cut short their extravehicular activity (EVA) and return to the Lunar Module (LM). Armstrong carefully folded the sample bag and tucked it into a pocket on his right thigh for safekeeping.

Twelve minutes after collecting the soil sample, Armstrong removed the TV camera from the Modular Equipment Stowage Assembly (MESA) and conducted a panoramic sweep of the lunar landscape. He then mounted the camera on a tripod, although the camera cable remained partly coiled and posed a tripping hazard throughout their time on the lunar surface.

In addition to the live TV broadcast, still photography was a crucial part of documenting the mission. Armstrong used a Hasselblad camera, which he could operate either handheld or mounted on his Apollo space suit. This camera captured iconic images of the Moon's surface and the activities of the astronauts during their historic EVA.

Shortly after, Buzz Aldrin joined Armstrong on the lunar surface. Reflecting on the awe-inspiring sight before them, Aldrin famously described the lunar landscape as "Magnificent desolation." This phrase succinctly captured the stark beauty and barrenness of the Moon's surface, as seen through the eyes of the first humans to set foot there.

Neil Armstrong remarked that moving in the lunar gravity, which is one-sixth of Earth's, felt "even perhaps easier than the simulations... It's absolutely no trouble to walk around." This reduced gravity made locomotion surprisingly effortless for the astronauts compared to their training on Earth.

Buzz Aldrin joined Armstrong on the lunar surface and experimented with different movement methods, including two-footed kangaroo hops. Despite the bulky Portable Life Support System (PLSS) backpacks causing a slight backward tipping tendency, both astronauts managed to maintain balance effectively. They eventually adopted a loping gait as their preferred method for traversing the lunar terrain method. Planning movements in advance became crucial, with the astronauts needing to anticipate their steps six or seven moves ahead due to the slippery nature of the fine lunar soil.

Transitioning from sunlight into the shadow of the Lunar Module (Eagle) did not significantly alter the internal temperature of their suits. However, Aldrin noted that his helmet felt warmer in sunlight, making the shadow feel cooler by comparison.

The Modular Equipment Stowage Assembly (MESA), intended to provide a stable work platform, proved inadequate as it was positioned in shadow, which slowed their work. As they moved and worked on the lunar surface, the astronauts kicked up gray dust, which adhered to the outer parts of their suits, leaving visible traces of their activities on the Moon.

On the lunar surface, Neil Armstrong and Buzz Aldrin planted the Lunar Flag Assembly, which held the flag of the United States. The TV camera captured this iconic moment for viewers on Earth. Buzz Aldrin later recalled that among all the tasks he had to perform on the Moon, he wanted the flag raising to proceed smoothly.

However, the astronauts encountered difficulty with the telescoping rod used to erect the flag. They could only insert the pole about 2 inches (5 cm) into the hard lunar surface. Aldrin was concerned that the flag might topple over in front of the live TV audience, but he managed to give "a crisp West Point salute" after they secured it.

Before Aldrin could photograph Armstrong with the flag, President Richard Nixon initiated a telephone radio transmission to them. This communication, which Nixon described as "the most historic phone call ever made from the White House," briefly interrupted their activities. Initially prepared with a long speech, Nixon heeded advice from Frank Borman, NASA's liaison at the White House during Apollo 11, to keep his remarks concise.

Aldrin salutes the deployed United States flag on the lunar surface.

Nixon: "Hello, Neil and Buzz. I'm talking to you by telephone from the Oval Room at the White House. And this must be the most historic telephone call ever made from the White House. I can't tell you how proud we all are of what you have done. For every American, this has to be the proudest day of our lives. And for people worldwide, I am sure that they, too, join with Americans in recognizing what an immense feat this is. Because of what you have done, the heavens have become a part of man's world. And as you talk to us from the Sea of Tranquility, it inspires us to redouble our efforts to bring peace and tranquility to Earth. For one priceless moment in the whole history of man, all the people on this Earth are truly one: one in their pride in what you have done, and one in our prayers that you will return safely to Earth."

Armstrong: "Thank you, Mr. President. It's a great honor and privilege for us to be here, representing not only the United States but men of peace of all nations, with interest and curiosity, and men with a vision for the future. It's an honor for us to be able to participate here today."

Nixon: "Thank you very much, and I look forward, all of us look forward, to seeing you on the Hornet on Thursday."

During their historic mission, Neil Armstrong and Buzz Aldrin conducted a series of meticulous scientific tasks on the lunar surface, marking a pivotal moment in human exploration. Following the deployment of the Early Apollo Scientific Experiment Package (EASEP), which included a Passive Seismic Experiment to measure moonquakes and a retroreflector array for lunar laser ranging, Armstrong ventured 196 feet (60 m) from the Lunar Module (LM). His objective was to capture photographs at the rim of Little West Crater, while Aldrin concurrently collected core samples.

Equipped with a geologist's hammer, Aldrin attempted to drive in sampling tubes, achieving a depth of only 6 inches (15 cm), highlighting the unexpected challenges of lunar soil. Undeterred, the astronauts employed scoops and tongs on extension handles to gather diverse rock specimens. Despite time constraints that necessitated halting documentation halfway through the scheduled 34 minutes, Aldrin meticulously packed 6 kilograms (13 lb) of lunar soil alongside the collected rocks.

Their geological findings were profoundly significant. Analysis revealed two primary types of rocks: basalt and breccia. The astronauts discovered three new minerals previously unknown to science within these samples: armalcolite, tranquillityite, and pyroxferroite. Armalcolite, in particular, was named in honor of Armstrong, Aldrin, and Collins, signifying their indelible contribution to lunar exploration. Remarkably, subsequent studies have identified these minerals on Earth, underscoring the broader scientific implications of their lunar discoveries.

Upon touching down on the lunar surface, the Apollo crew embarked on a symbolic and historic gesture. They unveiled a plaque affixed to the Lunar Module's strut behind the ladder, reading aloud its inscription to a global television audience. The plaque proudly declared:

"HERE, MEN FROM THE PLANET EARTH FIRST SET FOOT UPON THE MOON JULY 1969, A.D. WE CAME IN PEACE FOR ALL MANKIND."

This significant marker bore the signatures of the three Apollo astronauts and President Richard M. Nixon, emphasizing their mission's unity and peaceful intentions.

Following this momentous act, Commander Neil Armstrong removed the television camera from the LM's descent stage. With it, he captured a panoramic view of the lunar landscape, strategically placing the camera on its tripod to document the subsequent extravehicular activities.

Armstrong discovered another plaque mounted on the LM's ladder during their exploration. This plaque featured intricate drawings of Earth's Western and Eastern Hemispheres, accompanied by the inscription:

"Here men from the planet Earth first set foot upon the Moon July 1969, A. D. We came in peace for all mankind."

At the Nixon administration's request to acknowledge a higher authority, NASA added "A.D." to the date, signifying "Anno Domini" or "in the year of our Lord." This detail underscored the cultural and historical significance of the Apollo 11 mission as a milestone in human exploration and global unity and cooperation.

During Neil Armstrong's historic moonwalk, Mission Control utilized a coded phrase to alert him that his metabolic rates were elevated and advised him to slow down. Armstrong, driven by the urgency of completing tasks within a tight schedule, moved rapidly from one objective to the next as time ticked away. Interestingly, despite initial concerns, Armstrong's and Buzz Aldrin's metabolic rates remained lower than anticipated during their lunar excursion.

Mission Control granted them an additional 15 minutes to extend their time on the lunar surface in response to the efficient use of resources and the astronauts' continued well-being. This extension allowed for more comprehensive scientific observations and sample collection.

In a retrospective interview in 2010, Armstrong disclosed that NASA had initially imposed strict limits on the duration and scope of the first moonwalk due to uncertainties about the consumption rate of cooling water in the astronauts' Portable Life Support Systems (PLSS). These backpack-like systems were crucial for managing their body heat while working in the harsh lunar environment. The decision underscored NASA's cautious approach to ensure the safety and success of this pioneering mission, reflecting the meticulous planning and adaptive decision-making that characterized the Apollo program.

The lunar module pilot and mission commander successfully deployed the solar wind composition experiment on the lunar surface, positioning it to the north of the Lunar Module (LM) under direct sunlight. Both astronauts marveled at their enhanced mobility, which surpassed all initial expectations. Furthermore, their metabolic rates were significantly lower than originally predicted, a favorable outcome contributing to their operational efficiency.

At 110:09:43 into the mission timeline, the crew ceremoniously planted a three-by-five-foot United States flag using an eight-foot aluminum staff. This symbolic act marked their presence on the lunar surface and underscored the national pride associated with their historic achievement.

Astronaut Buzz Aldrin on the moon

Shortly after, at 110:16:30, President Richard M. Nixon initiated a conversation with the LM crew from the White House, extending congratulations and good wishes to the astronauts. This direct communication highlighted the monumental significance of the Apollo 11 mission to the nation and the world.

During subsequent environmental assessments, the lunar module pilot emphasized maintaining careful balance due to his center of mass dynamics. Notably, he observed that the LM's shadow had minimal impact on the temperature of his backpack, a vital consideration for thermal management in the lunar environment. Additionally, he remarked on his enhanced agility, which exceeded initial predictions, demonstrating the adaptability and effectiveness of the astronauts' training and equipment design in lunar conditions.

Aldrin's bootprint; part of an experiment to test the properties of the lunar regolith

After completing initial surface tasks, the mission commander collected a bulk sample comprising various surface materials and rock fragments, carefully storing them in a designated sample return container. Subsequently, the entire crew thoroughly inspected the Lunar Module (LM), examining its quads, struts, skirts, and antennas, all in excellent condition, ensuring the integrity of their return journey.

Scientific experiments continued deploying the Passive Seismic Experiment Package (PSEP) and the lunar laser ranging retroreflector to the south of the LM. The PSEP provided valuable seismic data, while the retroreflector facilitated precise measurements of lunar distances from Earth.

Upon analysis at the Lunar Receiving Laboratory in Houston, the total weight of the collected samples was confirmed to be 47.52 pounds (21.55 kg). The crew traveled 3,300 feet (1 km) across the lunar surface throughout their extravehicular activities. The commander ventured as far as 200 feet (60 m) from the LM, reaching a crater with a diameter of 108 feet (33 m) near the conclusion of their extravehicular period.

Aldrin steps from LM ladder onto lunar surface (NASA AS11-40-5869).

Armstrong and Aldrin set up U.S. flag on lunar surface

William Safire, the presidential speechwriter, had meticulously drafted an "In Event of Moon Disaster" announcement for President Nixon, outlining a protocol in case the Apollo 11 astronauts became stranded on the lunar surface. This contingency plan was detailed in a memo addressed to Nixon's Chief of Staff, H. R. Haldeman. Safire's proposal envisioned a solemn procedure if communication with the Lunar Module (LM) was lost.

Aldrin next to the Passive Seismic Experiment Package with the Lunar Module Eagle in the background

According to Safire's memo, Mission Control would cease communications with the LM, marking the tragic conclusion of the mission. A clergyman would then deliver a public ritual, akin to a burial at sea, to "commend their souls to the deepest of the deep." This ceremony was crafted to honor the astronauts' sacrifice and recognize their historical place.

Armstrong on the lunar surface at the MESA packing the bulk sample

During the Apollo 17 mission alone, astronauts retrieved more than 250 pounds of moon rock over the course of three moon walks in 1972.

The final line of the prepared text included an allusion to Rupert Brooke's poignant World War I poem, "The Soldier," adding a poetic and solemn touch to the occasion's solemnity. While fortunately never enacted, this contingency plan reflects the gravity and risks associated with humanity's first ventures beyond Earth, underscoring the careful planning and emotional preparation undertaken during the Apollo missions.

During their activities inside the Lunar Module (LM), Buzz Aldrin accidentally damaged a critical circuit breaker essential for arming the main engine that would propel them off the Moon's surface. This mishap raised concerns about their ability to initiate liftoff, potentially stranding them on the lunar surface. In a resourceful move, Aldrin ingeniously used the nonconductive tip of a Duro felt-tip pen to activate the switch, ensuring the engine could be fired when needed.

Aldrin prepares to deploy experiments from LM

After more than 21 hours of exploring the lunar terrain and conducting scientific experiments, the Apollo 11 astronauts left behind a series of poignant tributes and symbolic items.

These included an Apollo 1 mission patch in memory of astronauts Roger Chaffee, Gus Grissom, and Edward White, who tragically perished in a fire aboard their command module during a ground test in January 1967.

Two memorial medals commemorating Soviet cosmonauts Vladimir Komarov and Yuri Gagarin, who lost their lives in separate incidents in 1967 and 1968, respectively.

And a memorial bag containing a gold replica of an olive branch symbolizing peace and reconciliation.

Aldrin deploys PSEP and LR3

A silicon message disk bearing goodwill statements from Presidents Eisenhower, Kennedy, Johnson, and Nixon and messages from leaders of 73 countries worldwide. The disk also included listings of the US Congress leadership, members of the committees overseeing NASA legislation, and the names of NASA's senior management at the time.

After a rest period of about seven hours, Houston roused the crew of Apollo 11 to begin preparations for their journey back to Earth. Two and a half hours later, precisely at 17:54:00 UTC, Neil Armstrong and Buzz Aldrin lifted off in the ascent stage of the Lunar Module Eagle, leaving the Moon's surface to rendezvous with Michael Collins aboard the Command Module Columbia in lunar orbit.

Eagle's ascent stage approaching Columbia

Footage captured from the LM ascent stage during liftoff revealed a poignant moment: the American flag, planted approximately 25 feet (8 m) from the LM's descent stage, fluttered vigorously in the exhaust of the ascent stage engine. Buzz Aldrin recalled witnessing the flag topple as the LM ascent stage separated from the lunar surface: "The ascent stage of the LM separated ... I was concentrating on the computers, and Neil was studying the attitude indicator, but I looked up long enough to see the flag fall over." This iconic image captured the bittersweet departure from the Moon's surface, symbolizing the conclusion of their historic mission.

Subsequent Apollo missions adjusted their flag placements, planting them farther away from the LM to prevent such occurrences and ensure that these symbols of national pride and human achievement remained upright and visible on the lunar landscape.

During his solo orbits around the Moon aboard the Command Module Columbia, Michael Collins experienced a profound connection to the mission despite being alone. Often described as experiencing the most solitude any human has known since Adam, Collins felt integral to the journey. He emphasized each crew member's essential role in his autobiography, stating, "This venture has been structured for three men, and I consider my third to be as necessary as either of the other two."

During the 48 minutes of each orbit, when Columbia passed over the far side of the Moon, out of radio contact with Earth, Collins did not feel fear or loneliness. Instead, he described awareness, anticipation, satisfaction, confidence, and almost exultation. This unique perspective underscored his deep involvement in the mission's success.

One of Collins' initial tasks was to locate the Lunar Module Eagle on the lunar surface. Mission Control informed him that they estimated Eagle had landed approximately 4 miles (6.4 km) off its target location. During his orbits over the suspected landing site, Collins diligently searched for the module but initially could not spot it among the vast lunar landscape.

While passing over the far side of the Moon during his first orbits, Collins attended to various maintenance tasks aboard Columbia. These included managing excess water produced by the fuel cells and ensuring the cabin was prepared for Armstrong and Aldrin's return from the lunar surface. These activities highlighted Collins' critical role in supporting the lunar landing and ensuring a safe journey back to Earth for the entire Apollo 11 crew.

Just before entering the dark side of the Moon during his third orbit, Mission Control alerted Michael Collins to a coolant temperature issue aboard Columbia. Concerned that overly cold temperatures could lead to system freezing, Mission Control instructed Collins to switch to manual control and initiate Environmental Control System Malfunction Procedure 17. Instead, Collins opted to flick the coolant system switch from automatic to manual and back to automatic while monitoring the temperature closely. By the time Columbia reemerged on the near side of the Moon, Collins reported that the problem had been successfully resolved.

During subsequent orbits over the far side of the Moon, Collins described his experience as "relaxing," having effectively managed the system issue and resumed normal housekeeping tasks aboard the spacecraft. Meanwhile, while Neil Armstrong and Buzz Aldrin conducted their historic Extra-Vehicular Activity (EVA) on the lunar surface, Collins took the opportunity to rest and sleep, ensuring he was well-rested for the upcoming rendezvous.

While the mission plan anticipated Eagle's ascent stage to rendezvous with Columbia, Collins prepared for contingencies. He was ready to maneuver Columbia to meet Eagle if necessary, demonstrating his readiness to adapt to changing circumstances and ensure the safe reunion of the Apollo 11 crew in lunar orbit. This flexibility and meticulous preparation underscored Collins' critical role in the mission's overall success, facilitating the safe return of all astronauts to Earth.

After a successful lunar surface mission, Eagle's ascent stage rendezvoused with Columbia at 21:24 UTC on July 21, and the two spacecraft docked at 21:35. Shortly after that, at 23:41, Eagle's ascent stage was intentionally jettisoned into lunar orbit.

However, Eagle's fate remained a subject of interest and uncertainty. Initial reports indicated its orbit decayed, impacting the lunar surface in an unspecified location. Recent calculations in 2021 suggest that the Eagle might still be in orbit around the Moon, challenging previous assumptions about its final resting place.

The ascent stage liftoff from the lunar surface occurred precisely at 17:54:00 GMT on July 21, marking 21 hours, 36 minutes, and 20.9 seconds since it first touched down. Achieving an initial orbit of 48.0 by 9.4 nautical miles at 124:29:15.67, 434.88 seconds post-liftoff, required several precise maneuvers to facilitate the rendezvous and docking with Columbia.

Critical maneuvers included a 47.3-second coelliptic orbit adjustment at 125:19:34.70 and a 17.8-second delta height maneuver at 126:17:49.6 to refine the orbit to 47.4 by 42.1 nautical miles. Further adjustments were made with a 22.7-second terminal phase initiate maneuver at 127:03:51.8, achieving an orbit of 61.7 by 43.7 nautical miles. The final 28.4-second terminal phase maneuver at 127:46:09.8 stabilized the ascent stage in a 63.0 by 56.5 nautical mile orbit, enabling its successful docking with the Command and Service Module (CSM) Columbia at 128:03:00.0.

This intricate series of orbital adjustments and maneuvers ensured the safe and precise rendezvous of Eagle's ascent stage with Columbia, concluding the historic Apollo 11 mission and securing the return of Neil Armstrong, Buzz Aldrin, and Michael Collins to Earth.

Meticulous precision was crucial during the intricate maneuver to dock the Lunar Module (LM) with Apollo's Command and Service Module (CSM). Navigating the LM into the precise docking attitude and avoiding direct sunlight glaring through the forward window required astute control. However, an unexpected challenge arose when the platform inadvertently reached gimbal lock, momentarily causing the LM to tumble. Swift action stabilized the craft, utilizing the abort guidance system for precise attitude control.

Once docked, the crew and lunar samples were successfully transferred to the CSM. At 130 hours, 9 minutes, and 31.2 seconds into the mission, with the LM's ascent stage no longer needed, it was jettisoned 61.6 nautical miles above the lunar surface. The CSM was then prepared for transearth injection, a critical maneuver to begin the journey back to Earth.

A precise 7.2-second separation maneuver at 130 hours, 30 minutes, and 1.0 second, achieved an orbit of 62.7 by 54.0 nautical miles for the CSM, while the LM's ascent stage remained in lunar orbit indefinitely.

Transearth injection occurred at 135 hours, 23 minutes, and 42.28 seconds at an altitude of 52.4 nautical miles. The maneuver, lasting 151.41 seconds, successfully propelled the spacecraft onto a trajectory homeward. Transearth phase operations mirrored the precision of translunar flight, requiring just one midcourse correction—a 10.0-second, 4.8-feet-per-second adjustment—at 150 hours, 29 minutes, and 57.4 seconds.

Passive thermal control maintained optimal spacecraft temperatures throughout the coast back to Earth. This leg of the journey spanned 30 lunar orbits, totaling 59 hours, 30 minutes, and 25.79 seconds, achieving transearth injection at a velocity of 8,589.0 feet per second.

On July 23, as they prepared to return to Earth, the Apollo astronauts took to television once more. In a poignant broadcast, Michael Collins reflected on the Saturn V rocket that propelled them into orbit, praising its flawless performance and acknowledging the countless individuals whose dedication made their journey possible. His gratitude extended beyond the visible crew to the thousands whose efforts remained unseen but essential.

View of Moon after departure from an altitude of

Buzz Aldrin echoed this sentiment, viewing their mission not just as a national endeavor but as a testament to humanity's universal curiosity and quest for exploration. He drew inspiration from the Psalms, contemplating the vastness of the heavens and mankind's insignificance in the grand scheme of the cosmos. Their messages resonated with the world, underscoring the collaborative spirit and indomitable human spirit that defined the Apollo missions.

As they neared their return to Earth, Neil Armstrong expressed profound gratitude in a final broadcast from Apollo 11. He credited the success of their mission to the enduring legacy of scientific pioneers, the unwavering will of the American people, and the steadfast support of multiple administrations and Congresses. Armstrong highlighted the teams behind the spacecraft—Saturn, Columbia, Eagle, and the EMU spacesuit—thanking them for their dedication, craftsmanship, and commitment to excellence.

In a touching anecdote, Armstrong recounted a critical moment during the return journey when a bearing at the Guam tracking station failed, jeopardizing communications for the final leg. Station director Charles Force faced a daunting challenge and turned to an unlikely hero: his ten-year-old son, Greg. With small hands able to reach into the tight housing, Greg packed it with grease, effectively repairing the crucial equipment. Armstrong later acknowledged Greg's pivotal role, underscoring the collective effort and ingenuity that characterized the Apollo 11 mission.

As their spacecraft continued homeward bound, Armstrong concluded with heartfelt thanks to all who had contributed, sending a message of appreciation and goodwill to those listening and watching around the world. He bid farewell with a simple yet powerful message: "God bless you. Good night from Apollo 11."

Due to adverse weather conditions in the initially planned recovery area, Apollo 11's splashdown point was shifted 215 nautical miles downrange. The new location offered favorable conditions, with clear visibility extending up to 12 miles, waves measuring up to 3 feet, and a moderate wind speed of 16 knots.

At 194 hours, 49 minutes, and 12.7 seconds into the mission, the service module was jettisoned, marking the beginning of the return sequence. Following an automatic entry profile, the Command Module (CM) reentered Earth's atmosphere at an altitude of 400,000 feet and a velocity of 36,194.4 feet per second after a transearth coast lasting 59 hours, 36 minutes, and 52.0 seconds.

The parachute system deployed successfully, guiding the CM to splashdown in the Pacific Ocean precisely at 16:50:35 GMT (12:50:35 EDT) on July 24th. The mission's total duration was 195 hours, 18 minutes, and 35 seconds. The splashdown occurred approximately 1.69 nautical miles from the target point and 13 nautical miles from the recovery ship, the U.S.S. Hornet, positioned at an estimated latitude of 13.30° north and longitude of 169.15° west.

Following the successful Command Module (CM) splashdown from Apollo 11, the spacecraft initially assumed an apex-down flotation position. Still, it was swiftly reoriented to its normal flotation posture within 7 minutes and 40 seconds. The inflatable bag uprighting system facilitated this adjustment.

Upon landing in the Pacific Ocean, the crew promptly donned biological isolation garments as a precaution against potential lunar contaminants, being scrubbed down with an iodine solution to mitigate any risks of "lunar germs." They then disembarked into a rubber boat and were airlifted by helicopter to the primary recovery ship, the USS Hornet, arriving just 63 minutes after splashdown.

Meanwhile, the CM, estimated to weigh 10,873 pounds at splashdown and having traveled approximately 828,743 nautical miles during its mission, was recovered 125 minutes later. The crew, accompanied by a recovery physician, technician, and precious lunar samples, was directed into the Mobile Quarantine Facility aboard the Hornet. This specialized quarantine unit was crucial for ensuring containment during transport to the Lunar Receiving Laboratory in Houston.

The USS Hornet, under the command of Captain Carl J. Seiberlich, had been designated as the primary recovery ship for Apollo 11, having embarked from Long Beach, California, and stopped at Pearl Harbor to load essential recovery equipment. This included Sikorsky SH-3 Sea King helicopters, specialized divers, a NASA recovery team, and media representatives. Most of Hornet's air wing was left behind to accommodate these needs, making space for necessary gear and personnel. Additionally, a training boilerplate command module was onboard for operational preparedness.

Apollo 11 crew in raft while waiting for helicopter

After offloading in Hawaii at 00:15 GMT on July 27, the Mobile Quarantine Facility was loaded onto a C-141 aircraft, arriving safely in

Houston at 06:00 GMT on July 28. This meticulous process ensured that both crew and lunar samples were safely contained and transported, marking the conclusion of Apollo 11's historic journey to the Moon.

On July 12, 1969, while Apollo 11 remained poised on the launch pad, the USS Hornet set sail from Pearl Harbor toward the designated recovery area in the central Pacific, near coordinates 10°36′N 172°24′E. This area was carefully chosen for its strategic position to retrieve the returning astronauts and spacecraft.

Simultaneously, a historic presidential party was underway. President Richard Nixon, accompanied by astronaut Frank Borman, Secretary of State William P. Rogers, and National Security Advisor Henry Kissinger, departed for Johnston Atoll aboard Air Force One. From there, they transferred to the command ship USS Arlington, continuing their journey aboard Marine One.

After spending the night aboard the USS Arlington, the presidential entourage proceeded to the USS Hornet via Marine One for ceremonial events. Upon arrival, they were warmly greeted by Admiral John S. McCain Jr., Commander-in-Chief, Pacific Command (CINCPAC), and Thomas O. Paine, the Administrator of NASA. Paine had flown to Hornet from Pago Pago aboard one of the carrier's onboard delivery aircraft, underscoring the significance and grandeur of the occasion.

In the days leading up to Apollo 11's splashdown, US Air Force Captain Hank Brandli played a crucial role. Armed with access to classified spy satellite images, Brandli detected a storm front advancing toward the designated recovery area. His assessment highlighted impending challenges: poor visibility that could hinder capsule location efforts and strong upper-level winds posing a serious risk of parachute damage.

Brandli promptly alerted Navy Captain Willard S. Houston Jr., commander of the Fleet Weather Center at Pearl Harbor, who possessed the necessary security clearance. Recognizing the potential threat, Captain Houston acted on Brandli's recommendation. He relayed the concerns to Rear Admiral Donald C. Davis, commander of Manned Spaceflight Recovery Forces, Pacific, who advised NASA to relocate the recovery area—a decision fraught with career risks for those involved.

The new recovery location, chosen 215 nautical miles northeast of the original, necessitated the mission's flight plan adjustments. A novel sequence of computer programs was employed, untested in previous missions. Typically, re-entry sequences involved trajectory events like P64 followed by P67. In this instance, anticipating an extended re-entry without a skip-out, the sequence skipped P66, directly transitioning from P65 to P67. The crew was forewarned not to experience a full-lift (heads-down) attitude during the critical P67 phase.

President Richard M. Nixon was in the central Pacific recovery area to welcome the Apollo 11 astronauts aboard the USS Hornet, prime recovery ship for the historic Apollo 11 lunar landing mission.

These adjustments were critical as the first program subjected the astronauts to 6.5 standard gravities of acceleration (64 m/s²), followed by 6.0 standard gravities (59 m/s^2) in the second phase. Such precise calculations and operational changes ensured the navigation of Apollo 11 through challenging atmospheric conditions, safeguarding both crew and spacecraft during their return to Earth.

In the early hours before dawn on July 24, the USS Hornet took decisive action for the Apollo 11 recovery operation. The ship launched a fleet of aircraft essential to the mission's success. Four Sea King helicopters were deployed, along with three Grumman E-1 Tracers. Two E-1s coordinated operations as "air boss" aircraft, while the third acted as a crucial communications relay.

Among the Sea Kings, two carried divers and specialized recovery equipment, poised to assist in retrieving the spacecraft and astronauts. A third Sea King was equipped with advanced photographic gear to document the historic event. At the same time, the fourth helicopter transported the decontamination swimmer and flight surgeon, crucial roles in ensuring the safety and health of the returning crew.

At 16:44 UTC (05:44 local time), Columbia's momentous descent began with the deployment of its drogue parachutes, an event observed closely by the hovering helicopters. Seven minutes later, Columbia impacted the ocean surface forcefully. The splashdown occurred approximately 2,660 kilometers (1,440 nautical miles) east of Wake Island, 380 kilometers (210 nautical miles) south of Johnston Atoll, and just 24 kilometers (13 nautical miles) from USS Hornet's position, specifically located at 13°19′N 169°9′W.

Weather conditions at the recovery site were reported as favorable, with a temperature of 82 °F (28 °C), seas reaching 6 feet (1.8 meters), and eastward winds blowing at 17 knots (31 km/h; 20 mph). Broken clouds at 1,500 feet (460 meters) altitude provided 10 nautical miles (19 kilometers; 12 miles) visibility, ensuring clear operational conditions for recovery efforts.

Aircraft dispatched to the original splashdown area confirmed the accuracy of earlier predictions by Hank Brandli and Captain Houston, reporting adverse weather conditions that reinforced the critical decision to relocate the recovery site.

During Apollo 11's splashdown, the Command Module Columbia landed upside down in the Pacific Ocean. However, within ten minutes of touching down, the astronauts inside activated flotation bags that swiftly righted the capsule, stabilizing its position in the water.

Hovering above in a Navy helicopter, a diver swiftly descended and attached a sea anchor to Columbia to prevent it from drifting away. Immediately afterward, more divers deployed from the helicopters, attaching flotation collars to further stabilize the module and positioning rafts strategically for the astronauts' extraction.

Once the module was secure, the divers passed biological isolation garments (BIGs) down to the astronauts through the open hatch of Columbia. These garments were crucial in preventing any potential lunar pathogens from contaminating Earth's environment. The astronauts, assisted by the divers, transferred into the life raft and underwent further precautionary measures.

While the likelihood of lunar pathogens was minimal, NASA took meticulous precautions at the recovery site. The astronauts were carefully rubbed down with a sodium hypochlorite solution. At the same time, Columbia itself was wiped down with Povidone-iodine to eliminate any traces of lunar dust that might have adhered to its exterior surfaces.

Once safely aboard the recovery helicopter, the astronauts remained in their BIGs until they reached isolation facilities aboard the USS Hornet. As a final measure, the raft containing the decontamination materials was intentionally sunk, ensuring that any potential contaminants remained isolated from Earth's biosphere.

After their touchdown aboard USS Hornet at 17:53 UTC, the Apollo 11 astronauts were swiftly transported by helicopter, lowered by elevator into the hangar bay. From there, they walked just 30 feet (9.1 meters) to the Mobile Quarantine Facility (MQF), where they would commence a rigorous 21-day quarantine upon Earth. This precautionary measure, implemented for Apollo 11 and the subsequent missions of Apollo 12 and Apollo 14, aimed to ensure containment against any potential lunar pathogens until the Moon's barrenness of life was definitively confirmed.

President Nixon personally welcomed the astronauts back to Earth, acknowledging the monumental achievement of Apollo 11: "[A]s a result of what you've done, the world has never been closer together before."

Once Nixon departed, the USS Hornet carefully maneuvered alongside the 5-short-ton (4.5 metric ton) Columbia Command Module. Using its crane, the ship lifted Columbia onto a dolly, transporting it to a position adjacent to the MQF. Here, Columbia was securely attached to the MQF using a flexible tunnel, facilitating the removal of lunar samples, film, data tapes, and other mission artifacts.

USS Hornet then sailed back to Pearl Harbor, where the MQF was loaded onto a Lockheed C-141 Starlifter aircraft. This aircraft airlifted the MQF directly to the Manned Spacecraft Center, later renamed Johnson Space Center, in Houston. Meanwhile, the astronauts arrived at the Lunar Receiving Laboratory at 10:00 UTC on July 28, initiating the careful process of analyzing and studying the lunar samples and data collected during their historic mission.

Columbia was transported to Ford Island for deactivation, where its pyrotechnics were safely disarmed. Subsequently, the Command Module was relocated to Hickham Air Force Base and flown to Houston aboard a Douglas C-133 Cargomaster aircraft, reaching the Lunar Receiving Laboratory on July 30.

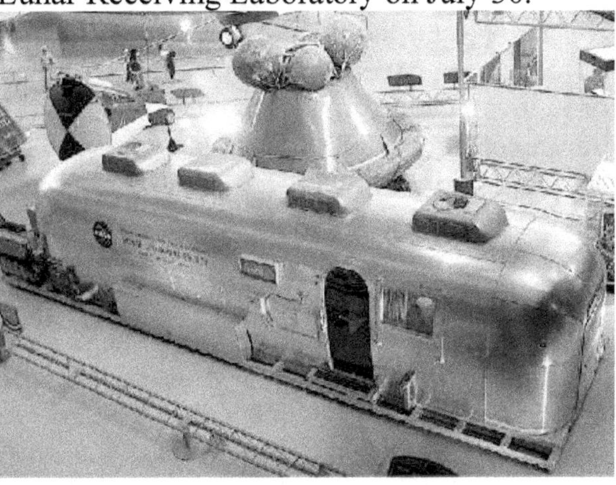

The Apollo 11 Mobile Quarantine Facility on display at the Steven F. Udvar-Hazy Center in 2009

Following NASA's Extra-Terrestrial Exposure Law, enacted on July 16 to formalize quarantine procedures, the astronauts remained in confinement throughout their quarantine period. After three weeks of isolation, initially aboard the Apollo spacecraft, then in their trailer on Hornet, and finally at the Lunar Receiving Laboratory, the astronauts received a clean bill of health.

On August 10, 1969, the Interagency Committee on Back Contamination convened in Atlanta. It officially lifted the quarantine on the astronauts and those who had shared quarantine with them—NASA physician William Carpentier and MQF project engineer John Hirasaki. Loose equipment from Columbia remained isolated until the lunar samples were deemed safe for study, marking the conclusion of a meticulously planned and executed mission that marked a significant milestone in human exploration of space.

Following the successful completion of the Apollo 11 mission, the culmination of a national endeavor to land humans on the Moon and return them safely to Earth by the end of the decade was achieved. Mission Director Chris Kraft and fellow NASA officials celebrated the milestone, marking the pinnacle of human achievement in space exploration.

The mission's success was underpinned by meticulous planning and the exceptional performance of spacecraft systems honed through preceding missions. The Apollo 11 crew executed the piloted lunar landing and surface exploration with remarkable skill and precision, reflecting the effectiveness of their rigorous pre-mission training.

Guidance, navigation, and control techniques employed during the descent were executed flawlessly, supported by the reliable performance of the landing radar. The crew's adaptation to lunar gravity was swift, facilitated by well-designed extravehicular mobility units that enabled easy movement across the lunar surface.

Critical operational phases, including pre-launch checks and the lunar ascent, proceeded smoothly, attesting to thorough planning and execution. All activities were well within the crew's operational capabilities throughout the mission timeline, from landing to departure from the Moon.

The meticulous quarantine operation, spanning from spacecraft landing to the release of crew members and lunar samples from the Lunar Receiving Laboratory, was flawlessly executed, ensuring containment without breach. Importantly, no extraterrestrial microorganisms were found among the crew or spacecraft, validating the effectiveness of quarantine measures.

Technical challenges encountered in earlier missions were mitigated, ensuring crew safety and mission success were never compromised. The Mission Control Center and Manned Space Flight Network demonstrated robust capabilities in monitoring and controlling all mission phases, from descent to ascent.

Apollo 11, the pinnacle of NASA's ambitious lunar exploration program, achieved its primary objective with the historic piloted lunar landing and safe return of its crew to Earth. Beyond this landmark achievement, the mission's secondary objectives encompassed a comprehensive exploration and scientific agenda.

Astronauts diligently conducted selenological inspections and sampling on the lunar surface. They collected contingency samples, studied lunar surface characteristics, and gathered bulk samples, providing crucial insights into the Moon's geological composition and history. Their assessments of the lunar environment's visibility were thorough and instrumental in enhancing our understanding of lunar conditions.

The mission also aimed to assess the capabilities of astronauts and their equipment in the challenging lunar environment. Astronauts successfully executed extravehicular operations and maneuvered with extravehicular mobility units, showcasing human adaptability to lunar gravity. They meticulously studied the effects of lunar landing on the module and evaluated potential contamination by lunar material, ensuring scientific integrity.

Photographic and television coverage from various vantage points—from the command module in orbit to the lunar surface itself—captured iconic images and critical data. This included sequences of the descent, lunar surface activities, and ascent, providing a visual chronicle of humanity's first steps on another celestial body.

Scientific experiments conducted during the mission yielded significant results. The passive seismic experiment, aimed at studying moonquakes, successfully gathered data. Despite time constraints, astronauts collected core tube and additional lunar samples, contributing to our understanding of lunar geology. Other experiments, such as laser ranging and cosmic ray detection, provided essential data on lunar environment and cosmic phenomena.

Wearing biological isolation garments, crew enters the Mobile Quarantine Facility aboard recovery ship U.S.S. Hornet (NASA S69-40753).

The launch vehicle performed flawlessly, precisely inserting the spacecraft into Earth's orbit before propelling it toward the Moon. Careful maneuvering ensured the spacecraft's safe separation from the S-IVB stage, averting potential collisions with Earth or the Moon.

In summary, Apollo 11 fulfilled its primary objective of landing and returning to the Moon and exceeded expectations in scientific exploration and technological achievement.

Ticker tape parade in New York City

Apollo 11 bulk rock samples collected during the mission (NASA S69-45519).

Apollo 12

Apollo 12 stands as a pivotal chapter in NASA's Apollo program, representing the agency's sixth crewed mission aimed at lunar exploration. Led by Commander Charles Conrad Jr., with Richard F. Gordon Jr. as Command Module Pilot and Alan L. Bean as Lunar Module Pilot, the mission commenced on November 14, 1969, lifting off from Kennedy Space Center's Launch Complex 39A atop the towering Saturn V rocket, designated SA-507.

The spacecraft, comprising the Command and Service Module (CSM) named Yankee Clipper, and the Lunar Module (LM) named Intrepid—crafted by North American Rockwell and Grumman respectively—embarked on a trajectory toward the Moon after a flawless launch. Four days into the mission, on November 18, Apollo 12 entered lunar orbit, preparing for the critical descent to the surface.

The chosen landing site was the Ocean of Storms, strategically close to the Surveyor 3 spacecraft, an earlier robotic probe that had touched down on the Moon two years prior. On November 19, 1969, Commander Conrad and Lunar Module Pilot Bean made their historic descent to the lunar surface aboard Intrepid. Over two extravehicular activities (EVAs), spanning 7 hours and 45 minutes, they traversed the rugged lunar terrain, deploying scientific instruments, collecting lunar samples, and conducting experiments.

The prime crew of the Apollo 12 lunar landing mission. From left to right they are: Commander, Charles "Pete" Conrad Jr.; Command Module pilot, Richard F. Gordon Jr.; and Lunar Module pilot, Alan L. Bean.

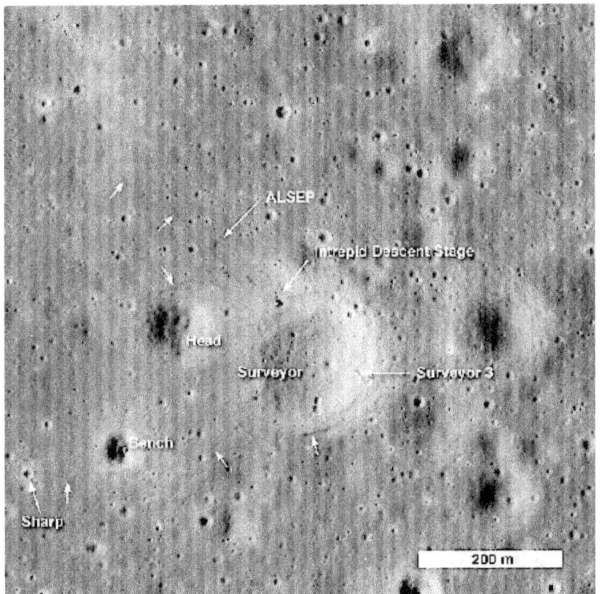

Meanwhile, Command Module Pilot Gordon orbited the Moon aboard Yankee Clipper, conducting observations and maintaining communications with Earth. Apollo 12's mission objectives extended beyond exploration; they included demonstrating pinpoint landing capabilities near previous robotic missions. The crew achieved these goals, amassing crucial lunar samples and deploying scientific instruments to study the Moon's geology and environment.

After approximately 31 hours on the lunar surface, Conrad and Bean rejoined Gordon in lunar orbit. On November 21, 1969, they bid farewell to the Moon, ascending aboard the LM ascent stage to rendezvous and dock with the CSM for their return journey to Earth.

Apollo 12 concluded its mission on November 24, 1969, splashing down in the South Pacific Ocean, where the crew was recovered by the USS Hornet. The mission's total duration spanned 10 days, 4 hours, 36 minutes, and 24 seconds, marking another triumph in NASA's lunar exploration endeavors and furthering human capabilities in space exploration, flying 622,268 miles.

The commander of the all-Navy Apollo 12 crew was Charles "Pete" Conrad Jr., who, at 39 years old during the mission, brought a wealth of experience and expertise to NASA's lunar exploration efforts. Conrad began his journey into aviation after earning a bachelor's degree in aeronautical engineering from Princeton University in 1953. Following his education, he became a naval aviator and completed training at the United States Naval Test Pilot School at Patuxent River Naval Air Station.

Selected as part of NASA's second group of astronauts in 1962, Conrad's career in spaceflight began with Gemini 5 in 1965, where he served as pilot. He later commanded Gemini 11 in 1966, demonstrating his capabilities and paving the way for his role as commander of Apollo 12.

Richard "Dick" Gordon, the mission's Command Module Pilot, was 40 years old at the time of Apollo 12. He, too, started as a naval aviator after earning a chemistry degree from the University of Washington in 1951 and completing test pilot school at Patuxent River. Selected in NASA's Group 3 of astronauts in 1963, Gordon had previously flown with Conrad on Gemini 11, establishing a strong rapport crucial for their Apollo mission.

Alan L. Bean, at 37 years old during the mission, was making his first journey into space as the Lunar Module Pilot. Graduating from the University of Texas in 1955 with a degree in aeronautical engineering, Bean also pursued a career as a naval aviator. Selected alongside Gordon in 1963, Bean's assignment to Apollo 12 was initially uncertain due to other obligations within the Apollo Applications Program. However, following the tragic loss of Clifton C. Williams Jr., Conrad successfully lobbied for Bean to join his crew, building on their prior relationship from test pilot school.

Together, Conrad, Gordon, and Bean formed a cohesive team that had previously served as backups for Apollo 9 earlier in 1969. Their combined skills, dedication, and camaraderie played pivotal roles in the success of Apollo 12, marking another significant achievement in NASA's ambitious lunar exploration program.

For the Apollo 12 mission, a diverse team of capsule communicators (CAPCOMs) and support crew members played crucial roles in supporting the astronauts during their journey to the Moon. Leading the CAPCOM team were military and civilian experts including Lt. Colonel Gerald Paul Carr (USMC), Edward George Gibson, Ph.D., Commander Paul Joseph Weitz (USN), and Don Leslie Lind, Ph.D. They communicated directly between mission control and the crew, ensuring smooth operations and relaying critical information.

Supporting the mission were also civilian backup CAPCOMs Dickie K. Warren, James O. Rippey, James L. Lewis, and Michael R. Wash, who stood ready to assist as needed. The support crew, consisting of Carr, Weitz, and Gibson, worked alongside the prime and backup crews, aiding in mission preparation and ensuring all operational aspects were meticulously managed.

The flight directors for Apollo 12 included Gerald D. Griffin, M. P. "Pete" Frank, Clifford E. Charlesworth, and Milton L. Windler, each responsible for overseeing different shifts during the mission. Their primary role was to make critical decisions ensuring crew safety and mission success, adhering to a succinct directive: "The flight director may take any actions necessary for crew safety and mission success."

Conrad and Bean rehearse their lunar surface activities before the mission

Additionally, the Apollo 12 backup crew, originally designated to support the mission but later assigned to Apollo 15, consisted of David R. Scott as commander, Alfred M. Worden as Command Module pilot, and James B. Irwin as Lunar Module pilot.

Apollo 12, originally poised as NASA's contingency for a lunar landing had Apollo 11 encountered issues, experienced a significant alteration in its timeline and mission scope following the success of Neil Armstrong's historic mission. The mission was postponed by two months, aligning with a more relaxed schedule for subsequent Apollo missions. This delay allowed for extensive geological training, emphasizing Conrad and Bean's preparation with multiple field trips focused on lunar geology—a crucial aspect for their upcoming lunar exploration.

The spacecraft and launch vehicle for Apollo 12 closely mirrored those of Apollo 11, with minor modifications such as including hammocks to enhance comfort for Conrad and Bean during their lunar stay.

The site selection process for Apollo 12 closely followed that of Apollo 11, integrating lessons learned and leveraging existing criteria for optimal landing sites on the Moon. Initially, the selection criteria were stringent and primarily focused on operational feasibility rather than scientific interest. Key considerations included proximity to the lunar equator, ensuring visibility from Earth, and a relatively flat terrain with minimal obstructions along the Lunar Module's flight path.

Sites were meticulously evaluated using photographs of Lunar Orbiter probes to confirm suitability. Additional factors included the availability of backup sites further west, ensuring favorable lighting conditions, and accommodating potential launch delays. Given the three-day recycling period required if a launch was scrubbed, only three out of the initially identified five suitable sites were designated as potential landing sites for Apollo 11.

The primary landing site for Apollo 11, the Sea of Tranquility, was selected as the easternmost option due to its alignment with mission requirements and operational logistics. The Apollo 11 and Apollo 12 crews trained extensively for these designated sites, ensuring they were well-prepared for any contingency, including a potential first lunar landing attempt by Apollo 12 had Apollo 11 encountered difficulties.

Gordon in the CM simulator

Following the triumphant success of Apollo 11, plans for Apollo 12 initially aimed to land in Sinus Medii, the next site west of the Sea of Tranquility. However, a compelling argument arose from NASA planning coordinator Jack Sevier and engineers at the Manned Spaceflight Center in Houston. They proposed a landing site near the location of the Surveyor 3 probe, which had touched down on the lunar surface in 1967. This site not only met operational criteria but also held significant scientific interest, offering the opportunity for astronauts to retrieve parts of the probe for analysis back on Earth.

Despite the appeal of this scientifically compelling site, concerns lingered within NASA. The precision of lunar landings still needed to be improved; Apollo 11 had landed several miles off its intended target. There was apprehension that Apollo 12 might also miss its mark, potentially leaving the astronauts unable to reach Surveyor 3 and causing embarrassment for the agency.

Nevertheless, prioritizing precision landings was pivotal for NASA's lunar exploration program. On July 25, 1969, Apollo Program Manager Samuel Phillips decided to designate the landing site near Surveyor 3, now known as Surveyor Crater. Although members of two site selection boards opposed this decision, it underscored NASA's commitment to advancing lunar exploration capabilities and executing scientific missions with increasing precision and effectiveness.

Conrad and Bean's preparation for the Apollo 12 mission involved an intensive training regimen and rehearsal, surpassing the standards set by previous missions. They dedicated extensive hours to mission-specific training, averaging five hours of training for every anticipated hour of flight—a total exceeding 1,000 hours per crew member. This surpassed the training hours allocated to Neil Armstrong and Buzz Aldrin for Apollo 11.

In addition to their primary mission training, Conrad and Bean had accumulated over 1,500 hours as backup crew members for Apollo 9, enhancing their readiness and familiarity with mission protocols. Specific to Apollo 12, their training regimen included more than 400 hours per crew member in simulators for both the Command Module (CM) and the Lunar Module (LM). These simulations were crucial, as they were conducted in real-time coordination with flight controllers at Mission Control, replicating mission scenarios and contingencies.

Conrad honed his skills by flying the Lunar Landing Training Vehicle (LLTV) to simulate lunar surface operations. This training continued despite a previous incident where Armstrong had to eject from a similar vehicle just before it crashed in 1968. This hands-on experience with the LLTV was essential for practicing the critical maneuvers required for landing and operating the Lunar Module on the Moon's surface.

Astronauts Charles Conrad Jr. (on left), commander; and Alan L. Bean, lunar module pilot, are shown in the Apollo Lunar Module Mission Simulator during simulator training at the Kennedy Space Center (KSC).

After assuming command of the Apollo 12 mission, Conrad took a decisive approach to lunar surface training, drawing from his past experiences and lessons learned. He clarified to NASA geologists that training for lunar activities would follow a structure similar to Apollo 11's. Still, with a strict directive, there would be minimal publicity and media involvement. Conrad had experienced intrusive media scrutiny during his Gemini missions and was determined to avoid distractions during Apollo 12's critical preparation phase.

The decision stemmed partly from the chaotic experience of Apollo 11's geology field trip, which had attracted a large media contingent, causing logistical challenges and distractions for the astronauts. Conrad emphasized the need for focused training without external disruptions, ensuring that simulations and preparations were conducted efficiently and without interference.

Following the successful return of Apollo 11 in July 1969, more time was allocated for geology training for the Apollo 12 crew. Despite this, Conrad and his team faced challenges in securing dedicated simulator time as the Apollo 11 crew continued to prioritize their own post-mission activities.

The core of the Apollo 12 geology training involved six field trips designed to replicate lunar conditions as closely as possible. Conrad and Bean practiced collecting and documenting samples with photographs while communicating with CAPCOM and geologists stationed out of sight in a nearby tent. This setup allowed for realistic training scenarios where their actions could be critiqued and refined afterward.

However, the training was fraught with frustrations. Conrad expressed dissatisfaction with the frequent changes in photo documentation procedures imposed by scientists. After multiple revisions, he eventually insisted on freezing the procedures to maintain consistency and clarity in the mission protocols.

Following Apollo 11's return, the Apollo 12 crew had the opportunity to examine lunar samples firsthand and receive detailed briefings from scientists. This direct interaction with the lunar material provided valuable insights and prepared them further for the scientific tasks awaiting them on the Moon.

Conrad's approach to training underscored his commitment to meticulous preparation and focus, ensuring that the Apollo 12 mission would proceed smoothly and effectively, with minimal distractions and maximum readiness.

For Apollo 11, the landing was targeted within a broad ellipse-shaped zone, which did not include specific plans for detailed geology traverses or designated tasks at predetermined sites. This approach contrasted with the preparations for Apollo 12, where Commander Conrad proactively engaged NASA's geology team well before the mission.

Conrad initiated discussions with the geology team, suggesting they collaborate on planning possible routes for exploration once he and Bean landed on the Moon. This proactive approach led to four specific traverses, each based on potential landing points for the Lunar Module (LM). These traverses were designed to guide Conrad and

Bean to scientifically interesting locations on the lunar surface, enhancing the mission's scientific objectives and maximizing the astronauts' efficiency in collecting samples and conducting experiments.

The introduction of these geology traverses marked the beginning of a systematic planning effort that would evolve significantly in subsequent missions. As Apollo missions progressed, geology traverse planning became a substantial collaborative effort involving multiple organizations. This systematic approach ensured that each mission could target specific geological features and conduct detailed scientific investigations, contributing to a deeper understanding of the Moon's composition and history.

The components of the Apollo 12 spacecraft were meticulously assembled and prepared for their mission to the Moon. The Lunar Module (LM-6) stages arrived at Kennedy Space Center (KSC) on March 24, 1969, and were joined together on April 28. Similarly, the Command Module (CM-108) and Service Module (SM-108) were delivered on March 28 and integrated on April 21. After thoroughly installing equipment and rigorous testing, the entire spacecraft, atop its launch vehicle, was moved to Launch Complex 39A on September 8, 1969.

SA-507 en route to the launch pad, September 1969

By November 1, 1969, the training schedule for the Apollo 12 crew was completed as planned, with subsequent activities serving as refresher sessions. The crew members expressed confidence that their training adequately prepared them for their mission to the Moon.

The Apollo 12 launch vehicle, designated SA-507, was a Saturn V rocket similar to the one used for Apollo 11, with no significant changes. It was also known by its Eastern Test Range designation, #2793. The Command and Service Module (CSM) was designated CSM-108 and bore the call-sign "Yankee Clipper." The Lunar Module (LM) was designated LM-6 and had the call-sign "Intrepid."

The entire launch vehicle, including the spacecraft, weighed 6,487,742 pounds (2,942,790 kg) at launch, slightly more than Apollo 11's 6,477,875 pounds (2,938,315 kg). The spacecraft itself weighed 110,044 pounds (49,915 kg), a slight increase from Apollo 11's 109,646 pounds (49,735 kg).

The Apollo 12 CSM on a test stand, June 30, 1969

The final countdown for Apollo 13 commenced at T-28 hours on November 13th, 02:00:00 GMT, marking a pivotal moment in NASA's ambitious lunar program. However, meticulous preparations were punctuated by challenges. On November 12th, during spacecraft preparations, engineers detected a leak in the CSM LH2 tank No. 2 during cryogenic loading, necessitating swift action. The tank was promptly drained and replaced with one from the Apollo 13 CSM, a critical substitution to ensure mission readiness.

As the countdown progressed, another hurdle emerged at T-17 hours on November 13th, 12:00:00 GMT, prompting an unscheduled hold. Crews swiftly reloaded cryogenics into the CSM, a meticulous process that took six hours to complete. Despite these setbacks, the countdown resumed at 19:00:00 GMT, with adjustments made to mitigate the impact on the launch schedule.

Outside, the weather posed its own challenges. A slow-moving cold front traversed the central region of Florida, casting rain showers and enveloping the launch pad in overcast conditions. The sky was dominated by stratocumulus clouds, stretching from a base of 2,100 feet. The temperature held steady at 68.0° F with a stifling 92 percent humidity. The barometric pressure stood at 14.621 lb/in², indicative of the atmospheric pressure at the launch site.

Amidst these meteorological conditions, winds gusted at 13.2 knots from 280° true north, as recorded by an anemometer positioned atop a light pole 60.0 feet above ground level. These precise environmental measurements underscored the meticulous planning and attention to detail required for each Apollo mission, where scientific precision and operational readiness were paramount in the quest for lunar exploration.

Apollo 12's historic launch from Kennedy Space Center Launch Complex 39, Pad A, on November 14, 1969, marked a pivotal moment in space exploration. Scheduled precisely between 16:22:00 GMT and 19:26:00 GMT, the launch aimed to synchronize with a sun elevation angle of 5.1° on the lunar surface, optimizing conditions for future mission phases.

Despite challenging weather conditions, including rainstorms that would typically halt such endeavors under NASA's strict safety protocols, Apollo 12 launched, defying conventional wisdom. The decision to waive the Manned Space Flight Center Launch Mission Rule 1-404, which traditionally prohibited launches during cumulonimbus cloud formations, highlighted the mission's urgency and NASA's confidence in its technological preparations.

Apollo 12 launches from Kennedy Space Center, November 14, 1969

During the ascent phase, the mission encountered a dramatic electrical disturbance just moments after liftoff. Between 000:00:12.8 and 000:00:32.3, the Saturn V rocket veered from its launch azimuth of 90° to a flight azimuth of 72.029°. At 000:00:36.5, lightning struck the vehicle itself, a rare event that resulted in a cascade of electrical failures. The discharge caused a voltage transient that disabled all three fuel cells in the service module, leaving the spacecraft reliant solely on battery power. A second lightning strike at 000:00:52 further compounded the situation by knocking out the "8-ball" attitude indicator, essential for navigation.

Despite these critical setbacks, the Saturn V's instrument unit guidance system remained unaffected, ensuring the rocket continued on its intended trajectory toward space. Inside Mission Control, telemetry streams became garbled, complicating the situation further;. At the same time, aboard the spacecraft, the crew faced a cockpit illuminated by caution and warning lights, signaling an unprecedented challenge.

The presence of President Richard Nixon and Vice President Spiro Agnew at the launch underscored the national significance of Apollo 12's mission. This launch aimed to continue humanity's exploration of the Moon and demonstrated NASA's resilience in overcoming unforeseen challenges, setting the stage for subsequent lunar landings and scientific achievements.

In Mission Control, John Aaron, the Electrical, Environmental, and Consumables Manager (EECOM), played a crucial role during Apollo 12's tumultuous launch. Drawing on his sharp memory of a previous test scenario, Aaron quickly recognized the telemetry failure pattern caused by the power loss to the Command Module's signal conditioning electronics (SCE). Without hesitation, he recommended a solution: "Flight, EECOM. Try SCE to Aux."

This directive puzzled Flight Director Gerald Griffin, CAPCOM Gerald P. Carr, Commander Charles Conrad, and Command Module Pilot Richard Gordon. However, Lunar Module Pilot Alan Bean, as the spacecraft's engineer, knew exactly where to locate the obscure switch and promptly toggled it. Miraculously, telemetry was restored, revealing no significant system malfunctions. Bean swiftly reactivated the fuel cells, restoring power to the spacecraft and allowing the mission to proceed.

Once safely in Earth parking orbit, the crew meticulously inspected their craft before preparing for the next critical maneuver: igniting the S-IVB third stage for trans-lunar injection. Despite the lightning strikes during ascent, which initially raised concerns about potential damage to the explosive bolts used for deploying the Command Module's parachutes, subsequent checks confirmed no lasting harm.

Amidst these challenges, a decision was made not to disclose the potential parachute damage to the crew. The mission continued as planned, understanding that aborting due to this uncertainty wouldn't mitigate the risk since failure during re-entry, whether from an Earth orbit or return from the Moon, would have dire consequences regardless. Fortunately, the parachutes deployed flawlessly at the mission's conclusion, affirming the team's cautious optimism and perseverance in the face of unexpected adversity.

Following the separation of the Lunar Module (LM), the third stage of the Saturn V, known as the S-IVB, was intended to be propelled into solar orbit. However, due to an operational error, the S-IVB passed the Moon at too high an altitude to achieve escape velocity from Earth's gravity. Consequently, it entered a semi-stable Earth orbit until eventually escaping Earth's gravitational pull in 1971, only to return to Earth orbit 31 years later briefly. Discovered by amateur astronomer Bill Yeung, it was initially designated J002E3 before its artificial origin was confirmed. As of 2021, it remains in solar orbit, with the potential to be recaptured by Earth's gravity in the future, though only expected at least the 2040s.

The S-IVBs from subsequent lunar missions were intentionally crashed into the Moon to generate seismic events, providing valuable data on the Moon's internal structure through seismometer readings left behind by previous missions. This dual purpose of lunar impact and scientific experimentation underscored NASA's commitment to maximizing the scientific returns from each Apollo mission despite unexpected challenges and setbacks.

Ground camera data, telemetered data, and analysis from launch computers unequivocally confirmed that Apollo 12 was struck by lightning during its ascent. This event had virtually no discernible impact on the launch vehicle during the subsequent disturbance. The detailed investigation pointed to atmospheric electrical conditions and the vehicle's limited capacitance, which indicated that the initial discharge was likely self-induced. A second, lesser discharge followed shortly after that, but neither event resulted from the Saturn V flying through cumulonimbus clouds, thus validating the waiver of Rule 1-404.

Importantly, the launch vehicle's hardware and software remained resilient, suffering no significant effects from the lightning strikes. Consequently, the mission continued without delay or alteration to its planned trajectory. The S-IC engine shutdown occurred precisely at 000:02:41.74, followed by the separation of S-IC/S-II stages and the ignition of the S-II engine at 000:09:12.34. Subsequently, the S-II engine shutdown at 000:09:12.34 facilitated the separation from the S-IVB stage, which ignited at 000:09:16.16.

Throughout these critical maneuvers, Apollo 12 maintained remarkable precision, with deviations from its planned trajectory limited to just -1.9 ft/sec in velocity and 0.2 nautical miles in altitude at the time of S-IVB's first engine cutoff at 000:11:33.91. This exceptional performance underscored NASA's meticulous planning and the robustness of its mission protocols, ensuring that even unexpected challenges like lightning strikes were managed without compromising the mission's objectives or safety protocols.

After the successful launch of Apollo 12, the various stages of the Saturn V rocket followed their planned trajectories, culminating in controlled impacts in the Atlantic Ocean. The first stage, S-IC, completed its mission at 000:09:14.5, impacting the ocean at approximately latitude 30.273° North and longitude 73.895° West. This location was about 365.2 nautical miles from the Kennedy Space Center launch site.

Electrical discharge between douds and ground 36.5 seconds after liftoff

Subsequently, the second stage, S-II, carried the mission further into space before completing its task. It impacted the Atlantic Ocean at 000:20:21.6, positioned much farther away at latitude 31.465° North and longitude 34.214° West, approximately 2,404.4 nautical miles from the launch site. These impacts marked the successful conclusion of the rocket stages' roles in facilitating Apollo 12's journey toward the Moon, highlighting the precision and complexity of mission planning and execution in space exploration.

During Apollo 12's ascent, the spacecraft encountered challenging atmospheric conditions. At 46,670 feet, maximum wind speeds peaked at 92.5 knots, blowing from an azimuth of 245° true north. This significant wind shear was further underscored by a maximum shear rate of 0.0183 seconds^{-1} at 46,750 feet, highlighting the dynamic and demanding environment through which the mission navigated.

After meticulous systems checks in Earth orbit, prompted by the earlier lightning strikes, Apollo 12 proceeded with the critical trans-lunar injection burn at 02:47:22.80 into the mission. This precise maneuver, powered by the S-IVB stage, propelled the spacecraft onto its trajectory toward the Moon.

Following this burn, approximately an hour and twenty minutes later, the Command and Service Module (CSM) separated from the S-IVB stage. Commander Charles Conrad then initiated the transposition, docking, and extraction maneuver. This complex sequence involved maneuvering the CSM to dock with the Lunar Module (LM) and subsequently separating the combined craft from the spent S-IVB stage.

Unlike Apollo 11, where the Service Module's Service Propulsion System (SPS) engine was used to distance itself from the S-IVB, Apollo 12's S-IVB stage fired its engines to depart the spacecraft's vicinity. This maneuver was intended to set the stage on a trajectory toward solar orbit, marking a strategic departure from Earth's gravitational influence and clearing the way for the continued journey of the lunar-bound Apollo 12 spacecraft.

Due to initial concerns over potential damage from lightning strikes, Commander Charles Conrad and Lunar Module Pilot Alan Bean performed an unscheduled inspection of the Lunar Module (LM) on the first day of Apollo 12's flight. Their thorough examination, earlier than planned, revealed no issues with the spacecraft, reassuring the crew and mission control alike.

As Apollo 12 continued its translunar coast, the only midcourse correction necessary occurred at 30:52.44.36 into the mission. This adjustment positioned the spacecraft onto a hybrid, non-free-return trajectory—a departure from previous crewed missions to lunar orbit, following a free-return trajectory. This allowed a safe return to Earth without requiring engine firings to enter lunar orbit.

Apollo 12 thus became the first crewed spacecraft to adopt this hybrid trajectory. Unlike the free-return path, this trajectory would require an additional burn to return to Earth, but it offered increased mission flexibility. Notably, it enabled Apollo 12 to launch during daylight hours and arrive at its intended lunar landing site on schedule, demonstrating the strategic advantages of the hybrid trajectory in mission planning.

However, using this hybrid trajectory meant that Apollo 12's journey from trans-lunar injection to lunar orbit took approximately 8 hours longer than missions on a free-return trajectory. Despite this extended duration, the mission proceeded smoothly, leveraging the Lunar Module's Descent Propulsion System (DPS) as a backup to the Command Module's Service Propulsion System (SPS) in case of engine failure, ensuring redundancy and safety throughout the lunar approach phase.

Apollo 12 achieved lunar orbit with a trajectory of 170.2 by 61.66 nautical miles (315.2 by 114.2 km; 195.9 by 70.96 mi) after a critical 352.25-second burn of the Service Propulsion System (SPS) at mission time 83:25:26.36. This burn effectively positioned the spacecraft to begin its lunar exploration.

During the first lunar orbit, a significant television transmission provided high-quality video footage of the lunar surface, offering unprecedented views back to Earth. This broadcast was a momentous event, capturing the attention of audiences worldwide and enhancing the public's engagement with the mission.

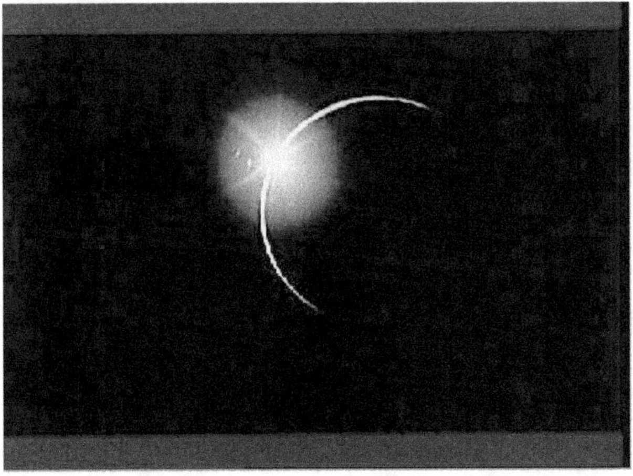

A solar eclipse seen from Apollo 12

On the third lunar orbit, another crucial burn was executed to circularize the spacecraft's orbit, adjusting it to 66.1 by 54.59 nautical miles (122.4 by 101.1 km; 76.07 by 62.82 mi). This circularization maneuver prepared the craft for subsequent operational phases, particularly the lunar landing preparations.

At 107:54:02.3 into the mission, the Command and Service Module (CSM) and Lunar Module (LM) were undocked in preparation for the landing sequence. Thirty minutes later, a precise 14.4-second burn by the CSM's thrusters facilitated a controlled separation, positioning the two spacecraft approximately 2.2 nautical miles (4.1 km; 2.5 mi) apart. This maneuver allowed the LM into a lower orbit in readiness to descend to the lunar surface, marking a pivotal step toward Apollo 12's historic lunar landing mission.

The Lunar Module's (LM) Descent Propulsion System initiated a crucial 29-second burn at 109:23:39.9, maneuvering the spacecraft to a lower orbit in preparation for its descent to the lunar surface. This positioning set the stage for the next pivotal phase of Apollo 12's mission.

View of Earth taken en route to the Moon

At 110:20:38.1, the LM began its powered descent to the lunar surface, marking a tense and critical moment for Commander Charles Conrad and Lunar Module Pilot Alan Bean.

Conrad, expecting to see a distinct pattern of craters known as "the Snowman" during the LM's descent "pitchover," was relieved and amazed to witness the anticipated landmark, with the Surveyor Crater prominently centered. This visual confirmation indicated their trajectory was precisely on course, boosting confidence in the mission's success.

Taking manual control of the LM, Conrad navigated toward the designated landing site near Surveyor Crater, an area affectionately dubbed "Pete's Parking Lot" in honor of his nickname. However, the terrain proved rougher than anticipated, requiring Conrad's expert piloting skills to adjust the LM's descent path.

Finally, at 110:32:36.2 (06:54:36 UT on November 19, 1969), Conrad successfully landed the LM on the lunar surface just 535 feet (163 m) from the Surveyor probe. This achievement fulfilled a key objective of the mission: executing a precision landing near the Surveyor spacecraft, demonstrating the advanced capabilities and precision of the Apollo lunar landing program.

Upon achieving parking orbit at 000:11:43.91 into the mission, Apollo 12 meticulously documented its spacecraft conditions. The orbit parameters were carefully recorded: an apogee (the farthest point from Earth) of 100.1 nautical miles and a perigee (the closest point to Earth) of 97.8 nautical miles. The orbit's inclination, the angle between the orbital plane and Earth's equatorial plane, was measured at 32.540°. The orbital period, the time taken to complete one orbit around Earth, was precisely calculated at 88.16 minutes. The spacecraft maintained a velocity of 25,565.9 feet per second relative to Earth.

These measurements were computed based on a spherical Earth model with a radius of 3,443.934 nautical miles. Such accurate orbital calculations were essential for planning and executing the complex maneuvers required throughout Apollo 12's mission, ensuring precise navigation from Earth orbit to lunar orbit and ultimately to the successful landing on the Moon.

Upon achieving orbit, the components of the Apollo 12 spacecraft were assigned international designations for clear identification and tracking purposes. The Command and Service Module (CSM) received the designation 1969-099A, marking its role as the primary vehicle housing the crew during their journey to and from the Moon. The S-IVB stage, which played a crucial role in the initial phases of the mission, was designated as 1969-099B.

Following the undocking at the Moon, the Lunar Module (LM) ascent stage was identified as 1969-099C, responsible for returning the astronauts from the lunar surface to the CSM. The LM descent stage, designated 1969-099D, remained on the Moon after its purpose was fulfilled, maintaining distinct operational distinctions throughout Apollo 12's lunar mission phases.

The Apollo 12 spacecraft was meticulously engineered, comprising various components to ensure mission success and crew safety. These included Command Module 108 and Service Module 108, collectively called CSM-108, which provided essential living space, navigation controls, and communication equipment. The CSM-108 served as the mission's primary command center, overseeing critical maneuvers and facilitating all lunar operations with precision and reliability.

The Lunar Module 6 (LM-6), crucial for the lunar landing phase, was securely housed within the Spacecraft-Lunar Module Adapter 15 (SLA-15). Serving as a protective cover and a structural interface between the Saturn V rocket and the LM, the SLA-15 ensured the safe transportation and deployment of the lunar module during the entirety of the Apollo 12 mission.

Integral to crew safety, the Launch Escape System (LES) featured three rocket motors designed to swiftly propel the Command Module to a safe distance in the event of an abort scenario shortly after launch. This system provided a robust mechanism for crew escape, enhancing the overall safety protocols of the mission.

While SLA-15 mirrored the configuration used successfully in Apollo 11, ensuring reliability, the LES incorporated upgrades such as a more dependable motor igniter. These enhancements further optimized the spacecraft's abort capabilities and bolstered its resilience throughout the mission.

The Apollo 12 mission distinguished itself not only through its crew's chosen call signs — "Yankee Clipper" for the Command and Service Module (CSM) and "Intrepid" for the Lunar Module (LM) — but also through subtle yet significant modifications made to its spacecraft compared to its predecessor, Apollo 11.

The names "Yankee Clipper" and "Intrepid" were carefully chosen by the all-Navy crew of Apollo 12 from a vast array of proposed names submitted by employees of the prime contractors responsible for building the spacecraft modules. George Glacken, a flight test engineer at North American Aviation involved in the Command and Service Module (CSM), proposed "Yankee Clipper." His inspiration came from the historic ships known as Clippers, which symbolized American pride and maritime excellence.

On the other hand, Robert Lambert, a planner at Grumman Aerospace who contributed to the Lunar Module (LM), suggested the name "Intrepid." This name was chosen to embody the nation's steadfast dedication to space exploration and emphasize the astronauts' bravery and determination to overcome challenges.

Regarding technical enhancements, the CSM of Apollo 12 featured a hydrogen separator to prevent gaseous hydrogen from entering the potable water tank. This separator addressed the discomfort experienced by the Apollo 11 crew due to hydrogen in the water causing severe flatulence. This separator was relocated from the water dispenser in the Command Module's cabin, where it had been on Apollo 11. Another improvement was made to the recovery loop attached to the CSM following splashdown, eliminating the need for swimmers to attach an auxiliary loop during recovery operations.

Changes to the LM included structural modifications to accommodate scientific experiment packages intended for deployment on the lunar surface. Additionally, two hammocks were added to enhance astronaut comfort during rest periods on the Moon. A notable upgrade was replacing the black-and-white television camera used on Apollo 11 with a color television camera for Apollo 12, improving the quality and detail of visual data transmitted from the lunar surface.

The Apollo Lunar Surface Experiments Package (ALSEP) represented a pivotal component of NASA's lunar exploration strategy, comprising a suite of scientific instruments designed for deployment on the lunar surface by Apollo astronauts. Developed to operate autonomously, these instruments would transmit valuable data back to Earth, contributing significantly to our understanding of lunar geology and environment.

NASA's decision to develop ALSEP was partly in response to scientific debates about the feasibility and cost-effectiveness of crewed lunar missions compared to robotic exploration. By demonstrating the unique capabilities of human astronauts in deploying and managing such complex scientific equipment, ALSEP aimed to justify the human presence on the Moon for scientific exploration.

In 1966, Bendix Corporation was awarded the contract to design and build the ALSEPs. The initial deployment of lunar experiments began with the Early Apollo Surface Experiment Package (EASEP) during Apollo 11, which consisted of a limited set of instruments due to time constraints. However, Apollo 12 marked the first full deployment of an ALSEP, setting a precedent for subsequent lunar missions.

The ALSEP deployed during Apollo 12 included various instruments strategically placed at least 300 feet (91 meters) from the Lunar Module (LM). This distance was crucial to shield the instruments from debris generated during the LM's ascent back to lunar orbit after completing its surface mission. Each ALSEP package varied across missions, tailored to specific scientific objectives and advancements in instrumentation.

Throughout the Apollo program, ALSEPs played a vital role in collecting seismic data, measuring lunar heat flow, studying lunar magnetic fields, and conducting other experiments that provided invaluable insights into the Moon's geological and environmental characteristics.

Apollo 12's ALSEP was equipped with diverse scientific instruments designed to study various aspects of the lunar environment and phenomena. These instruments included:

Lunar Surface Magnetometer (LSM): Measured the magnetic field at the Moon's surface, providing insights into lunar magnetic properties and their implications for lunar geology.

Lunar Atmosphere Detector (LAD, Cold Cathode Gauge Experiment): Intended to measure the density and temperature variations of the thin lunar atmosphere, helping scientists understand its composition and dynamics.

Lunar Ionosphere Detector (LID, Suprathermal Ion Detector Experiment or SIDE): Studied charged particles in the lunar atmosphere, contributing to our knowledge of lunar ionospheric conditions and interactions with solar radiation.

Solar Wind Spectrometer: Measured the strength and direction of the solar wind at the lunar surface, crucial for understanding how solar particles interact with the Moon's exosphere and surface materials.

Solar Wind Composition Experiment: This standalone experiment measured the composition of the solar wind, providing critical data on the elemental makeup and characteristics of particles emanating from the Sun. The astronauts later returned it to Earth for detailed analysis.

Dust Detector: Monitored lunar dust accumulation on the equipment, essential for assessing its impact on lunar surface operations and equipment longevity.

In addition to these instruments, Apollo 12's ALSEP included the Passive Seismic Experiment (PSE), which deployed a seismometer to measure moonquakes and other seismic activities within the Moon's crust. The PSE was calibrated using data from the planned impact of the LM ascent stage, which was expected to produce seismic signals equivalent to the explosive force of one ton of TNT upon impacting the lunar surface.

Apollo 12's Passive Seismic Experiment

The Apollo 12 ALSEP (Apollo Lunar Surface Experiments Package) was a sophisticated array of scientific instruments left on the lunar surface, connected to a central station that served as a hub for data transmission and power distribution. The central station housed essential components, including a transmitter, receiver, timer, data processor, and control equipment for managing the experiments.

Powering these experiments was the SNAP-27 (Systems for Nuclear Auxiliary Power) radioisotope thermoelectric generator (RTG), developed by the Atomic Energy Commission. This marked the first use of atomic energy on a crewed NASA spacecraft, though similar systems had been employed on certain NASA and military satellites before. The RTG contained plutonium and was securely transported from Earth in a cask attached to an LM landing leg. This design ensured the safety of the plutonium core in case of a mission abort, a scenario deemed unlikely by NASA. Remarkably, the cask survived an unplanned re-entry during the ill-fated Apollo 13 mission, sinking intact into the Tonga Trench of the Pacific Ocean without any radioactive leakage reported.

Activated remotely from Earth on November 19, 1969, the ALSEP experiments commenced operations on the lunar surface. However, the Lunar Atmosphere Detector (LAD) encountered an early setback when its power supply failed shortly after activation, resulting in limited data return. Despite this setback, other experiments, like the Lunar Surface Magnetometer (LSM), continued to operate, contributing valuable scientific insights.

Over the years, budgetary constraints led to the deactivation of the powered ALSEP experiments. The LSM, operational until June 14, 1974, and its counterpart deployed during Apollo 15 provided crucial data on lunar magnetic fields. Finally, on September 30, 1977, all remaining active ALSEP experiments were deactivated, marking the end of their scientific operations on the lunar surface.

During the Earth orbit phase of Apollo 12, meticulous systems checks were conducted, marked by heightened scrutiny due to the previous lightning strikes. These checks ensured the spacecraft's readiness for its journey toward the Moon. At 002:47:22.7 mission time, the second firing of the S-IVB engine commenced a crucial 341.24-second translunar injection maneuver. The engine shutdown occurred precisely at 002:53:03.94, followed ten seconds later by the actual translunar injection at a velocity of 35,419.3 feet per second.

This maneuver occurred after completing 1.5 Earth orbits, totaling 2 hours, 41 minutes, and 30.03 seconds. Notably, Apollo 12 adopted a high-pericynthion free-return translunar profile for the first time, enhancing its capability for a safe return to Earth within the reaction control velocity correction capacity. Termed a "hybrid" non-free-return trajectory, this new profile offered significant mission planning flexibility. It enabled a daylight launch aligned with the intended landing site and provided a larger performance margin for the service propulsion system.

Despite its flexibility, the hybrid profile was designed with constraints, ensuring that a safe return using the descent propulsion system remained feasible even if the vehicle failed to enter lunar orbit.

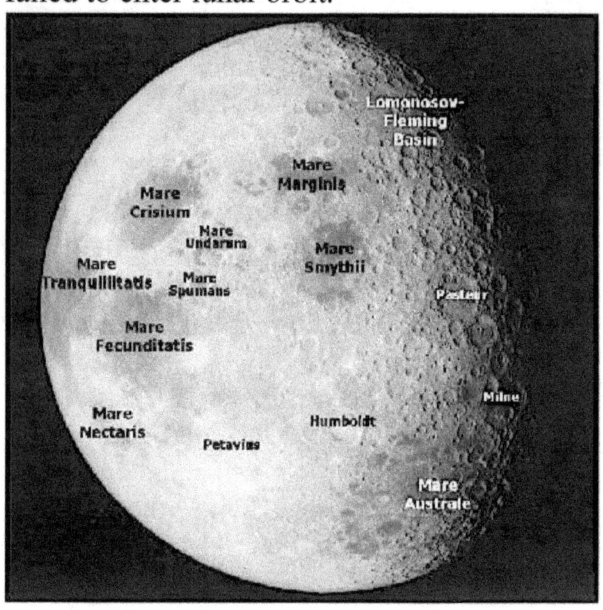

East side of the Moon showing the Lomonosov Fleming basin near the terminator. Major features are marked for orientation. North is toward the top.

At 003:18:04.9, the Command and Service Module (CSM) completed its separation from the S-IVB stage, underwent transposition, and successfully docked with the Lunar Module (LM) by 003:26:53.3. This critical maneuver was captured and transmitted via onboard television from 003:25 to 004:28, providing clear visual confirmation of the docking process.

Following the successful docking, the docked spacecraft were ejected from the S-IVB stage at 004:13:00.9. At 004:27, an evasive maneuver was executed using the S-IVB's auxiliary propulsion system, also observed via television transmission. A ground command initiated the propulsive venting of residual propellants, intended to direct the S-IVB beyond the Moon and into a solar orbit. However, due to an unexpectedly prolonged ullage engine burn, the resulting trajectory did not provide sufficient energy for the S-IVB to escape the Earth-Moon system completely.

Bean places the fuel element into the SNAP-27 RTG

As a consequence, the S-IVB was placed into an elliptical orbit around Earth and the Moon. Despite this deviation from the original plan, the primary mission objectives of ensuring the S-IVB did not impact the spacecraft, Earth, or the Moon were successfully achieved. The closest approach of the S-IVB to the Moon occurred at a distance of 3,082 nautical miles at 085:48 mission time. This incident highlighted the complexities and contingencies involved in space mission planning and execution, underscoring NASA's ability to adapt and manage unexpected situations effectively during the Apollo 12 mission.

During the Apollo missions, meticulous preparation was crucial to ensuring the safety and functionality of the Lunar Module (LM) amid the electrical transients experienced during launch. On this particular mission, the commander and lunar module pilot took early action, entering the LM at 007:20, ahead of schedule. Their primary objective was to conduct thorough housekeeping and systems checks, verifying the integrity of critical LM systems post-launch.

The checks confirmed that all LM systems were operating satisfactorily, a critical reassurance before embarking on the journey toward the Moon. As the spacecraft continued its translunar coast, a precise midcourse correction became necessary at 030:52:44.36. Lasting just 9.19 seconds and providing a velocity change of 61.8 feet per second, this maneuver ensured the spacecraft remained precisely on course, steering it toward a carefully plotted hybrid, non-free-return circumlunar trajectory.

The crew's activities were closely monitored throughout the mission and broadcast back to Earth. Detailed television coverage, spanning 47 minutes starting at 030:18, captured the meticulous preparations leading up to the crucial burn. Subsequently, a longer 56-minute transmission commencing at 062:52 delivered vivid color images, showcasing the command module, intravehicular transfer procedures, the interior of the LM, and fleeting glimpses of Earth and the Moon.

As the spacecraft approached its lunar destination, another pivotal moment unfolded at 083:25:23.36. Positioned at 82.5 nautical miles above the lunar surface, the service propulsion engine fired for an extensive 352.25 seconds. This maneuver successfully inserted the spacecraft into a lunar orbit measuring 168.8 by 62.6 nautical miles—a precise orbital placement critical for subsequent lunar exploration activities.

Crescent view of Earth on the way to the Moon (NASA

The journey from Earth to lunar orbit had spanned an impressive 80 hours, 38 minutes, and 1.67 seconds of translunar coasting. Along the way, astronauts were treated to awe-inspiring views, including the breathtaking crescent view of Earth—an unforgettable sight against the backdrop of deep space as they ventured closer toward their lunar objective.

When Pete Conrad, known as the shortest man among the initial groups of astronauts, joyfully set foot on the lunar surface during Apollo 12, his exuberant first words echoed through space and history: "Whoopie! Man, that may have been a small one for Neil, but that's a long one for me." Conrad's quip was no spontaneous remark; it was the culmination of a playful bet he had made with journalist Oriana Fallaci, who had questioned whether NASA scripted Neil Armstrong's famous line. Conrad had wagered $500 that he would utter these words, but he humorously lamented later that he could never collect his winnings.

Earthrise over lunar surface following lunar orbit insertion (NASA)

Apollo 12 aimed to improve its predecessor's achievements, including enhancing television coverage from the lunar surface. Unlike Apollo 11's monochrome camera, Apollo 12 carried a color camera for better visual fidelity. However, as astronaut Alan Bean moved to set up the camera near the Lunar Module (LM), a fateful mistake occurred: he inadvertently pointed it directly into the Sun's harsh glare, damaging its sensitive Secondary Electron Conduction (SEC) tube. This mishap abruptly ended television coverage of the mission, a setback that highlighted the challenges and risks of lunar exploration.

Despite this setback, Apollo 12's lunar orbit phase commenced with notable achievements. During the first orbit around the Moon, the crew delivered clear and detailed descriptions of lunar features, accompanying sharp pictures transmitted back to Earth. Approximately 33 minutes of high-quality television coverage began at 084:00, allowing viewers on Earth a remarkable glimpse of the lunar landscape and the astronauts' activities.

As the mission progressed, crucial maneuvers were executed to refine the spacecraft's orbital path. At 087:48:48.08, a precisely timed 16.91-second burn was performed, circularizing the orbit at approximately 66.1 by 54.3 nautical miles—a critical adjustment enabling further exploration and scientific endeavors on the lunar surface by the crew of Apollo 12.

During the next revolution around the Moon, the Lunar Module (LM) crew embarked on critical tasks, transitioning to the LM to conduct essential housekeeping and communication checks. The sequence of operations unfolded with precision timing: at 104:20, the mission commander entered the LM, with the lunar module pilot following at 105:00, both preparing meticulously for the imminent descent to the lunar surface.

At 107:54:02.3, with the spacecraft at an altitude of 63.0 nautical miles, the LM and Command Module (CM) were undocked, initiating the next phase of the mission. Shortly after that, at 108:24:36.9, a brief 14.4-second separation maneuver ensured a safe distance between the two vehicles. The descent toward the lunar surface proceeded with calculated maneuvers: at 109:23:39.9, a 29.0-second descent orbit insertion maneuver placed the LM into a precise orbit of 60.6 by 8.1 nautical miles.

The pivotal moment of powered descent initiation followed at 110:20:38.1, when the LM's descent engine fired for 717.0 seconds, guiding the craft toward its historic landing. Touchdown occurred precisely at 06:54:36 GMT (01:54:36 a.m. EST) on November 19, 110:32:36.2 mission time, with the engine shutting down a mere 1.1 seconds before landing.

The landing site, located at 3.01239° S latitude and 23.42157° W longitude in the Oceanus Procellarum region (Ocean of Storms), held significance beyond its geographical coordinates. The descent caused high-velocity sandblasting on the nearby Surveyor III probe, a previous robotic mission to the Moon. Interestingly, this process inadvertently removed more lunar dust than it deposited, revealing a tan hue on the probe's surface and lightening areas directly exposed to the abrasive effects of the LM's landing.

Charles Conrad Jr., Apollo 12 Commander, examines the unmanned Surveyor III spacecraft during the second extravehicular activity (EVA-2). The Lunar Module (LM) "Intrepid" is in the right background.

One of the mission's primary objectives was achieved with remarkable precision: a precise landing near Surveyor III, which had touched down on April 20, 1967. The LM landed an impressive 535 feet from the robotic probe, demonstrating NASA's capability to conduct targeted lunar landings with increasing accuracy.

As the mission continued, reports from the Command Module (CM) orbiting overhead provided critical visual confirmations: during subsequent revolutions, the commander and command module pilot spotted both the CSM and Surveyor III, reinforcing the success and strategic importance of Apollo 12's lunar exploration objectives.

LM following separation. Large crater in foreground is Ptolemaeus (NASA AS12-51-7507).

Three hours after touching down on the lunar surface, the crew of Apollo 12 began their meticulous preparations for their historic extravehicular activity (EVA). Commanding the mission, Pete Conrad led the way as he exited the LM's hatch, deploying the modularized equipment stowage assembly and activating a newly installed color television camera, relaying live images of their actions back to Earth.

Before descending, Conrad reported spotting Surveyor III approximately 600 feet away while noting that the LM had settled just 25 feet from the edge of a crater. Stepping onto the lunar soil at 115:22:22 mission time, Conrad described the surface as soft and loosely packed, causing his boots to sink in with each step. Shortly after, Alan Bean joined him on the lunar surface at 115:51:50, marking a pivotal moment in human exploration.

Initial activities included setting up equipment essential for scientific endeavors and national symbolism: deploying the S-band erectable antenna, establishing the United States flag at 116:19:31, and initiating the solar wind composition experiment. Despite an initial hiccup with the radioisotope thermoelectric generator fuel element, the Apollo Lunar Surface Experiments Package (ALSEP) deployment proceeded smoothly. Positioned approximately 600 to 700 feet from the LM, the ALSEP site would serve as a crucial scientific outpost, transmitting data such as the crew's own footsteps recorded by the passive seismometer.

Conrad erects U.S. flag at landing site

As they ventured farther from the LM, the crew embarked on a productive geological survey, collecting a core tube sample and various surface samples during their exploratory traverse. They covered a remarkable distance of 3,300 feet (about 1 kilometer). They gathered an estimated 36.8 pounds (16.7 kg) of lunar material—a testament to their efficiency and determination in conducting scientific research on the Moon.

Following nearly four hours of groundbreaking exploration, Conrad and Bean re-entered the LM, sealing the hatch at 119:06:36. This marked the conclusion of their first extravehicular activity period, characterized by unprecedented scientific accomplishments and paving the way for further human endeavors beyond Earth's boundaries.

At 119:47:13.23, the Command and Service Module (CSM) executed a critical plane change maneuver lasting 18.23 seconds. This maneuver was designed to adjust the spacecraft's orbital trajectory, shifting it to a new path measuring 62.5 by 57.6 nautical miles around the Moon. Such maneuvers were essential for optimizing orbital dynamics and facilitating subsequent mission objectives.

After rest, the crew commenced their second extravehicular activity (EVA) period at 131:33, demonstrating their resilience and readiness for further lunar exploration. Their first task involved troubleshooting and preparing for equipment retrieval: they disengaged the cable and carefully stored the malfunctioning LM TV camera in the equipment transfer bag, intending to return it to Earth for detailed failure analysis.

Commander Pete Conrad then proceeded to the Apollo Lunar Surface Experiments Package (ALSEP) site, focusing on checking and adjusting the lunar atmosphere detector's positioning. During this process, the instrument unexpectedly recorded higher atmospheric readings, initially puzzling the crew. Further analysis revealed that the anomaly was caused by the outgassing from Conrad's spacesuit as he approached the delicate instrument, a phenomenon indicative of the unique challenges and meticulous precautions inherent in lunar exploration.

After proudly planting the U.S. flag on the lunar surface, Pete Conrad and Alan Bean shifted their focus to the meticulous deployment of the Apollo Lunar Surface Experiments Package (ALSEP) during the remainder of their first extravehicular activity (EVA). This mission phase was challenging, as the astronauts encountered minor difficulties with some aspects of the deployment process.

One notable hurdle arose when Alan Bean had trouble extracting the Radioisotope Thermoelectric Generator (RTG) plutonium fuel element from its protective cask. To overcome this obstacle, the astronauts resorted to using a hammer to carefully dislodge the fuel element, ensuring that the critical power source could be properly deployed and utilized for scientific experiments on the lunar surface.

Despite these initial challenges, Conrad and Bean successfully deployed all the components of the ALSEP package. This included positioning various scientific instruments designed to study lunar seismic activity, magnetic fields, and other environmental parameters. These experiments were crucial for advancing our understanding of the Moon's geology and its interaction with the space environment.

Throughout their EVA, the Passive Seismic Experiment (PSE) monitored the astronauts' movements, which detected their footprints as they traversed back to the Lunar Module (LM). They also conducted further scientific sampling, securing a core tube filled with lunar material and collecting additional diverse samples from the Moon's surface.

The first extravehicular activity period lasted three hours, 56 minutes, and 3 seconds—a period marked by groundbreaking scientific accomplishments and meticulous attention to detail. During this historic EVA, Conrad and Bean's efforts expanded our knowledge of lunar science and set the stage for continued exploration and discovery during subsequent Apollo missions.

Bean removes the RTG fuel element from its cask

The movement of astronauts on the lunar surface during Apollo 12 was meticulously recorded using advanced scientific instruments, including the passive seismometer and the lunar surface magnetometer. These instruments provided valuable data on the disturbances caused by human activities and environmental interactions on the Moon.

During their exploration, Commander Pete Conrad conducted an intriguing experiment near Head Crater, approximately 300 to 400 feet from the passive seismometer. He rolled a grapefruit-sized rock down the crater wall to observe any seismic response. Despite the significant force exerted, no notable seismic activity was detected across any of the four axes monitored by the seismometer, highlighting the Moon's relatively quiet geological environment.

They conducted a comprehensive scientific sampling campaign as the crew embarked on their geological traverse toward Surveyor III. This included capturing panoramic photographs of the lunar landscape, utilizing stereo photography techniques for detailed surface mapping, and collecting multiple core samples—two single and one double—and an eight-inch-deep trench sample.

Their collection efforts extended to lunar environment samples and various geological specimens, encompassing rocks, dirt, bedrock, and even samples described as "molten," reflecting the diverse geological composition of the lunar surface.

A notable observation made during these activities was the discovery of fine dust accumulation around larger rocks and a noticeable lightening of soil color as they dug deeper into the lunar regolith. These findings provided valuable insights into the lunar soil's characteristics and its interaction with the lunar environment over time, shedding light on processes such as micrometeorite impacts and solar radiation effects on the Moon's surface.

During their mission on the lunar surface, the Apollo crew undertook a series of meticulously planned activities that expanded our understanding of lunar exploration. One of their pivotal tasks was examining and interacting with the Surveyor III spacecraft. Positioned nearby, Surveyor III provided a unique opportunity for a close-up investigation. The crew meticulously photographed the spacecraft, noting its weathered appearance, with distinct brown hues across its structure. They observed that while the glass components showed signs of warping, they remained intact, mounted securely within their framework, and opted not to retrieve them.

Bean holds a special environmental sample container filled with lunar soil. Conrad's reflection can be seen in Bean's visor

After completing their examination, the crew embarked on a return traverse, retrieving crucial scientific experiments and samples. They collected the solar wind composition experiment, which had been exposed for nearly 19 hours, contributing valuable data to our understanding of lunar surface conditions. Utilizing the Apollo lunar surface close-up camera, they captured stereo images of the area surrounding their lunar module during their final moments on the surface.

Conrad stands by Surveyor III. Note the lunar module on the horizon

Before re-entering the lunar module (LM), the crew removed any accumulated lunar dust on their suits, ensuring a clean transfer back into the spacecraft. The Lunar Module Pilot initiated transferring gathered samples, equipment, and parts to the Commander inside the LM. This careful exchange allowed for efficient retrieval and preservation of mission-critical materials.

Throughout their extravehicular activities, the crew demonstrated exceptional mobility and operational proficiency with their portable life support systems, mirroring the success of their predecessors on Apollo 11. Their second extravehicular period spanned nearly 3 hours and 49 minutes, during which they covered a distance of approximately 4,300 feet (1.3 km) and collected an estimated 38.8 pounds (17.6 kg) of lunar samples.

Upon reentering the LM and jettisoning expendable equipment, the crew prepared for their return journey. The mission's total time spent outside the LM amounted to 7 hours, 45 minutes, and 18 seconds, covering a total distance of 7,600 feet (2.3 km). They gathered 75.73 pounds (34.35 kg) of lunar material, carefully documented and analyzed upon their return to Earth by the Lunar Receiving Laboratory in Houston.

Their mission concluded with a sense of achievement, marking the completion of the second successful human exploration of the Moon.

Following the landing of the Lunar Module (LM) by Pete Conrad between planned geologic traverse points, scientists at Mission Control in Houston adapted swiftly to optimize the astronauts' exploration efforts. During the first Extravehicular Activity (EVA) and the subsequent rest period, these scientists merged two originally planned traverses into a single route that Conrad and Alan Bean could follow starting from their landing site.

Conrad uses tongs to pick up rock from lunar surface

The resulting traverse path resembled a rough circle, strategically designed to maximize scientific observations and sample collection across varied lunar terrain. Approximately 13 hours after concluding their first EVA, Conrad and Bean embarked on their second exploration phase. Their initial stop was Head Crater, roughly 100 yards (91 meters) from the LM.

Alan Bean made a significant discovery at Head Crater when he observed Pete Conrad's footprints revealing lighter material underneath. This finding suggested the presence of ejecta—material ejected from the impact site—originating from Copernicus crater, located some 230 miles (370 kilometers) to the north. This discovery was a crucial confirmation of hypotheses formed by scientists studying overhead photographs of the lunar surface before the mission.

Samples collected from the Head Crater during Apollo 12 provided invaluable insights for lunar geologists. Analysis of these samples allowed scientists to establish the age of the impact that formed the Copernicus crater, estimating it to be approximately 810 million years ago based on initial dating methods.

After exploring Head Crater, Pete Conrad and Alan Bean continued their geologic traverse across the lunar surface. They visited several notable landmarks, including Bench Crater and Sharp Crater, before reaching Halo Crater. Their ultimate destination was Surveyor Crater, where the Surveyor 3 probe had landed previously on the Moon.

Approaching Surveyor Crater cautiously due to concerns about footing and the stability of the probe, Conrad and Bean descended into the shallow crater from a safe distance. Then it followed a contour to reach the probe. Fortunately, they found the terrain solid and the Surveyor probe stable. During their visit to Surveyor Crater, they collected various components of the probe, including pieces of the television camera, and retrieved rocks previously studied via television imagery.

One memorable objective for Conrad and Bean was to capture a photograph with the Surveyor probe using an automatic timer for their Hasselblad cameras, which they had brought without informing Mission Control. However, amidst the lunar samples they had collected and placed in their Hand Tool Carrier, they couldn't locate the timer when the opportune moment arrived.

Bean prepares to step onto the lunar surface

Before returning to the vicinity of their Lunar Module (LM), Conrad and Bean also explored Block Crater, situated within Surveyor Crater itself. Their second Extravehicular Activity (EVA) lasted 3 hours, 49 minutes, and 15 seconds, during which they covered a distance of 4,300 feet (1,300 meters) on foot. Throughout their EVAs, Conrad and Bean ventured as far as 1,350 feet (410 meters) from the LM, demonstrating their mobility and capability to conduct extensive scientific exploration on the Moon.

They collected an impressive 73.75 pounds (33.45 kilograms) of lunar samples during their mission, contributing substantially to our understanding of lunar geology and scientific exploration beyond Earth. The achievements of Conrad and Bean during their time on the lunar surface underscored the meticulous planning and execution that characterized the Apollo 12 mission, solidifying its place in the annals of space exploration history.

During their Lunar Module (LM) stay on the lunar surface, the Command Module Pilot aboard the Command and Service Module (CSM) conducted the S-158 lunar multispectral photography experiment. This comprehensive photographic survey captured detailed imagery of three notable lunar features: the Wall of Theophilus, and prospective Apollo landing sites at Fra Mauro and Descartes. These targets of opportunity provided crucial data for future lunar missions and scientific study.

After completing their scientific tasks, the crew prepared for their ascent back to lunar orbit. The ascent stage engine fired precisely at 142:03:47.78, initiating liftoff after the LM had spent 31 hours, 31 minutes, and 12.0 seconds on the lunar surface. The ascent engine burn lasted 423.2 seconds, slightly exceeding the planned duration by 1.2 seconds. This maneuver positioned the spacecraft into an initial orbit with dimensions of 46.3 by 8.8 nautical miles at 142:10:59.9.

Precision maneuvers were then executed to facilitate rendezvous and docking with the orbiting CSM. At 143:01:51.0, a 41.1-second coelliptic orbit adjustment raised the LM's orbit to 51.0 by 41.5 nautical miles. This was followed by a 13.0-second maneuver at 144:00:02.6 to refine the orbit to 44.4 by 40.4 nautical miles. Further adjustments included a 26-second terminal phase initiate maneuver at 144:36:26, placing the ascent stage into an orbit of 60.2 by 43.8 nautical miles. Finally, a 38.0-second burn at 145:19:29.3 fine-tuned the orbit to 62.3 by 58.3 nautical miles in preparation for docking with the CSM at 145:36:20.2, positioned at an altitude of 58.1 nautical miles.

Throughout the intricate docking sequence, which spanned 37 hours, 42 minutes, and 17.9 seconds from undocking to rendezvous, the CSM transmitted high-quality television footage for 24 minutes. This visual documentation provided a captivating view of the final stages of the rendezvous and docking process, showcasing the crew's precision and expertise in executing complex orbital maneuvers.

Evidence that Surveyor III bounced when it landed is the footpad imprint seen to the right

After Pete Conrad and Alan Bean completed their activities on the lunar surface and returned safely to the Lunar Module (LM), Richard Gordon remained in orbit aboard the Command Module (CM), tasked with monitoring their progress and preparing for their rendezvous after their Moon exploration. Initially focused on supporting the lunar landing operation, Gordon's communications with Mission Control were occasionally challenged by shared bandwidth with Conrad and Bean, who were using the same communication circuit.

Once Conrad and Bean re-entered the LM, Gordon executed a crucial plane change maneuver to adjust the orbit of the Command and Service Module (CSM). This maneuver compensated for the Moon's rotation and ensured that Gordon would be in the correct position to rendezvous with the LM when it launched from the lunar surface.

During his solo orbit, Gordon engaged in the Lunar Multispectral Photography Experiment using four Hasselblad cameras mounted in a ring and aimed through one of the CM's windows. Each camera was equipped with a different color filter, allowing simultaneous photography of lunar features across various points on the spectrum. These multispectral images provided insights into lunar surface composition and revealed colors and details that were invisible to the naked eye or detectable with conventional color film.

The experiment focused on capturing detailed imagery of potential future Apollo landing sites, providing valuable data for mission planning and scientific analysis. Gordon's efforts in orbit significantly expanded our understanding of lunar geology and paved the way for future explorations of the Moon.

After a successful mission on the lunar surface, the LM Intrepid lifted off from the Moon at 143:03:47.78 mission time, or 14:25:47 UT on November 20, 1969. Following several maneuvers, the Command and Service Module (CSM) and LM docked approximately three and a half hours later at 147:59:31.6. Shortly after that, at 147:59:31.6, the LM ascent stage was jettisoned, and the CSM maneuvered away. Controlled from Earth, the remaining propellant in the LM was depleted in a controlled burn that impacted the Moon 39 nautical miles (72 km; 45 mi) from the Apollo 12 landing site. The vibrations from this impact were recorded by the seismometer left on the lunar surface, registering them for more than an hour.

The crew remained in lunar orbit for another day, capturing photographs of the lunar surface, including potential landing sites for future Apollo missions. A second plane change maneuver was executed at 159:04:45.47, lasting 19.25 seconds, to adjust the trajectory of the CSM.

For the journey back to Earth, the trans-Earth injection burn occurred at 172:27:16.81, lasting 130.32 seconds, to propel the CSM Yankee Clipper toward home. Along the way, two short midcourse correction burns were performed to refine the spacecraft's trajectory. During this return journey, the crew also conducted a final television broadcast, answering questions submitted by the media.

The return voyage allowed ample time for rest, punctuated by memorable events such as photographing a solar eclipse. This celestial event occurred when the Earth passed between the spacecraft and the Sun, offering the astronauts a spectacular view, which Bean described as one of the most memorable sights of the mission.

After transferring the crew and lunar samples to the Command and Service Module (CSM), the LM's ascent stage was jettisoned at 147:59:31.6, and preparations began for the transearth injection burn. The ascent stage was then remotely maneuvered to impact the lunar surface. A 5.4-second maneuver at 148:04:30.9 separated the CSM from the ascent stage, placing the CSM in an orbit of 62.0 by 57.5 nautical miles.

Next, an 82.1-second deorbit firing of the ascent stage took place at 149:28:14.8, performed at 57.6 nautical miles. This burn depleted the ascent stage's propellants, leading to its impact on the lunar surface at 149:55:16.4. The impact point was estimated to be at approximately latitude 3.94° south and longitude 21.20° west, about 40 nautical miles east-southeast of the Apollo 12 landing site and 5 nautical miles from the intended target.

During the final lunar orbits before departing for Earth, extensive tracking and photography of landmarks on the lunar surface were conducted. A 500 mm long-range lens was specifically used to capture detailed mapping and training data essential for planning future lunar missions.

During Apollo's return journey, meticulous maneuvers and adjustments ensured a precise path back to Earth. Approximately 88 hours and 58 minutes after beginning their journey from the Moon, the Command and Service Module (CSM) executed a critical transearth injection at a velocity of 8,351 feet per second. Following 45 orbits around the lunar surface, this maneuver marked a pivotal moment in the mission's timeline.

Before the injection, at 159 hours and 4 minutes into the mission, a plane change maneuver lasting 19.25 seconds adjusted the CSM's orbit, aligning it to intersect Earth's path at precisely the right angle. Subsequently, at 172 hours and 27 minutes, another precise adjustment at an altitude of 63.3 nautical miles fine-tuned the trajectory, culminating in the transearth injection at 172 hours and 29 minutes.

Following this crucial maneuver, the crew captured approximately 38 minutes of high-quality television footage, showcasing the receding lunar surface and the spacecraft's interior. This broadcast, which commenced about 20 minutes post-injection, provided a remarkable view of the Moon as they departed its orbit.

Throughout the transearth phase, minor corrections were made to ensure the CSM's trajectory remained on course. At 188 hours and 27 minutes, a brief midcourse correction of 4.4 seconds and 2.0 feet per second was executed, strategically timed after a period of crew rest to optimize performance. This adjustment was part of the meticulous planning to guarantee a safe and accurate return.

Later in the journey, at 241 hours and 21 minutes, a final midcourse correction lasting 5.7 seconds and adjusting the velocity by 2.4 feet per second was conducted. These precision maneuvers underscored NASA's commitment to maintaining a precise trajectory for the crew's return to Earth.

As the mission approached its conclusion, the crew engaged in a final televised broadcast, lasting approximately 37 minutes. This broadcast included views of the spacecraft's interior and an interactive question-and-answer session with scientists and the press. This final transmission provided insights into life aboard the spacecraft and a poignant conclusion to their extraordinary journey from lunar exploration back to Earth.

Upon reentry, the final stages of Apollo's journey unfolded with the jettisoning of the service module at 244 hours and 7 minutes into the mission. This marked the separation of the command module (CM), which began its descent toward Earth's atmosphere at an altitude of 400,000 feet, entering at a staggering velocity of 36,116 feet per second. The return journey, spanning 71 hours, 52 minutes, and 52 seconds since departing the Moon, culminated in this critical phase of the mission.

Following the separation, the service module's reaction control system was fired until depletion. However, despite these efforts, neither radar nor visual contact was established by either the crew or ground recovery teams. It was later determined that the service module likely became unstable during the depletion phase, failing to achieve the necessary velocity adjustment to continue on its intended trajectory out of Earth's atmosphere into a high-apogee orbit as planned. Instead, it is believed that the module reentered Earth's atmosphere and impacted before detection could be made.

Apollo 12 crew in raft following egress from CM (l. to r.): Conrad, Bean and Gordon

On November 24, 1969, Apollo 12's Command Module (CM), known as Yankee Clipper, faced a tumultuous return to Earth, landing in the Pacific Ocean at 20:58 GMT (3:58 p.m. EST). The sea-state conditions were notably rough, contributing to an exceptionally hard splashdown that subjected the crew to forces reaching about 15 times the force of Earth's gravity (15 g). The impact was so severe that parts of the heat shield were dislodged, and the 16 mm sequence camera, detached from its bracket, struck Lunar Module Pilot (LMP) Alan Bean above his right eye.

Despite these challenges, the splashdown occurred approximately 2.0 nautical miles from the intended target point and 3.91 nautical miles from the recovery ship, the USS Hornet, positioned in the Pacific. The estimated coordinates of the splashdown site were latitude 15.78° south and longitude 165.15° west.

Apollo 12 about to impact the surface of the Pacific

Immediately after splashdown, the CM assumed an apex-down orientation, but within just 4 minutes and 26 seconds, the inflatable bag uprighting system successfully restored it to its normal flotation position. This rapid recovery process demonstrated the effectiveness of the safety systems designed aboard the spacecraft.

Once recovered by the USS Hornet, the astronauts were swiftly transferred to the Mobile Quarantine Facility (MQF) as a precautionary measure against potential lunar contaminants. Meanwhile, lunar samples and parts of the Surveyor spacecraft were airlifted to Houston's Lunar Receiving Laboratory (LRL).

Following the Hornet's docking in Hawaii, the MQF was offloaded and flown to Ellington Air Force Base near Houston on November 29. From there, it was transported to the LRL, where the astronauts remained under quarantine until their release on December 10. This meticulous quarantine procedure underscored NASA's commitment to safeguarding against any potential extraterrestrial pathogens, ensuring a cautious conclusion to the historic Apollo 12 mission.

Following their dramatic splashdown, the Apollo 12 crew—Commander Charles Conrad, Lunar Module Pilot Alan Bean, and Command Module Pilot Richard Gordon—emerged safely from their Command Module (CM) and were promptly rescued by a helicopter. Remarkably, within just 60 minutes of splashdown, they were aboard the recovery ship, USS Hornet. Immediate precautions were taken to ensure biological isolation, similar to those implemented during the Apollo 11 mission.

The recovery operation continued swiftly, with the CM retrieved just 48 minutes later. During the splashdown, the CM weighed 11,050 pounds, completing an impressive journey covering approximately 828,134 nautical miles.

The Mobile Quarantine Facility (MQF) and the CM were offloaded from the Hornet at 02:18 GMT on November 29 in Hawaii. Shortly after, the MQF was loaded onto a C-

141 aircraft bound for Ellington Air Force Base in Houston, Texas, where it arrived at 11:50 GMT. Two hours later, the crew entered the Lunar Receiving Laboratory (LRL), beginning their quarantine period as a precaution against potential lunar contaminants.

Meanwhile, the CM was transported to Hickam Air Force Base in Hawaii for deactivation. Upon completing this process at 14:15 GMT on December 1, the CM was flown to Ellington Air Force Base aboard a C-133 aircraft and delivered to the Lunar Receiving Laboratory at 19:30 GMT on December 2. This meticulous handling ensured that the spacecraft and crew remained isolated until all potential risks were mitigated.

The crew's quarantine concluded on December 10, shortly after that the CM was released. On January 11, the CM was transferred to the North American Rockwell Space Division facility in Downey, California, where it underwent detailed postflight analysis, contributing crucial insights to NASA's ongoing exploration efforts.

The Apollo 12 mission was a testament to the advancements in lunar exploration capabilities, paving the way for future missions to the Moon. This mission's success, characterized by precision in both lunar landing and operational execution, yielded significant scientific discoveries and operational insights.

Key conclusions drawn from the meticulous analysis of post-mission data include:

Precision Lunar Landing: The mission validated the effectiveness of crew training, mission planning, and real-time navigation capabilities from ground control. This precision enabled a landing near a previously visited Surveyor spacecraft and well within the designated landing area.

In its Translunar Profile Apollo 12 introduced a hybrid non-free-return translunar trajectory, showcasing enhanced maneuvering capabilities necessary for future lunar landings at higher latitudes and different surface conditions.

Wives of Apollo 12 crew greet them when the mobile quarantine facility arrives at Ellington AFB,

During its extended lunar surface operations, the mission demonstrated that crew members, supported by portable life support systems, could effectively manage timeline activities and metabolic demands during extended scientific explorations on the lunar surface.

For the first time, Apollo 12 deployed the Apollo Lunar Surface Experiments Package (ALSEP), despite encountering minor operational challenges. Despite these issues, ALSEP provided valuable scientific data across various research disciplines.

Apollo 12, NASA's second manned lunar landing mission, achieved a series of groundbreaking objectives that expanded human understanding and capability in lunar exploration.

The primary goals of the mission were multifaceted. Firstly, Apollo 12 successfully conducted a detailed inspection, survey, and sampling in a designated mare area on the lunar surface. This included deploying the Apollo Lunar Surface Experiments Package (ALSEP), which consisted of several critical scientific instruments such as a passive seismic experiment, a lunar surface magnetometer, a solar wind spectrometer, a suprathermal ion detector, and a cold cathode ion gauge. These instruments provided unprecedented data on lunar geological and environmental conditions.

Apollo 12 crew aboard recovery ship U.S.S. Hornet enter the mobile quarantine facility (NASA

One of the mission's key technical achievements was the demonstration of precise point landing capabilities. This involved navigating and landing the Lunar Module (LM) near a specific target, showcasing advancements in crew training, mission planning, and real-time navigation from ground control.

Apollo 12 also focused on enhancing human capabilities to operate effectively in the lunar environment. Astronauts successfully conducted extravehicular activities (EVAs), collected selected samples, and managed the portable life support systems necessary for extended lunar surface operations. Despite some challenges, including partial achievements in high-resolution stereoscopic photography and landmark tracking due to time constraints, the mission provided crucial insights into the practicalities of sustained human presence on the Moon.

Scientific experiments conducted during Apollo 12 yielded significant discoveries. Lunar field geology studies, analysis of solar wind composition, multispectral photography of the lunar surface, and evaluations of lunar dust dynamics all contributed valuable data to our understanding of the Moon's composition and environment.

From the perspective of launch vehicle objectives, Apollo 12 successfully launched into Earth orbit, executed a precise translunar trajectory following S-IVB restart and managed spacecraft attitude control throughout critical maneuvers. Although the planned slingshot trajectory using lunar gravity to propel the spent S-IVB stage into solar orbit was unsuccessful, safe venting and disposal of remaining propellants were planned.

Apollo 12 CM Yankee Clipper on display at the Virginia Air and Space Center in Hampton, Virginia

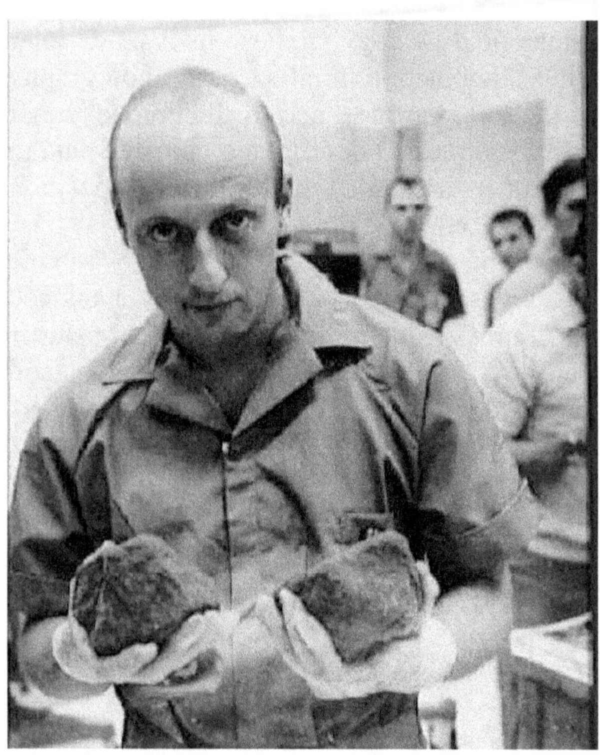

Astronaut Charles Conrad Jr., commander of the Apollo 12 lunar landing mission, holds two lunar rocks which were among the samples brought back from the Moon by the Apollo 12 astronauts.

Apollo 13

Apollo 13 was meticulously planned as a Type H mission, designed for precision piloted lunar landing demonstration and systematic lunar exploration. Its ambitious objectives spanned spacecraft operations, scientific exploration, and precise mission milestones. Its primary goals included conducting detailed selenological surveys of the Fra Mauro formation, deploying the Apollo Lunar Surface Experiments Package (ALSEP) for scientific measurements, advancing human capabilities for lunar operations, and capturing essential photographs of potential landing sites.

The mission also aimed to achieve detailed objectives such as broadcasting live television coverage from space, collecting contingency and selected samples from the lunar surface, refining techniques for pinpointing lunar landings, and documenting lunar sites with photography under various lighting conditions. Extravehicular communication tests, studies on lunar soil mechanics, and updates to lunar mapping references were also planned to enhance scientific understanding.

Scientific experiments planned for Apollo 13 included deploying instruments to measure lunar seismic activity (ALSEP III), conducting geological studies (S-059), analyzing solar wind composition (S-080), and performing radar observations of the Moon (S-164, S-170). Additional experiments aimed to study unique lunar phenomena such as the gegenschein and capture detailed close-up images of the lunar surface (S-184).

The launch vehicle objectives were equally ambitious. They required precise insertion into Earth orbit, successful navigation toward the Moon, maintaining spacecraft orientation during critical maneuvers, executing evasive actions post-ejection, and attempting a controlled impact on the lunar surface within specified parameters. The mission also included tasks like accurately determining impact locations and safely venting residual propellants and gases from the launch vehicle.

The original prime crew designated for Apollo 13 was initially intended to serve as the backup crew for Apollo 10. This team included Gordon Cooper as commander, Donn F. Eisele as command module pilot (CMP), and Edgar Mitchell as lunar module pilot (LMP). However, internal considerations at NASA led to a change in plans. Cooper's training approach was deemed too relaxed, and Eisele had faced challenges during his tenure on Apollo 7, compounded by personal issues including an extramarital affair.

Deke Slayton, NASA's Director of Flight Crew Operations, assigned them to the backup crew primarily due to the absence of other experienced astronauts at the time. Originally, Slayton had intended for Alan Shepard, Stuart Roosa, and Edgar Mitchell to lead Apollo 13. However, Shepard required additional training time after returning to active duty following surgery for an inner ear disorder, having not flown since 1961. This necessitated a shift in crew assignments, with Slayton selecting Jim Lovell as commander, Fred Haise as LMP, and Ken Mattingly as CMP.

Originally slated as the CMP for Apollo 13's backup crew, Jack Swigert found himself in the prime crew position under extraordinary circumstances. Just seven days before the scheduled launch, Charles Duke, the lunar module pilot (LMP), contracted rubella from a contact with his son's friend. This exposure affected both the prime and backup crews who had been training together, posing a significant risk as only Ken Mattingly, the original CMP of the prime crew, lacked immunity to rubella.

According to standard protocol, if a prime crew member became unfit for flight due to illness, the backup crew would typically replace the entire crew. However, this standard procedure could not be implemented due to the simultaneous exposure of both crews. With only two days left until launch, NASA made the critical decision to replace Mattingly with Swigert, who was immune to rubella, thereby minimizing the risk of illness affecting the mission's success.

This newly configured crew, previously backups for Apollo 11 and scheduled for Apollo 14, thus became the prime crew for Apollo 13. The final crew composition just before the mission's launch included Captain James Arthur Lovell, Jr. as commander, John Leonard "Jack" Swigert, Jr. as command module pilot (following the replacement of Ken Mattingly), and Fred Wallace Haise, Jr. as lunar module pilot.

James Arthur Lovell, Jr., an accomplished astronaut and seasoned naval officer, found himself at a crucial juncture aboard Apollo 13, a mission defining his storied space exploration career. Selected as an astronaut in 1962, Lovell had already amassed an impressive array of achievements by the time of this mission. He was embarking on his fourth spaceflight, marking his second journey to the Moon, a distinction unmatched by any of his contemporaries.

Lovell's career trajectory began with significant contributions to the Gemini program, where he first served as the pilot of Gemini 7 and then commanded Gemini 12. These missions showcased his proficiency and leadership in the demanding realm of spaceflight. However, his pivotal role as the command module pilot during Apollo 8 catapulted him into history. This mission, in 1968, marked humanity's inaugural manned orbit of the Moon, a milestone that underscored Lovell's invaluable experience and capability.

The original prime crew of the Apollo 13 mission post for a publicity photo. From left: James A. Lovell Jr., commander; Thomas K. Mattingly II, command module pilot; and Fred W. Haise Jr., lunar module pilot.

Born on March 25, 1928, in Cleveland, Ohio, Lovell was 42 years old during the Apollo 13 mission. His educational background included earning a Bachelor of Science from the U.S. Naval Academy in 1952, laying the foundation for his illustrious aviation and space exploration career. By the time of Apollo 13, Lovell had been NASA's most seasoned astronaut in terms of cumulative time spent in space, totaling an impressive 572 hours across his missions.

Jim Lovell, Apollo 13's mission commander, epitomized the spirit of a pioneer in both military aviation and the burgeoning field of space exploration. His journey began post-graduation from the United States Naval Academy, where he distinguished himself as a naval aviator and later as a test pilot, roles that equipped him with the skills necessary for the challenges of space travel. Selected as part of NASA's second group of astronauts in 1962, Lovell had already left an indelible mark on the Gemini program, flying missions with renowned peers such as Frank Borman and Buzz Aldrin. His pivotal role in Apollo 8, the historic 1968 mission that achieved lunar orbit, solidified his reputation as one of NASA's most seasoned and capable astronauts.

The Apollo 13 lunar landing mission prime crew from left to right are: Commander, James A. Lovell Jr., Command Module pilot, John L. Swigert Jr., and Lunar Module pilot, Fred W. Haise Jr.

Jack Swigert, appointed as the command module pilot (CMP) for Apollo 13 at the age of 38, brought a wealth of expertise and experience to the mission. His background in mechanical engineering and aerospace science was complemented by a distinguished career in the Air Force and various roles in state Air National Guards. Before joining NASA's fifth astronaut group in 1966, Swigert had established himself as an accomplished engineering test pilot, demonstrating both technical proficiency and operational readiness crucial for manned spaceflight missions.

Fred Haise, the lunar module pilot (LMP) at 35 years old, contributed a wealth of experience to Apollo 13. Haise's journey to becoming an astronaut began with his education in aeronautical engineering and his Marine Corps fighter pilot service. His career path intersected with NASA when he joined as a civilian research pilot, paving the way for his selection as part of Group 5 astronauts.

In the Apollo program, NASA introduced a novel concept: the support crew, a pivotal addition alongside the prime and backup crews established during earlier Mercury and Gemini missions. This innovation, championed by Deke Slayton in response to insights from James McDivitt, set to command Apollo 9, addressed the complex logistics of mission coordination spread across multiple U.S. facilities.

Support crew members, typically junior in seniority compared to their prime and backup counterparts, assumed critical roles in assembling and maintaining mission protocols, flight plans, and meticulous checklists. Their primary responsibilities encompassed updating operational guidelines and ensuring meticulous preparation to ensure flawless mission execution.

During Apollo 13, the support crew included Vance D. Brand, Jack Lousma, and either William Pogue or Joseph Kerwin, selected based on mission-specific requirements and crew availability.

Alongside the crews, the mission was overseen by a team of dedicated directors led by Gerald D. Griffin. Each operational shift—manned by M.P. "Pete" Frank III, Clifford E. Charlesworth, and Milton L. Windler—ensured round-the-clock vigilance and management of mission operations.

During the mission itself, the guidance and oversight from mission control were entrusted to a team of experienced flight directors. Gene Kranz led the White team as the lead flight director, supported by Glynn Lunney (Black team), Milton Windler (Maroon team), and Gerry Griffin (Gold team). These directors provided round-the-clock support in managing mission operations and responding to any contingencies that arose.

Communication between mission control and the spacecraft was facilitated by capsule communicators (CAPCOMs), who were astronauts responsible for maintaining voice communications with the crew. For Apollo 13, the CAPCOMs included Joseph Kerwin,

Vance Brand, Jack Lousma, John Young, and Ken Mattingly, ensuring clear and effective communication channels throughout the mission.

The Apollo 13 mission insignia, designed by artist Lumen Martin Winter, embodies profound symbolism reflecting its goals and challenges. At its center is Apollo, the Greek god of the Sun, depicted riding his chariot across the face of the Moon, pulled by three horses. This imagery metaphorically represents the Apollo missions as bringing the light of knowledge to all humanity, with the Earth visible in the distant background. The mission motto, "Ex luna, scientia" ("From the Moon, knowledge"), encapsulates the mission's purpose of expanding human understanding through lunar exploration. Jim Lovell, influenced by the motto of his alma mater, the Naval Academy, adapted it to inspire the mission's significance.

The mission patch prominently features the Roman numerals "Apollo XIII" and notably lacks the crew's names, a rarity among Apollo mission insignias. This design characteristic meant that the patch did not require modification when Jack Swigert replaced Ken Mattingly just three days before launch due to health concerns. This abrupt crew change also necessitated the melting down and reminting of the Apollo 13 Robbins medallions flown aboard the mission, ensuring they accurately reflected the final crew composition and omitted references to a lunar landing date that had become unachievable.

Jim Lovell's thoughtful selection of call signs further enriched the mission's narrative. "Aquarius," chosen for the lunar module, derived from the constellation associated with the bringer of water, aligning symbolically with the mission's journey. Contrary to some media reports, the name was not inspired by the musical "Hair" but rather by its astronomical connotations. The command module's call sign, "Odyssey," not only evoked Homer's epic tale but also paid homage to Stanley Kubrick's film "2001: A Space Odyssey," resonating with Lovell's appreciation for its thematic exploration of human destiny and discovery.

Following U.S. President John F. Kennedy’s challenge to his nation to land an astronaut on the Moon by the decade's end and ensure a safe return to Earth, this ambitious goal had spurred NASA's gradual progression through Projects Mercury and Gemini, paving the way for the Apollo program. The culmination came with Apollo 11, which achieved a lunar landing on July 20, 1969. Neil Armstrong and Buzz Aldrin ventured onto the lunar surface while Michael Collins orbited in Command Module Columbia. The historic mission concluded on July 24, 1969, marking the fulfillment of Kennedy's vision.

CSM-109 Odyssey in the Operations and Checkout Building

To support this endeavor, NASA had commissioned fifteen Saturn V rockets, and it was still being determined how many missions would be required at the outset. The triumph of Apollo 11, achieved with the sixth Saturn V launch, left nine rockets available for future missions, originally aiming for ten lunar landings. However, publicity enthusiasm waned following Apollo 11, compounded by budget cuts from Congress, leading to the cancellation of Apollo 20.

Despite the success of lunar landings, the missions remained perilous; astronauts faced such risks that life insurance was unattainable, highlighting the dangers involved in space exploration.

Even before the first U.S. astronaut's flight, NASA envisioned a centralized facility for spacecraft communication and performance monitoring, championed by Christopher C. Kraft Jr., NASA's inaugural flight director. Kraft's pivotal role during John Glenn's Mercury Friendship 7 flight solidified the flight director's authority, establishing a rule that his decisions were final for mission safety and success.

Lovell practices deploying the flag

The establishment of Houston's Mission Control Center in 1965, designed in part by Kraft and now named in his honor, became integral to Apollo missions. Here, flight controllers monitored spacecraft telemetry while collaborating with specialists in the Staff Support Room, ensuring meticulous oversight of spacecraft systems.

Apollo 13, designated as the second H mission, aimed for precision lunar landings and scientific exploration. With Kennedy's initial challenge met and Apollo 12 showcasing precision landing capabilities, Apollo 13 underscored a broader scientific agenda, particularly in geology. This shift was encapsulated in the mission's motto, "Ex luna, scientia" (From the Moon, knowledge), emphasizing scientific aspirations beyond mere landing achievements.

The crew selected for Apollo 13 immersed themselves in an exhaustive training regimen, accumulating well over 1,000 hours tailored to their mission. This preparation amounted to more than five hours of training for every hour planned for their ten-day journey into space. At Kennedy Space Center (KSC) and in Houston, each member of the prime crew dedicated upwards of 400 hours inside simulators meticulously crafted to replicate both the Command Module (CM) and, for Lovell and Haise, the Lunar Module (LM).

These simulation sessions were not mere rehearsals but critical exercises that involved direct interaction with the flight controllers stationed at Mission Control. They meticulously simulated various scenarios, from routine procedures to handling emergencies. Flight controllers, in particular, underwent rigorous training alongside the crew, engaging in numerous simulations designed to hone their ability to respond swiftly and effectively to any potential spacecraft malfunctions.

Beyond KSC and Houston, the crew utilized specialized simulators at different facilities, each designed to replicate specific aspects of their mission. These diverse training environments ensured that every aspect of the mission was thoroughly rehearsed, preparing crew and ground control for the challenges in space.

The training regimen for the Apollo missions demanded a balance between technical proficiency and scientific exploration, a delicate equilibrium that evolved over successive missions. Apollo 11's crew faced a tight timeline, just six months from assignment to launch, leaving minimal room for extensive geology training. Their focus was primarily on higher-priority mission tasks.

Lovell (left) and Haise during geology training in Hawaii, January 1970

By the time Apollo 12 prepared for its lunar journey, the approach to geology training had expanded. Astronauts participated in field exercises, working with a CAPCOM and a simulated team of scientists who awaited their reports from the field. Harrison Schmitt, a scientist-astronaut deeply committed to enhancing geological understanding, recognized the need for inspiration. He facilitated a meeting between Lovell, Haise, and his former professor, Lee Silver from Caltech. This encounter led to an informal geology field trip financed by the astronauts, which concluded with Lovell appointing Silver as their geological mentor for Apollo 13's planning.

Meanwhile, Farouk El-Baz spearheaded the geology training for Mattingly and Swigert, incorporating aerial surveys where astronauts described and photographed simulated lunar features. Mattingly's enthusiasm during these exercises inspired others, such as Apollo 14's Command Module Pilot, Roosa, to seek out El-Baz for additional instruction.

These initiatives underscored a growing recognition within NASA of the pivotal role geological understanding played in lunar exploration. As missions progressed, integrating scientific expertise alongside technical training became increasingly essential, shaping a more comprehensive approach to lunar exploration.

Significant changes were implemented beginning with Apollo 13 in response to the challenges encountered during Apollo 11's lunar descent, during which the LM Eagle nearly ran out of propellant. Unlike Apollo 11 and 12, where the LM performed the crucial burn to lower its orbit before landing, mission planners decided that the Command and Service Module (CSM) would now undertake this task. This adjustment ensured that more propellant remained available for the LM during critical hover phases, especially as future missions targeted rougher lunar terrains.

For Apollo 13, the mission plan allocated the first of two planned four-hour lunar surface extravehicular activities (EVAs) to setting up the Apollo Lunar Surface Experiments Package (ALSEP), a collection of scientific instruments. During the second EVA, Lovell and Haise were scheduled to explore Cone crater, located near their intended landing site. Before these EVAs, both astronauts meticulously rehearsed their procedures in simulated conditions, including approximately 20 walk-throughs in their spacesuits. These sessions focused on sample collection techniques, tool usage, and equipment operation, often conducted in simulated microgravity aboard the "Vomit Comet" or in lunar gravity simulations.

Lovell underwent training on the Lunar Landing Training Vehicle (LLTV) to further prepare for the descent to the lunar surface, complemented by helicopter training to refine his piloting skills. Despite previous accidents involving similar vehicles (the LLTV and Lunar Landing Research Vehicle), which resulted in crashes, mission commanders emphasized the invaluable experience gained from these training tools. Their advocacy convinced hesitant NASA management to retain these vehicles, underscoring their critical role in preparing astronauts for the intricate maneuvers required during lunar landings.

The Apollo 13 mission was conducted using the formidable Saturn V launch vehicle, identified explicitly as SA-508. This rocket played a pivotal role in NASA's ambitious lunar exploration program, tasked with propelling astronauts and their spacecraft beyond Earth's atmosphere toward the Moon. SA-508 closely resembled its predecessors used in Apollo missions 8 through 12, marking a continuation of proven technology and design.

Weighing an impressive 2,949,136 kilograms (6,501,733 lb), including the spacecraft, SA-508 stood as NASA's heaviest launch vehicle to date. Its power was harnessed through multiple stages, beginning with the S-IC first stage, equipped with engines generating 440,000 newtons (100,000 lbf) with less thrust than Apollo 12, yet it remained well within operational specifications. Notably, the S-II second stage introduced a new method for insulating cryogenic tanks, replacing earlier affixed panels with sprayed-on insulation—an innovation intended to enhance efficiency and performance.

The Apollo 13 spacecraft itself comprised the Command Module (CM) and Service Module (SM), collectively known as CSM-109 and christened "Odyssey." This module served as the crew's living quarters and operational nerve center during their journey to and from the Moon. It housed essential navigation, communication, and life support systems critical for the astronauts' well-being and mission success.

Accompanying the CSM was the Lunar Module (LM), designated LM-7 and nicknamed "Aquarius." This specialized module was instrumental for lunar descent and surface operations, engineered to safely transport astronauts to the lunar surface and back into lunar orbit. It featured distinct descent and ascent stages tailored for precise landing and rendezvous maneuvers essential to lunar exploration missions.

Preparations for Apollo 13 began with the arrival of its spacecraft components at Kennedy Space Center in June 1969, followed by the Saturn V stages in June and July of the same year. Rigorous testing and meticulous assembly culminated in the rollout of the fully integrated launch vehicle on December 15, 1969. Originally slated for launch on March 12, 1970, the mission's schedule was adjusted to April 11, 1970, allowing ample time for planning and accommodating NASA's strategic decision to extend the timeline between Apollo missions in response to budgetary considerations and the cancellation of Apollo 20.

Launch preparations for Apollo missions involved meticulously monitoring weather conditions to ensure optimal launch conditions. As the countdown for Apollo 13 began at T-28 hours on April 10th at 05:00:00 GMT, scheduled holds were planned at T-9 hours for 9 hours and 13 minutes and another at T-3 hours 30 minutes for one hour.

At launch, a cold front originating from a low-pressure system in the North Atlantic extended and became stationary, stretching through northern Florida along the Gulf Coast to a low-pressure area in southern Louisiana. In northern Florida, the front was relatively weak, but its intensity increased as it moved northwestward toward the Gulf of Mexico and Louisiana.

Haise practices removing the fuel capsule from its transport cask mounted on the LM. The real cask sank unopened into the Pacific Ocean with its radioactive contents.

Surface winds at Kennedy Space Center were light and variable, typical for the area, with a tendency for the sea breeze to influence winds to shift to the east-southeast by early afternoon. The sky was partly covered with altocumulus clouds, accounting for 40 percent coverage at a base of 19,000 feet. Cirrostratus clouds were also present, forming at a higher altitude with full coverage at 26,000 feet.

Temperature readings showed 75.9° F with a relative humidity of 57 percent. The barometric pressure was measured at 14.676 lb/in². Anemometer readings at the launch site, positioned 60.0 feet above ground level, recorded winds blowing at 12.2 knots from 105° true north.

Apollo 13 commenced its ascent from Kennedy Space Center Launch Complex 39, Pad A, at precisely 19:13:00 GMT on April 11, 1970 (2:13:00 p.m. EST). The launch window was extended until 22:36:00 GMT to optimize the sun's elevation angle on the lunar surface at 10.0°.

During the initial phase, between 000:00:12.6 and 000:00:32.1, the spacecraft transitioned from a launch pad azimuth of 90° to a flight azimuth of 72.043°. The S-IC engine shutdown occurred at 000:02:43.60, triggering the separation of the S-IC and S-II stages and the ignition of the S-II engine.

However, a critical anomaly arose when the center engine of the S-II stage shut down prematurely at 000:05:30.64, approximately 2 minutes and 12 seconds earlier than scheduled. This early shutdown was attributed to severe pogo oscillations, intense vibrations in the propulsion and structural systems. As a result, the vehicle deviated significantly from its planned trajectory, with the altitude at shutdown 10.7 nautical miles lower and velocity 5,685.3 feet per second slower than anticipated.

Despite these challenges, the remaining engines of the S-II stage and the S-IVB third stage compensated by burning longer. This adjustment enabled Apollo 13 to achieve a near-circular parking orbit of approximately 190 kilometers (100 nautical miles). Approximately two hours later, a translunar injection (TLI) maneuver successfully positioned the mission on course for the Moon.

Post-flight investigations revealed that the affected engine was perilously close to catastrophic failure due to the pogo oscillations, exacerbated by an interaction with turbopump cavitation. Although a solution to prevent pogo had been developed, time constraints prevented its integration into the Apollo 13 vehicle.

Following the initial phases of ascent, Apollo 13 continued its journey with subsequent critical maneuvers and events. The S-II engine shutdown occurred at 000:09:52.64, slightly delayed, followed by the separation from the S-IVB stage, which ignited at 000:09:56.90, also 34 seconds behind schedule. Despite these delays, deviations from the planned trajectory were minimal, with a velocity variance of only -1.9 ft/sec and an altitude variation of 0.2 nautical miles.

Apollo 13 launches from Kennedy Space Center, April 11, 1970

The first cutoff of the S-IVB engine occurred at 000:12:29.83, 44 seconds later than scheduled. However, this again resulted in minor deviations, with velocity differing by -1.9 ft/sec and altitude by 0.2 nautical miles.

After completing its role in the mission, the S-IC stage impacted the Atlantic Ocean at 000:09:06.9, approximately 355.3 nautical miles from the launch site, while the S-II stage followed suit at 000:20:58.1, impacting at a significant distance of 2,452.6 nautical miles from the launch site.

During ascent, maximum wind conditions were encountered at 44,540 feet, measuring 108.13 knots from 252° true north, with a maximum wind shear of 0.0166 sec^{-1} at 50,610 feet.

Despite the early shutdown of the S-II center engine, conditions at parking orbit insertion at 000:12:39.83 were nearly nominal, with apogee and perigee measurements of 100.3 by 99.3 nautical miles, a period of 88.19 minutes, an inclination of 32.547°, and a velocity of 25,565.9 ft/sec.

All launch vehicle and spacecraft systems underwent verification following orbital insertion, preparing for the critical translunar injection. Onboard television operations commenced at 001:35, lasting approximately five-and-a-half minutes, providing early visuals from the mission. The decisive 350.75-second translunar injection maneuver, executed by the second firing of the S-IVB stage, occurred precisely at 002:35:46.30. The S-IVB engine shutdown followed at 002:41:37.15, with translunar injection confirmed just ten seconds later, marking the culmination of 1.5 Earth orbits spanning 2 hours, 29 minutes, and 7.3 seconds, achieving a velocity of 35,562.7 ft/sec.

Following the Trans-Lunar Injection (TLI), Jack Swigert skillfully executed the separation and transposition maneuvers, positioning the Command and Service Module (CSM) Odyssey for its crucial rendezvous with the Lunar Module (LM) Aquarius. With precision, the spacecraft disengaged from the S-IVB stage, a pivotal step captured on board through a televised broadcast that showcased the intricate docking process and provided viewers with panoramic views of both the interior and exterior of the Command Module (CM).

At 003:19:08.8, the transposition and docking maneuver successfully united the CSM and the LM, marking a significant achievement in the mission. Subsequently, at 004:01:00.8, the docked spacecraft were gracefully ejected from the S-IVB stage, followed by an 80.2-second separation maneuver initiated by the S-IVB auxiliary propulsion system at 004:18:00.6. Meanwhile, ground controllers meticulously directed the now-unneeded S-IVB stage on a precise trajectory toward the Moon's surface, aiming for impact within the operational range of Apollo 12's seismometer.

For Apollo 13, a departure from previous missions saw the S-IVB stage intentionally redirected toward the Moon rather than into solar orbit. This strategic decision aimed to generate seismic data by impacting the lunar surface within the detection range of Apollo 12's seismometer. The maneuver was meticulously planned, with a precise lunar impact maneuver executed at 005:59:59.5, setting the stage for scientific observation.

The S-IVB struck the Moon's surface at 077:56:39.7, generating seismic signals of unprecedented strength. These vibrations were so intense that they necessitated adjustments to the Apollo 12 seismometer's gain settings to prevent saturation of the recording equipment. Concurrently, the suprathermal ion detector recorded a significant spike in ion counts, surging from zero at impact to 2,500 before returning to baseline. This phenomenon was attributed to lunar surface particles potentially reaching heights of up to 200,000 feet, where they were ionized by sunlight.

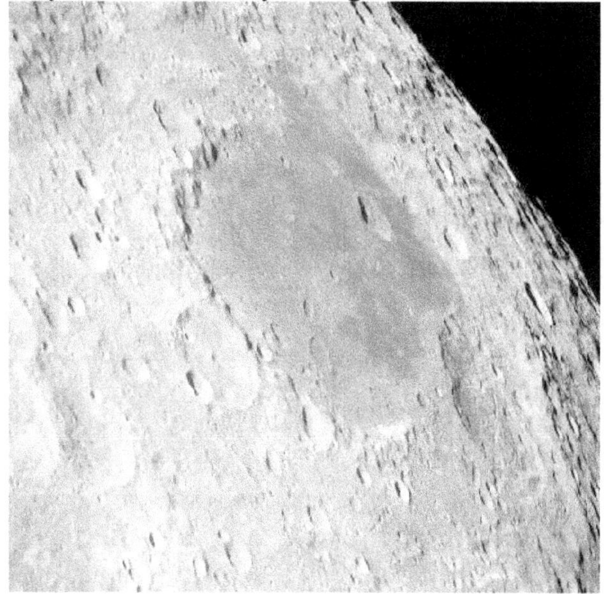

The Apollo 13 crew photographed the Moon out of the Lunar Module.

The impact occurred at coordinates latitude 2.5° south and longitude 27.9° west, approximately 35.4 nautical miles from the intended target point and 75 nautical miles from the Apollo 12 seismometer. At the moment of impact, the S-IVB stage weighed 29,599 pounds and traveled at a staggering speed of 8,465 feet per second.

Throughout these critical maneuvers and scientific observations, comprehensive television coverage provided real-time insights into the mission's progress, including the preparations and execution of the second midcourse correction burn, broadcasted for 49 minutes starting at 030:13.

As Apollo 13 settled into its three-day journey toward the Fra Mauro region of the Moon, a pivotal moment occurred 30 hours, 40 minutes, and 50 seconds into the mission. With the television camera broadcasting, the crew executed a critical burn to transition from a free-return trajectory to a hybrid trajectory. Unlike a free-return path, which would automatically loop back toward Earth if no further maneuvers were made, a hybrid trajectory allowed Apollo 13 to target sites at higher lunar latitudes, such as Fra Mauro.

This adjustment expanded the mission's scientific potential, enabling exploration beyond the lunar equator. Meanwhile, communication with mission control took an unexpected turn when Jack Swigert realized, to the amusement of everyone involved, that he had forgotten to file his federal income tax return, originally due on April 15. In a light-hearted moment amidst the seriousness of space travel, Swigert inquired about obtaining an extension, prompting laughter and swift assistance from mission controllers. Ultimately, it was determined that Swigert qualified for a 60-day extension due to his status as being out of the country—albeit quite far from any terrestrial borders.

On the third day of their mission toward Fra Mauro, the Apollo 13 crew prepared for significant operational adjustments. Originally scheduled to enter the Lunar Module (LM) for system tests at 58 hours into the mission, they were surprised to learn upon waking that the timeline had been accelerated by three hours and subsequently by another hour. Despite these changes, Commander Jim Lovell maintained a steady presence during a planned television broadcast for 55 hours into the mission. As emcee, Lovell guided viewers through virtual tours of the Command and Service Module (CSM) Odyssey and the Lunar Module Aquarius, showcasing the spacecrafts' interiors.

However, the broadcast reached a limited audience as none of the television networks were airing the transmission, prompting Marilyn Lovell, Jim's wife, to relocate to the VIP room at Mission Control for a glimpse of her husband and his crewmates.

During the early stages of their translunar coast, the crew also engaged in scientific activities, capturing photographs of Earth to support atmospheric wind analysis. Meanwhile, a crucial midcourse correction maneuver at 030:40:49.65 further refined the spacecraft's trajectory. This adjustment lowered the closest point of approach to the Moon to just 60 miles, a departure from the previous free-return trajectory that would have automatically directed the spacecraft back toward Earth without requiring further intervention.

Around 56 hours into the Apollo 13 mission, approximately six and a half minutes after a television broadcast, the spacecraft was positioned about 180,000 nautical miles (210,000 miles or 330,000 kilometers) away from Earth. Inside the Lunar Module (LM) Aquarius, Fred Haise completed the shutdown procedures after testing its systems, while Commander Jim Lovell stowed the television camera. Meanwhile, Jack Lousma, the Capsule Communicator (CAPCOM), relayed minor instructions to Jack Swigert, including adjustments to the spacecraft's attitude to facilitate the photography of Comet Bennett.

Earlier, a pressure sensor in one of the Service Module's (SM) oxygen tanks had shown signs of malfunctioning. Sy Liebergot, the Electrical, Environmental, and Consumables Manager (EECOM) responsible for monitoring the CSM's electrical system, requested the activation of the tank stirring fans. Typically, this stirring process helped redistribute the tank contents to ensure accurate pressure readings, normally performed once daily. Flight Director Gene Kranz instructed Liebergot to wait briefly after the television broadcast for the crew to settle down before initiating the action.

Subsequently, Lousma communicated Liebergot's request to Swigert, who promptly activated and deactivated the tank fans after a brief period. This action aimed to address the off-scale high readings observed on the quantity gauge of oxygen tank 2. This gauge anomaly had first been noted around 46 hours into the mission when, upon activating the fans in oxygen tank 2, the gauge suddenly indicated over 100 percent capacity, suggesting a possible electrical fault such as a short circuit.

As preparations continued, the crew readied the LM for activation and inspection. At 053:27, Commander Lovell and Lunar Module Pilot Haise entered the LM to commence an in-flight inspection, initiating a television transmission of the spacecraft's interior from 055:14 to 055:46. The crew subsequently returned to the Command Module (CM), closing the LM hatch at 055:50.

At 055:52:31, a master alarm on the CM caution and warning system alerted the crew to a low-pressure indication in cryogenic hydrogen tank 1, which had previously neared its operational limits. Flight controllers swiftly instructed the crew to activate the cryogenic system's fans and heaters to stabilize the situation. The response was prompt, with the command module pilot acknowledging and initiating the necessary actions to manage the power and stabilization systems.

Ninety-five seconds after Jack Swigert activated the switches controlling the oxygen tank fans, a significant event jolted the Apollo 13 spacecraft. The astronauts heard a loud "bang," with noticeable electrical fluctuations and the firing of attitude control thrusters. This sudden disturbance caused a brief interruption in communications and telemetry with Earth, lasting 1.8 seconds. The onboard systems swiftly rectified this by automatically switching the high-gain S-band antenna from narrow-beam to wide-beam mode, restoring contact with mission control.

The incident occurred precisely at 55 hours, 54 minutes, and 53 seconds into the mission (03:08 UTC on April 14, 10:08 PM EST, April 13). Jack Swigert promptly reported the issue to Houston 26 seconds later, stating, "Okay, Houston, we've had a problem here." Commander Jim Lovell echoed this message at 55:55:42, informing mission control of a "Main B Bus undervolt."

William Fenner, the guidance officer (GUIDO), was the first in the control room to detect and report the anomaly to Flight Director Gene Kranz. Approximately 90 seconds after the initial disturbance, at 55:54:53.555, telemetry from Apollo 13 experienced a brief but almost complete loss for 1.8 seconds. During this brief interval, the spacecraft's caution and warning system alerted the crew to a low voltage condition on DC main bus B, corroborating the crew's firsthand observation of a problem within the spacecraft.

Upon hearing the loud noise, Jim Lovell initially suspected Fred Haise had activated the LM's cabin-repressurization valve, a playful act known to startle crewmates. However, seeing Haise's confusion ruled out this possibility. Meanwhile, Jack Swigert considered the prospect of a meteoroid impact on the LM. Still, both he and Lovell quickly dismissed this idea upon confirming there was no indication of a leak.

The critical issue arose from the "Main Bus B undervolt" alarm, signaling that the Service Module's (SM) second electric power distribution system lacked sufficient voltage from its three hydrogen-oxygen-fueled fuel cells. Nearly all systems in the Command and Service Module (CSM) relied on this power. Although the voltage briefly returned to normal, both Bus A and Bus B soon experienced voltage shortages. Haise checked the status of the fuel cells and discovered that two of them had failed. Mission protocols strictly prohibited entering lunar orbit unless all fuel cells were operational.

Overall view of the Mission Operations Control Room in the Mission Control Center at the Manned Spacecraft Center, during the fourth television transmission from the Apollo 13 spacecraft while en route to the Moon. Eugene F. Kranz (foreground, back to camera), one of four Apollo 13 Flight Directors, views the large screen at front of MOCR. Astronaut Fred W. Haise Jr., lunar module pilot, is seen on the screen. The fourth television transmission from the Apollo 13 mission was on the evening of April 13, 1970. Shortly after the transmission ended and during a routine procedure that required the crew to flip a switch that stirred one of the cryogenic liquid oxygen tanks, an explosion occurred that ended any hope of a lunar landing and jeopardized the lives of the three crew members.

At the moment of the incident, Lovell was in the lower equipment bay stowing the television camera, while Haise was in the tunnel between the CSM and LM, returning to the CSM. Swigert, the command module pilot, monitored spacecraft performance from the left-hand couch. Due to the master alarm indicating low voltage on DC main bus B, Swigert moved to the right-hand couch to monitor the voltage readings more closely. He reported that voltages initially appeared stable at 55:56:10, but shortly after, fuel cell 3 failed 90 seconds later. He also noted fluctuations in the quantity reading of oxygen tank 2, which then returned to an off-scale high position.

Shortly after, at 56:09:07, Lovell reported, "...We're venting something... into space..." followed by Haise's report at 56:09:58 that fuel cell 1 was offline. Within half an hour, Haise reported that fuel cell three had also failed. As fuel cells 1 and 3 registered zero electrical output, ground controllers could not confirm their operational status, focusing on resolving the escalating electrical issues.

Several alarming indicators surfaced aboard Apollo 13 in the aftermath of the accident. The telemetry revealed that oxygen tank 2 was showing empty, while the pressure in oxygen tank 1 was slowly declining. Additionally, the spacecraft's computer had reset, and there were issues with the high-gain antenna.

Initially, Sy Liebergot, monitoring the electrical systems, had focused on oxygen tank 1, believing its readings would provide a reliable guide to the state of tank 2, as did his support team in the control room. This led to him initially missing the concerning signs from tank 2 following the stir operation. When Flight Director Gene Kranz questioned Liebergot about this discrepancy, Liebergot suggested it might be due to an instrumentation issue. This response would later become a source of good-natured teasing among colleagues.

Meanwhile, Jim Lovell, observing from the spacecraft window, reported seeing "a gas of some sort" venting into space, underscoring the seriousness of the situation.

Within five minutes of the initial incident, controllers directed the crew to connect fuel cell 3 to DC main bus B to confirm its operational status. Upon realizing that fuel cells 1 and 3 were not functioning, they instructed the crew to initiate an emergency powerdown to reduce the load on the remaining fuel cell. Fuel cell two was subsequently shut down at 58:00 into the mission, followed by the powerdown of the Command Module (CM) computer and platform ten minutes later.

As the pressure in Oxygen Tank 1 continued to decline rapidly, controllers directed the crew to switch power to the Oxygen Tank 2 instrumentation. Upon doing so, it became evident that Oxygen Tank 2 had also failed, highlighting the gravity of the situation. Efforts were then focused on preserving the remaining oxygen in tank 1, but despite several attempts, the pressure continued to decrease, escalating the urgency of the situation aboard Apollo 13.

Following the depletion of Oxygen Tank 1, which supplied the fuel cells essential for generating power aboard Apollo 13, the spacecraft faced a critical situation. The remaining operational fuel cell, already deprived of adequate oxygen, relied on the CM's surge tank. Recognizing the urgency, Flight Director Gene Kranz ordered the isolation of the surge tank to conserve its remaining oxygen. However, this action meant that the fuel cell would cease functioning within approximately two hours, as the oxygen from Tank 1 was either consumed or lost due to leaks.

Adding to the complexity, the space around the spacecraft was littered with countless small debris particles from the earlier accident, complicating attempts to use star sightings for navigation.

With the primary goal now shifted to ensuring the astronauts' safe return to Earth, it became evident within about 90 minutes of the incident that the leak in Oxygen Tank 1 could not be contained. Soon, the decision was made to utilize the Lunar Module (LM) as a "lifeboat" for the remainder of the mission.

With the fuel cells rendered inoperative due to the lack of oxygen, the CM's batteries became the sole power source, typically reserved for reentry. The remaining oxygen was contained in the surge tank, and depressurization packages were used for the CM.

With its sufficient electrical power and oxygen supplies, the LM became instrumental in facilitating a safe return to Earth. Consequently, the decision was made to abort the Apollo 13 mission. By 58 hours and 40 minutes into the mission, the LM had been activated, the guidance system transferred from the CSM to the LM, and the systems aboard were powered down.

Following the critical incident on Apollo 13, the remainder of the mission focused on two primary objectives: planning and executing the necessary maneuvers to return the spacecraft safely to Earth and managing the consumables aboard the Lunar Module (LM) to support the extended duration and additional crew members.

Initially, propulsion options were evaluated to facilitate a return to a free-return trajectory and perform required midcourse corrections. Typically, these maneuvers would be handled by the Service Module (SM) propulsion system. However, due to concerns about the SM's condition after the accident, including its structural integrity and the high electrical power demands of its engine, the decision was made to use the LM's descent engine instead.

On 61 hours, 29 minutes, and 43.49 seconds into the mission, the LM's descent engine was fired for 34.23 seconds to maneuver the spacecraft back onto a free-return trajectory. This trajectory would allow the spacecraft to loop behind the Moon and return toward Earth.

During the period when Apollo 13 was out of contact with Earth tracking stations (from 77 hours, 8 minutes, and 35 seconds to 77 hours, 33 minutes, and 10 seconds), lasting 24 minutes and 35 seconds, the LM utilized its charged batteries and full oxygen tanks, initially intended for lunar surface operations. This shift to using the LM as a "lifeboat" was a contingency plan developed by LM flight controllers after simulations and training exercises.

The outcome would have been dire if the accident had occurred during the return journey or after the LM had been jettisoned following a successful lunar landing. The LM became essential for its propulsion capabilities and life support systems, ensuring the crew's survival during the journey back to Earth. This scenario highlighted the critical importance of contingency planning and the adaptability of mission protocols in the face of unforeseen emergencies in space exploration.

After the critical incident aboard Apollo 13, one of the crucial decisions made by Mission Control under the leadership of Gene Kranz was the choice of return trajectory. Initially, a "direct abort" was considered using the Service Module's main engine (Service Propulsion System or SPS) to return to Earth before reaching the Moon. However, concerns over the condition of the SPS following the accident, coupled with the need for the fuel cells to sustain power for at least another hour, prompted Kranz to opt for an alternative route.

Instead of a direct abort, which would have required a fully operational SPS, Kranz directed the spacecraft to swing around the Moon. Apollo 13 was originally on a hybrid trajectory aimed toward Fra Mauro on the lunar surface. The new plan involved returning the spacecraft to a safer free-return trajectory, which could guide it back to Earth without requiring additional propulsion maneuvers.

The Lunar Module's Descent Propulsion System (DPS) was selected to execute this maneuver. While not as powerful as the SPS, the DPS could adjust the trajectory to a free-return path. However, this required new software to be developed for Mission Control's computers, as it had never been anticipated that the LM would need to maneuver the combined CSM/LM spacecraft.

During this critical phase, as the Command Module (CM) was being powered down to conserve resources, Commander Jim Lovell manually transferred the guidance system's orientation information from the CM to the LM. This involved hand calculations to ensure the LM's guidance system could accurately control the spacecraft's trajectory. Mission Control verified Lovell's calculations to ensure accuracy before proceeding with the DPS burn.

At 61 hours, 29 minutes, and 43.49 seconds into the mission, the DPS burn of 34.23 seconds successfully maneuvered Apollo 13 back onto a free-return trajectory. This maneuver was a pivotal step in the mission's success, ensuring the spacecraft would safely navigate around the Moon and begin its journey back to Earth despite the unprecedented challenges and uncertainties faced during the mission.

After the decision to use the Lunar Module's Descent Propulsion System (DPS) to maneuver Apollo 13 back onto a free-return trajectory, the focus turned to optimizing the return path to Earth. Initially, the spacecraft was projected to splash down in the Indian Ocean, an area with limited recovery resources. Flight Dynamics Officers (FIDOs), including Jerry Bostick, were eager to shorten the travel time and relocate the splashdown to the Pacific Ocean, where NASA's primary recovery forces were stationed.

Several options were considered to achieve this goal. One proposal involved a maneuver that would shave off 36 hours from the return time, but it required jettisoning the Service Module (SM). However, exposing the Command Module's (CM) heat shield to space during the return journey was a concern, as it was not designed to endure such conditions.

Flight controllers evaluated various return options based on mission constraints and recovery capabilities. They determined that the minimum practical return time for Apollo 13 was 133 hours to the Atlantic Ocean, while the maximum would be 152 hours to the Indian Ocean. Given the deployment of recovery forces in the Pacific, a decision was made to aim for a splashdown in that ocean at 142 hours and 40 minutes into the mission.

To expedite the return journey, a crucial transearth injection maneuver was executed at 079:27:38.95 mission elapsed time. This maneuver utilized the LM descent propulsion system to increase the spacecraft's velocity by 860.5 feet per second (ft/sec) after it had swung around the far side of the Moon. This burn was instrumental in shortening the travel time and ensuring a Pacific Ocean splashdown where recovery operations could be efficiently conducted.

During the critical moments preceding the burn, the Apollo 13 crew received confirmation that the S-IVB stage had successfully impacted the lunar surface as planned. This small victory prompted Commander Lovell to remark wryly, "Well, at least something worked on this flight."

In Mission Control, Gene Kranz and his White team of mission controllers, who had tirelessly supported various aspects of the mission and developed crucial procedures for the astronauts' safe return, focused on the upcoming PC+2 procedure. This maneuver was essential for adjusting their trajectory back toward Earth.

Typically, the accuracy of such a burn could be verified by aligning Lovell's transferred data in the LM's computer with the position of a navigational star. However, the spacecraft's surroundings were cluttered with debris, making this standard check impractical. Consequently, the crew relied on the position of the Sun, the only celestial body not obscured by debris, for their navigational fix.

Meanwhile, Houston informed the crew that the Moon would be perfectly centered in the commander's Lunar Module (LM) window during the burn, ensuring precise alignment. The maneuver itself, executed at 79 hours, 27 minutes, and 38.95 seconds into the mission, lasted for four minutes and 23 seconds.

Following the burn, the crew took proactive measures to conserve the LM's consumables, shutting down most of its systems. This strategic decision was crucial as they continued their journey back to Earth under increasingly challenging conditions.

The Apollo 13 mission faced daunting challenges as it returned from lunar orbit to Earth. Unlike the Command Module (CM), which generated electricity and water through fuel cells, the Lunar Module (LM) relied on silver-zinc batteries that did not produce water as a byproduct. This meant electrical power and water—essential for equipment cooling and drinking—were critical concerns.

Apollo13 - view of the crippled Service

The LM's consumption was minimized to its lowest feasible level to conserve power. Jack Swigert managed to transfer some water from the CM's reserves into drinking bags. Still, with personal consumption rationing, concerns arose that they might run out of water for cooling several hours before reentry.

However, historical data from Apollo 11 provided a glimmer of hope. Despite the lack of water after jettisoning in lunar orbit, its LM systems continued functioning for seven to eight hours. This precedent offered a degree of reassurance during Apollo 13's dire situation.

In the end, Apollo 13 returned safely to Earth with 12.8 kilograms (28.2 lb) of water still onboard. Each astronaut's daily ration was limited to just 0.2 liters (6.8 fl oz), resulting in a total weight loss of 14 kilograms (31 lb) among the crew. Fred Haise unfortunately developed a urinary tract infection, likely exacerbated by reduced water intake. The combination of microgravity and cosmic radiation may have further compromised his immune system's ability to combat the infection.

Throughout these challenges, the successful adaptation of the LM's environmental control systems by creating the makeshift "mailbox" device underscored the resourcefulness and collaboration between NASA's ground control and the mission crew. This innovative solution was pivotal in managing the spacecraft's atmosphere and ensuring the crew's survival during their extraordinary journey back from the Moon.

As the mission progressed, additional challenges arose. An error in guidance during the transearth injection maneuver required a midcourse correction using the LM's descent propulsion system. This adjustment, made at 105 hours, 18 minutes, and 42 seconds into the mission, was crucial for aligning the spacecraft's reentry trajectory within safe limits.

The spacecraft was maneuvered into a passive thermal control mode throughout the transearth coast phase to manage heat dissipation. The critical consumables—water for cooling systems, battery power for both the CSM and LM, oxygen for breathing, and LiOH canisters for carbon dioxide removal—initially appeared dangerously low following the accident. However, the situation improved significantly by conserving power by shutting non-essential systems in the LM.

By the time of splashdown, ample reserves of each consumable remained available, ensuring the crew's safe return despite the harrowing challenges encountered during Apollo 13's mission.

Following the dramatic events of the Apollo 13 mission, the Command Module (CM) faced unprecedented challenges as it prepared for reentry into Earth's atmosphere. With the spacecraft in a powered-down state for much of the journey, new procedures had to be devised to ensure a safe return.

Swigert with the rig improvised to adapt the CM's lithium hydroxide canisters for use in the LM

At one point during the mission, the CM was briefly powered up to evaluate the operational status of critical systems. The entry batteries, crucial for reentry, were recharged using umbilical connectors that had previously supplied power from the Lunar Module (LM) while the CM remained inactive.

The crew observed the Service Module (SM) during this assessment, revealing significant damage. An entire panel near the S-band high-gain antenna was missing, exposing internal components and systems to space. The fuel cells, vital for generating electrical power, were visibly tilted on their shelf above the oxygen supply area. The high-gain antenna itself sustained damage, which could potentially affect communications. Additionally, a substantial amount of debris was observed, further complicating the spacecraft's condition.

Inside the dimly lit spacecraft, the temperature plummeted to as low as 3°C (38 °F), creating an uncomfortable environment for the Apollo 13 crew. Commander Lovell contemplated having the crew wear their spacesuits to stay warm but realized this would cause overheating. Instead, they improvised with available gear: Lovell and Haise wore their lunar EVA boots, while Swigert added an extra coverall for insulation. Despite these efforts, all three astronauts remained chilly throughout the ordeal. Swigert was particularly affected after getting his feet wet while filling water bags and lacking lunar overshoes since he hadn't been slated for a moonwalk.

A unique challenge arose concerning waste management. To maintain the spacecraft's trajectory stability, the crew refrained from discharging urine into space and instead collected it in bags. This conservation measure added to the cramped and uncomfortable conditions inside the module.

Despite these hardships, the crew rarely voiced complaints. Any condensation that formed on the spacecraft walls, though minimal, posed no operational issues, thanks in part to enhanced electrical insulation measures implemented following the tragic Apollo 1 fire. Their stoicism and adherence to procedures underscored their professionalism and determination to overcome adversity during one of NASA's most challenging missions.

Flight controller John Aaron, alongside Mattingly and a team of engineers and designers, faced the unprecedented challenge of devising a procedure to power up the Command Module (CM) from a complete shutdown—a scenario never anticipated for in-flight operations, especially given Apollo 13's severe power and time constraints.

The procedure ensured the CM's systems were operational before re-entry. Despite the daunting circumstances, the astronauts executed the startup without apparent difficulty. Gene Kranz later attributed their success to their background as test pilots, accustomed to making critical decisions in high-stakes environments where their lives were on the line.

As Apollo 13 pressed onward, the challenges faced by the crew intensified. Mission Control closely monitored the effects of cold temperatures and fatigue on the astronauts' ability to execute crucial startup procedures for the Command Module (CM). Recognizing the urgency at 133 hours into the mission, they authorized Jim Lovell to fully power up the Lunar Module (LM) to increase cabin warmth. This involved restarting the LM's guidance computer, a critical step that allowed Lovell to perform essential navigational tasks and calibrate the LM's Inertial Measurement Unit (IMU).

As the dramatic events of Apollo 13 unfolded, Mission Control faced a pivotal decision. With temperatures plummeting and the crew's fatigue jeopardizing critical startup procedures for the Command Module (CM), they authorized Commander Jim Lovell to power up the Lunar Module (LM) fully. This included restarting its guidance computer, which was essential for navigation tasks, and calibrating the LM's Inertial Measurement Unit (IMU). Utilizing the LM's positional data relative to Earth, Lovell employed a novel process to streamline the CM's computer setup, enhancing the accuracy of the reentry guidance system controlled by the Primary Guidance, Navigation, and Control System (PGNCS).

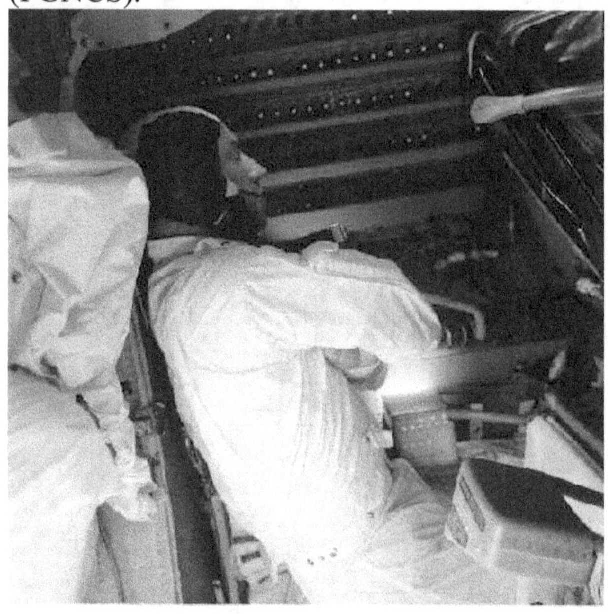

Lovell tries to rest in the frigid spacecraft

Despite the precise execution of their trajectory adjustment, Apollo 13 began to drift off course following the critical PC+2 burn. This deviation necessitated immediate corrective actions. With the LM's guidance system deactivated after the burn, the crew resorted to an unconventional navigation method—using Earth's terminator, the boundary between night and day, as a reference. This technique, typically employed in low Earth orbit, had never been attempted for a return from the Moon.

A brief 14-second Delta-V burn was executed at 105 hours, 18 minutes, and 42 seconds using the LM's Descent Propulsion System (DPS) to correct their path. This maneuver successfully realigned their trajectory, but further adjustments were required. Approximately 32 hours later, at 137 hours, 40 minutes, and 13 seconds, a 21.5-second burn using the LM's reaction control system (RCS) further refined their course.

Shortly after, the crew jettisoned the damaged Service Module (SM), giving them their first visual assessment of the extensive damage inflicted during the mission. They observed a missing panel, visibly tilted fuel cells, and damage to the high-gain antenna. Debris scattered around the spacecraft underscored the severity of the situation. Concerns about potential damage to the SM's engine bell reinforced Flight Director Gene Kranz's decision to avoid using the Service Propulsion System (SPS) engine for maneuvers.

Following these evaluations, the crew transitioned back to the Command Module (CM), reactivating its life support systems as they prepared for the final leg of their perilous journey home.

Approximately six hours before reentry, they deactivated the passive thermal control mode and conducted a final midcourse correction using the LM's RCS. This 21.5-second maneuver, performed at 137 hours, 40 minutes, and 13 seconds, adjusted their trajectory by 3.0 feet per second, ensuring they remained on course for a safe return to Earth.

Shortly after that, at 138 hours, 1 minute, and 48 seconds, the damaged SM was jettisoned, allowing the crew to visually inspect and photograph the damage caused by the oxygen tank failure. The LM remained attached until 141 hours, 30 minutes, and 0.2 seconds—approximately 70 minutes before entry into Earth's atmosphere—to minimize power usage from the CM's electrical systems. During undocking, normal tunnel pressure facilitated the separation of the two spacecraft.

One of the final challenges was safely separating the Lunar Module (LM) from the CM just before reentry. Since the usual method involving the SM's RCS was not feasible after its jettison, engineers from Grumman and a team led by senior scientist Bernard Etkin from the University of Toronto collaborated to devise an alternative solution. They successfully utilized air pressure to push the modules apart, a technique implemented by the astronauts under intense pressure as they prepared for the critical reentry phase.

The LM subsequently reentered Earth's atmosphere and was deliberately directed to impact the Tonga Trench in the Pacific Ocean, one of its deepest points. This decision was crucial as it ensured the safe landing of the cask containing plutonium oxide intended for the SNAP-27 Radioisotope Thermoelectric Generator (RTG), overseen by the Atomic Energy Commission. The cask descended 10 kilometers (6 mi) to the ocean floor, where subsequent helicopter surveys confirmed no radioactive leakage.

Apollo 13's final midcourse correction not only adjusted the spacecraft's trajectory but also addressed critical safety concerns, underscoring the meticulous planning and ingenuity that characterized the mission's successful return despite the daunting challenges encountered in space.

During Apollo 13's reentry into Earth's atmosphere, the ionization of the air around the Command Module (CM) typically caused a four-minute communications blackout. However, due to the mission's shallow reentry path, this blackout extended unexpectedly to six minutes, leading mission controllers to fear a potential failure of the CM's heat shield.

Fortunately, communication was reestablished when Odyssey, the CM, regained radio contact. It splashed down safely in the South Pacific Ocean at coordinates 21°38′24″S 165°21′42″W, southeast of American Samoa and just 6.5 kilometers (3.5 nautical miles) from the recovery ship, USS Iwo Jima.

After reentering Earth's atmosphere at an altitude of 400,000 feet and a velocity of 36,210.6 feet per second, the Command Module (CM) of Apollo 13 completed its 63-hour, 8-minute, and 42.9-second transearth coast. According to projected trajectory data, some pieces of the Lunar Module (LM) survived entry, striking the open sea between Samoa and New Zealand.

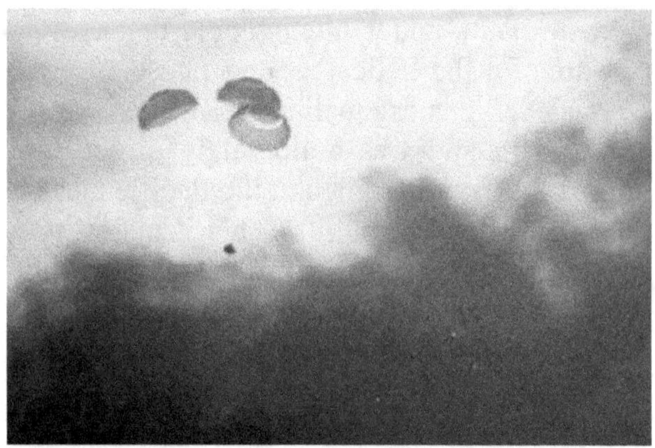

The CM deployed its parachute system, leading to a controlled splashdown in the Pacific Ocean precisely at 18:07:41 GMT (01:07:41 p.m. EST) on April 17th. The entire mission lasted 142 hours, 54 minutes, and 41 seconds. Following splashdown, the CM assumed an apex-up flotation attitude.

Recovery efforts were swift: the crew was retrieved by helicopter and safely aboard the recovery ship, USS Iwo Jima, just 45 minutes after splashdown. Lovell exited through the hatch, followed by Swigert and Haise, as captured in NASA image S70-35610. The CM itself was recovered 43 minutes later.

The splashdown site was located at approximately latitude 21.63° south and longitude 165.37° west, approximately 1.0 nautical mile from the target point and 3.5 nautical miles from the USS Iwo Jima. The estimated weight of the CM at splashdown was 11,133 pounds, and the total distance traveled during the mission was approximately 541,103 nautical miles.

Despite their ordeal, the crew was in relatively good health upon recovery, except for Fred Haise, who had developed a serious urinary tract infection likely exacerbated by limited water intake during the mission.

The incident involving Apollo 13 sparked a renewed global interest in the Apollo program, capturing the attention of millions around the world. Television coverage reached unprecedented levels, with broadcasts seen by millions of viewers. The significance of the situation was underscored by the response from various nations, including Soviet ships heading toward the potential landing area in case assistance was needed. Offers of help came from other countries in case the spacecraft needed to splash down elsewhere.

After President Nixon canceled appointments to attend to the unfolding events personally, demonstrating the gravity of the situation. He contacted the astronauts' families and visited NASA's Goddard Space Flight Center to oversee the mission's coordination of tracking and communications.

Apollo 13's rescue effort garnered more public attention than any previous spaceflight, second only to Apollo 11's historic Moon landing. The incident dominated headlines worldwide, prompting people to gather around their televisions for updates. It became a global event of concern and unity, with widespread prayer gatherings held in various countries, including a congregation led by Pope Paul VI and a religious festival in India.

In the United States, an estimated 40 million people watched Apollo 13's splashdown live on all three major networks, with an additional 30 million catching parts of the 6½-hour telecast. The event was also closely watched internationally. Commentators reflected that Apollo 13, despite its near-tragic outcome, had a unifying effect on the world that possibly exceeded the impact of a successful Moon landing, highlighting the shared concern and global fascination with human space exploration.

Apollo 13 splashes down in the South Pacific on April 17, 1970

After departing from the USS Iwo Jima by aircraft on April 18 at 18:20 GMT, the Apollo 13 crew arrived in Houston at 03:30 GMT on April 20. The USS Iwo Jima, carrying the Command Module (CM), reached Hawaii at 19:30 GMT on April 24. Deactivation procedures for the CM were finalized by April 26. The CM was subsequently transported to the North American Rockwell Space Division facility in Downey, California, for postflight analysis, arriving there at 14:00 GMT on April 27. This marked the beginning of extensive examinations and evaluations to understand the mission's challenges and ensure lessons learned for future space missions.

President Richard Nixon awarding the Apollo 13 astronauts the Presidential Medal of Freedom

The Apollo 13 mission stands as a testament to human ingenuity and resilience in the face of near-catastrophe. Following the unexpected loss of primary oxygen in the service module, the crew and ground support teams displayed exceptional skill and determination, ultimately ensuring the astronauts' safe return.

Analysis of post-mission data yielded several critical conclusions. Firstly, the mission had to be aborted due to a complete loss of primary oxygen caused by an unforeseen switch design and pre-mission procedure incompatibility. This issue, compounded by an abnormal detanking process before launch, led to an in-flight short circuit and rapid oxidation within one of the two oxygen storage tanks. The subsequent loss of pressure integrity in both tanks necessitated using emergency systems to sustain life aboard the spacecraft.

Notably, the concept of a backup crew proved its worth when the backup command module pilot stepped in just days before launch, replacing the prime crew member exposed to rubella. This substitution highlighted the foresight in NASA's crew preparation protocols, ensuring mission readiness despite unforeseen medical challenges.

The lunar module, initially intended for a moon landing, instead played a crucial role in ensuring the crew's survival. Its systems, designed for a shorter operational lifespan, supported the astronauts for double the duration they were originally intended to endure. This emergency operational capability underscored the lunar module's meticulous planning and robust engineering.

The mission's success, though marred by its failure to land on the moon, nonetheless achieved significant milestones. A lunar flyby was executed, allowing the completion of planned experiments, including the study of lightning phenomena, Earth photography, and the impact of the S-IVB stage on the lunar surface. Data collected during these operations contributed invaluable insights into the lunar module's capabilities and furthered scientific understanding.

Throughout the ordeal, the effectiveness of rigorous pre-mission training, coupled with seamless coordination between astronauts and ground personnel, shone brightly. The crew's adept response to the emergency underscored the importance of thorough preparation and teamwork in NASA's missions.

Following the near-catastrophic Apollo 13 mission, NASA Administrator Thomas O. Paine convened the Apollo 13 Review Board on April 17, 1970, appointing Edgar M. Cortright, director of NASA's Langley Research Center, as its chairman. Comprising eight members, the board meticulously investigated the events surrounding the mission over nearly two months, culminating in their final report submitted on June 15.

Among the board's members was Neil Armstrong, renowned as the commander of Apollo 11, the historic mission that first landed humans on the moon. William Anders, the lunar module pilot of Apollo 8 and executive secretary of the National Aeronautics and Space Council, served as an observer alongside two others. Their findings concurred with an electrical short circuit with arcing as the initiating incident.

The critical sequence of events began approximately 2.7 seconds after the activation of fans within the Service Module's oxygen tanks. This activation caused a sudden spike of 11.1 amperes in current and a simultaneous drop in voltage within the spacecraft's electrical system. Almost immediately, there was a corresponding reduction in current drawn from the fuel cells, consistent with losing power to one of the fans. Notably, no other concurrent changes in spacecraft power configuration were noted at that precise moment, and the heaters in the oxygen tanks, which operated at low power, were confirmed to be inactive.

The subsequent critical event recorded during the Apollo 13 mission was the onset of a pressure increase in oxygen tank 2, thirteen seconds after the initial electrical anomalies.

This delay was consistent with the potential low-level combustion that could occur under such circumstances. The investigation indicated a high likelihood that an electrical short circuit with arcing had occurred, specifically within the fan motor or its associated wiring, thereby triggering the sequence of events leading to the accident.

The energy released from this short circuit was estimated to be between 10 to 20 joules, which subsequent tests confirmed was sufficient to ignite the Teflon insulation in the oxygen tank. It was also revealed that the electrical wiring within the tank may have been damaged during abnormal tanking procedures at Kennedy Space Center before launch, further exacerbating the risk of electrical initiation.

While precise details regarding the exact mechanism by which the oxygen tank 2 system lost its integrity were not fully determinable from the available data, extensive analyses and tests conducted during the investigation strongly suggested that localized combustion within the pressure vessel led to thermal heating and subsequent failure at the closure of the pressure vessel. This critical point, situated at the upper end of the quantity probe where the Inconel conduit entered the vessel containing Teflon-insulated wires, likely became a focal point for the combustion progression.

It is hypothesized that the combustion propagated along the insulation of the wires until reaching the junction point where all wires converged. This, potentially compounded by the ignition of the metal at the upper end of the probe, contributed to the weakening and failure of either the closure mechanism or the conduit, or possibly both. Such a failure would have immediately led to pressurization of the tank dome, which included a ruptured disc rated at approximately 75 pounds per square inch (psi). Subsequent rupture of this disc or the entire dome would have resulted in the release of oxygen, accompanied by combustion byproducts, into bay 4 of the spacecraft.

Following the critical events on Apollo 13, spacecraft accelerations were observed, likely triggered by the release of oxygen into bay 4. As oxygen began pressurizing the oxygen shelf space, the conditions were such that if the holes formed in the pressure vessel were sufficiently large and developed rapidly, the escaping oxygen alone could have been enough to blow off the bay 4 panel. However, the situation was compounded by the possibility that the oxygen release was accompanied by the combustion of Mylar and Kapton, extensively used as thermal insulation within the oxygen shelf compartment and tank dome. This combustion would have further increased the internal pressure beyond what the oxygen alone could generate. Recorded slight temperature increases at various locations within the Service Module indicated that external combustion likely occurred.

The panel's ejection from bay 4 subsequently struck the high-gain antenna, disrupting spacecraft communications for approximately 1.8 seconds.

The sequence of events leading to this critical situation can be attributed to several factors:

The Oxygen Tank 2, which flew aboard Apollo 13, initially appeared satisfactory after assembly and acceptance testing at Beech Aircraft Corporation. However, it was later discovered that the tank contained inadequate thermostatic switches on the heater assembly and subsequently failed during ground tests at Kennedy Space Center (KSC).

Additionally, it is likely that the tank had a loosely fitted fill tube assembly. This assembly may have been displaced during handling incidents, including an incident at the prime contractor's facility, possibly exacerbating the subsequent issues.

While the displaced fill tube assembly itself was not initially considered a significant concern, it necessitated improvised detanking procedures at KSC. These makeshift procedures, undertaken due to difficulties in detanking through the compromised fill tube assembly, inadvertently set the stage for the accident.

Despite Beech Aircraft not encountering issues during detanking in their acceptance tests, normal detanking procedures proved unsuccessful at KSC due to gas leakage through the displaced fill tube assembly. This unforeseen complication was a critical factor in the subsequent events aboard Apollo 13.

The unique detanking procedures implemented at Kennedy Space Center (KSC) involved prolonged heater operation and pressure cycling, practices that had yet to be previously utilized. Though within the operational specifications governing heater use at KSC, these procedures subjected the oxygen tank to conditions for which it had not been formally qualified through testing.

Before the flight, NASA, North American Rockwell (NR), and Beech Aircraft Corporation's comprehensive review of these procedures did not anticipate potential damage due to overheating. Importantly, key officials involved in the review needed to be made aware of the extended duration of the heater operation. Nevertheless, it was presumed that adequate thermostatic switches would safeguard the tank against overheating issues.

The root cause of the inadequate thermostatic switches can be traced to several factors. Initially, the original 1962 specifications issued by NR to Beech for the tank and heater assembly specified the use of a 28 V DC power supply, consistent with spacecraft requirements. However, in 1965, NR revised its specifications to recommend using a 65 V DC power supply for heater operation during tank pressurization, a measure aimed at reducing the time required for pressurization at KSC. Despite this update, Beech continued to order thermostatic switches based on the original 28 V DC specifications intended for Block II tanks without aligning them with the new 65 V DC requirement.

This discrepancy was noticed during the review processes conducted by NASA, NR, and Beech and during the testing phases. Qualification and acceptance testing protocols did not include the necessary cycling of switches under load conditions, which could have identified the compatibility issue with the ground support equipment at KSC. This oversight represented a significant lapse in oversight shared across all involved parties.

In summary, the Apollo 13 incident highlighted critical gaps in communication, testing protocols, and adherence to updated specifications during the development and testing phases of the oxygen tank and heater assembly.

The failure of the thermostatic switches in oxygen tank 2 during the special detanking procedures at Kennedy Space Center (KSC) had profound consequences for the Apollo 13 mission. These switches, designed to operate under 28 V DC conditions, were ill-equipped to handle the 65 V DC power used during tank pressurization at KSC. While they could manage the higher voltage without issue while closed and cool, they were not designed to open under these conditions safely. During the extended detanking process, as the switches reached their upper-temperature limits and attempted to open, they were welded shut by arcing, rendering them permanently closed and ineffective as protective devices.

The failure of these switches went unnoticed at KSC, where crucial switch operation checks, such as observing heater current readings on the control panel, were not performed. Although tank temperature readings indicated that the heaters had reached their limit, signaling that switch opening should have been anticipated, this critical indication was not recognized then.

Subsequent tests revealed that the thermostatic switches' failure allowed the heater tube assembly to reach temperatures as high as approximately 1000°F in localized spots during the continuous eight-hour heater operation. This extreme heating severely damaged the Teflon insulation on the fan motor wires near the heater assembly. Consequently, from this point onward, including during the launch preparations, oxygen tank 2 was in a hazardous state when filled with oxygen and powered electrically.

The consequences of these earlier failures manifested nearly 56 hours into the mission. The fan motor wiring movement, possibly agitated by the fan's operation, resulted in a short circuit that ignited the insulation via an electric arc within the oxygen tank. The ensuing combustion within the tank likely overheated and compromised the wiring conduit where it entered the tank, potentially damaging a portion of the tank structure itself.

The rapid release of high-pressure oxygen that followed, possibly augmented by the combustion of insulation materials in the tank's vicinity, forcefully expelled the outer panel of bay 4 in the Service Module (SM). This event caused a leak in the high-pressure system of oxygen tank 1, damaged the high-gain antenna, inflicted various other damages, and ultimately forced the mission to be aborted.

These events underscored critical failures in component design, testing protocols, and operational oversight, revealing vulnerabilities in the systems supporting manned space missions. The lessons from Apollo 13 significantly influenced subsequent missions, leading to enhanced safety protocols and rigorous testing regimes to mitigate risks in future space exploration endeavors.

Significant redesigns were implemented for Apollo 14 and subsequent missions to enhance spacecraft safety and reliability, especially focusing on the oxygen system following the lessons learned from Apollo 13. The oxygen tank underwent crucial upgrades, primarily in the thermostats, which could now handle the correct voltage levels to prevent failures seen in previous missions.

To mitigate risks further, the stirring fans, which had caused issues in Apollo 13 due to their unsealed motors, were removed. This change, however, necessitated adjustments since the absence of the fans affected the accuracy of the oxygen quantity gauge. As a solution, a third oxygen tank was introduced and positioned in Bay 1 of the Service Module, strategically isolated from the other two tanks by an isolation valve. This setup ensured that this tank could independently supply oxygen to the Command Module's environmental system even in emergencies if needed.

Moreover, all electrical wiring in Bay 4 was now encased in stainless steel to improve safety, reducing the risk of electrical shorts or fires in oxygen-rich environments. The fuel cell oxygen supply valves were redesigned to prevent contact between Teflon-coated wiring and oxygen, further enhancing overall safety protocols.

Additional enhancements included modifications to spacecraft and Mission Control monitoring systems to provide more immediate and visible alerts for any anomalies detected. Emergency provisions were also expanded, with a reserve of 19 liters of water stored in the Command Module and an emergency battery, similar to those used in the Lunar Module's descent stage, installed in the Service Module.

Furthermore, the Lunar Module was adapted to facilitate easier power transfer to the Command Module, improving operational flexibility and redundancy. These comprehensive upgrades reflected NASA's commitment to learning from past challenges and ensuring the safety and success of future space missions.

On February 5, 1971, the Lunar Module Antares of Apollo 14 successfully touched down on the Moon, precisely at the intended Fra Mauro landing site, which Apollo 13 had aimed to explore. Commanded by Alan Shepard and accompanied by Edgar Mitchell, this mission marked NASA's return to lunar exploration after the harrowing events of Apollo 13.

During the descent to the lunar surface, astronaut Fred Haise served as Capsule Communicator (CAPCOM), maintaining crucial communication with the astronauts aboard Antares. Shepard and Mitchell conducted two Extra-Vehicular Activities (EVAs), notably exploring near the prominent Cone crater, conducting experiments, and collecting lunar samples.

For the astronauts of Apollo 13—Jim Lovell, Jack Swigert, and Fred Haise—the mission marked their last spaceflight opportunities. Lovell retired from both NASA and the Navy in 1973, venturing into the private sector. Swigert, initially scheduled for the Apollo-Soyuz Test Project, faced a different trajectory due to fallout from the Apollo 15 postal covers incident, eventually pursuing politics until his untimely death from cancer before taking office in 1982. Fred Haise was set to command Apollo 19, which was later canceled, and concluded his NASA career after contributing to the Space Shuttle Approach and Landing Tests in 1979.

Despite Apollo 13 not achieving its intended lunar landing due to the oxygen tank incident, several experiments were successfully conducted during the mission. The Saturn V's S-IVB stage, instead of being left in solar orbit as in previous missions, was deliberately crashed into the Moon. This impact, observed from a distance, provided valuable seismic data, shedding light on the Moon's internal structure and impact history. Additionally, experiments measuring atmospheric electrical phenomena and testing cloud height determination from Earth's orbit were carried out, yielding significant scientific insights despite the mission's challenges.

Apollo 13, despite its harrowing ordeal, left a lasting legacy in space exploration history. Dubbed a "successful failure" by mission commander Jim Lovell, it showcased NASA's resilience and capability under pressure. The mission, originally intended to land on the Moon, faced a critical setback when an oxygen tank explosion jeopardized the lives of its crew—Jim Lovell, Jack Swigert, and Fred Haise.

The response to the crisis became a testament to NASA's ingenuity and teamwork. Mission control teams worked tirelessly to devise improvised solutions to bring the crew safely back to Earth. William R. Compton highlighted the heroic efforts of these teams, emphasizing how their real-time improvisation played a crucial role in the crew's survival. Rick Houston and Milt Heflin underscored that Apollo 13 demonstrated the unwavering commitment and capability of mission control to manage emergencies in space.

Former NASA chief historian Roger D. Launius noted that the successful recovery of Apollo 13 solidified public confidence in NASA's abilities. It was a pivotal moment that showcased the agency's dedication to astronaut safety and mission success despite the inherent risks of space exploration.

Despite the accolades and acknowledgments of NASA's achievements, the incident also prompted serious reflection within the agency and among policymakers. Some officials, including Manned Spaceflight Center director Gilruth, expressed concerns over the risks associated with continued Apollo missions. Budgetary pressures and political considerations further influenced decisions, leading to the cancellation of two remaining lunar missions after Apollo 13. The Apollo program eventually concluded with Apollo 17 in December 1972.

Today, Apollo 13 is remembered for its challenges and the triumph of human ingenuity and teamwork that brought its crew safely home. It remains a symbol of NASA's capability to overcome adversity and continue pushing the boundaries of space exploration.

In the space race of the Cold War, in September 1970, the Soviet Union expanded its expertise in space engineering with the introduction of the Civilian Engineer Group. This group welcomed Anatoli Demyanenko, Valeri Makrushin, Dmitri Yuyukov, and others who focused on crucial technical advancements essential for Soviet space missions. Their work underscored the nation's dedication to engineering excellence in advancing its ambitious space exploration endeavors.

By April 27, 1970, the Soviet Air Force announced Group 5, which included Anatoli Berezovoi, Vladimir Dzhanibekov, and several other skilled aerospace professionals. This group underwent rigorous training aimed at enhancing the USSR's capabilities in crewed spaceflight operations, reflecting the country's strategic focus on maintaining leadership in space exploration.

The year 1971 witnessed the formation of the 1971 Scientific Group in the Soviet Union, which included Gurgen Ivanyan. This group was crucial in advancing scientific research within the Soviet space program, contributing to groundbreaking discoveries and technological innovations.

In May of the same year, China entered the international space arena by establishing the Shuguang Group 1970. Notable individuals like Wang Fuquan and Yu Guilin were part of this group, marking China's proactive steps toward developing its space capabilities. This initiative laid the foundation for China's future achievements in space exploration, highlighting its growing presence and contributions to global space endeavors.

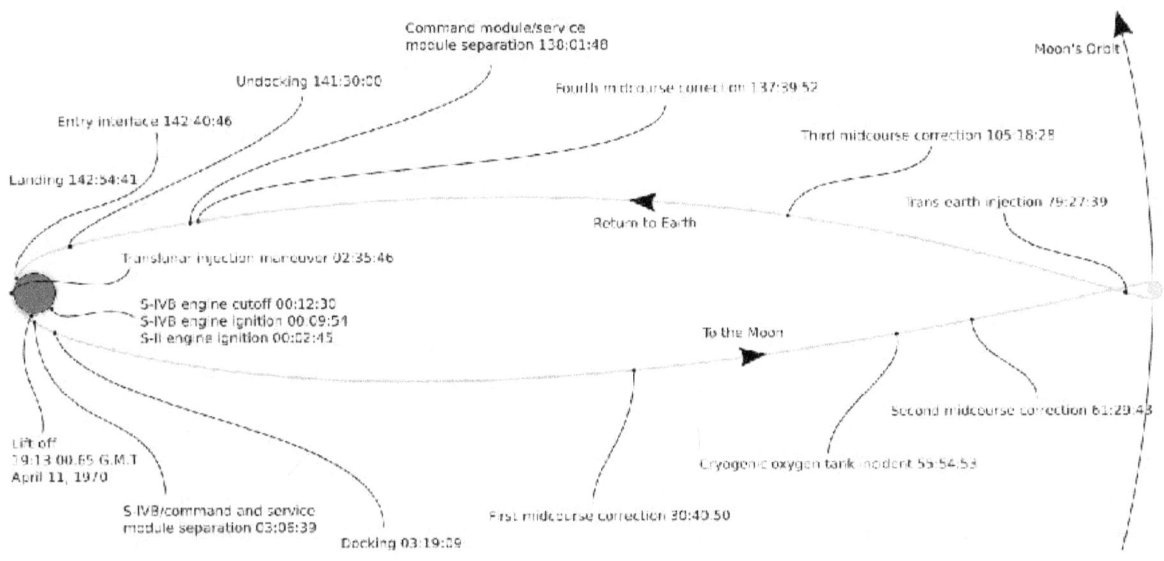

The circumlunar trajectory followed by Apollo 13, drawn to scale. The accident occurred about 56 hours into the mission.

Apollo 14

The Eighth Crewed Mission

Apollo 14, designated as a crewed lunar landing mission (H) by NASA, embarked on its journey on January 31, 1971. Its focused mission was to conduct precision piloted lunar landings and systematic lunar exploration, flying 1,150,321 miles. It marked NASA's third successful lunar landing, following previous missions in the Apollo program.

The mission was originally scheduled for 1970 but was postponed because of the investigation into Apollo 13's failure to reach the Moon's surface and the need for modifications to the spacecraft.

The primary objectives of Apollo 14 were ambitious and scientifically significant. These included conducting detailed selenological inspections, surveys, and sample collections from a carefully chosen region within the Fra Mauro formation on the lunar surface. Another key goal was to deploy and activate the Apollo Lunar Surface Experiments Package (ALSEP) to gather data on lunar seismicity, solar wind, and other environmental factors. Additionally, the mission sought to enhance human capabilities for lunar operations and to capture detailed photographs of potential future exploration sites.

Notably, Apollo 14 aimed to surpass the achievements of its predecessor, Apollo 13, by significantly increasing the amount of lunar material collected and scientific data returned to Earth. This was made possible in part by innovative equipment such as the Modular Equipment Transporter (MET), a collapsible two-wheeled cart designed to transport tools, cameras, a portable magnetometer, and lunar samples across the lunar terrain. The MET facilitated greater mobility and efficiency during lunar surface activities.

The mission also incorporated critical improvements in response to the oxygen tank failure experienced during Apollo 13. These enhancements included modifications to the internal construction of the Command and Service Module (CSM) oxygen tanks, the addition of a third oxygen tank, and the installation of an auxiliary battery.

The prime crew of the Apollo 14 lunar landing mission. From left to right they are: Command Module pilot, Stuart A. Roosa, Commander, Alan B. Shepard Jr. and Lunar Module pilot Edgar D. Mitchell. The Apollo 14 mission emblem is in the background.

Commanded by Captain Alan Bartlett Shepard, Jr., Apollo 14 was poised to make history with its crew: Major Stuart Allen Roosa as command module pilot and Commander Edgar Dean Mitchell as lunar module pilot. Shepard, a seasoned astronaut selected in 1959, had already made his mark as the first American in space during the Mercury program's inaugural suborbital flight. Despite setbacks due to Ménière's syndrome, an inner ear disorder, Shepard's determination and successful experimental surgery in 1968 paved his way back to flight status. At 47, he would become the oldest to set foot on the Moon.

Stuart Roosa, born in Durango, Colorado, and aged 37 at the time of Apollo 14, brought a unique background as a former smoke jumper turned Air Force pilot. He earned his wings and later distinguished himself at the Aerospace Research Pilot School before joining NASA's astronaut corps in 1966. Edgar Mitchell, born in Hereford, Texas, was 40 years old during the mission. His career as a Navy pilot and subsequent academic pursuits, including aerospace engineering, prepared him well for his role as lunar module pilot.

Initially slated for Apollo 13, Shepard and his crew were redirected to Apollo 14 by NASA's management, who sought additional training time for Shepard. This decision reshuffled crew assignments, leading to the now-famous Apollo 13 team, including Jim Lovell, Ken Mattingly, and Fred Haise, taking on the ill-fated mission instead.

Apollo 14's backup crew, originally comprised of Eugene A. Cernan as commander, Ronald E. Evans Jr. as command module pilot (CMP), and Joe H. Engle as lunar module pilot (LMP), played a crucial role as the contingency team for the mission. In a significant turn of events, this backup crew was reassigned to become the prime crew for Apollo 17, with Harrison Schmitt replacing Joe Engle as the lunar module pilot. This decision was driven by NASA's strategic aim to include a trained scientist on a lunar mission, with Schmitt, a geologist, fulfilling this role admirably.

Joe Engle, known for his pioneering flights in the X-15 rocket plane that reached the edge of outer space, would later contribute to NASA's Space Shuttle program. In 1981, he flew as commander on STS-2, the second flight of the Space Shuttle, showcasing his versatility and experience in manned spaceflight. Thus, while Engle did not land on the Moon with Apollo, his career continued to make significant contributions to human space exploration.

During Apollo 14, a team of capsule communicators (CAPCOMs), including Major Charles Gordon Fullerton, Lt. Commander Bruce McCandless II, Fred Wallace Haise, Jr., and Commander Ronald Ellwin Evans, handled communication with the astronauts in space. They were pivotal in relaying instructions and crucial information between Mission Control and the crew aboard the spacecraft.

Supporting the mission's operations on the ground were a dedicated support crew comprising McCandless, Lt. Colonel William Reid Pogue, Fullerton, and Phillip Kenyon Chapman, Sc.D. These astronauts played a vital role in crafting and updating mission protocols, flight plans, and checklists, ensuring the smooth execution of Apollo 14's objectives.

Overseeing the mission from Mission Control was a team of experienced flight directors, each responsible for critical decision-making during different shifts. M.P. "Pete" Frank and Glynn S. Lunney led the first shift, Milton L. Windler on the second shift, Gerald D. Griffin on the third shift, and Glynn S. Lunney again on the fourth. Their mandate was clear: ensure crew safety and mission success by making any necessary operational decisions in real time.

Apollo 14's flight directors upheld this responsibility with unwavering commitment, navigating the mission through various phases, including crucial Extra-Vehicular Activities (EVAs). Fred Wallace Haise, Jr., who had previously trained for Apollo 13, brought invaluable expertise to the mission, particularly during the EVAs, which aimed for the same lunar landing site as the previous mission. Although Haise did not walk on the Moon during Apollo 14, Edgar Dean Mitchell, a member of Group 5 astronauts, achieved this milestone, solidifying his place in history.

The prime and backup crews for both Apollo 13 and Apollo 14 were officially announced on August 6, 1969, marking a significant milestone in NASA's planning for these missions. Originally, Apollo 14 was slated for launch in July 1970. However, early in 1970, NASA faced budget constraints that led to the cancellation of Apollo 20 and a decision to reduce the number of Apollo missions per year to two. This adjustment meant Apollo 13 was scheduled for April 1970, with Apollo 14 tentatively planned for October or November of the same year.

The events took a turn following the investigation into the accident that aborted Apollo 13's lunar landing attempt. On May 7, 1970, NASA Administrator Thomas O. Paine announced that Apollo 14's launch would be delayed, aiming for by December 3, 1970, with a landing near the originally targeted site of Apollo 13. The Apollo 14 astronauts continued their rigorous training regimen throughout this period in preparation for their mission.

Further adjustments became necessary following the release of the accident report and subsequent NASA reviews of spacecraft modifications required for safety and mission readiness. On June 30, 1970, NASA announced that the launch of Apollo 14 would be postponed further, now targeting by January 31, 1971.

The crew of Apollo 14 underwent an extensive 19-month training period after being assigned to the mission, making it the longest training duration for any Apollo crew up to that point. This extended training period was not only a response to the complexities of lunar exploration but also involved overseeing critical modifications to the Command and Service Module (CSM) following the Apollo 13 incident. Commander Alan Shepard delegated much of the responsibility for managing these changes to Command Module Pilot Stuart Roosa.

Reflecting on the gravity of their mission, Lunar Module Pilot Edgar Mitchell later emphasized the intense pressure felt by the crew. They understood that the success of Apollo 14 was pivotal for the continuation of the Apollo program itself. Mitchell articulated their mindset, stating, "We realized that if our mission failed—if we had to turn back—that was probably the end of the Apollo program. There was no way NASA could stand two failures in a row. We figured a heavy mantle was on our shoulders to ensure we got it right."

Before the unfortunate abort of the Apollo 13 mission, NASA had planned for Apollo 14 to target a landing near Littrow Crater, located in Mare Serenitatis on the Moon. This area was known for its volcanic features, which scientists were eager to explore.

However, following the safe return of Apollo 13, a reassessment of scientific priorities led NASA to prioritize the Fra Mauro region for Apollo 14's landing site. The Fra Mauro formation, created from ejecta resulting from the impact that formed Mare Imbrium, offered a unique opportunity to retrieve samples from deeper layers beneath the Moon's surface. Of particular interest was Cone Crater, formed by a relatively recent and substantial impact, potentially providing materials that could be dated to gain insights into lunar history.

Choosing Fra Mauro over Littrow also facilitated orbital photography of another promising landing site, the Descartes Highlands, which eventually became the landing site for Apollo 16. While Littrow remained unvisited by the Apollo missions, a nearby area known as Taurus-Littrow would later become famous as the landing site for Apollo 17.

Shepard in front of the Lunar Landing Research Vehicle, flown to simulate the landing

For Apollo 14 specifically, the chosen landing site was slightly closer to Cone Crater than the original target designated for Apollo 13.

The decision to shift Apollo 14's landing site from Littrow to Fra Mauro had significant implications for the astronauts' geological training. Originally, the crew had been prepared with visits to volcanic sites on Earth, anticipating a landing near volcanic features on the Moon like those at Littrow. However, the change to the Fra Mauro region necessitated a shift in training focus to crater sites, reflecting the geological composition and objectives of the new landing site.

Shepard (left) and Mitchell during geological training

Instead of volcanic landscapes, the astronauts now trained at sites such as the Ries Crater in West Germany and an artificial crater field in Arizona's Verde Valley. This adjustment aimed to simulate the geological conditions and challenges they would encounter at Fra Mauro, emphasizing impact cratering processes and sample collection from impact ejecta.

Commander Alan Shepard's demeanor influenced the effectiveness of this training, which set the tone for the crew, including Lunar Module Pilot Edgar Mitchell. According to Harrison Schmitt, who observed Shepard closely, the commander seemed preoccupied with the challenges of returning to space after a decade-long absence and ensuring the mission's success following the near-disaster of Apollo 13. This mindset potentially affected the crew's engagement during training activities.

Meanwhile, Command Module Pilot Stuart Roosa undertook specialized training for his role alone in lunar orbit. He worked closely with geologist Farouk El-Baz, inspired by the effective training El-Baz had provided to Apollo 13's Command Module Pilot Ken Mattingly. Roosa and El-Baz studied lunar maps extensively to prepare for Roosa's observations and photography tasks during lunar orbit. While Shepard and Mitchell conducted their geological field trips on Earth,

Roosa practiced making observations from an airplane, simulating the conditions of lunar orbit to refine his observational skills.

Following the incident with Apollo 13, where last-minute crew changes were necessitated by exposure to a communicable disease, NASA implemented stringent measures to safeguard the health of the Apollo 14 crew. This initiative, known as the Flight Crew Health Stabilization Program, was designed to minimize the risk of any health-related disruptions close to launch.

Starting 21 days before the scheduled launch date, the Apollo 14 crew members began residing in dedicated quarters at Kennedy Space Center (KSC) in Florida. During this period, their interactions were strictly limited to essential personnel such as their spouses, the backup crew, mission technicians, and others directly involved in their training. All individuals in contact with the crew underwent thorough physical examinations and received necessary immunizations to prevent any potential illness transmission. Movement of the crew within KSC and its vicinity was carefully controlled to reduce exposure risks.

Apollo 14 marked a significant evolution in the Apollo spacecraft design, influenced heavily by the lessons learned from the Apollo 13 mission and the expanded lunar activities planned for this mission. Command Module (CM) 110 and Service Module (SM) 110, collectively known as CSM-110 and named Kitty Hawk by astronaut Stuart Roosa, were pivotal components. The call sign paid homage to the birthplace of aviation, where the Wright Brothers made their historic flight in 1903. Accompanying CSM-110 was Lunar Module 8 (LM-8), christened Antares by astronaut Edgar Mitchell, named after a star in the Scorpius constellation, crucial for navigation during lunar descent.

Extensive modifications were implemented in response to the Apollo 13 incident when an oxygen tank failure jeopardized the mission. The oxygen tanks underwent a redesign, integrating upgraded thermostats capable of handling proper voltage to prevent similar issues. A third oxygen tank was installed in the Service Module's Bay 1, equipped with an emergency isolation valve to safeguard the Command Module's environmental system. Upgrades from aluminum to stainless steel for the quantity probes in each tank further enhanced reliability.

Bay 4, where the Apollo 13 oxygen tank explosion occurred, saw significant safety improvements. Electrical wiring was encased in stainless steel to prevent potential damage, and fuel cell oxygen supply valves were redesigned to isolate Teflon-coated wiring from oxygen exposure. Both onboard and at Mission Control, monitoring systems were enhanced to provide more immediate and visible alerts for any anomalies, ensuring proactive response capabilities.

Critical provisions were also added to enhance crew safety and mission resilience. In response to the power and water shortages experienced during Apollo 13, Apollo 14 was equipped with emergency supplies: a 5 US gallon water reserve in the Command Module and an emergency battery, identical to those in the Lunar Module's descent stage, in the Service Module. Modifications were made to facilitate power transfer between the Lunar Module and Command Module.

Technical improvements extended to the Lunar Module as well, with the installation of anti-slosh baffles in the descent stage's propellant tanks. These baffles prevented premature triggering of the low fuel indicator, a recurring issue during previous missions. Structural adaptations were made to accommodate new lunar surface equipment, including the Modular Equipment Transporter, ensuring efficient mobility and operational flexibility during surface activities.

Regarding spacecraft preparations, the Command and Service Modules arrived at

KSC on November 19, 1969. Shortly after, on November 21, the ascent stage of the Lunar Module (LM) was delivered, followed by the descent stage three days later. Subsequently, extensive checkout, testing, and equipment installations were conducted to ensure the readiness of all components for the upcoming mission.

The Saturn V rocket chosen for Apollo 14, designated SA-509, continued the legacy of its predecessors used in Apollo missions from 8 to 13. At a staggering weight of 6,505,548 pounds (2,950,867 kg), SA-509 surpassed its predecessor for Apollo 13 by 3,814 pounds (1,730 kg), making it the heaviest launch vehicle ever flown by NASA up to that point.

Following the challenges experienced during Apollo 13, several modifications were implemented to enhance reliability and safety. These included measures to prevent pogo oscillations, which had prematurely shut down the center J-2 engine on the S-II second stage of Apollo 13. Key enhancements for Apollo 14 included the installation of a helium gas accumulator in the liquid oxygen (LOX) line of the center engine, a backup cutoff device for added redundancy, and the simplification of the propellant utilization valve to a two-position setup across all five J-2 engines.

On November 9, 1970, the fully assembled launch vehicle, consisting of the spacecraft mounted atop the Saturn V rocket, was ceremoniously rolled out from NASA's iconic Vehicle Assembly Building to Launch Pad 39A, marking a significant milestone in the preparations for the upcoming mission.

The Apollo 14 launch vehicle is rolled out from the Vehicle Assembly Building, November 9, 1970

The mission's Command and Service Module (CSM), designated CSM-110 and named "Kitty Hawk," along with the Lunar Module (LM), designated LM-8 and named "Antares," were integral components of this historic launch, underscoring NASA's meticulous planning and commitment to advancing lunar exploration capabilities.

The final countdown for the Apollo 14 mission commenced at T-28 hours on January 30, 1971, at 06:00:00 GMT. Scheduled holds were observed, including one at T-9 hours for 9 hours and 23 minutes and another at T-3 hours 30 minutes for one hour. As launch time approached, a cold front extended across northern Florida, with scattered rain showers positioned to the south. These showers gradually moved toward the launch area just before the scheduled liftoff.

About 30 minutes before launch, a band of cumulus congestus clouds with showers formed along a line stretching from Orlando toward the northern Merritt Island Launch

Area (MILA). This weather condition and lightning concerns necessitated a hold at T-8 minutes and 40 seconds until the showers had sufficiently cleared the launch complex.

Despite earlier rain, the launch pad was dry at liftoff, although the Saturn V rocket ascended through layers of clouds. Surface winds around Cape Canaveral were light and westerly, with cumulus clouds covering 70 percent of the sky at a base of 4,000 feet and altocumulus clouds covering 20 percent at 8,000 feet. The temperature was 71.1°F, relative humidity was 86 percent, and barometric pressure measured 14.652 lb/in².

Wind speeds were recorded at two levels: at 60 feet above ground level, the anemometer measured 9.7 knots blowing from 255° true north; at 530 feet, winds were 16.5 knots blowing from 275° true north. Due to weather conditions, the flight azimuth had to be adjusted from 72.067° to 75.5579° east of north.

Apollo 14 embarked on its historic journey from Launch Complex 39-A at Kennedy Space Center at 4:03:02 pm EST (21:03:02 UTC) on January 31, 1971. This launch followed a delay of 40 minutes and 2 seconds due to weather conditions, marking the first such delay in the Apollo program. The original launch time of 3:23 pm was at the onset of a narrow launch window lasting just under four hours; missing this window would have necessitated postponing the mission until March.

Launch of Apollo 14

The decision to delay was influenced by lessons learned from Apollo 12, which had launched during adverse weather and encountered lightning strikes. As a precaution, stricter launch criteria were implemented. Among the distinguished guests witnessing the launch were U.S. Vice President Spiro T. Agnew and the future King Juan Carlos I of Spain.

To compensate for the delayed launch, Apollo 14 adopted a faster trajectory to the Moon, allowing the mission to catch up on lost time during flight. Approximately two days after liftoff, mission timers were adjusted ahead by 40 minutes and 3 seconds to ensure that subsequent events aligned with the scheduled flight plan.

During the ascent phase, Apollo 14 transitioned smoothly through critical stages. The Saturn V rocket executed a roll maneuver from a launch pad azimuth of 90° to a flight azimuth of 75.558° between 000:00:12.814 and 000:00:28.000. The first stage (S-IC) engine shutdown occurred at 000:02:44.094, followed by separation from the second stage (S-II) and ignition of its engine. The S-II engine shutdown at 000:09:19.05, followed by separation from the third stage (S-IVB), which ignited at 000:09:23.4. The first S-IVB engine cutoff took place at 000:11:40.56, with minor deviations from the planned trajectory.

During ascent, the maximum wind conditions encountered were 102.6 knots at 255° from true north at 43,270 feet, with a maximum wind shear of 0.0201 sec^{-1} at 43,720 feet.

After achieving orbit, the S-IVB third stage of the Saturn V shut down, and the Apollo 14 astronauts proceeded with spacecraft checks in preparation for the crucial Translunar Injection (TLI) burn. This burn was essential to propel the spacecraft toward the Moon along the planned trajectory.

Following the TLI burn, the Command and Service Module (CSM) separated from the S-IVB stage. Stuart Roosa, the Command Module Pilot, executed the transposition maneuver to turn the CSM around for docking with the Lunar Module (LM), named Antares. Roosa, who had practiced this maneuver extensively, aimed to achieve the record for the minimal propellant used during docking.

However, despite several careful attempts over the next two hours, the docking mechanism failed to activate upon attempting to dock. This unexpected issue prompted intense discussions among mission controllers, who eventually proposed retracting the docking probe and retrying the maneuver, hoping that physical contact would engage the docking latches. This adjustment proved successful, allowing the joined spacecraft to separate from the S-IVB stage within an hour.

The now-unmanned S-IVB stage was redirected on a trajectory to impact the Moon, a planned event that occurred just over three days later. Seismometers left by previous missions monitored this impact, with Apollo 12's seismometer detecting vibrations for more than three hours.

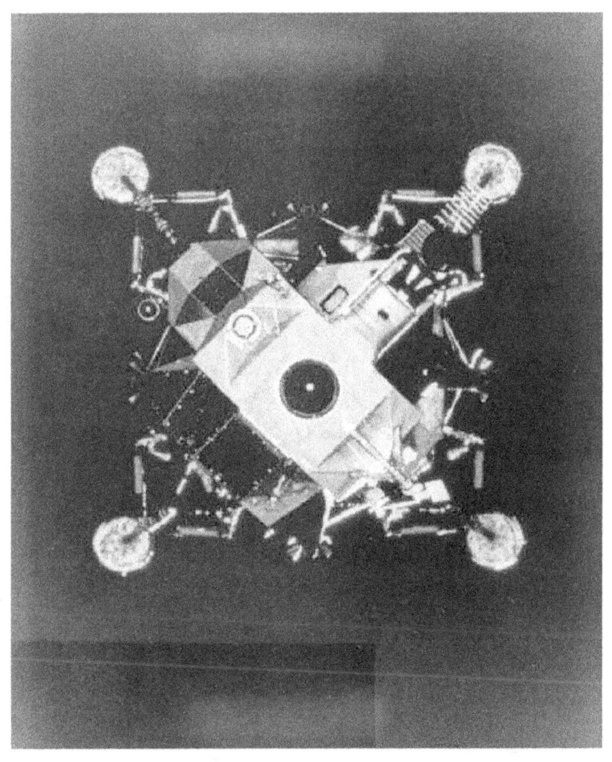

Antares as seen from Kitty Hawk

Upon achieving orbit, the conditions were recorded at insertion with an apogee and perigee of 100.1 by 98.9 nautical miles (n mi), an inclination of 31.120°, and a velocity of 25,565.9 feet per second (ft/sec). The CSM was designated 1971-008A internationally upon achieving orbit, while the S-IVB stage was designated 1971-008B.

After completing in-flight systems checks, the crucial 350.84-second Translunar Injection maneuver was executed at 002:28:32.40. The S-IVB engine shutdown occurred at 002:34:23.24, followed by Translunar Injection ten seconds later, achieving a velocity of 35,541.0 ft/sec. This maneuver occurred after completing 1.5 Earth orbits lasting 2 hours, 22 minutes, and 42.68 seconds, marking a critical phase in Apollo 14's journey toward the Moon.

During the Translunar Phase of Apollo 14's mission, several critical maneuvers and activities unfolded as the spacecraft journeyed toward the Moon. At 003:02:29.4, the Command and Service Module (CSM) successfully separated from the S-IVB stage. This separation was followed by the routine transposition maneuver, ensuring the spacecraft were properly oriented for further operations.

However, achieving docking between the CSM and the Lunar Module (LM) named Antares proved to be more challenging than anticipated. It took six attempts before a successful docking was achieved at 004:56:56.0. The docked spacecraft were then ejected from the S-IVB stage through a 6.9-second maneuver at 005:47:14.4, followed by an 80.2-second separation maneuver performed at 006:04:01.7. Subsequent examination of the docking probe revealed no issues, leading to the conclusion that the capture-latch assembly might not have been properly engaged during the initial attempts.

Throughout the translunar journey, the crew engaged in various activities essential to mission success. They conducted star and Earth horizon calibration sightings, crucial for navigation exercises during the transearth coast phase. Dim-light photography of Earth provided additional data. At 030:36:07.91, a 10.19-second midcourse correction maneuver was performed to refine the spacecraft's trajectory.

At 060:30 Ground Elapsed Time, Commander Shepard and Lunar Module Pilot Mitchell transferred to the LM for two hours of housekeeping and systems checks. During this time, they photographed a wastewater dump from the CSM, contributing to a particle contamination study intended for the Skylab program.

Further course adjustments were made with a second midcourse correction at 076:58:11.98, lasting 0.65 seconds, to achieve the desired trajectory for lunar orbit insertion. With these preparations complete, the crew settled in for the continued journey toward the Fra Mauro landing site on the lunar surface.

pollo 14 landing site, 2009 photograph by Lunar Reconnaissance Orbiter

During the Apollo 14 mission, meticulous timing and precise maneuvers were crucial as the spacecraft journeyed toward the Moon. At 054:33:36 into the mission timeline, a critical adjustment was made to the onboard clock, compensating for a delay in the launch countdown due to weather conditions. This update, adding precisely 40 minutes and 2.9 seconds, would prove invaluable for the command module pilot during lunar orbit, eliminating the need for frequent manual corrections in his flight log.

As Apollo 14 approached the lunar sphere, another pivotal event unfolded at 081:56:40.70 mission time. Positioned at an altitude of 87.4 nautical miles above the lunar surface, the spacecraft's service propulsion engine roared to life, burning for 370.84 seconds. This burn successfully maneuvered the spacecraft into a lunar orbit with an apocynthion (farthest point from the Moon) of 169.0 nautical miles and a pericynthion (nearest point to the Moon) of 58.1 nautical miles. The journey from Earth to this point, known as the translunar coast, spanned an impressive 79 hours, 28 minutes, and 18.30 seconds.

Like Apollo 13's mission, the S-IVB stage was targeted to impact the lunar surface to provide seismic data deliberately. This objective was accomplished with a precise 252.2-second auxiliary propulsion system lunar impact maneuver at 009:04:11.2. The S-IVB impacted the Moon at 082:37:52.17, striking at latitude 8.07° south and longitude 26.04° west, approximately 159 nautical miles from the target point and 94 nautical miles southwest of Apollo 12's seismometer. The seismometer recorded the impact 37 seconds later and detected vibrations for over three hours.

Shortly after that, at 086:10:52.97 mission time, another critical maneuver shaped the mission's trajectory. A brief, 20.81-second burn of the service propulsion system fine-tuned the spacecraft's orbit to 58.8 nautical miles by 9.1 nautical miles in preparation for the upcoming separation of the Lunar Module (LM), named Antares.

Apollo 14 marked a milestone as the first mission where the Command and Service Module (CSM) propelled the LM into a lower lunar orbit, a maneuver intended to maximize hover time above the Moon's rugged terrain. This safety measure, planned meticulously in advance, underscored NASA's commitment to ensuring the astronauts' safety as they prepared for their lunar descent.

In a strategic departure from previous Apollo missions, Apollo 14 implemented a significant change in its lunar descent strategy. Rather than relying solely on the Lunar Module's descent propulsion system for the critical descent orbit insertion maneuver, engineers opted to conserve LM propellant. This adjustment aimed to provide a larger margin for potential landings in rugged lunar terrain, a decision that underscored NASA's commitment to mission flexibility and safety.

At 101:20 mission time, Commander Alan Shepard and Lunar Module Pilot Edgar Mitchell entered the LM, initiating thorough system checks and preparations for the imminent undocking procedure. This meticulous process ensured all systems were primed and ready for the challenges ahead.

Undocking occurred at 103:47:41.6 following a precise 2.7-second firing of the service module's reaction control system. This maneuver effectively separated the Command Module (CM) from the LM, placing it into an initial orbit measuring 60.2 nautical miles by 7.8 nautical miles above the lunar surface.

Subsequently, at 105:11:46.11 mission time, a critical 4.02-second burn of the service module's engines was executed to circularize the CM's orbit. This adjustment transformed the orbit into a stable path measuring 63.9 nautical miles by 56.0 nautical miles, providing an optimal operational position for continued lunar operations.

After the Lunar Module (LM) Antares separated from the Command Module in lunar orbit, two critical issues threatened the mission's success. The first problem arose from the LM computer receiving an erroneous ABORT signal due to a faulty switch. NASA suspected that a small ball of solder had dislodged and was intermittently causing the switch to close the circuit, triggering the false signal. Tapping on the panel near the switch temporarily resolved the issue, but it soon recurred. If the ABORT signal persisted after the descent engine fired, the computer could have initiated an auto-abort, separating the ascent stage from the descent stage and potentially aborting the landing attempt.

In response, NASA and software teams from the Massachusetts Institute of Technology (MIT) urgently devised a workaround. Since the LM's software was hard-wired and couldn't be updated remotely, they simulated an abort event to reset the system. This solution tricked the computer into believing an abort had already occurred, preventing it from reacting to further false signals to abort. While this safeguard allowed the astronauts to continue piloting the spacecraft, they would need to initiate an abort if necessary during critical phases manually.

As preparations continued, ground personnel detected an unexpected abort command from a computer input channel just before the ignition despite the crew not activating the abort switch. Further investigation pinpointed the issue to the faulty abort switch. A procedural workaround was swiftly implemented and followed to preempt any unintended aborts during the powered descent.

The powered descent proceeded as planned, with the LM's engines firing for 764.61 seconds, initiating at 108:02:26.52 mission time at 7.8 nautical miles above the lunar surface. However, approximately six minutes into the descent, a complication arose with the landing radar system switching unexpectedly to the low-range scale. This issue would have prevented the radar from acquiring essential data for a safe landing. Quickly, the crew manually cycled the circuit breaker, resolving the problem and restoring the radar to full functionality.

During the powered descent of Apollo 14's Lunar Module (LM) Antares, a critical challenge emerged when the landing radar failed to lock onto the Moon's surface automatically. This failure deprived the navigation computer of essential altitude and descent speed data for a safe landing. Acting swiftly, the astronauts cycled the breaker of the landing radar system, successfully restoring functionality just as the LM approached approximately 22,000 feet (6,700 meters) above the lunar terrain.

Mission protocols mandated an abort if the radar was non-functional below 10,000 feet (3,000 meters), a crucial safety measure to ensure the crew's safety in case of inadequate landing data. However, Commander Alan Shepard considered attempting a manual landing without the radar's assistance, showcasing the astronauts' skill and decision-making under pressure.

With the landing radar operational again, Shepard skillfully guided the LM to a pinpoint landing—the closest to the intended target among the six missions that successfully touched down on the Moon. The historic touchdown occurred precisely at 09:18:11 GMT (04:18:11 a.m. EST) on February 5, at mission time 108:15:09.30. The descent engine shut down a mere 1.83 seconds after first contact with the lunar surface.

Antares landed within the designated landing site in the Fra Mauro highlands, positioned at latitude 3.64530° south and longitude 17.47136° west. The site was originally targeted for Apollo 13's mission before its course was altered due to its abort. The LM settled on a gentle slope of approximately 7°, providing a stable base for the upcoming lunar exploration activities.

Approximately 70 seconds of descent engine firing time remained upon landing, underscoring the precision of the descent phase. Preparation for the initial lunar surface exploration commenced promptly, with cabin depressurization starting at 113:39:11 mission time.

Panorama of the Apollo 14 landing site taken in 1971

However, the start of the first extravehicular activity (EVA) encountered a delay of 49 minutes due to intermittent communication issues with the Portable Life Support System (PLSS). These disruptions were traced to a configuration problem within the LM's systems. The issue was swiftly resolved by cycling the audio circuit breaker, restoring clear and reliable communications necessary for the astronauts' safety and mission objectives.

Following their successful landing, Commander Alan Shepard stepped onto the lunar surface at 113:47 mission time, marking a historic moment as he became the first astronaut to set foot on the Moon during the

Apollo 14 mission. Eight minutes later, Lunar Module Pilot Edgar Mitchell followed suit, immediately proceeding with the crucial task of collecting the contingency sample—a precautionary measure in case an unexpected need arose to return to Earth hastily.

Shepard's initial steps onto the lunar surface were captured in an iconic image, where he shaded his eyes from the stark lunar sunlight, reflecting on the monumental journey that had brought them there.

During the first extravehicular activity (EVA), which commenced at 9:42 am EST (14:42 UTC) on February 5, 1971—delayed by earlier communications issues—the astronauts embarked on a series of tasks essential to the mission's scientific objectives and operational goals. They deployed critical equipment such as the television camera, S-band antenna for communications, and the solar wind experiment. Additionally, they set up and loaded the Modularized Equipment Transporter (MET) for mobility and sample collection.

The crew meticulously documented their activities, capturing panoramic views and detailed photographs of the lunar terrain and equipment setup. At 115:46 mission time, Shepard and Mitchell embarked on a journey to deploy the Apollo Lunar Surface Experiments Package (ALSEP) approximately 500 feet west of the LM. This included positioning the laser-ranging retro-reflector to aid in precise measurements from Earth.

Shepard, reflecting on the moment's significance, remarked, "And it's been a long way, but we're here," encapsulating the spirit of exploration and achievement driving the Apollo missions.

One of the key scientific activities conducted during this EVA was Mitchell's deployment of the Active Seismic Experiment's (ASE) geophone lines. These extended 310 feet (94 meters) from the ALSEP's Central Station and were crucial for gathering data on lunar subsurface structure and composition through seismic vibrations generated by thumper explosives. Despite encountering challenges with some thumpers failing to fire, the experiment provided valuable insights into the lunar regolith.

Throughout their exploration, Shepard and Mitchell collected lunar samples, meticulously documenting each find and photographing the lunar landscape for further analysis back on Earth. The first EVA concluded after 4 hours, 47 minutes, and 50 seconds, showcasing the astronauts' endurance and adaptability in the demanding lunar environment.

The Apollo Lunar Surface Experiments Package (ALSEP) deployed during Apollo 14 included a comprehensive array of scientific instruments designed to explore various aspects of the lunar environment. These instruments were critical for advancing our understanding of the Moon's geology, atmosphere, and space environment.

Passive Seismic Experiment (PSE): Originally flown on Apollo 12 and 13, the PSE was designed to detect and measure lunar seismic activity, providing insights into the Moon's interior structure and geological processes.

Active Seismic Experiment (ASE): Also flown on Apollo 13, the ASE involved the deployment of geophones and thumper devices to generate seismic waves. These waves helped scientists study the Moon's subsurface composition and structure in greater detail.

Suprathermal Ion Detector Experiment (SIDE): Flown previously on Apollo 12, SIDE aimed to measure and analyze suprathermal ions—charged particles moving at high speeds in the lunar exosphere—providing data on the Moon's interaction with the solar wind.

Cold Cathode Ion Gauge Experiment (CCIG): Deployed on Apollo 12 and 13, the CCIG measured the density of gases in the lunar atmosphere, helping scientists understand lunar exosphere dynamics.

Charged Particle Lunar Environment Experiment (CPLEE): This experiment focused on studying the flux and characteristics of charged particles near the lunar surface, contributing to our understanding of space weather and its effects on lunar conditions.

Additionally, two other experiments were part of Apollo 14's scientific payload but were not integrated into the ALSEP:

Laser Ranging Retroreflector (LRRR or LR3): Flown on Apollo 11, the LRRR was deployed near the ALSEP to enable precise measurements of lunar distance from Earth using laser technology, aiding in tests of gravitational theories and lunar motion studies.

Lunar Portable Magnetometer (LPM): Introduced for Apollo 14, the LPM resembled equipment used on Apollo 12 and was designed to be carried by the astronauts during their second extravehicular activity (EVA). It aimed to measure magnetic fields on the lunar surface, providing insights into the Moon's magnetic history and geological processes.

Following the loss of the ALSEP components on Apollo 13 due to the LM's re-entry into Earth's atmosphere, deploying the ALSEP and these scientific instruments became integral objectives of the Apollo 14 mission.

The Passive Seismic Experiment (PSE) carried aboard Apollo 14 was a seismometer designed to detect and measure lunar seismic activity, similar to its counterpart left on the Moon during the Apollo 12 mission. This instrument was crucial in studying the Moon's interior structure and seismic events.

A unique aspect of the Apollo 14 PSE was its calibration method. After deployment on the lunar surface, the instrument would be calibrated using the impact of the Lunar Module's (LM) ascent stage, which would be intentionally jettisoned to impact the Moon. This impact would involve an object of known mass and velocity striking a known location on the lunar surface, providing a precise calibration reference for the seismometer.

Astronaut Edgar D. Mitchell, Apollo 14 Lunar Module pilot, moves across the lunar surface as he looks over a traverse map during extravehicular activity (EVA). Lunar dust can be seen clinging to the boots and legs of the space suit. I

The Apollo 12 seismometer was also designed to detect impacts, particularly from the spent S-IVB booster stage of Apollo 14. This booster, directed by mission operations, would impact the Moon after Apollo 14 entered lunar orbit. Both seismometers—Apollo 12's and Apollo 14's—working in tandem, along with those deployed on subsequent Apollo missions, formed a network of instruments distributed across different lunar locations. This network enabled scientists to gather comprehensive data on lunar seismic activity, offering insights into the Moon's internal structure and geological processes.

The Active Seismic Experiment (ASE) carried by Apollo 14 had a dual-part design to study seismic waves and their transmission through the Moon's regolith. This experiment was a crucial component of the mission's scientific objectives.

The first part of the ASE involved one of the crew members deploying three geophones.

These geophones were positioned at distances up to 310 feet (94 meters) from the ALSEP's Central Station. As the astronaut returned from placing the furthest geophone, they periodically fired thumper devices placed every 15 feet (4.6 meters). Thumpers were small explosive devices designed to generate controlled seismic waves upon detonation, providing data on the subsurface structure of the Moon.

The second part of the ASE included four mortars, each equipped with different properties and set to impact at varying distances from the experiment site. These mortars were intended to generate additional seismic waves upon impact, further enhancing the experiment's data collection capabilities.

However, due to concerns about potential damage to other delicate experiments deployed as part of the ALSEP, the mortar shells were never fired during the mission. It was decided that firing them would pose too great a risk to the integrity of the nearby scientific instruments.

A similar experiment was successfully conducted during the Apollo 16 mission. The mortars were deployed and fired, providing valuable seismic data that contributed to our understanding of lunar subsurface geology and seismic activity.

The ASE on Apollo 14, despite not firing its mortars, played a significant role in advancing lunar science by studying the propagation of seismic waves through the Moon's regolith, laying the groundwork for future lunar exploration missions and scientific endeavors.

The Lunar Portable Magnetometer (LPM) was designated to accompany the astronauts during Apollo 14's second extravehicular activity (EVA). It would measure the Moon's magnetic field at various locations on the lunar surface, gathering data crucial for understanding the Moon's magnetic properties and geological history.

The Suprathermal Ion Detector Experiment (SIDE) on Apollo 14 measured ions on the lunar surface, particularly those originating from the solar wind. This experiment provided insights into the interaction between the Moon and the solar wind, contributing to our understanding of space weather effects on planetary bodies.

Accompanying the SIDE, the Cold Cathode Ion Gauge Experiment (CCIG) monitored the lunar atmosphere to detect any variations over time. By measuring the density of gases around the Moon, the CCIG helped scientists study the Moon's exosphere dynamics and its interaction with the space environment.

The Charged Particle Lunar Environment Experiment (CPLEE) measured the energies of protons and electrons emitted by the Sun, which reached the lunar surface. This experiment was crucial for studying the solar particle environment near the Moon and its potential effects on lunar surface materials.

The Laser Ranging Retroreflector (LRRR or LR3) deployed during Apollo 14 acts as a passive target for laser beams from Earth, enabling precise measurements of the Earth-Moon distance over time. Alongside the LRRRs left by Apollo 11 and 15, those from Apollo 14 are the only experiments on the Moon still actively providing data. These measurements are essential for testing theories of gravitational dynamics and refining our understanding of the Earth-Moon system's orbital mechanics.

Apollo 14 introduced several new innovations and adaptations to enhance astronaut safety and mission effectiveness during their lunar exploration activities.

One notable addition was the Buddy Secondary Life Support System (BSLSS), a set of flexible hoses designed to allow astronauts Alan Shepard and Edgar Mitchell to share cooling water in the event of a Primary Life Support System (PLSS) failure. If one astronaut's PLSS backpack malfunctioned, they could rely on the BSLSS to provide cooling without consuming additional oxygen from their backup Oxygen Purge System (OPS) cylinder. This innovative system extended the operational capability of the OPS by ensuring that oxygen was preserved solely for breathing, a critical safety feature for lunar surface operations.

The OPS units used on Apollo 14 were also modified from previous missions by removing unnecessary internal heaters. This adjustment helped streamline the OPS design while maintaining its reliability in the lunar environment.

Additionally, Apollo 14 astronauts carried water bags known as "Gunga Dins" to the lunar surface. These bags were inserted into their helmets, allowing them to sip water during extravehicular activities (EVAs). Although similar water bags had been flown on Apollo 13, Shepard and Mitchell were the first to use them while actually walking on the Moon, enhancing their hydration and comfort during their lunar explorations.

Regarding spacesuit design, Alan Shepard became the first astronaut on the lunar surface to wear a spacesuit adorned with commander's stripes. These distinctive red stripes were applied to his spacesuit's arms, legs, and helmet. This addition was implemented following difficulties in distinguishing astronauts from each other in photographs during previous missions. Commander's stripes helped visually identify Shepard as the mission's leader, facilitating easier recognition and coordination during their activities on the lunar surface.

After the successful deployment of the Apollo Lunar Surface Experiments Package (ALSEP), the first data began transmitting back to Earth at 116:47:58, marking a significant milestone in the mission's scientific objectives. However, the deployment process encountered several challenges that required astronaut intervention and troubleshooting.

One of the initial issues was the difficulty in releasing the Boyd bolt on the Suprathermal Ion Detector, which temporarily hindered its proper deployment. Additionally, stiffness in the cable connecting the Suprathermal Ion Detector to the Cold Cathode Ion Gauge caused the latter to topple over initially. Adjustments were made to stabilize these instruments to ensure their functionality.

Further complications included low transmitter strength at the Central Station, leading to potential data transmission issues, and noisy data received from the Suprathermal Ion Detector experiment, which required careful calibration and monitoring. Additionally, five of the thumper initiators for the Active Seismic Experiment failed to fire as planned, impacting the seismic data collection objectives.

Despite these challenges, the Laser Ranging Retro-Reflector was successfully set up during the first Extravehicular Activity (EVA-1), providing a passive target for precise laser measurements of the Earth-Moon distance.

Throughout the EVA-1 period, communication with Earth remained generally stable, although the resolution of the television picture gradually degraded toward the end of the extravehicular session.

After completing their tasks, the crew re-entered the Lunar Module (LM), and the cabin was repressurized at 118:27:01, concluding a productive and eventful first EVA period. The astronauts covered a distance of approximately 3,300 feet (1 km) and collected an estimated 45.2 pounds (20.5 kg) of lunar samples during their exploration.

During Apollo 14's second extravehicular activity (EVA-2), the crew explored the area around Cone Crater, located approximately 0.7 nautical miles (1.3 km) east-northeast of their lunar landing site in the Fra Mauro highlands. This journey was significant as it aimed to investigate geological features and gather samples from a unique lunar formation.

The preparations for EVA-2 began with cabin depressurization at 131:08:13, slightly earlier than planned, followed by Commander Shepard exiting the Lunar Module (LM) at 131:13 GMT. Lunar Module Pilot Mitchell followed seven minutes later. Their primary objective was to reach Cone Crater using the Modular Equipment Transporter (MET), a two-wheeled handcart specifically designed for Apollo 14. The MET allowed them to transport tools, equipment, and lunar samples more efficiently without carrying them manually, a significant advantage over previous missions.

However, the crew encountered challenges navigating the lunar terrain, particularly the steep slopes leading toward Cone Crater. These difficulties caused them to fall approximately 30 minutes behind schedule. Despite this setback, they reached a point within 50 feet (15 m) from the rim of Cone Crater, achieving the key objectives of this excursion.

During EVA-2, the crew deployed and used the MET effectively at Station C-Prime, located in a field of small boulders near Cone Crater. This station was pivotal for collecting geological samples and conducting scientific experiments near the crater.

The MET used on Apollo 14 was approximately 86 inches (220 cm) long, 39 inches (99 cm) wide, and 32 inches (81 cm) high when deployed on the lunar surface. It featured pressurized rubber tires, a pioneering design developed by Goodyear known as the Experimental Lunar Tire (XLT). These tires were crucial for providing traction and stability on the Moon's surface and were inflated with nitrogen to about 1.5 pounds per square inch (10 kPa).

Fully loaded with equipment and samples, the MET weighed around 165 pounds (75 kg) and included two legs combined with wheels to ensure stationary stability. This innovation paved the way for future lunar exploration vehicles, such as the Lunar Roving Vehicle (LRV), used in subsequent Apollo missions to enhance mobility and productivity during lunar surface operations.

During Apollo 14's second extravehicular activity (EVA), astronauts Shepard and Mitchell encountered unexpected challenges on the lunar surface. They were taken aback by the rugged terrain, which differed significantly from the flat landscape they had anticipated based on orbital maps. Navigational landmarks, originally identified from overhead photographs, appeared distorted and difficult to recognize amidst the undulating ground.

As they embarked toward Cone crater with the Modular Equipment Transporter (MET) in tow, Shepard and Mitchell's journey proved more strenuous than anticipated. Through CAPCOM Fred Haise, Mission Control monitored their progress anxiously from afar. However, the live television feed, fixed near the Lunar Module (LM), offered no visual confirmation of their struggle. Meanwhile, the astronauts' exertion became evident through their heavy breathing and elevated heart rates, raising concerns at Mission Control as time ticked away on their EVA duration.

Shepard and the Modular Equipment Transporter

Their trek was marked by deceptive ridges that promised the crater's rim but revealed more challenging terrain beyond. Despite Mitchell's strong intuition that they were near their target, physical exhaustion set in, prompting Haise to advise them to collect samples and commence their return to the LM.

Subsequent analysis of photographs taken during the EVA revealed that Shepard and Mitchell had ventured within approximately 65 feet (20 meters) of Cone Crater's rim. Detailed images from the Lunar Reconnaissance Orbiter (LRO) later confirmed their path, showing the distinct tracks of the astronauts and the MET coming tantalizingly close, within 30 meters of the crater's edge.

This experience underscored the critical need for improved lunar surface transportation and navigation systems. Lessons learned from Apollo 14 directly contributed to the development of the Lunar Roving Vehicle, slated for use during subsequent missions starting with Apollo 15. This innovative vehicle would provide astronauts with enhanced mobility and navigation capabilities, crucial for exploring the rugged and unpredictable lunar terrain.

Once back in view of the LM's television camera, Commander Shepard seized a moment that would forever mark Apollo 14 in the annals of space exploration. Prepared for this historic occasion, Shepard unveiled a Wilson six-iron golf club head, ingeniously attached to the handle of a contingency sample tool, along with two golf balls he had brought along.

Despite the constraints of his bulky EVA suit, Shepard managed several one-handed swings, showcasing his enthusiasm in the low lunar gravity. He jubilantly reported that the second ball soared "miles and miles and miles," a testament to the unique physics of lunar golfing. Not to be outdone, astronaut Mitchell improvised by throwing a lunar scoop handle javelin-style. His "javelin" and one of Shepard's golf balls found themselves nestled together in a nearby crater, with Mitchell's toss slightly farther afield.

In subsequent interviews, Shepard recalled that one of his shots landed near the Apollo Lunar Surface Experiment Package (ALSEP), while the other, digitally estimated decades later, traveled approximately 24 and 40 yards, respectively. This impromptu display of skill and humor amidst the solemnity of their mission captured the public imagination, cementing Apollo 14's legacy beyond its scientific achievements.

The second EVA, which included these memorable events, lasted 4 hours, 34 minutes, and 41 seconds. Shepard later donated the modified golf club to the USGA Museum in New Jersey and arranged for a replica to reside in the National Air and Space Museum. In 2021, on the 50th anniversary of Apollo 14, imaging specialist Andy Saunders utilized advanced techniques to produce enhanced images, providing new insights into the final resting places of Shepard's historic golf shots on the lunar surface.

Upon returning to the Lunar Module (LM), the Apollo 14 crew conducted additional scientific activities to maximize their mission's scientific yield. Magnetometer measurements were taken at two designated sites along their traverse route, providing valuable data on lunar magnetic fields. At one site, they dug an approximately 1.5-foot trench to extract soil samples, including containerized samples, after an unsuccessful attempt to retrieve a triple core tube sample.

Before preparing to re-enter the LM, the astronauts made critical adjustments to improve the signal strength of the ALSEP Central Station's antenna. This alignment tweak boosted signal reception by approximately 1/2 dB, enhancing data transmission to the Manned Space Flight Network ground stations, although the signal remained detectable by the 30-foot antenna.

The second extravehicular activity (EVA) period lasted 4 hours, 34 minutes, and 41 seconds, during which the crew covered 9,800 feet (about 3 kilometers). They collected an estimated 49.2 pounds (22.3 kilograms) of lunar samples, crucial for scientific analysis back on Earth.

Upon re-entering the LM, the cabin was repressurized at 135 hours, 42 minutes, and 54 seconds into the mission, marking the conclusion of Apollo 14's third piloted lunar exploration mission. This comprehensive effort advanced our understanding of lunar geology and magnetic properties and contributed significantly to the overall success of the Apollo program's scientific objectives.

While the Apollo 14 landing crew conducted their historic activities on the lunar surface, Command Module Pilot Stuart Roosa remained aboard the Kitty Hawk, performing critical scientific tasks from lunar orbit. Over nearly two days, Roosa executed an intensive program of observations and experiments originally intended for the ill-fated Apollo 13 mission.

One of Roosa's primary tasks was orbital photography of the lunar surface. Using a Hasselblad camera, he captured detailed images of the Descartes Highlands, a proposed landing site under consideration for Apollo 16. Despite technical issues with the Lunar Topographic Camera (Hycon), which suffered a shutter malfunction that hindered its use, Roosa successfully obtained photographs confirming the suitability of Descartes as a landing location.

Roosa also documented the impact point of Apollo 13's S-IVB stage near Lansburg B crater, providing valuable imagery for post-mission analysis. He conducted a series of maneuvers with the Service Propulsion System (SPS), adjusting the Command and Service Module's (CSM) orbit to approximately 60 nautical miles (110 kilometers) above the lunar surface. Additionally, Roosa performed a plane change maneuver to compensate for the Moon's rotation, optimizing the CSM's orbital path.

In addition to lunar surface photography, Roosa engaged in astronomical observations. He photographed the Gegenschein phenomenon and the Lagrangian point (L2) of the Sun-Earth system, exploring theories about the origin of Gegenschein's illumination from reflections off particles at L2.

Roosa's scientific duties extended to experimental radar activities as well. He conducted bistatic radar experiments, using Kitty Hawk's transmitters to emit signals toward the Moon's surface. By analyzing the echoes received back on Earth, these experiments aimed to provide insights into the lunar regolith's depth and composition.

Roosa's efforts significantly contributed to lunar science and mission planning throughout his solo operations aboard the Kitty Hawk. His detailed observations and data collection enhanced our understanding of lunar geology, orbital dynamics, and astronomical phenomena, advancing the goals of NASA's Apollo program beyond the constraints of lunar surface operations.

During Apollo 14, the astronauts spent 9 hours, 22 minutes, and 31 seconds outside the Lunar Module (LM), covering a distance of 13,100 feet (approximately 4 kilometers) across the lunar surface. They meticulously gathered a significant scientific haul, collecting 93.21 pounds (42.28 kilograms, as verified by the Lunar Receiving Laboratory in Houston) of lunar samples.

The farthest point reached from the LM during their exploration was 4,770 feet. As their lunar surface mission concluded, the LM's ascent stage engine ignited precisely at 18:48:42 GMT (13:48:42 EST) on February 6, 1971, after a total duration of 33 hours, 30 minutes, and 31 seconds on the lunar surface.

Antares, the ascent stage, lifted off from the Moon at 13:48:42 EST (18:48:42 UTC) on February 6, 1971. Following a direct rendezvous on the first lunar landing mission, the command and service module Kitty Hawk successfully docked with Antares an hour and 47 minutes later. Despite early concerns over docking issues and a failure of the LM's Abort

Guidance System just before docking, the operation was completed successfully on the initial attempt.

Kitty Hawk in lunar orbit

After transferring crew members, equipment, and lunar samples to Kitty Hawk, the ascent stage of Antares was jettisoned and deliberately impacted the lunar surface. The impact generated seismic waves recorded by seismometers deployed during previous Apollo missions, further contributing to the scientific understanding of lunar geology and structure.

Following the successful liftoff from the lunar surface, the ascent stage of Apollo 14 executed a precisely timed 432.1-second engine burn, placing it into an initial orbit measuring 51.7 by 8.5 nautical miles (n mi). This marked the Apollo program's first use of a direct lunar orbit rendezvous. However, a minor 12.1-second vernier adjustment was necessary at 141:56:49.4 to fine-tune the orbit to 51.2 by 8.4 n mi.

Later, a critical 3.6-second terminal phase initiation maneuver at 142:30:51.1, along with two small midcourse corrections, further refined the ascent stage's trajectory to an orbit of 60.1 by 46.0 n mi. To facilitate docking with the Command and Service Module (CSM), a final 26.7-second adjustment maneuver at 143:13:29.1 set the stage into a precise 61.5 by 58.2 n mi orbit. The actual docking occurred at 143:32:50.5, with both spacecraft at an altitude of 58.6 n mi, after 39 hours, 45 minutes, and 8.9 seconds of separation.

During the approach for docking, telemetry indicated a failure in the abort guidance system, although no caution or warning signals were triggered. Despite attempts to resolve the issue by cycling circuit breakers and switches, the problem persisted. Fortunately, the docking process itself proceeded smoothly without any probe/drogue malfunctions, and all equipment was carefully returned for post-flight analysis.

Once the crew and lunar samples were successfully transferred to the Command Module (CM), the ascent stage was jettisoned at 145:44:58.0, clearing the way for preparations for the journey back to Earth. Remote-controlled maneuvers guided the ascent stage to impact the lunar surface. A brief 15.8-second maneuver at 145:49:42.5 separated the CM from the ascent stage, leaving it in a final orbit of 63.4 by 56.8 n mi.

To begin the journey home, a crucial 76.2-second deorbit firing at an altitude of 57.2 n mi utilized the remaining propellant of the ascent stage, leading to its deliberate impact on the lunar surface at 147:42:23.4. The impact point was precisely identified at latitude 3° 25' 12" south and longitude 19° 40' 1" west, located 36 n mi west of the Apollo 14 landing site, 62 n mi from the Apollo 12 landing site, and only seven n mi from the targeted impact location.

On Apollo 14, special dust control procedures were used to effectively decrease the amount of lunar surface dust in the cabins. On previous missions, dust adhering to equipment being returned to Earth had created a problem.

Following a 149.23-second maneuver at 148:36:02.30, transearth injection was achieved at 148:38:31.53 at a velocity of 8,505 ft/sec after 34 lunar orbits lasting 66 hours 35 minutes 39.99 seconds.

On February 6, at 8:39:04 pm EST (February 7, 01:39:04 UTC), during Kitty Hawk's 34th lunar revolution, the crucial trans-Earth injection burn commenced, lasting 350.8 seconds. This burn was pivotal in initiating the return journey of Apollo 14 back to Earth.

During the subsequent trans-Earth coast phase, several important tests and exercises were conducted aboard the Command Module (CM). These included tests of the oxygen system to ensure functionality under low tank oxygen densities and at high flow rates, which would be essential for planned in-flight extravehicular activities (EVAs) on future missions like Apollo 15. Additionally, a navigation exercise simulated a scenario where communication with ground control was lost, testing the crew's ability to navigate independently.

During his rest periods on the journey home, astronaut Edgar Mitchell conducted unauthorized ESP experiments, attempting to transmit images of cards to four individuals on Earth telepathically. Mitchell later reported varying success, with two recipients scoring significantly above chance.

As they neared Earth, the crew held a press conference, during which the Capsule Communicator (CAPCOM) read them pre-submitted questions, providing insights into their experiences and observations during the mission.

During the transearth phase, a minor 3.0-second midcourse correction using the service module's reaction control system was executed at 165:34:56.69, fine-tuning their trajectory. Scientific experiments during this phase included televised demonstrations of electrophoretic separation, liquid transfer, heat flow, and convection studies, and composite casting under microgravity conditions.

As the mission approached its conclusion, the service module was jettisoned at 215:32:42.2, followed by the CM reentering Earth's atmosphere at 215:47:45.3, approximately 400,000 feet above the Earth's surface, traveling at a velocity of 36,170 feet per second. After a transearth coast lasting 67 hours, 9 minutes, and 13.8 seconds, the CM successfully splashed down in the Pacific Ocean at 21:05:00 GMT (16:05:00 EST) on February 9.

However, the service module's expected reentry into Earth's atmosphere and subsequent splashdown in the Pacific Ocean, approximately 650 nautical miles southwest of the CM's splashdown point, did not have confirmed radar data or visual sightings, indicating a potential deviation from the planned trajectory.

After a successful mission, the Apollo 14 Command Module (CM), with astronauts Alan B. Shepard Jr., commander; Stuart A. Roosa, command module pilot; and Edgar D. Mitchell, lunar module pilot, aboard, splashed down in the South Pacific Ocean on February 9, 1971, at 21:05 UTC, approximately 900 miles (1,400 km) south of American Samoa. The USS New Orleans swiftly recovered the crew, who were transferred to the ship's Mobile Quarantine Facility. From there, they were flown to Pago Pago International Airport in Tafuna, American Samoa, then onward to Honolulu, and finally to Ellington Air Force Base near Houston aboard a specially equipped aircraft containing another Mobile Quarantine Facility trailer. This stringent quarantine procedure was followed until their release on February 27, 1971, marking the end of their isolation in the Lunar Receiving Laboratory.

Apollo 14 landing in the South Pacific

Apollo 14 astronauts were the last lunar explorers to undergo quarantine upon returning from the Moon, and uniquely, they were the only Apollo crew quarantined both before and after their mission.

The mission's total duration was 216 hours, 1 minute, and 58.1 seconds. The splashdown occurred approximately 0.6 nautical miles from the target point and 3.8 nautical miles from the recovery ship USS New Orleans. The splashdown site was located at approximately latitude 27.02° south and longitude 172.67° west. Following splashdown, the command module assumed an apex-up flotation attitude, and the crew was swiftly retrieved by helicopter, boarding the recovery ship just 48 minutes after splashdown. The command module itself was recovered 76 minutes later.

The command module weighed 11,481.2 pounds at splashdown and traveled approximately 1,000,279 nautical miles during the mission. Upon reaching Hawaii, the CM and its quarantine facility were offloaded on February 17, 1971, and subsequently transported to Houston for further examination and analysis in the Lunar Receiving Laboratory.

After the Apollo 14 mission, Stuart Roosa, who had a background in forestry, carried several hundred tree seeds with him on the flight. These seeds later germinated upon returning to Earth and became known as the "Moon trees." They were distributed around the world as part of a commemorative effort. Some of these seedlings were given explicitly to state forestry associations in 1975 and 1976 to mark the United States Bicentennial.

NASA's Apollo 14 Mission Splashes Down in Pacific Ocean,

Following the splashdown, the command module (CM) was transported to Hickam Air Force Base, Hawaii, for deactivation. Once deactivated, it was flown via C-133 aircraft to Ellington Air Force Base, Houston, where it arrived on February 22, 1971, at 21:45 GMT. The crew and medical support personnel were released from quarantine on February 26. Subsequently, the CM and lunar samples were released on April 4.

Tests conducted on the lunar samples showed no evidence of lunar microorganisms at the exploration sites, leading NASA to discontinue quarantine procedures for future missions. On April 8, 1971, the CM was delivered to the North American Rockwell Space Division facility in Downey, California, for postflight analysis.

In the aftermath of Apollo 13, rigorous mission data analysis yielded pivotal insights that shaped future endeavors. The modifications to the cryogenic oxygen system, precipitated by the mission's harrowing failure, were rigorously tested and found to meet all safety benchmarks for subsequent missions, including critical extravehicular activities.

Apollo 14's mission underscored the indispensable role of human expertise in spaceflight. Crew members adeptly diagnosed and circumvented hardware malfunctions that could have otherwise jeopardized the mission. Their adaptability ensured mission continuity and showcased the irreplaceable value of piloted space missions.

Navigating the lunar surface emerged as one of the most challenging tasks due to various factors, including poor visibility in both sunlit and shadowed areas and difficulties in discerning distances and small-scale features. These challenges underscored the need for precise navigational aids and training for lunar explorers.

The "Big Bertha" rock (Lunar Sample 14321) was the third largest rock collected during the Apollo program.

Apollo 14 also marked a milestone by successfully demonstrating lunar ascent rendezvous within a single orbit. This advancement significantly reduced the time between liftoff from the lunar surface and docking with the command module, cutting it by approximately two hours compared to earlier missions. However, it demanded highly compressed timelines for mission activities.

Addressing previous issues with lunar dust contamination, Apollo 14 employed specialized procedures and equipment that effectively mitigated dust ingress into returning spacecraft. This innovation alleviated a persistent problem encountered during earlier lunar missions.

During the journey back to Earth, the mission successfully validated onboard navigation capabilities independent of ground-based communications. This capability, proven effective during the transearth phase, now stood ready as a reliable contingency for future missions.

Furthermore, Apollo 14's launch through cumulus clouds, with cloud tops reaching heights of 10,000 feet, set a safe precedent for missions, demonstrating that such conditions did not pose a risk of triggered lightning during liftoff. This finding established critical safety parameters for future launch operations under similar atmospheric conditions.

Apollo 14 embarked on a meticulously planned mission with comprehensive objectives to advance lunar exploration and scientific discovery. Central to its primary goals was the thorough inspection, survey, and collection of samples from the Fra Mauro formation—a crucial geological target that promised insights into lunar history and composition. This objective was successfully fulfilled, marking a pivotal achievement in lunar science.

Another primary objective, the deployment and activation of the Apollo Lunar Surface Experiments Package (ALSEP), was pivotal in expanding our understanding of lunar geophysics and environmental conditions. This suite of experiments encompassed passive and active seismology, solar wind composition analysis, and the study of lunar dust and charged particles. Each experiment provided valuable data that contributed significantly to our knowledge of the Moon's environment and its interaction with space.

Apollo 14 also aimed to enhance human capabilities in the lunar environment, demonstrating advancements in extravehicular activity (EVA) operations, navigational accuracy, and modular equipment transporters. These achievements underscored the mission's success in refining operational procedures critical for future lunar missions.

Despite encountering challenges such as camera malfunctions and visibility limitations during high-sun angles, the mission achieved substantial photographic documentation of lunar features and geological formations. These visuals, albeit partially obtained, provided crucial insights into potential future exploration sites and lunar topography.

Operational tests conducted during the mission included classified experiments for the Department of Defense, demonstrating Apollo's dual-use capabilities in advancing both civilian and military technologies. These tests encompassed radar tracking, acoustic measurements, and ionospheric disturbance studies, highlighting the program's broader scientific and strategic significance.

Furthermore, Apollo 14 flawlessly achieved all launch vehicle objectives, from a precise Earth parking orbit insertion to a successful lunar impact of the S-IVB stage, further validating NASA's capability in precise trajectory and operational control in deep space.

Apollo 15

The Ninth Crewed Mission

The prime crew of Apollo 15; from left to right: Commander, David R. Scott, Command Module pilot, Alfred M. Worden and Lunar Module pilot, James B. Irwin.

Apollo 15, launched from Kennedy Space Center on July 26, 1971, marked a significant milestone in the United States' Apollo program. It was the ninth crewed mission and the fourth to achieve a lunar landing successfully. This mission, known as the first of the J missions, was characterized by its extended duration on the Moon and a robust focus on scientific exploration compared to its predecessors, which flew 1,274,137 miles.

Colonel David Randolph Scott, selected as an astronaut in 1963, commanded the historic Apollo 15 mission with Major Alfred Merrill Worden serving as command module pilot and Lt. Colonel James Benson Irwin as lunar module pilot. Scott, born in San Antonio, Texas, in 1932, brought a wealth of experience to the mission, having previously piloted Gemini 8 and served as command module pilot for Apollo 9. His educational background included a Bachelor of Science from the U.S. Military Academy in 1954 and a Master of Science in aeronautics and astronautics from MIT in 1962.

Major Alfred Worden, born in Jackson, Michigan, in 1932, shared a West Point education with Scott, graduating in 1955, and earned dual master's degrees in astronautical and aeronautical engineering and instrumentation engineering from the University of Michigan in 1963. Irwin, born in Pittsburgh in 1930, graduated from the U.S. Naval Academy in 1951, pursued further studies at Michigan, and was selected alongside Worden in NASA's fifth astronaut group in 1966.

Their mission, Apollo 15, marked Worden's and Irwin's first and only spaceflight, distinguished by their roles as pivotal members of the all-Air Force crew.

During the Apollo 15 mission, communication between the astronauts and Mission Control was facilitated by a distinguished group of capsule communicators (CAPCOMs), each selected for their expertise and experience. Among them were notable figures such as Joseph Percival Allen IV, Ph.D., Major Charles Gordon Fullerton (USAF), Karl Gordon Henize, Ph.D., Commander Edgar Dean Mitchell (USN/Sc.D.), Robert AlaB Ridley Parker, Ph.D., Harrison Hagan "Jack" Schmitt, Captain Alan Bartlett Shepard, Jr. (USN), Richard F. Gordon Jr., and Vance D. Brand. This team of CAPCOMs, all fellow astronauts, played a crucial role in relaying instructions, updates, and critical information to the Apollo 15 crew throughout their mission.

Supporting the mission on the ground were astronauts Karl G. Henize, Joseph P. Allen, and Robert A. Parker, serving in the support crew role. These scientist-astronauts, selected in 1967, provided essential assistance with scientific tasks and mission operations, ensuring the success of the lunar exploration objectives.

The mission directors overseeing Apollo 15 included Gerald D. Griffin, Milton L. Windler, Glynn S. Lunney, and Eugene F. Kranz, who led the Gold, Maroon, Black, and White teams respectively. These seasoned flight directors coordinated every aspect of the mission from launch to lunar landing, ensuring meticulous planning and decision-making during critical mission phases.

During the early Apollo missions, the push to integrate scientific exploration faced significant hurdles as technical and operational goals often took precedence over scientific inquiry. Scientist-astronauts such as Harrison Schmitt were instrumental in advocating for a stronger emphasis on scientific training and objectives during lunar missions. However, their advocacy initially met resistance or indifference from some of their astronaut colleagues.

Schmitt recognized the crucial need for specialized training in lunar geology and sought to inspire his fellow astronauts. He enlisted Caltech geologist Lee Silver to impart his expertise, introducing him to Apollo 13's crew members, Jim Lovell and Fred Haise. Lovell and Haise eagerly embraced the opportunity to enhance their mission preparations by participating in field expeditions with Silver, thereby significantly integrating geology into their training regimen.

Meanwhile, geologist Farouk El-Baz was pivotal in preparing Ken Mattingly, Apollo 13's command module pilot, to observe and document lunar features from orbit effectively. However, the crew's newfound geological skills went largely unutilized due to the mission's unforeseen abort caused by an onboard explosion.

As efforts continued into Apollo 14, ongoing initiatives continued to embed geological exploration into missions. Command module pilot Stuart Roosa demonstrated enthusiasm for geological activities, contrasting with mission commander Alan Shepard's comparatively limited interest in scientific pursuits. This underscored the persistent challenges in balancing technical mission goals with the imperative for scientific discovery within the Apollo program.

Already familiar with the spacecraft from their experience as backup crew for Apollo 12, David Scott, Alfred Worden, and James Irwin could focus more of their training time as the prime crew for Apollo 15 on geology and lunar sampling techniques. Scott, in particular, was driven by a strong determination to maximize the scientific yield of their mission. In April 1970, he initiated geological planning sessions with Caltech geologist Lee Silver to enhance their geological training.

Harrison Schmitt's role as the backup lunar module pilot for Apollo 15 allowed him to engage closely with mission preparations and foster healthy competition between the prime and backup crews. The cancellation of two Apollo missions in September 1970 resulted in Apollo 15 being designated as a "J mission," which entailed an extended stay on the lunar surface and the introduction of the Lunar Roving Vehicle (LRV). This transformation was enthusiastically welcomed by Scott, who, as described by David West Reynolds, saw himself not just as a pilot but as an explorer driven to exploit the opportunities afforded by the J mission format fully.

Throughout their rigorous 20-month training period, the Apollo 15 crew engaged in regular geology field trips, which were pivotal in preparing them for their lunar mission. Initially, these trips were educational excursions led by Caltech geologist Lee Silver to geological sites in Arizona and New Mexico. As training progressed and the mission drew nearer, the field trips evolved into highly realistic simulations.

Closer to launch, the crews simulated lunar surface conditions by wearing mock-ups of the lunar backpacks they would use and hiking near locations such as the Rio Grande Gorge. They communicated via walkie-talkies with a CAPCOM stationed in a tent nearby. The CAPCOM, accompanied by a geologist unfamiliar with the area, relied solely on the astronauts' descriptions to interpret their geological findings. This exercise helped the crew hone their skills in accurately describing landscapes and geological features to those who could not visually observe them.

David Scott, known for his dedication and enthusiasm, embraced these field geology exercises, considering himself a serious amateur. The practical experience gained during these simulations proved invaluable in preparing the crew for the complexities and challenges they would face during their lunar exploration mission.

In September 1970, the decision to select the landing site for Apollo 15 was a critical milestone. The Site Selection Committee had narrowed the options to two promising locations: Hadley Rille, a deep channel on the edge of Mare Imbrium near the Apennine mountains, and the crater Marius, surrounded by low, potentially volcanic domes. Although the final decision did not rest solely with him, the mission commander's influence was significant.

Backup Apollo 15 Commander Richard Gordon (right) and backup Lunar Module Pilot Harrison Schmitt during geology training in Taos, New Mexico. NASA photo number S71-23768

For David Scott, the choice was clear-cut. He favored Hadley Rille because of its diverse geological features and an intangible sense of exploration spirit. In his words, "Hadley had more variety... It looked beautiful, and usually, when things look good, they are good." This subjective assessment played a crucial role in the site selection process.

Interestingly, NASA lacked high-resolution images of the Hadley landing site at the time of selection. The terrain was considered too rugged to risk sending earlier Apollo missions for detailed imaging. Despite this limitation, the committee decided based on available data and Scott's compelling rationale.

The proximity of the Apennine mountains posed a unique challenge for the landing trajectory. Unlike earlier Apollo missions that used a 15-degree approach, Apollo 15 required a steeper 26-degree descent path. This adjustment was necessary to safely navigate the rugged lunar terrain and achieve a successful landing at Hadley.

Apollo 15 performed geology training in Hawaii. Commander David R. Scott is seen here taking a photograph during training carrying map and sample bag.

During Apollo 15's preparation, the expanded mission scope required Alfred Worden to devote much of his time to training at North American Rockwell's facilities in Downey, California, where the command and service module (CSM) was under construction. Unlike his crewmates, who focused on field geology, Worden embarked on specialized geology training with Farouk El-Baz.

El-Baz, renowned for his expertise in lunar geology, guided Worden through studying maps and photographs of the lunar surface, particularly the craters he would pass over while orbiting alone in the CSM. This training aimed to equip Worden with the skills to effectively describe lunar features during his solo orbits, ensuring that his transmissions back to Earth would provide valuable scientific data.

Dave Scott sampling at the boulder on the rim of Hadley Rille at Station 9a. Jim Irwin's reflection can be seen in the visor.

Worden found El-Baz to be informative and an enjoyable and inspiring teacher. Additionally, he participated in his crewmates' geology field trips, albeit from a different perspective. While they explored the terrain on foot, Worden often observed from an airplane overhead, simulating the speed and perspective from which he would view the lunar landscape from orbit.

The intensive training regimen took its toll on both Worden's and James Irwin's marriages, leading them to seek advice from their commander, David Scott. Concerned that marital issues might jeopardize their mission roles, Scott consulted with Deke Slayton, Director of Flight Crew Operations at NASA. Slayton emphasized that the astronauts' foremost priority was to perform their duties effectively, reassuring them that personal challenges would not affect their mission assignments.

Ultimately, while the Irwins were able to overcome their marital difficulties, Worden's marriage ended in divorce before the mission began. Despite these personal challenges, both astronauts remained committed to their professional responsibilities, ensuring they were fully prepared for the demands of the Apollo 15 mission.

Apollo 15 was equipped with Command and Service Module CSM-112, named Endeavour after HMS Endeavour, the ship captained by James Cook during the first purely scientific sea voyage. The choice of name reflected Apollo 15's pioneering role as the first lunar landing mission to emphasize scientific exploration heavily. A small piece of wood from Cook's Endeavour was carried aboard the spacecraft to honor this connection.

Accompanying the CSM was Lunar Module LM-10, designated Falcon, named after the United States Air Force Academy mascot. Falcon symbolized the crew's Air Force background, carrying two falcon feathers to the Moon as a tribute.

While preparing for Apollo 15 at the Kennedy Space Center, technicians encountered challenges with the instruments housed in the Service Module's Scientific Instrument Module (SIM) bay. Delays in instrument deliveries and the need for further testing or minor adjustments by principal investigators and NASA contractors were common. These instruments, designed for space operations, had to undergo rigorous testing on Earth to simulate the space environment. For instance, the 7.5-meter booms used for the mass and gamma-ray spectrometers required specialized equipment to replicate conditions they would encounter in space. Despite efforts, issues like the mass spectrometer boom occasionally failing to retract fully were observed during testing.

Modifications were made to the Lunar Module (LM) to enhance its capabilities for the extended J mission of Apollo 15. The descent and ascent stages saw enlarged fuel and oxidizer tanks, while the engine bell on the descent stage was extended. Additional batteries and solar cells were installed to boost electrical power. These enhancements increased the Lunar Module's weight to 36,000 pounds (16,000 kilograms), making it 4,000 pounds (1,800 kg) heavier than earlier models.

Had Apollo 15 been designated as an H mission, it would have utilized CSM-111 and LM-9. However, these specific spacecraft components were later repurposed for the Apollo-Soyuz Test Project in 1975. The Lunar Module LM-9 now resides at the Kennedy Space Center Visitor Complex. At the same time, the Command and Service Module CSM-111 was eventually transferred to the National Museum of the United States Air Force at Wright-Patterson Air Force Base in Dayton, Ohio, after being part of the Smithsonian's collection since December 1974.

For the Apollo 15 mission, the astronauts were outfitted with newly redesigned space suits known as the "A7LB". These suits significantly improved over previous Apollo missions, incorporating several key modifications to enhance functionality and comfort during lunar operations.

Unlike earlier suits with connectors arranged in two parallel rows, the A7LB suits featured connectors positioned in triangular pairs. This redesign, along with a diagonal entry zipper running from the right shoulder to the left hip, facilitated easier donning and doffing within the confined spaces of the spacecraft. Moreover, a new waist joint allowed greater mobility, enabling astronauts to bend fully and sit comfortably on the lunar rover—a critical improvement for surface exploration.

The Apollo 15 command and service module in lunar orbit, photographed from Falcon

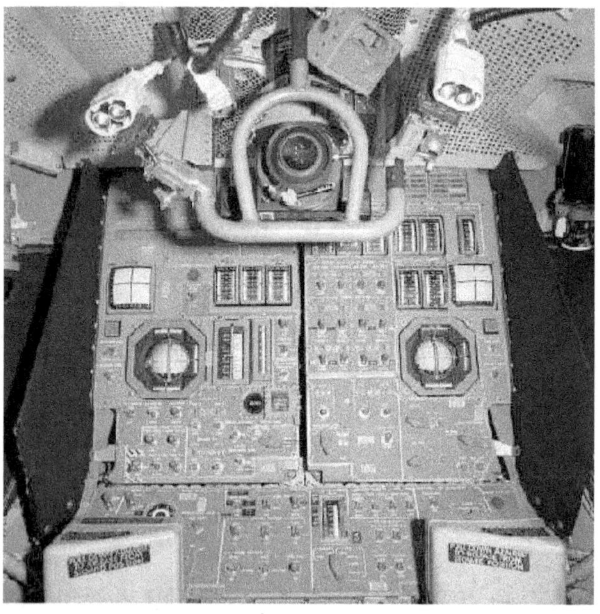

The interior of Falcon

The suits were equipped with upgraded backpacks that supported longer-duration moonwalks, essential for extended missions on the lunar surface. Additionally, as a visual identification feature introduced since Apollo 13, the commander's suit prominently displayed a red stripe on the helmet, arms, and legs.

Command Module Pilot Alfred Worden wore a suit similar to those worn during Apollo 14, albeit modified to interface specifically with Apollo 15's equipment. Notably, gear required solely for lunar surface Extra-Vehicular Activities (EVAs), such as the liquid cooling garment, should have been included in Worden's suit configuration. His planned EVA task was limited to retrieving film cartridges from the Scientific Instrument Module (SIM) bay during the return journey to Earth.

On the evening of July 24, 1971, the final countdown for the Apollo launch commenced at Cape Kennedy, marking the beginning of meticulous preparations for a historic journey. The countdown sequence, initiated at T-28 hours, unfolded with precision amid carefully scheduled holds—first at T-9 hours for 9 hours and 34 minutes, followed by another pause at T-3 hours 30 minutes, lasting an additional hour. These deliberate breaks ensured that every detail was scrutinized and any necessary adjustments could be made without haste.

The Saturn V rocket, designated SA-510, chosen for the Apollo 15 mission, represented the pinnacle of engineering achievements in space exploration. As the tenth flight-ready model of its kind, it underwent significant modifications to accommodate a larger payload. Given the mission's objectives and NASA's ambitious goals, these adjustments were crucial.

To enhance its capability, several key changes were implemented. Firstly, the launch trajectory was altered to a more southerly direction, ranging between 80 to 100 degrees azimuth. This adjustment, coupled with lowering the Earth parking orbit to 166 kilometers (90 nautical miles), enabled an additional 1,100 pounds (500 kg) to be carried into space. Propellant reserves were reduced to achieve this, and the number of retrorockets on the S-IC first stage was halved from eight to four. This meant that the four outboard engines of the S-IC burned longer, with the center engine also contributing more thrust.

The astronauts pose before the VAB as the Saturn V is rolled out

Moreover, modifications were made to the S-II stage to mitigate pogo oscillations, ensuring smoother and more stable flight dynamics during ascent. These refinements underscored NASA's commitment to optimizing the performance and reliability of its most powerful rocket.

Following the meticulous installation of all major systems, the Saturn V was transported from the Vehicle Assembly Building to Launch Complex 39A. However, during late June and early July 1971, the rocket and its towering Launch Umbilical Tower (LUT) encountered a series of lightning strikes—four in total. Despite these electrifying encounters, the vehicle sustained no damage of consequence, with only minor impacts noted on ground support equipment.

Worden, Scott, Slayton and Schmitt eat the pre-launch breakfast

As the launch approached, Cape Kennedy was under the influence of fair weather conditions attributed to a high-pressure system extending westward from the Bermuda High, encompassing central Florida. Above, the sky was adorned with cirrus clouds covering 70 percent of its expanse, their bases hovering at 25,000 feet. The temperature stood at 85.6°F, with a relative humidity of 68 percent, and the barometric pressure was recorded at 14.788 lb/in². Winds were measured meticulously at two heights: at ground level, 60.0 feet above, an anemometer recorded a steady 9.9 knots blowing from 156° true north, while at 530 feet above the launch site, winds maintained a slightly stronger pace of 10.5 knots originating from 158° true north.

On July 26, 1971, at 9:34 AM EDT, Apollo 15 embarked on its journey to the Moon from Kennedy Space Center Launch Complex 39, Pad A, on Merritt Island, Florida. The launch was meticulously timed to commence at the very beginning of a critical two-hour, 37-minute window. This precision was crucial to ensure the spacecraft would reach the Moon under optimal lighting conditions for landing at Hadley Rille. Any delay beyond the next available window on July 27 would have necessitated postponing the mission until late August, highlighting the stringent scheduling constraints faced by NASA.

The Apollo 15 launch vehicle during rollout

Earlier that morning, the astronauts—awakened five and a quarter hours before launch—had been prepared by Deke Slayton. After breakfast and suiting up, they were transported to Pad 39A, the launch site for all crewed lunar landing attempts, arriving approximately three hours before liftoff. The countdown proceeded flawlessly, with no unplanned delays reported.

Apollo 15's Saturn V rocket performed as planned during the ascent phase. The vehicle rolled from a launch pad azimuth of 90° to a flight azimuth of 80.088° between 12.21 and 23.02 seconds after liftoff. The S-IC first-stage engine shutdown occurred at 2 minutes and 39.56 seconds into the flight, followed by the separation of the spent S-IC stage and ignition of the S-II second-stage engine. The S-II engine operated until 9 minutes and 9.06 seconds, after which it separated from the S-IVB third stage, which ignited at 9 minutes and 13.20 seconds. The first cutoff of the S-IVB engine happened at 11 minutes and 34.67 seconds, with minor deviations from the planned trajectory—only -2.0 feet per second in velocity and 0.4 nautical miles in altitude.

Throughout ascent, the spacecraft encountered maximum wind conditions of 36.2 knots blowing from 63° true north at an altitude of 45,110 feet, with a maximum wind shear of 0.0110 seconds at 36,830 feet. Upon achieving parking orbit at 11 minutes and 44.67 seconds after liftoff (S-IVB cutoff plus 10 seconds), the Command and Service Module (CSM) exhibited an orbit with an apogee and perigee of 91.5 by 89.6 nautical miles, an inclination of 29.679°, a period of 87.84 minutes, and a velocity of 25,602.7 feet per second.

Designated internationally as 1971-063A, the CSM continued its journey toward lunar orbit, while the S-IVB stage, designated 1971-063B, played its role in facilitating the mission's trajectory toward the Moon.

Following a thorough series of in-flight systems checks, Apollo 15 initiated its crucial Translunar Injection (TLI) maneuver. This maneuver, involving the second firing of the S-IVB stage's engine, lasted 350.79 seconds and commenced precisely at 002:50:02.90 into the mission timeline. The S-IVB engine shutdown followed at 002:55:53.61, concluding the TLI maneuver. Ten seconds later, the spacecraft achieved translunar injection, propelling it toward the Moon at 35,606.5 feet per second.

Before this maneuver, Apollo 15 had completed 1.5 Earth orbits, totaling 2 hours, 44 minutes, and 19.02 seconds. Initially placed in a planned parking orbit around Earth following the first S-IVB engine cutoff at 000:11:36, the mission spent 2 hours and 40 minutes in this orbit. This period allowed the crew aboard and mission control in Houston to meticulously assess and confirm the spacecraft's operational readiness through telemetry.

The restart of the S-IVB engine for TLI marked a critical transition point in the mission, redirecting Apollo 15 on its trajectory toward lunar orbit.

After achieving Translunar Injection (TLI), which set Apollo 15 on a trajectory toward the Moon, the Command and Service Module (CSM) and Lunar Module (LM) remained attached to the nearly-spent S-IVB booster stage. Following TLI, explosive charges separated the CSM from the booster. Command Module Pilot Alfred Worden then maneuvered the CSM using its thrusters to dock with the LM, which was mounted at the end of the S-IVB stage. Once docked, the combined spacecraft was detached from the S-IVB stage through another set of explosive charges.

Upon separation, the now-empty S-IVB booster executed a planned maneuver to move away from the spacecraft. Approximately an hour after Apollo 15 entered lunar orbit, the S-IVB booster impacted the Moon. However, due to a navigational error, it landed 79 nautical miles (146 km) from its target point. Despite this deviation, the impact of the S-IVB booster was detected by seismometers left on the lunar surface by the Apollo 12 and Apollo 14 missions, providing valuable scientific data for lunar geology and seismic studies.

At 003:22:27.2 into the mission timeline, the CSM separated from the S-IVB stage, maneuvered into a transposition maneuver, and successfully docked with the LM at 003:33:49.50. Color television onboard was initiated to broadcast coverage of the docking procedure back to Earth. The docked spacecraft were then ejected from the S-IVB at 004:18:01.2, followed by an 80.2-second separation maneuver performed at 004:40:01.8 to refine their trajectory.

Later, at 005:46:00.7, the tanks of the now-empty S-IVB stage were vented, and its auxiliary propulsion system was fired for 241.2 seconds to target a deliberate impact on the lunar surface. An additional 71-second maneuver was executed at 010:00:01, approximately 30 minutes later than originally planned. This adjustment provided extra tracking time to account for any trajectory deviations caused by liquid oxygen and liquid hydrogen venting. Ultimately, the S-IVB impacted the Moon at 079:24:41.55 mission elapsed time, landing at latitude 0.99° south and longitude 11.89° west—83 nautical miles from its target, 191 nautical miles from the Apollo 12 seismometer, and 102 nautical miles from the Apollo 14 seismometer. At the moment of impact, the S-IVB weighed 30,880 pounds and traveled at a velocity of 8,455 feet per second.

During the Apollo 15 mission, a malfunctioning light on the Service Propulsion System (SPS) prompted the astronauts to troubleshoot extensively. Concerned that the light indicated a potential risk of the SPS unexpectedly firing, they limited the use of the affected control bank, activating it only for critical burns and managing it manually. Upon the mission's return, subsequent investigation revealed that the issue stemmed from a tiny wire fragment trapped within a switch, causing the malfunction.

Around 028:40:00 into the mission, the astronauts performed a test burn of the SPS, which also served as a midcourse correction maneuver. This adjustment aimed to refine their trajectory and ensure proper course toward their destination. Despite the initial concern with the malfunctioning light, the SPS operated effectively during critical maneuvers, using the redundant system (bank B) for automatic control while manually overseeing bank A.

During the translunar flight, two minor midcourse corrections were necessary to optimize the lunar orbit injection. The first correction, executed at 028:40:22.00, lasted 0.8 seconds and adjusted the velocity by 5.3 feet per second. The second correction involved troubleshooting an intermittent short in the SPS bank A system, which could still be operated satisfactorily in manual mode despite the issue.

In preparations for lunar operations, the Lunar Module (LM) crew entered the LM at 033:56 for initial checkout, slightly ahead of schedule. Communications checks were conducted between 034:21 and 034:45, utilizing the Goldstone tracking station in California, although initial configuration issues briefly affected down-voice backup checks. Although a momentary loss of downlink carrier locks approximately 15 minutes later, data loss was minimized due to alternate tracking stations providing continuous support.

After purging and refreshing the Lunar Module's (LM) atmosphere to ensure it was free from any contaminants, the Apollo 15 astronauts entered the LM approximately 34 hours into their mission. Their primary task was to inspect the condition of the LM's equipment and transfer essential items needed for their lunar activities. Much of this activity was broadcast live to Earth, with Command Module Pilot Worden operating the camera.

During their inspection, the crew discovered a broken outer cover on the Range/Range Rate tapemeter inside the LM. This was concerning not only because the tapemeter provided critical distance and rate of approach information but also because shards of the glass cover were floating inside the LM's cabin. Originally designed to operate in a helium atmosphere, the tapemeter was now exposed to the LM's oxygen environment due to the breakage.

Ground testing confirmed that the tapemeter could still function properly despite the broken cover. The crew carefully removed most of the glass fragments using a vacuum cleaner and adhesive tape, ensuring the cabin remained safe and operational.

Between 034:55 and 035:46, a televised transmission of the CSM and LM interiors was broadcast, capturing the crew's activities inside the spacecraft. Despite variations in picture quality due to lighting conditions, the camera operation was generally nominal.

Following the discovery of the broken tapemeter cover, the crew initiated intravehicular transfer and LM housekeeping tasks at 056:26, slightly ahead of the scheduled timeline. They meticulously cleaned the LM cabin to eliminate any remaining glass debris from the damaged tapemeter.

Based on the data from the first midcourse correction burn test, it was decided to utilize only bank B for all Service Propulsion System (SPS) maneuvers, except for lunar orbit insertion and transearth injection. For these critical maneuvers, a modified procedure allowed for dual-bank burns with automatic start and shutdown using bank B, ensuring redundancy and reliability during key mission phases.

Around 61 hours and 15 minutes into the mission, on the evening of July 28, according to Houston time, astronaut Dave Scott discovered a water system leak while preparing to chlorinate the water supply onboard. The source of the leak was initially unclear, raising concerns about its potential impact on the mission.

Experts at Mission Control in Houston quickly devised a solution, which they relayed to the crew. With their guidance, the astronauts successfully implemented the fix onboard. They used towels to mop up the leaked water, which was then dried in the tunnel connecting the Command Module (CM) and Lunar Module (LM). Scott humorously remarked that it resembled someone's laundry hanging out.

At 073:31:14 into Apollo 15's mission, the crew executed a critical midcourse correction, a swift adjustment that altered their trajectory by 5.4 feet per second using the spacecraft's propulsion system. This correction marked the second of only two needed following Trans Lunar Injection (TLI), underscoring the precision of mission planning and execution.

Approaching the Moon on July 29, Apollo 15 faced a pivotal moment as it entered lunar orbit. The Lunar Orbit Insertion (LOI) burn, conducted on the far side of the Moon and out of direct radio contact with Earth, was essential. Failure to execute this burn would have prematurely brought the spacecraft into radio visibility, signaling a missed maneuver. The successful 398.36-second burn commenced at 078:31:46.7 into the mission, occurring at an altitude of 86.7 nautical miles above the lunar surface. This burn placed Apollo 15 into an elliptical lunar orbit, measuring 170.1 by 57.7 nautical miles.

Maturate operational adjustments continued during the journey. At 074:06:47.1, the crew jettisoned the scientific instrument bay door, capturing photographs of its slow drift away from the Command and Service Module (CSM), eventually settling into a heliocentric orbit. This action cleared the way for subsequent operations and observations.

As the spacecraft settled into lunar orbit, further maneuvers were required to refine its path. At 082:39:49:09, a 24.53-second service propulsion system maneuver established a descent orbit of 58.5 by 9.6 nautical miles in preparation for the Lunar Module (LM) undocking. A subsequent 30.40-second orbit trim maneuver at 095:56:44.70 adjusted the orbit to 60.3 by 8.8 nautical miles, fine-tuning their trajectory for upcoming lunar operations.

However, a significant challenge arose during the 12th lunar revolution. The scheduled CSM/LM undocking and separation maneuver faced a delay due to a loose or disconnected umbilical between the two modules. The Command Module Pilot undertook a spacewalk into the tunnel connecting the modules, identifying and re-securing the loose umbilical plug. This critical intervention restored proper communications and alignment, allowing the planned undocking and separation to proceed approximately 25 minutes behind schedule at 100:39:16.2 and an altitude of 7.4 nautical miles.

With the spacecraft now maneuvering independently, a final 3.67-second adjustment at 101:38:58.98 circularized the CSM's orbit at 65.2 by 54.8 nautical miles. This preparation set the stage for the crew's intensive scientific activities and data collection during their lunar surface operations.

Beginning with Apollo 14, a crucial change in lunar landing strategy was implemented to conserve Lunar Module fuel. Instead of the LM decoupling and performing the Descent Orbit Insertion (DOI) burn independently, the Service Propulsion System (SPS) in the Service Module now executed this burn. At the same time, the LM remained docked to the Command and Service Module (CSM). This maneuver, critical for shaping the descent trajectory, was pivotal for Apollo 15 as well.

Apollo 15 found itself in an initial lunar orbit where its highest point (apocynthion) passed over the planned landing site at Hadley. A DOI burn was scheduled at the orbit's opposite point to prepare for the lunar descent, ensuring Hadley would transition beneath the craft's lowest point (pericynthion). Executed precisely at 082:39:49.09, this 24.53-second burn by the SPS adjusted the orbit parameters to an apocynthion of 58.5 nautical miles and a pericynthion of 9.6 nautical miles.

As the crew rested overnight between July 29 and 30, Mission Control observed the lunar orbit becoming increasingly elliptical due to gravitational anomalies or "mascons" on the Moon's surface. By the crew's awakening on July 30, the pericynthion had decreased to 7.6 nautical miles, raising concerns about the accuracy of the landing site altitude. A meticulous adjustment was made using the Reaction Control System (RCS) thrusters to optimize the trajectory. At 095:56:44.70, a 30.40-second maneuver raised the pericynthion to 8.8 nautical miles and the apocynthion to 60.2 nautical miles, ensuring a more stable and precise path for the upcoming lunar operations.

As preparations intensified for the lunar descent, the crew of Apollo 15 continued their detailed observations of the Moon, meticulously documenting the terrain, including crucial visuals of the landing site at Hadley. These observations were essential for mission planning and provided captivating television footage of the lunar surface to audiences back on Earth.

Commander Scott and Lunar Module Pilot Irwin began their final preparations by entering the Lunar Module (LM) in readiness for the historic landing attempt. However, as the scheduled undocking time approached at 100:13:56, an unexpected issue arose—nothing happened when separation was attempted over the far side of the Moon.

Quick analysis by the crew in tandem with Mission Control pointed toward a probable cause: a loose or disconnected probe instrumentation umbilical between the Command and Lunar Modules. Command Module Pilot Worden undertook the critical task of investigating the connection within the tunnel linking the modules. His inspection confirmed the issue, and he secured the umbilical plug to be more firmly in place.

With this pivotal problem resolved, the Lunar Module Falcon successfully separated from the Command Module Endeavour at 100:39:16.2, approximately 25 minutes behind schedule. The separation occurred at an altitude of 5.8 nautical miles, marking a crucial milestone toward the planned lunar landing.

Meanwhile, aboard Endeavour, Command Module Pilot Worden executed a precise burn using the Service Propulsion System (SPS) at 101:38:58.98. This maneuver adjusted Endeavour's orbit to 65.2 nautical miles by 54.8 nautical miles, positioning the spacecraft optimally for Worden's upcoming scientific activities and observations from lunar orbit

On board the Lunar Module Falcon, astronauts Scott and Irwin prepared for the critical Powered Descent Initiation (PDI) — the burn that would bring them from lunar orbit to the surface. With clearance from Mission Control, they initiated PDI at 104:30:09.4, beginning their descent from an altitude of 5.8 nautical miles, slightly higher than intended.

Initially, during the descent phase, Falcon was positioned so that the astronauts were lying on their backs, unable to view the lunar surface below directly. However, a pivotal pitch over maneuver reoriented the craft, allowing Scott and Irwin to see the rugged lunar landscape directly ahead of them. This new perspective revealed a stark reality: the lunar surface differed significantly from their training simulations.

Commander Scott, responsible for guiding the landing, encountered unexpected challenges. First, Mission Control informed them of an approximately 3,000 feet (910 meters) error in their descent trajectory prior to the pitch-over. The lunar surface conditions also made it difficult to discern familiar landmarks from the simulator, such as the distinctive Hadley Rille. Scott realized they were likely heading toward overshooting their planned landing site.

Once he identified Hadley Rille, Scott adjusted the spacecraft's trajectory, manually guiding it back toward the computer's designated landing target. Simultaneously, he scanned the lunar surface for a suitable, relatively smooth area to touch down safely.

The powered descent of the Lunar Module Falcon commenced precisely at 104:30:09.4, initiating from an altitude of 5.8 nautical miles above the lunar surface. This crucial engine firing spanned 739.2 seconds, culminating just 0.7 seconds before the historic touchdown at 22:16:29 GMT (06:16:29 p.m. EDT) on July 30, 1971, marking Apollo 15's arrival on the Moon at 104:42:29.3.

The Lunar Module (LM) "Falcon" is photographed against the barren lunarscape during the third Apollo 15 lunar surface extravehicular activity (EVA) at the Hadley-Apennine landing site on the lunar nearside.

The chosen landing site for this mission was strategically located near the base of the Montes Apenninus (Apennine Mountains) and adjacent to Hadley Rille. These geological features provided a rich scientific landscape for exploration and investigation.

Apollo 15's mission objectives were as ambitious as they were groundbreaking. They aimed to unlock new insights into the lunar landscape and expand the capabilities of human space exploration. Positioned near the Montes Apenninus (Apennine Mountains) and alongside Hadley Rille, the chosen landing site provided a strategic vantage point for scientific exploration.

The primary goals included conducting meticulous selenological inspections and surveys and meticulously examining and sampling lunar materials and surface features within the Hadley-Apennine region. Additionally, the mission aimed to deploy and activate various surface experiments to study lunar phenomena directly on the Moon's surface.

Apollo 15 also sought to evaluate the endurance and adaptability of Apollo equipment, enabling extended stays on the lunar surface. This involved increasing extravehicular activities (EVAs) and improving surface mobility, pushing the boundaries of what was possible for human exploration.

Moreover, the mission encompassed conducting in-flight experiments and capturing detailed orbital photographs. These tasks aimed to provide comprehensive insights into lunar geology and surface conditions, further advancing our understanding of Earth's celestial neighbor.

As the Lunar Module Falcon descended toward the lunar surface during Apollo 15's historic landing attempt, Commander Scott faced unexpected challenges due to the thick cloud of lunar dust that the Falcon's powerful engine kicked up. Below 60 feet (18 meters), visibility became severely limited, hindering Scott's ability to gauge the landing site accurately.

Falcon, equipped with a larger engine bell to manage its heavier payload, posed additional risks during descent. Mission planners had stressed the critical importance of shutting down the engine upon initial contact with the lunar surface to avoid the potential hazard of "blowback." This phenomenon, where engine exhaust reflects off the lunar terrain and re-enters the engine, could lead to catastrophic consequences such as an explosion.

When Lunar Module Pilot Irwin called out "Contact," signifying that one of the landing leg probes had touched the surface, Scott swiftly responded by cutting off the engine. This decision allowed Falcon to freefall the remaining distance to the lunar surface gently. Descending at approximately 0.5 feet (0.15 meters) per second, Falcon dropped from a height of 1.6 feet (0.49 meters). However, due to Scott's speed and the obscured terrain, Falcon made what was likely the hardest lunar landing of any crewed mission, impacting at about 6.8 feet (2.1 meters) per second.

The unexpected jolt caused Lunar Module Pilot Irwin to exclaim, "Bam!" as Falcon settled onto the lunar surface. Scott had inadvertently landed Falcon on the rim of a small crater obscured by lunar dust, resulting in the lander coming to rest at a tilted angle of 6.9 degrees, leaning to the left by 8.6 degrees.

Reflecting on the intense landing, Irwin later described it as the most challenging touchdown he had experienced. Concerned that the tilted position could lead to a tip-over, potentially requiring an immediate mission abort, the astronauts remained vigilant as they prepared for their historic exploration of the Moon's surface.

Upon landing at 104:42:29.3 (22:16:29 GMT on July 30), Lunar Module Falcon touched down in the Montes Apenninus (Apennine Mountains), near Hadley Rille, at coordinates approximately 26.13222° north latitude and 3.63386° east longitude. The landing site was about 1,800 feet (550 meters) from the initially planned target, with approximately 103 seconds of fuel remaining after touchdown.

After Lunar Module Pilot Irwin's exclamation confirming contact with the lunar surface, Commander Scott radioed back to Houston, announcing, "Okay, Houston. The Falcon is on the Plain at Hadley." This marked the successful initiation of their lunar exploration mission.

Once settled within the designated landing zone, the crew utilized the enhanced mobility provided by the Lunar Roving Vehicle (LRV), determining that no further maneuvering was necessary. This versatile vehicle would prove instrumental in expanding their range and efficiency during surface operations.

At 106:42:49, two hours post-landing, Commander Scott depressurized the cabin and opened the LM's top hatch. He embarked on a "stand-up EVA" (SEVA), taking 33 minutes and 7 seconds to capture panoramic photographs of the surrounding terrain. These initial images provided crucial visual data of the lunar landscape directly surrounding the landing site.

The first formal extravehicular activity (EVA) on the lunar surface commenced at 119:39:17 as the crew prepared for more extensive explorations. Commander Scott descended the ladder first, deploying the Modularized Equipment Stowage Assembly (MESA) and activating the television camera. The live broadcast of his descent offered vivid footage of the historic moment.

Following his commander, Lunar Module Pilot Irwin exited Falcon to join Scott on the lunar surface. While Scott set up the television camera on a tripod to document their activities, Irwin immediately began collecting a contingency sample of lunar soil, a crucial task in case of an unexpected early mission termination.

When Falcon settled onto the lunar surface, Commander Scott prioritized maintaining their accustomed circadian rhythm despite the excitement. The landing, timed for late afternoon Houston time, dictated their schedule: sleep first, then lunar exploration. Nevertheless, Scott seized a rare opportunity, opening Falcon's top hatch—the usual docking port—for a thirty-minute survey. Inspired by Lee Silver's teachings on reconnaissance, this decision defied initial concerns from Deke Slayton and other mission managers over oxygen loss. Scott's stand-up extravehicular activity (EVA) through the hatch marked a historic moment, allowing him to meticulously plan the next day's activities on the lunar terrain.

Scott does geology work near Hadley Rille

During this unique maneuver, Scott invited Irwin to join in observing their lunar surroundings, though technical constraints prevented it without rearranging life support connections—an offer Irwin declined. Following their return to cabin pressure, Scott and Irwin undertook another unprecedented action: removing their space suits to rest, making them the first humans to do so on the Moon.

Mission Control in Houston detected a gradual oxygen decline during their scheduled sleep period, prompting Scott and Irwin to wake up early. Investigation revealed an unexpected culprit: an open valve on the urine transfer device. The crew swiftly addressed this incident, prompting Scott to recommend immediate awakening protocols for future missions facing similar issues, a precaution from their lunar experience.

Once the technical glitch was resolved, the astronauts refocused on their mission objectives.

After suiting up and depressurizing the cabin, Scott and Irwin embarked on their historic first full Extravehicular Activity (EVA), stepping onto the lunar surface as the seventh and eighth humans to do so. Their immediate task was to deploy the lunar rover, stowed folded within Falcon's descent stage. However, the slanted terrain posed an unexpected challenge during extraction.

Unlike earlier Apollo missions, Apollo 15 aimed to delve deeper into lunar science, leveraging advanced technologies and modified spacecraft configurations. One of the most iconic additions was the Lunar Roving Vehicle (LRV), a compact and agile rover that allowed astronauts to traverse the lunar surface efficiently. This innovation expanded the exploration reach, enabling astronauts to cover greater distances and collect diverse samples.

The mission was strategically outfitted with enhanced scientific instrumentation, including a dedicated Scientific Instrument Module (SIM) installed in the service module. This module facilitated comprehensive data collection both from lunar orbit and during the lunar module's stay on the surface. These upgrades were pivotal in broadening our understanding of lunar geology, surface composition, and cosmic phenomena.

Apollo 15's lunar module underwent significant modifications to accommodate the LRV and increase payload capacity, enabling longer lunar surface stays and more extensive exploration. Moreover, the mission included the deployment of a scientific subsatellite into lunar orbit, augmenting observational capabilities from above.

At 120:18:31 into their mission, Scott and Irwin began offloading and deploying the Lunar Roving Vehicle (LRV), a critical component for extended lunar surface exploration. Despite encountering initial difficulties during deployment and checkout, including a non-functional front steering mechanism and issues with battery readings on LRV #2, the crew resolved these challenges through troubleshooting efforts.

The Lunar Roving Vehicle (LRV), a groundbreaking innovation designed for the Apollo missions, was poised to revolutionize lunar exploration across three planned missions. Crafted with meticulous engineering, the LRV was a four-wheeled marvel strategically fashioned to expand astronauts' exploratory reach on the Moon's rugged terrain.

Weighing 462 pounds on Earth and measuring 10 feet 2 inches long with a height of 44.8 inches, the LRV boasted a robust 7.5-foot wheelbase. Its design integrated five pivotal systems: mobility, crew station, navigation, power, and thermal control. Each element is meticulously crafted to withstand the extremes of the lunar environment while facilitating comprehensive scientific exploration.

Two 36-volt batteries, capable of independently powering all LRV systems, were central to its operation. Its cargo capacity, totaling 1,080 pounds, accommodated vital equipment such as astronaut life support gear, communication apparatus, scientific instruments, photographic tools, and lunar sample storage, each crucial for mission success.

Stowed compactly within the Lunar Module's Quad 1 during the journey to the Moon, the LRV awaited deployment upon landing. Astronauts undertook the meticulous task of unfurling and configuring the vehicle, preparing it for crucial cargo loading and operational readiness.

During the initial Extravehicular Activity (EVA-1), despite initial challenges with the LRV's front steering, astronauts Scott and Irwin adeptly adapted their plans to navigate the lunar landscape effectively. Subsequent EVAs saw improved maneuverability as technical glitches were resolved, enabling smoother operations and enhanced mobility.

Guided by mission control in Houston, astronauts maneuvered the rover into action, overcoming initial instrumentation hiccups. Scott, after completing a meticulous system check, famously inaugurated the LRV's lunar journey with the historic words, "Okay. Out of detent; we're moving." This milestone marked humanity's first dialogue while operating a vehicle on the Moon, underscoring the Apollo program's triumphs in lunar exploration and technological prowess.

At 121:44:55, astronauts David Scott and James Irwin embarked on a pioneering journey with the Lunar Roving Vehicle (LRV) Equipped with a remote-controlled television camera overseen by NASA's Ed Fendell in Houston, the LRV provided a live broadcast, albeit with lower resolution than still photographs. This real-time view allowed geologists on Earth to participate virtually in Scott and Irwin's lunar exploration, bridging the vast distance between worlds through pioneering scientific exploration.

As Scott and Irwin traversed the undulating lunar terrain in the LRV, they encountered a significant geological feature visible from their initial landing site: the rille, a distinctive lunar trench. Spotting Elbow Crater in the distance, they navigated toward it, knowing its coordinates would assist Mission Control in precisely pinpointing their historic landing site.

Their mission was meticulously planned: to collect vital soil and rock specimens from the Moon's surface, offering a window into its geological history. With methodical precision, they documented samples and provided vivid descriptions of their lunar surroundings, painting a detailed picture of the alien terrain.

Continuing their scientific expedition near St. George Crater, Scott and Irwin gathered additional samples and meticulously captured photographs, leveraging the LRV's advanced navigation system to navigate the lunar landscape. Their thorough documentation of these lunar features aimed to expand our understanding of the Moon's composition and environmental conditions.

Returning to their Lunar Module (LM), appropriately named Falcon, the astronauts carefully deposited their collected samples, preparing to deploy the Apollo Lunar Surface Experiments Package (ALSEP). This sophisticated suite of scientific instruments was designed to operate autonomously after their departure, conducting vital experiments that would continue contributing to NASA's knowledge of lunar science.

Mission Control in Houston during the third Apollo 15 EVA, August 2, 1971. CAPCOM Joe Allen is to left (pointing) with Dick Gordon next to him.

During these tasks, Scott encountered technical difficulties while attempting to drill holes for the heat flow experiment. Despite this setback, their inaugural Extravehicular Activity (EVA) lasted an impressive 6 hours and 32 minutes, allowing them to achieve substantial scientific and operational milestones on the lunar surface. Throughout their EVA, the crew provided enthusiastic and informative commentary on the lunar landscape, broadcast globally via television feeds facilitated by mission control.

In a moment captured live on television, Scott conducted a legendary experiment: dropping a falcon feather and hammer to demonstrate Galileo's theory on gravitational acceleration. This iconic display, suggested by CAPCOM Joe Allen, illustrated that objects of different masses fall simultaneously in a vacuum like the Moon's. The feather, likely from a symbolic gyrfalcon associated with the United States Air Force Academy, descended alongside the hammer, reaching the lunar surface simultaneously due to the Moon's negligible atmosphere. This demonstration echoed Alan Shepard's historic golf swing, resonating as a celebrated moment in scientific history.

Following these extraordinary scientific endeavors, Scott maneuvered the LRV from the LM to a vantage point ideal for capturing the LM liftoff on camera. In a poignant tribute observed later, he discreetly placed a small aluminum statuette, known as Fallen Astronaut, alongside a plaque honoring fallen astronauts and cosmonauts. Initially reported to Mission Control as routine cleanup, this solemn memorial underscored the crew's reverence for those who had sacrificed in pursuing space exploration.

Scott later disclosed the existence of this memorial during a post-flight news conference, further emphasizing its significance. Before leaving the rover for the final time to re-enter the LM, he placed a Bible on its control panel, marking a personal moment of reflection and tribute.

View of Commemorative plaque left on moon at Hadley-Apennine landing site. A close-up view of a commemorative plaque left on the Moon at the Hadley-Apennine landing site in memory of 14 NASA astronauts and USSR cosmonauts, now deceased. Their names are inscribed in alphabetical order on the plaque. The plaque was stuck in the lunar soil by Astronauts David R. Scott and James B. Irwin during their Apollo 15 lunar surface extravehicular activity. The tiny, man-like object represents the figure of a fallen astronaut/cosmonaut.

Their initial EVA concluded after 4 hours, 49 minutes, and 50 seconds, during which Scott and Irwin spent 18.5 hours outside the LM, collecting approximately 170 pounds of lunar samples. These meticulously gathered and cataloged samples would prove invaluable for scientific analysis upon their return to Earth, shedding light on the Moon's geological composition and history.

During the second Extravehicular Activity (EVA), which commenced at 142:14:48 with the cabin depressurization, Scott and Irwin focused on further exploration and scientific tasks on the lunar surface. They began by thoroughly checking the Lunar Roving Vehicle (LRV). They successfully addressed issues with its front steering by recycling circuit breakers, ensuring it was fully operational for their second traverse.

Setting out at 143:10:43, the astronauts traveled south toward the Apennine Front, east of their initial traverse route. They made strategic stops at significant landmarks such as Spur Crater and various points along the base of the Apennine Front, collecting documented samples, a core sample, and a comprehensive soil sample. Throughout their journey, they meticulously documented their findings through photography, enhancing the scientific record of their expedition.

Television transmissions from the lunar surface during this period were clear and provided real-time views of their activities, allowing audiences on Earth to share in the exploration experience. The return trip closely followed their outbound route, with additional stops at Dune Crater enriching their scientific haul.

During this EVA, Lunar Module Pilot Irwin engaged in soil mechanics tasks, contributing to their understanding of lunar surface properties. Meanwhile, Commander Scott attempted to drill for a deep-core sample but halted the effort due to time constraints, ensuring that critical mission objectives remained achievable within their operational timeframe.

After returning to the Lunar Module (LM) at 148:32:17, Scott and Irwin revisited the Apollo Lunar Surface Experiments Package (ALSEP) site to complete unfinished tasks from earlier. Commander Scott successfully drilled the second hole required for the heat flow experiment, placed the probe, and retrieved a core tube sample, leaving the drill core stems at the site for later retrieval during their third Extravehicular Activity (EVA-3).

Following these scientific tasks, the crew deployed the United States flag at the LM site, symbolizing their nation's achievement in lunar exploration. They stowed their collected lunar samples and film in the LM, ensuring their safe return to Earth for extensive scientific study.

NASA's Apollo 14 Mission Splashes Down in Pacific Ocean,

The second Extravehicular Activity period lasted 7 hours, 12 minutes, and 14 seconds, during which Scott and Irwin traveled 41,000 feet (12.5 km) in the Lunar Roving Vehicle (LRV). They spent 1 hour and 23 minutes driving the LRV, with an additional 2 hours and 34 minutes parked for scientific activities and sample collection.

Approximately 76.9 pounds (34.9 kg) of lunar samples were gathered during this EVA, contributing significantly to scientific understanding of the Moon's geology and history. The crew re-entered the LM at 149:27:02, and the cabin was repressurized, marking the conclusion of their productive second extravehicular excursion.

The crew spent almost 14 hours in the LM before the cabin was depressurized for the third extravehicular period Apollo IS~ at 163:18:14. The third extravehicular activity During their third and final Extravehicular Activity (EVA-3), Scott and Irwin encountered delays that pushed their start time 1 hour and 45 minutes later than planned, due to adjustments in the surface activities timeline. As a result, their intended trip to the North Complex had to be canceled.

During their third and final moonwalk, EVA 3, astronauts David Scott and James Irwin faced a formidable challenge: retrieving a core sample. Due to technical setbacks, this crucial scientific objective had eluded them previously. The extraction process proved demanding, requiring significant effort to dislodge and break the core into transportable pieces for return to Earth.

Complicating matters further, an improperly mounted vise on their rover hindered their attempts to segment the core as planned, forcing them to transport it back intact, albeit with one segment longer than anticipated. Scott, questioning the value of the core given the effort expended, received reassurance from CAPCOM Joe Allen regarding its scientific significance.

Indeed, the core sample would be one of the mission's most significant scientific discoveries, offering profound insights into the Moon's geological history. However, the time spent on this task necessitated a change in plans: their scheduled visit to the North Complex hills had to be canceled. Instead, they redirected their exploration to the northwest edge of Hadley Rille, expanding their scientific observations beyond the immediate landing site.

After their successful lunar surface activities, the Apollo 15 crew prepared to depart from the Moon's surface. They returned to the Lunar Module (LM) and carefully unloaded the Lunar Roving Vehicle (LRV). Positioned strategically at 167:35:24 elapsed mission time, the LRV was parked closer to the LM than initially planned. This adjustment allowed for optimal ground-controlled television coverage of the LM's ascent back to lunar orbit.

The commander selected a site with slightly elevated terrain, enhancing the visibility for televised coverage. This final parking spot marked a significant moment for the first lunar rover vehicle, poised to capture the liftoff of the Apollo 15 Lunar Module ascent stage.

Mission Control in Houston as Falcon takes off from the Moon

During their third extravehicular activity (EVA), lasting 4 hours, 49 minutes, and 50 seconds, the astronauts traversed a remarkable distance of 16,700 feet (5.1 km) aboard the LRV. The actual drive time amounted to 35 minutes, with the vehicle parked for 1 hour and 22 minutes. Throughout this EVA, they gathered approximately 60.2 pounds (27.3 kg) of lunar samples, crucial for scientific study back on Earth.

Following their productive EVA, the crew re-entered the LM and repressurized the cabin at 168:08:04, concluding Apollo 15's fourth manned lunar exploration mission. Across the entire mission, their total extravehicular time tallied up to an impressive 18 hours, 34 minutes, and 46 seconds. They covered a cumulative distance of 91,500 feet (27.9 km) in the LRV, spending a total drive time of 3 hours. The vehicle remained parked for 5 hours and 10 minutes during their EVAs. They collected 170.44 pounds (77.31 kg) of lunar samples, as verified by the Lunar Receiving Laboratory in Houston.

Their exploration took them as far as 16,470 feet from the LM, reaching notable lunar features like Crater La Hire A—a prominent bowl-shaped crater with a distinctive ridge to the south.

While the Lunar Module (LM) conducted its surface operations, the Command Module Pilot (CMP) of Apollo 15, Alfred Worden, carried out a series of critical tasks aboard the Command and Service Module (CSM) in lunar orbit. Throughout 34 orbits around the Moon, Worden performed extensive scientific experiments using the Scientific Instrument Module (SIM) bay and operated cameras to gather crucial data about the lunar surface and its environment.

One of Worden's primary responsibilities was photographing the sunlit lunar terrain, capturing high-resolution images that would contribute to detailed mapping efforts. These images provided essential data for mapping the chemical composition of the lunar surface and determining the topography along the spacecraft's orbital path.

Additionally, Worden visually surveyed various lunar regions to help identify the geological processes responsible for shaping lunar features. He also collected atmospheric data from the Moon. He conducted surveys using gamma-ray and X-ray detectors, aiming to detect and analyze sources of these radiations on the lunar surface.

With its SIM bay exposed, the CSM was equipped with advanced scientific instruments. These included a gamma-ray spectrometer mounted on a boom, an X-ray spectrometer, and a laser altimeter, though the latter experienced technical issues mid-mission. The SIM bay also housed cameras, including a stellar camera and a metric camera forming the mapping camera system, complemented by a panoramic camera derived from classified spy technology. These cameras enabled precise determination of the time and location of each photograph taken.

Apollo 15 SM SIM bay

Further instruments included an alpha particle spectrometer to detect traces of lunar volcanic activity and a mass spectrometer mounted on a boom to minimize contamination from the spacecraft. Worden sometimes encountered challenges with retracting the boom, complicating data collection efforts.

During his solo operations in lunar orbit, Worden executed an 18.31-second plane change maneuver to adjust the CSM's orbit, resulting in an elliptical path around the Moon with a periselene of 53.6 nautical miles and an aposelene of 64.5 nautical miles. His dedicated scientific observation and data acquisition efforts significantly enhanced our understanding of the Moon's geological and atmospheric characteristics.

Worden and the CSM Endeavour continued their mission independently, supported by a separate CAPCOM and flight team as the LM Falcon departed the lunar surface. This phase marked a critical transition in Apollo 15's mission profile, with Worden continuing his vital role in lunar orbit while his colleagues began their journey back to Earth.

During his solitary orbit around the Moon aboard the Command and Service Module (CSM) Endeavour, Alfred Worden faced a series of challenges and responsibilities that underscored the complexity and isolation of his mission. Scheduled to pass directly over the Apollo 15 landing site at the moment of Falcon's planned liftoff, Worden initially struggled to spot his returning crewmates. It wasn't until a subsequent orbit that he finally located Falcon and could observe its ascent back into lunar orbit.

While in orbit, Worden diligently maintained his physical health by exercising regularly to prevent muscle atrophy, a crucial regimen for extended space missions. Meanwhile, mission control in Houston provided him with continuous updates on Scott and Irwin's activities on the lunar surface, ensuring he remained informed and connected despite the vast distances involved.

The panoramic camera aboard the CSM, though not without its flaws, managed to capture sufficient images of the lunar surface. Despite its imperfections, no special adjustments were deemed necessary due to the volume of usable data it provided. Additionally, Worden used the command module's windows to take numerous photographs, systematically documenting the Moon at regular intervals throughout his orbital journey.

However, Worden encountered technical setbacks during his mission. A malfunctioning mission timer in the Lower Equipment Bay of the command module posed challenges, as its circuit breaker had popped en route to the Moon, limiting his ability to precisely time observations and tasks.

Endeavour, with the SIM bay exposed, as seen from the Lunar Module Falcon

Worden's meticulous observations and extensive photographic documentation played a pivotal role in shaping future lunar exploration missions. His data and insights were instrumental in selecting the Taurus-Littrow Valley as the landing site for Apollo 17, where subsequent missions would search for evidence of lunar volcanic activity.

Endeavour's orbital path occasionally led it to pass over the Moon's far side, causing temporary communication blackouts with mission control on Earth. Each time contact was reestablished, Worden greeted Earth with the message "Hello, Earth. Greetings from Endeavour," a gesture he and geology instructor Farouk El-Baz had devised together. El-Baz assisted Worden in compiling translations of the greeting in various languages.

The experiments conducted using the Scientific Instrument Module (SIM) bay aboard Apollo 15's Command and Service Module (CSM) Endeavour yielded significant findings that deepened our understanding of the lunar environment. Data collected by the X-ray spectrometer revealed a higher-than-expected flux of fluorescent X-rays, indicating unexpected concentrations of certain elements. Notably, the lunar highlands were richer in aluminum compared to the darker maria regions.

The landing area is shown in an image taken by the mapping camera

Endeavour's orbital path, set at a more inclined orbit than previous manned missions, provided Alfred Worden with unique vantage points of the Moon. He made valuable contributions to lunar science by capturing photographs and offering detailed descriptions of features previously unknown or poorly understood.

As Scott and Irwin prepared to depart from the lunar surface in the Lunar Module (LM), Worden faced the challenge of maintaining alignment between the LM's planned ascent trajectory and Endeavour's orbital plane. The Moon's rotation had caused Endeavour's orbit to drift, necessitating a precise plane change maneuver to synchronize their paths. Using the Service Propulsion System (SPS), Worden executed an 18-second burn to adjust Endeavour's orbit, ensuring optimal rendezvous conditions for the LM's return.

The Apollo 15 mission concluded with a meticulously orchestrated sequence of events as the Lunar Module (LM) prepared to depart from the lunar surface. At precisely 17:11:23 GMT on August 2, 1971 (01:11:23 p.m. EDT), the ascent stage engine ignited, signaling the start of the journey back to lunar orbit. By this time, the LM had spent 66 hours, 54 minutes, and 53.9 seconds on the lunar surface.

The engine firing lasted 431.0 seconds, achieving an initial lunar orbit of 42.5 by 9.0 nautical miles (approximately 78.7 by 16.7 kilometers). Several rendezvous maneuvers were then required to align the ascent stage's orbit with the Command and Service Module (CSM) Endeavour orbiting the Moon. A terminal phase initiate maneuver lasting 2.6 seconds at 172:29:40.0 adjusted the ascent stage's orbit to 64.4 by 38.7 nautical miles (approximately 119.3 by 71.7 kilometers).

Docking between the ascent stage and the CSM occurred at 173:36:25.5, with the two spacecraft reuniting at 57.0 nautical miles (approximately 105.6 kilometers). Since the LM first separated from the CSM, they had been undocked for 72 hours, 57 minutes, and 9.3 seconds.

Following a successful transfer of crew members and collected lunar samples from the LM to the CSM, the ascent stage was jettisoned at 179:30:01.4. This step was delayed by one revolution due to difficulties in verifying the spacecraft tunnel sealing and ensuring astronaut pressure suit integrity. Once jettisoned, the LM was intentionally crashed into the lunar surface, its impact recorded by seismometers left behind by previous Apollo missions (Apollo 12, 14, and 15).

After these intense and critical maneuvers, Flight Director Deke Slayton suggested that the astronauts take sleeping pills to rest before returning to Earth. However, Mission Commander David Scott declined, believing it unnecessary. During their extravehicular activities (EVAs), medical personnel had noted irregularities in both Scott's and Jim Irwin's heartbeats. However, this information was withheld from the crew until after their return to Earth. Irwin later experienced heart problems and passed away in 1991 from a heart attack, prompting Scott to reflect on the need for transparency regarding biomedical readings during the mission.

NASA doctors at the time hypothesized that these irregular heart readings might have been related to potassium deficiency, stemming from the astronauts' strenuous activities on the lunar surface and inadequate replenishment of electrolytes.

After concluding their lunar operations, the crew of Apollo 15 focused on conducting scientific experiments while in lunar orbit. This included further detailed observations of the Moon and the deployment of a subsatellite from the Command and Service Module (CSM) Endeavour's Scientific Instrument Module (SIM) bay. On August 4, Endeavour departed lunar orbit with a precise burn of the Service Propulsion System (SPS) engine lasting 2 minutes and 21 seconds at 21:22:45 GMT.

The following day, during their journey back to Earth, Alfred Worden conducted a historic Extravehicular Activity (EVA). Lasting 39 minutes, this "deep space" EVA took place approximately 171,000 nautical miles (197,000 miles or 317,000 kilometers) from Earth, making it the first of its kind in history. Worden ventured out into the vacuum of space to retrieve film cassettes from Endeavour's SIM bay, assisted by Jim Irwin who remained at the command module's hatch. This milestone EVA remains one of only three conducted during the Apollo J missions under similar deep space conditions. Later that day, the crew also set a record for the longest Apollo flight up to that point.

Meanwhile, the ascent stage of the Lunar Module was deliberately maneuvered to impact the lunar surface. Using its remaining fuel, the engine fired until depletion, leading to impact at 03:03:37 GMT on August 3 (11:03:37 p.m. EDT on August 2). The impact site, located at latitude 26° 21' north and longitude 0° 15' east, was approximately 12.7 nautical miles (23.5 kilometers) from the intended target and 50 nautical miles (93 kilometers) west of the Apollo 15 landing site. Seismic stations left behind by previous missions (Apollo 12, 14, and 15) recorded the event, providing valuable scientific data.

In preparation for deploying the subsatellite into lunar orbit, a precise orbit-shaping maneuver adjusted Endeavour's path at 221:20:48.02. The subsatellite was successfully ejected from the SIM bay at 222:39:29.1 during the 74th revolution, entering an orbit of 76.3 by 55.1 nautical miles (141.2 by 102.1 kilometers) at an inclination of -28.7°. Equipped with instruments to measure plasma and energetic-particle fluxes, vector magnetic fields, and lunar gravitational anomalies, the subsatellite operated flawlessly, providing crucial data throughout its mission.

Following these operations, a final 140.90-second maneuver achieved transearth injection at 223:51:06.74. This critical maneuver propelled the spacecraft back toward Earth at a velocity of 8,272.4 feet per second, marking the end of 74 lunar orbits totaling 145 hours, 12 minutes, and 41.68 seconds.

Irwin with the Lunar Roving Vehicle on the Moon. Mons Hadley is in the background.

During the transearth phase of the mission at 241:57:12, Command Module Pilot Al Worden conducted a crucial Extravehicular Activity (EVA). This 39-minute, 0-second EVA was broadcast live on television, showcasing Worden's retrieval of film cassettes from the Scientific Instrument Module (SIM) bay onboard the Command and Service Module (CSM). Throughout the EVA, Worden made three excursions to the bay. The first and second trips were dedicated to retrieving the panoramic and mapping camera film cassettes. On his third trip, Worden inspected the overall condition of the instruments, focusing particularly on the mapping camera.

The Moon as seen from the departing Apollo 15 spacecraft

During his inspection, Worden reported that he found no evidence to explain the failure of the mapping camera's extend/retract mechanism, which was stuck in the extended position. He also observed no visible cause for the failure of the pan camera's velocity/altitude sensor. Additionally, Worden noted that the mass spectrometer boom was not fully retracted, indicating a minor operational issue.

The transearth EVA concluded at 242:36:19, marking the end of a total extravehicular activity time for the mission of 19 hours, 46 minutes, and 59 seconds. The data collected during this EVA, including seismometer readings, were carefully analyzed at Mission Control to ensure all mission objectives were met.

To fine-tune the spacecraft's trajectory for Earth reentry, a 22.30-second midcourse correction was performed at 291:56:49.91. This adjustment, achieving a velocity change of 5.6 feet per second, ensured that the CSM remained on a precise course for its return to Earth.

As the Apollo 15 mission approached Earth on August 7, the service module was jettisoned at 294:43:55.2, marking the start of the final phase. Following a normal reentry profile, the command module (CM) entered Earth's atmosphere at 294:58:54.7, descending from an altitude of 400,000 feet at a velocity of 36,096 feet per second. This reentry followed a transearth coast lasting 71 hours, 7 minutes, and 48.0 seconds.

Upon atmospheric entry, the parachute system deployed, with two main parachutes inflating properly and one collapsing, likely due to damage sustained as the spacecraft vented fuel. Despite this, the two functioning parachutes were sufficient to ensure a safe splashdown of the CM in the North Pacific Ocean at 20:45:53 GMT (04:45:53 p.m. EDT) on August 7. The impact point was approximately 1.0 nautical mile from the target and 5 nautical miles from the recovery ship, USS Okinawa.

The collapsed parachute contributed to one of the fastest entry-to-splashdown times in the Apollo program, just 778.3 seconds. The splashdown site was estimated to be near latitude 26.13° north and longitude 158.13° west. Immediately after splashdown, the CM assumed an apex-up flotation attitude. The crew was swiftly retrieved by helicopter, boarding the recovery ship within 39 minutes. The CM itself was recovered 55 minutes after splashdown.

At the time of splashdown, the estimated weight of the CM was 11,731 pounds, and the total distance traveled during the mission was approximately 1,107,945 nautical miles. Despite the parachute anomaly, all crew members emerged unharmed from the successful conclusion of their 12-day, 7-hour, 11-minute, and 53-second mission aboard Apollo 15.

The astronauts disembark their helicopter aboard the Okinawa

Apollo 15 stands as a pivotal achievement in the annals of space exploration. It marked the fourth successful lunar landing and a triumph of scientific endeavor. The mission achieved its primary objectives and yielded an unprecedented wealth of new knowledge about the Moon and its characteristics.

The Apollo system was central to the mission's success, which evolved beyond a mere transport vehicle to become a sophisticated scientific platform. Equipped with enhanced consumables and specialized scientific instruments, the Command and Service Module (CSM) was crucial in gathering crucial data. This capability enabled real-time collaboration between astronauts and ground-based scientists, facilitating informed decision-making and maximizing scientific returns.

Apollo 15 validated numerous advancements and modifications to the Apollo hardware. Enhanced configurations of the launch vehicle, spacecraft, and life support systems enabled the safe transport of larger payloads and extended lunar surface operations. Innovations such as modified pressure garments and portable life support systems significantly improved astronaut mobility and prolonged extravehicular activities on the lunar surface.

Key technological innovations, including the ground-controlled mobile television camera and the Lunar Roving Vehicle (LRV), expanded the mission's operational scope. The LRV notably increased exploration range and payload capacity, while a lunar communications relay unit ensured uninterrupted communication over extended distances.

The mission also highlighted the evolving role of astronauts as scientific observers and investigators. Apollo 15 demonstrated the crew's adaptability and innovation, enhancing scientific data collection through on-the-spot task adjustments and creative problem-solving. This dynamic underscored the irreplaceable value of human presence in space exploration, affirming the unique capabilities of human observation and cognitive flexibility.

Moreover, Apollo 15 reinforced the importance of rigorous training with flight-grade equipment, proving critical in mission preparedness and success. These insights were encapsulated in groundbreaking experiments, including efforts to cultivate plants in lunar soil—a pioneering step toward understanding extraterrestrial agriculture.

Scientific exploration formed the cornerstone of Apollo 15's ambitious mission profile, spanning various disciplines crucial to advancing our understanding of the Moon and beyond. The mission's scientific experiments, from the Passive Seismic Experiment to investigations into solar wind composition and soil mechanics, generated essential data across fields ranging from planetary science to astrophysics. Of particular note was the deployment of a Laser Ranging Retroreflector, a pivotal advancement that enabled precise measurements of lunar distance and set the stage for future scientific endeavors.

In addition to its scientific objectives, Apollo 15 undertook critical operational tests that underscored its dual role in advancing civil and defense-related aerospace capabilities. These tests included classified Department of Defense experiments and trials evaluating lunar module operations. They encompassed a wide spectrum of assessments, from acoustic measurements to radar tracking, highlighting the mission's broader implications for national defense and technological innovation.

Furthermore, the mission's launch vehicle objectives were executed flawlessly, ensuring the precise insertion of the spacecraft into Earth orbit and subsequent trajectory toward the Moon. The deliberate impact of the S-IVB stage on the lunar surface validated operational procedures and provided valuable insights into lunar dynamics and surface interactions.

Despite Apollo 15's significant scientific achievements, controversy overshadowed the careers of its crew due to an unauthorized deal involving postage covers taken to the Moon. Before the mission, astronauts David Scott, Alfred Worden, and James Irwin had arranged, through an intermediary, to carry around 400 postal covers to the lunar surface. This arrangement was made with West German stamp dealer Hermann Sieger, with each astronaut expecting compensation of about $7,000 intended for their children's future.

During the mission, Scott discreetly placed the covers inside the Lunar Module Falcon, where they remained throughout the astronauts' activities on the Moon. Upon their return to Earth, 100 of these covers were handed over to the intermediary, Walter Eiermann, who facilitated their transfer to Sieger, earning a commission.

The controversy arose because this arrangement was not approved by NASA management, specifically Flight Director Deke Slayton, as required by NASA policy. The unauthorized inclusion of the postal covers aboard the spacecraft raised ethical concerns about commercializing space missions and adherence to established protocols.

After the unauthorized postal covers aboard Apollo 15 were discovered, a series of events impacted the careers of Scott, Worden, and Irwin. The covers, intended for sale to collectors through Sieger, were priced at approximately $1,500 each and sold in late 1971. Upon receiving the payments as agreed, the astronauts returned the covers and declined any compensation.

In April 1972, Deke Slayton, then Director of Flight Crew Operations, learned about the unauthorized items aboard Apollo 15. Consequently, Scott, Worden, and Irwin were removed from their roles as the backup crew for Apollo 17. The matter became public in June 1972, resulting in a public reprimand for the astronauts for what was seen as a breach of trust. Importantly, this incident effectively ended their careers as active astronauts, with none flying in space again.

During the subsequent investigation, the astronauts surrendered any remaining covers. Alfred Worden later filed a lawsuit related to the situation. Ultimately, in 1983, the covers were returned to them, which some interpreted as a vindication of their actions, including Slate magazine.

The affair surrounding the postal covers of Apollo 15 highlighted the ethical standards and strict protocols maintained by NASA, emphasizing the consequences of disregarding these regulations. Despite their contributions to the mission's success, the incident left a lasting stain on the reputations and careers of the astronauts involved, marking a controversial chapter in the history of lunar exploration and astronaut conduct.

During the Apollo 15 mission, Commander David Scott left a poignant tribute on the lunar surface—a small statuette known as the Fallen Astronaut. Crafted by Belgian artist Paul Van Hoeydonck, the statuette was intended as a quiet homage to commemorate those who had lost their lives in pursuing space exploration. However, what began as a gesture of remembrance soon became entangled in a web of controversy and conflicting expectations.

Scott's initial arrangement with Van Hoeydonck was straightforward: a discreet memorial with no public replicas beyond one for display at the National Air and Space Museum. Yet, misunderstandings arose. Van Hoeydonck, viewing the statuette as a tribute to human achievement in space, expected recognition and sought to offer signed replicas for sale, a prospect that clashed with NASA's stringent policies against commercializing space missions.

The discord intensified when Scott, in a 2021 document titled "Memorandum for the Record," asserted that the figurine was, in fact, created by NASA personnel—a stark departure from his earlier testimony attributing its creation to Van Hoeydonck. This revelation added layers to the controversy surrounding the Fallen Astronaut, complicating its narrative within the annals of space history.

Beyond the statuette, another artifact from Apollo 15 garnered attention and scrutiny: a Bulova Chronograph watch worn by Scott on the lunar surface. This watch, a prototype gifted to Scott by Bulova, became the only privately owned timepiece ever to grace the Moon. Its appearance during Scott's third Extravehicular Activity, where he famously saluted the American flag against the backdrop of Hadley Delta, marked a unique intersection of private enterprise and historical exploration.

The presence of the Bulova watch on the lunar surface was not without its own controversy. Without disclosing his decision to NASA officials, Scott had opted to bring the Bulova timepieces alongside the NASA-issued Omega Speedmaster. When a crystal on the Omega malfunctioned during his second EVA, Scott switched to the Bulova, making it an inadvertent participant in one of humanity's most significant achievements.

In 2015, the Bulova watch achieved further renown when it fetched $1.625 million at auction, underscoring its status as a priceless relic of space exploration and one of the most valuable astronaut-owned artifacts ever sold. This auction highlighted the enduring allure of space memorabilia and cemented the Bulova watch's place among the most coveted and expensive timepieces in history.

Thus, the legacies of the Fallen Astronaut statuette and the Bulova watch from Apollo 15 continue to resonate, not merely as relics of a historic mission but as symbols of the complex intersections between art, commerce, and exploration in the realm beyond Earth.

To highlight the significance of Apollo 15 in the Cold War space race, on March 22, 1972, the Soviet Union established two new groups crucial for advancing its space program: Civilian Specialist Group 4 and Medical Group 3. These selections underscored the Soviet commitment to expanding its capabilities in specialized technical fields and medical expertise essential for sustaining prolonged space missions.

Civilian Specialist Group 4 included Boris Andreyev, Valentin Lebedev, Georgi Machinski, Valeri Polyakov, Lev Smirenny, and others. They brought specialized skills and knowledge aimed at enhancing the scientific and operational aspects of Soviet space missions. Their contributions were pivotal in advancing technological innovations and ensuring the success of complex space endeavors.

Simultaneously, Medical Group 3 comprised professionals dedicated to safeguarding cosmonaut health and well-being during space missions. Their expertise in space medicine and physiological research played a critical role in understanding the effects of spaceflight on the human body and developing effective countermeasures.

These developments in 1972 marked significant milestones in the Soviet Union's ongoing efforts to maintain leadership in space exploration. They emphasized the integration of diverse expertise to overcome the challenges of long-duration missions and further expand the frontiers of human spaceflight.

Apollo 16

The Tenth Crewed Mission

Apollo 16, launched on April 16, 1972, marked a pivotal moment in the United States' Apollo space program. It represented the tenth crewed mission and the fifth to achieve a successful lunar landing. This mission, part of the "J missions" series, was meticulously planned to extend human exploration and scientific investigation on the Moon. It flew 1,391,550 miles.

The prime crew of the Apollo 16 lunar landing mission. From left to right: Thomas K. Mattingly II, Command Module pilot; John W. Young, Commander; and Charles M. Duke Jr., Lunar Module pilot.

Under astronauts John Young, Ken Mattingly, and Charles Duke, Apollo 16 aimed to conduct an intensive study of the lunar surface, building upon the achievements of its predecessors. Equipped with advanced instrumentation and the groundbreaking Lunar Roving Vehicle (LRV), the mission was designed for enhanced mobility and scientific productivity.

The primary objectives of Apollo 16 were comprehensive: to meticulously inspect, survey, and collect samples from the Descartes region—a site selected for its geological diversity and scientific significance. This mission was not merely about landing on the Moon but extending human understanding of lunar geology and conducting ambitious experiments.

In addition to surface activities, Apollo 16 encompassed crucial in-flight experiments and extensive photographic documentation. From lunar orbit, the crew captured detailed imagery and data that would contribute significantly to our understanding of the Moon's composition and history.

Apollo 16 was crewed by three remarkable individuals who each brought unique experience and expertise to the mission. Captain John Watts Young of the US Navy was at the helm, serving as the mission commander. Born on September 24, 1930, in San Francisco, California, Young was 41 years old during Apollo 16. His illustrious career began in 1962 when he was selected as part of NASA's second astronaut group. Young's spaceflight resume was impressive; he flew aboard Gemini 3 in 1965 with Gus Grissom, making him the first astronaut from the second group to venture into space. He later piloted Gemini 10 in 1966 alongside Michael Collins. He served as the command module pilot for Apollo 10 in 1969—a mission crucial for testing the lunar module in lunar orbit ahead of the historic Apollo 11 landing. With Apollo 16, Young became only the second American, after Jim Lovell, to venture into space four times.

Joining Young was Lieutenant Commander Thomas Kenneth "Ken" Mattingly II, also of the US Navy, who served as the command module pilot. Born on March 17, 1936, in Chicago, Illinois, Mattingly was 36 years old during Apollo 16. Selected as an astronaut in 1966 as part of NASA's fifth astronaut group, Mattingly was initially designated to fly on Apollo 13 but was replaced just days before launch due to exposure to rubella. Despite never contracting the illness, his removal was precautionary. Mattingly had previous experience as a member of the support crews for Apollo 8 and Apollo 9.

Lieutenant Colonel Charles Moss Duke, Jr. of the US Air Force, who served as the lunar module pilot, completed the crew. Born on October 3, 1935, in Charlotte, North Carolina, Duke was also 36 years old during Apollo 16. Like Mattingly, Duke was selected as part of NASA's fifth astronaut group in 1966. Before Apollo 16, Duke contributed to the support crew for Apollo 10 and served as a capsule communicator (CAPCOM) for Apollo 11.

Apollo 16 was supported by a comprehensive team of professionals crucial to its success, extending beyond the prime and backup crews to CAPCOMs and flight directors who played vital roles in mission control.

The CAPCOMs, or capsule communicators, were the vital link between the astronauts in space and mission control on Earth. For Apollo 16, the CAPCOM team included Major Donald Herod Peterson, Major Charles Gordon Fullerton, Colonel James Benson Irwin, Fred Wallace Haise Jr., Stuart Allen Roosa, Edgar Dean Mitchell, Major Henry Warren Hartsfield Jr., and Anthony Wayne "Tony" England, Ph.D. These individuals provided essential guidance, relayed instructions, and facilitated communications during critical phases of the mission.

Backing up the prime crew were experienced astronauts Fred W. Haise Jr., Stuart A. Roosa, and Edgar D. Mitchell, forming the backup crew. Planned initially with different assignments before the cancellation of Apollo 18 and 19, these astronauts brought valuable experience from previous lunar missions, ensuring readiness to step in if needed.

The flight directors oversaw operations from mission control and made real-time decisions to ensure crew safety and mission success. The first shift was led by M.P. "Pete" Frank and Philip C. Shaffer, the second by Eugene F. Kranz and Donald R. Puddy, and the third by Gerald D. Griffin, Neil B. Hutchinson, and Charles R. Lewis. Their collective expertise and leadership were essential in coordinating the complex activities of Apollo 16, from launch through lunar exploration and return.

Additionally, the support crew, an innovation in the Apollo program, included Anthony W. England, Karl G. Henize, Henry W. Hartsfield Jr., Robert F. Overmyer, and Donald H. Peterson. These members played crucial roles in mission planning, rule development, and checklist management, ensuring that all operational aspects were meticulously prepared and executed.

Apollo 16's landing site selection marked a strategic shift toward exploring the lunar highlands, a region of immense scientific interest due to its potential to reveal the Moon's early geological history. As the second mission of the Apollo "J missions," which aimed to enhance scientific capability and exploration duration, Apollo 16 focused on expanding our understanding of the Moon's surface characteristics.

Unlike earlier missions that primarily explored lunar mare regions, Apollo 16 aimed to investigate the highlands untouched by previous landings. The mission aimed to collect samples from the Descartes region, specifically chosen for its geological diversity and the presence of ancient lunar material. Scientists hoped these samples would provide insights into the Moon's early formation, predating the volcanic activity that filled its low-lying areas with lava.

Previous Apollo missions, such as Apollo 14 and Apollo 15, had gathered lunar material from pre-mare regions, likely ejected by meteorite impacts. These samples offered valuable clues about the Moon's composition before volcanic activity altered its surface. However, direct exploration of the highlands remained a critical gap in lunar exploration.

Apollo 16 and its successor, Apollo 17, thus represented an opportunity to address these scientific uncertainties. By exploring the highlands and conducting detailed geological surveys, the mission aimed to advance our knowledge of lunar evolution and contribute significantly to our understanding of planetary geology.

Apollo 14 and Apollo 15 had focused their explorations on regions near the Mare Imbrium impact basin, providing valuable insights into the geological processes associated with these areas. However, scientists recognized that the lunar highlands farther from Mare Imbrium remained largely unexplored, presenting an opportunity to investigate different geological environments.

Scientist Dan Milton identified an intriguing area in the Descartes region of the Moon through analysis of Lunar Orbiter photographs. This area exhibited a notably high albedo, suggesting the presence of volcanic rock—a hypothesis that garnered widespread support within the scientific community. Researchers speculated that these lunar highlands might harbor geological formations similar to terrestrial volcanic regions, prompting excitement about the potential discoveries Apollo 16 could bring.

While some scientists advocated for a landing near the prominent Tycho crater, known for its distinct features and geological interest, logistical challenges ruled out this option. Tycho's location far from the lunar equator and its rough terrain significantly hindered the Lunar Module's approach and safe landing.

Ultimately, the decision to target the Descartes region for Apollo 16's landing site reflected a strategic choice to explore an area with unique geological characteristics and the potential for significant scientific discoveries. By venturing to this relatively uncharted territory on the Moon, NASA aimed to broaden our understanding of lunar geology and uncover clues about the Moon's early history that could shed light on planetary formation processes both on Earth and beyond.

The selection process for Apollo 16's landing site involved meticulous deliberation by the Ad Hoc Apollo Site Evaluation Committee, chaired by Noel Hinners of Bellcomm, during meetings held in April and May 1971. The committee's consensus emphasized the importance of exploring the lunar highlands, distinct from previous landing sites, to expand scientific knowledge.

Among the sites considered, the Descartes Highlands west of Mare Nectaris emerged as a promising candidate. This region was noted for its geological diversity and the potential to yield insights into the Moon's early history, untouched by the volcanic activity that characterized mare regions explored in earlier missions.

Astronauts John Young and Charles Duke participate in geology training at the Rio Grande Gorge near Taos, New Mexico in September 1971.

Another contender was the crater Alphonsus, which presented specific scientific objectives of great interest. Geologists identified three primary objectives: investigating the possibility of ancient, pre-Imbrium impact material preserved within the crater's walls, analyzing the composition of the crater's interior, and exploring potential evidence of past volcanic activity within smaller "dark halo" craters on its floor.

However, concerns arose about potential contamination of samples within Alphonsus by the Imbrium impact, which could complicate efforts to obtain pristine pre-Imbrium material. Given that analyses of Apollo 14 samples were ongoing and Apollo 15 had not yet provided definitive data, there was uncertainty about the necessity of revisiting these objectives.

Ultimately, the Descartes region was chosen for Apollo 16 due to its geological significance and the strategic advantage it offered for deploying a network of seismometers. This decision underscored NASA's commitment to advancing scientific exploration of the Moon's diverse terrains and deepening our understanding of its geological evolution.

Following the decision of the site selection committee on June 3, 1971, Apollo 16 was slated to land at the Descartes site on the Moon's surface. This site was chosen based on orbital photography obtained during the Apollo 14 mission, confirming its suitability for a crewed landing. The specific landing location was between two relatively young impact craters, North Ray and South Ray, measuring 1,000 meters (3,280 feet) and 680 meters (2,230 feet) in diameter, respectively.

One key advantage of the Descartes landing site was the presence of these impact craters, which acted as natural drill holes penetrating through the lunar regolith (soil). This geological feature exposed bedrock, allowing the Apollo 16 crew to sample and study materials from beneath the lunar surface.

Location of the Apollo 16 landing site

After the site was chosen, mission planners focused the scientific objectives on exploring the Descartes and Cayley formations, two geological units within the lunar highlands. These formations were of significant interest to the scientific community, as they were initially suspected to have formed due to lunar volcanism. However, findings from lunar samples collected during the mission would ultimately disprove this hypothesis.

Apollo 16's exploration of the Descartes site represented a critical phase in lunar exploration, contributing valuable data that refined our understanding of the Moon's geological history. The mission aimed to uncover new insights into lunar evolution and provide clues about planetary formation processes, marking another milestone in NASA's Apollo program.

In preparation for their mission, John Young and Charlie Duke, along with backup commander Fred Haise, underwent an intensive geological training program in addition to the standard Apollo spacecraft training. This specialized training was crucial for their role in analyzing lunar features and collecting samples while exploring the Descartes region.

The geological training program included multiple field trips designed to familiarize the astronauts with geological concepts and techniques applicable to lunar surface exploration. These trips were instrumental in helping them understand the unique geological characteristics they would encounter on the Moon. During these excursions, Young, Duke, and Haise provided scientific descriptions and analyses of various geological formations, simulating the conditions they would face on the lunar surface.

Initially, backup Lunar Module Pilot Edgar Mitchell was occupied with responsibilities related to Apollo 14 and could not participate in the early stages of the geological training. However, by September 1971, Mitchell joined the field trips, enriching his understanding of lunar geology alongside the primary and backup crew members.

During Mitchell's absence, Tony England, a member of the support crew and the Lunar EVA CAPCOM (Capsule Communicator), or other geologist trainers, filled in to ensure that backup commander Fred Haise remained well-versed in the geological aspects of the mission.

Young (right) and Duke training to drive the Lunar Roving Vehicle

During the preparatory phases of the Apollo missions, extensive geological training was crucial for the astronauts slated to explore the Moon's surface. Initially focused on volcanic terrain, the training pivoted when it became clear that the planned landing site, Descartes, was not volcanic as previously thought. Instead, the training encompassed diverse geological formations to prepare astronauts for the unexpected.

In July 1971, a significant training session occurred in Sudbury, Ontario, Canada, marking the first time American astronauts trained in Canada. The site was chosen for its remarkable Sudbury Basin, a 97 km wide crater formed nearly 1.8 billion years ago by a massive meteorite impact. This crater provided a unique opportunity to study shatter cone geology, a critical analog for lunar surface features.

During these exercises, astronauts engaged in fieldwork without space suits, focusing on radio equipment communication procedures. This simulated lunar surface operations, allowing them to practice coordination and scientific protocols under conditions akin to those they would face in space.

As the training progressed, these field trips evolved into complex exercises involving up to eight astronauts and numerous support personnel. Media coverage underscored the importance of these simulations, highlighting their rigorous nature and the strategic importance of the locations chosen. For instance, exercises at the Nevada Test Site simulated lunar craters with unprecedented accuracy, requiring security clearances and special permissions for participants.

In addition to their rigorous field geology training, astronauts Young and Duke undertook comprehensive preparations for their Apollo 16 mission, focusing on essential tasks such as adapting to lunar gravity, mastering EVA procedures in their space suits, and practicing the operation of the Lunar Roving Vehicle (LRV). Their prior assignment as backups for Apollo 13, originally slated for a lunar landing, afforded them approximately 40 percent of their training time dedicated to surface operations, ensuring thorough readiness for their upcoming mission.

Integral to their preparation was the study of lunar samples brought back by earlier missions, which provided crucial insights into the geological compositions and formations they might encounter. This knowledge proved invaluable when, upon landing, Young and Duke quickly discerned that the expected volcanic rocks were absent, contrary to initial expectations from Mission Control.

A significant portion of their training, approximately 350 hours, involved practicing tasks in their space suits. Young deemed this hands-on experience essential, enabling them to understand the equipment's capabilities and limitations in executing their assigned duties on the lunar surface.

Meanwhile, astronaut Mattingly underwent specialized training for his role in lunar orbit, including recognizing geological features from spacecraft and operating the Scientific Instrument Module. This training included flying over field areas in aircraft to simulate orbital observations, ensuring he was well-prepared to support the mission's scientific objectives from above.

These rigorous training efforts collectively equipped the Apollo 16 crew with the knowledge, skills, and adaptability necessary to effectively carry out their mission objectives on the Moon, despite encountering unexpected geological challenges.

Apollo 16 embarked on its journey to the Moon aboard a Saturn V rocket designated AS-511, marking the eleventh flight of this powerful launch vehicle and the ninth to carry a crewed mission. This particular Saturn V closely resembled the one used for Apollo 15, with a notable modification involving the reinstatement of four retrorockets to the S-IC first stage. This adjustment brought the total number of retrorockets to eight, a configuration seen in earlier Apollo missions like Apollo 14.

The decision to reintroduce these retrorockets was prompted by lessons learned from Apollo 15's flight. Initially omitted to reduce weight, their absence led to the unexpected closeness between the jettisoned S-IC stage and the ascending Saturn V. This raised concerns that relying solely on four retrorockets, with the risk of potential failure, could result in a collision scenario. Therefore, adding these retrorockets for Apollo 16 was a precautionary measure aimed at ensuring a safer separation and ascent trajectory for the Saturn V, thereby minimizing risks during critical phases of the mission.

Apollo 16's launch vehicle by the VAB, January 27, 1972

The preparations for Apollo 16 at Kennedy Space Center commenced in July 1970, with all spacecraft and launch vehicle components arriving by September 1971. Initially scheduled for launch on March 17, 1972, the mission faced delays due to several technical issues. These included a bladder failure in the Command Module's reaction control system during testing, concerns about the reliability of explosive cords for the Lunar Module's separation from the Command Module after lunar surface operations, and a problem with astronaut Duke's spacesuit.

In light of these issues, NASA postponed the launch to the next available window, setting a new date for April 16, 1972. The Saturn V launch vehicle initially rolled out to Launch Complex 39A from the Vehicle Assembly Building on December 13, 1971, was temporarily returned for adjustments on January 27, 1972, and was again successfully rolled out on February 9.

The official countdown for the April 16 launch began on Monday, April 10, 1972, six days before liftoff. During this phase, all three stages of the Saturn V rocket were powered up, and preparations, such as loading drinking water into the spacecraft, were completed. Meanwhile, the crew of Apollo 16 engaged in final training exercises, culminating in their last preflight physical examination on April 11.

Apollo 16 lift-off

As the countdown for Apollo 16 progressed smoothly, scheduled holds were executed without incident, ensuring meticulous readiness for the historic launch. The mission's Saturn V rocket, designated SA-511 and known as Eastern Test Range #1601, stood poised at Cape Kennedy's Launch Complex 39A.

The Command and Service Module (CSM-113), named "Casper," and the Lunar Module (LM-11), known as "Orion," were integral parts of this ambitious mission. The terminal countdown resumed at T-28 hours on April 15, 1972, with planned holds at T-9 hours for 9 hours and another hold at T-3 hours 30 minutes for one hour.

On the launch day, Cape Kennedy enjoyed fair weather conditions under a ridge of high pressure extending from the Atlantic Ocean across central Florida. Cumulus clouds covered 20 percent of the sky with a base at 3,000 feet. The temperature stood at 88.2°F with a relative humidity of 44 percent, and the barometric pressure measured 14.769 lb/in^2. Winds were measured at 12.2 knots at 269° from true north at 60.0 feet above ground level and 9.9 knots at 256° from true north at 530 feet above ground level.

On April 16, 1972, at 12:54 pm EST, Apollo 16 launched flawlessly from Kennedy Space Center's Launch Complex 39, Pad A. The Saturn V rocket, designated SA-511, initiated its ascent with precision, marking the beginning of a crucial journey to the Moon.

During the ascent phase, the rocket executed a series of critical maneuvers according to plan. Shortly after liftoff, between 12.7 and 31.8 seconds into the flight, the vehicle rolled from 90° on the launch pad to a flight azimuth of 72.034°. The first stage (S-IC) engine performed nominally until shutdown at 2 minutes and 41.78 seconds into the flight, followed by the separation of the S-IC and ignition of the second stage (S-II) engine.

The S-II stage operated as expected until its shutdown at 9 minutes and 19.54 seconds, initiating separation from the Saturn V Instrument Unit (S-IVB). The S-IVB stage ignited at 9 minutes and 23.60 seconds and executed its first engine cutoff at 11 minutes and 46.21 seconds, with minor deviations in velocity and exact altitude conforming closely to the mission plan.

During ascent, the maximum encountered wind conditions were 50.7 knots at 257° at 38,880 feet, with a maximum wind shear of 0.0095 sec^{-1} at 44,780 feet. These conditions posed no significant issues, ensuring a stable trajectory toward Earth orbit.

The spacecraft's parameters were meticulously recorded upon insertion into a parking orbit at 11 minutes and 56.21 seconds after launch. These included an apogee and perigee of 90.7 by 90.0 nautical miles, an inclination of 32.542°, a period of 87.84 minutes, and a velocity of 25,605.0 feet per second.

After achieving Earth orbit following a successful launch, the crew of Apollo 16 acclimated to the weightless environment and focused on readying their spacecraft for the critical trans-lunar injection (TLI) maneuver. This burn of the third-stage rocket, the S-IVB, would propel them out of Earth's orbit and on a trajectory toward the Moon.

While in Earth orbit, the crew encountered minor technical challenges, including issues with the environmental control system and the S-IVB's attitude control system. However, they resolved or mitigated these concerns through troubleshooting and adjustments, ensuring the spacecraft was prepared for its journey beyond Earth.

The international designations for the various mission components were assigned upon achieving orbit: the Command and Service Module (CSM) was designated 1972-031A, and the S-IVB stage was designated 1972-031B. These designations are used to track and identify objects in space as part of international agreements and conventions.

As preparations continued, the crew focused on their mission objectives, knowing that upon undocking at the Moon, further designations would mark the Lunar Module's ascent stage (1972-031C), descent stage (1972-031E), and the particles and fields subsatellite (1972-031D). Each designation plays a crucial role in tracking and managing the components and activities of the Apollo 16 mission, ensuring clarity and coordination in its execution.

The Apollo 16 Command and Service Module (CSM) "Casper" approaches the Lunar Module (LM). The two spacecraft were about to make their final rendezvous of the mission, on April 23, 1972. Astronauts John W. Young and Charles M. Duke Jr., aboard the LM, were returning to the CSM in lunar orbit after three successful days on the lunar surface. Astronaut Thomas K. Mattingly II was in the CSM.

During the translunar phase of the Apollo 16 mission, the crew proceeded through several critical maneuvers and observations as they journeyed toward the Moon. After completing in-flight systems checks, the spacecraft executed a precise 341.92-second translunar injection maneuver, also known as the second S-NB firing, at 002:33:36.50 into the flight. This maneuver accelerated the spacecraft to a velocity of 35,589.9 feet per second, marking their departure from Earth's orbit after completing 1.5 orbits lasting 2 hours 37 minutes 32.21 seconds.

NASA officials gathered around a console in the Mission Operations Control Room (MOCR) in the Mission Control Center (MCC) before making a decision on whether to land Apollo 16 on the moon or to abort the landing. They included Dr. Christopher C. Kraft Jr., Manned Spacecraft Center (MSC) Director, and Brig. Gen. James A. McDivitt (USAF), Manager, Apollo Spacecraft Program Office, MSC, Dr. Rocco A. Petrone, Apollo Program Director, Office Manned Space Flight (OMSF), NASA HQ.; Capt. John K. Holcomb (U.S. Navy, Ret.), Director of Apollo Operations, OMSF; Sigurd A. Sjoberg, Deputy Director, MSC; Capt. Chester M. Lee (U.S. Navy, Ret.), Apollo Mission Director, OMSF; Dale D. Myers, NASA Associate Administrator for Manned Space Flight; and Dr. George M. Low, NASA Deputy Administrator.

At 003:04:59.0, the Command and Service Module (CSM) separated from the S-NB stage, underwent transposition, and docked again at 003:21:53.4. This maneuver allowed the crew to transition into a new configuration for their journey, during which they transmitted color television for 18 minutes to Earth. The docked spacecraft then separated from the S-NB at 003:59:15.1, followed by an 80.2-second separation maneuver at 004:18:08.3 to finalize their trajectory toward the Moon.

Shortly after these maneuvers, the S-IVB rocket's third stage reignited for over five minutes, propelling the spacecraft toward its lunar destination at about 35,000 kilometers per hour (22,000 miles per hour). Six minutes after this burn, the CSM, housing the crew, separated from the S-IVB stage. They moved approximately 49 feet away before executing a turn to retrieve the Lunar Module (LM) from inside the expended rocket stage.

The view of Earth taken during the Translunar Coast from the distance of about 25.000 to 30.000 nautical miles from Earth. South America is to the left. Central America is in the upper center. The gulf of Mexico and Florida is seen to the lower right of Central America. North America is in the upper left. The north-west part of Africa is seen in the upper right.

NASA officials conferring on whether to allow the Apollo 16 landing, April 20, 1972

During this process, the crew noticed particles emanating from a section of the LM's exterior where the skin appeared torn or shredded. Duke estimated they were observing between five to ten particles per second. To investigate, Young and Duke entered the LM through the docking tunnel, connected it to the command module, and inspected its systems thoroughly. Fortunately, they identified no major issues then, ensuring confidence in their continued journey toward lunar orbit.

After setting their course toward the Moon, the crew of Apollo 16 initiated a crucial maneuver known as "barbecue mode." In this mode, the spacecraft rotated along its long axis three times per hour, a method designed to evenly distribute heat from the Sun around the spacecraft. This ensured that no single side of the spacecraft would overheat or freeze in the extremes of space.

With the spacecraft stabilized and systems in order, the crew proceeded with further preparations for their voyage. Approximately 15 hours after launch, they began their first sleep period of the mission, vital for maintaining their alertness and performance during the demanding journey ahead.

On the second day of the flight, Mission Control woke the crew. At the same time, the spacecraft was approximately 181,000 kilometers (98,000 nautical miles) from Earth, traveling at a speed of about 1.622 kilometers per second (5,322 feet per second). Despite reaching lunar orbit on the fourth day of flight, days two and three were dedicated to critical spacecraft maintenance and scientific experiments.

One significant experiment was an electrophoresis study previously performed during the Apollo 14 mission. This experiment aimed to demonstrate whether electrophoretic separation in a near-weightless environment could produce purer substances than possible on Earth. Using polystyrene particles of different sizes and colors, the crew successfully achieved separation. However, some challenges related to electro-osmosis in the experimental equipment prevented perfect clarity in the separation bands.

During the remainder of day two of the Apollo 16 mission, several critical activities and observations occurred as the spacecraft continued its journey toward the Moon. The day began with a minor mid-course correction burn using the Command and Service

Module's (CSM) service propulsion system (SPS) engine. This brief two-second burn was aimed at fine-tuning the spacecraft's trajectory to ensure precision in reaching lunar orbit.

Later in the day, the astronauts again entered the Lunar Module (LM) to conduct a detailed inspection of its systems. During this inspection, they noted additional paint peeling from sections of the LM's outer aluminum skin. Despite these observations, all systems were reported to function normally, reassuring the crew and Mission Control.

Following the LM inspection, the crew reviewed checklists and procedures in preparation for upcoming milestones, including the critical Lunar Orbit Insertion (LOI) burn scheduled for the next day. However, during these preparations, Command Module Pilot Mattingly encountered a challenge known as "gimbal lock." This occurs when the system responsible for maintaining the spacecraft's attitude loses its frame of reference, often due to approaching a position where two gimbal axes parallel. Mattingly realigned the guidance system using the positions of the Sun and Moon as reference points to restore accuracy.

By the end of day two, Apollo 16 had traveled approximately 260,000 kilometers (140,000 nautical miles) away from Earth, steadily progressing toward its rendezvous with the Moon. These meticulous checks, adjustments, and preparations underscored the crew's dedication to ensuring a successful mission despite the occasional technical challenges encountered along the way.

As the crew of Apollo 16 awakened for flight day three, the spacecraft advanced to a distance of approximately 291,000 kilometers (157,000 nautical miles) from Earth. Still outside the lunar sphere of gravitational influence, the craft gradually decelerated. The early part of the day was primarily dedicated to routine housekeeping tasks, spacecraft maintenance, and exchanging status reports with Mission Control in Houston.

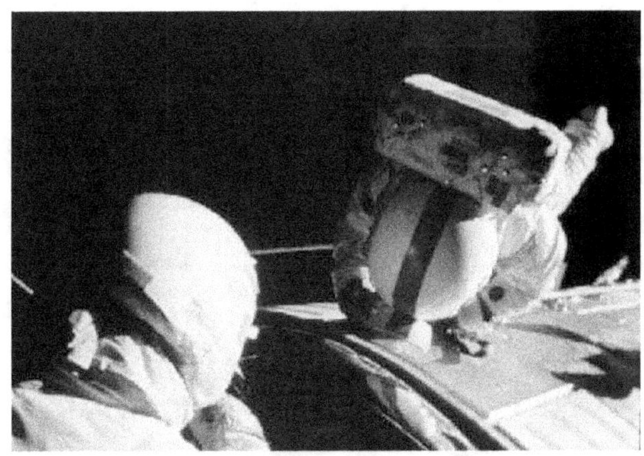

Ken Mattingly performing his deep-space EVA, retrieving film cassettes from the CSM's exterior

One significant experiment during this period was the Apollo light flash experiment, ALFMED. This experiment aimed to investigate the phenomenon of "light flashes" observed by astronauts during previous lunar missions when the spacecraft was in darkness, irrespective of whether their eyes were open. These flashes were hypothesized to be caused by cosmic ray particles penetrating the astronauts' eyes, and ALFMED sought to understand this occurrence further.

In the latter half of flight day three, astronauts Young and Duke again entered the Lunar Module to power it up and meticulously check its systems. All systems were confirmed to operate within expected parameters, reassuring the crew as they prepared for the upcoming lunar landing. Following these checks, the astronauts suited up in their space suits and rehearsed the procedures that would be crucial on the day of landing.

Charlie Duke's family portrait left on the surface of the moon. Cat Crater at Station 14 was named for sons Charles And Tom. Dot Crater at Station 16 was named for his wife.

As flight day three drew to a close, approximately 59 hours, 19 minutes, and 45 seconds after liftoff, Apollo 16 was positioned about 330,902 kilometers (178,673 nautical miles) from Earth and 62,636 kilometers (33,821 nautical miles) from the Moon. At this juncture, the spacecraft's velocity began to increase steadily as it entered the gravitational sphere of influence of the Moon, marking a critical phase as the mission neared its lunar destination.

On flight day four of Apollo 16, the crew focused on executing the Lunar Orbit Insertion (LOI) maneuver, a critical step in entering orbit around the Moon. They began their day with preparations for this maneuver, which involved breaking the spacecraft into lunar orbit.

As the spacecraft approached an altitude of 20,635 kilometers (11,142 nautical miles), the scientific instrument module (SIM) bay cover was jettisoned. This action prepared the spacecraft for the upcoming maneuvers without obstruction.

Approximately 74 hours into the mission, Apollo 16 passed behind the Moon, temporarily losing communication with Mission Control as it entered the far side. During this period, the Service Propulsion System (SPS) engine fired for 6 minutes and 15 seconds. This burn successfully slowed down the spacecraft, enabling it to enter lunar orbit with a pericynthion (low point) of 58.3 nautical miles (108.0 kilometers) and an apocynthion (high point) of 170.4 nautical miles (315.6 kilometers).

Following the Lunar Orbit Insertion, the crew immediately began preparations for the next maneuver, the Descent Orbit Insertion (DOI). This maneuver was designed to refine further the spacecraft's orbit, specifically reducing the pericynthion to 19.8 kilometers (10.7 nautical miles). The successful completion of DOI ensured that the spacecraft was on the correct trajectory for the upcoming phases of the mission.

The remainder of flight day four was dedicated to making observations of the lunar surface from orbit and finalizing preparations for activating the Lunar Module, which would be crucial for the subsequent undocking and landing operations scheduled for the following day.

Apollo 16's mission included a significant event to advance our understanding of the Moon's geological makeup. At 005:40:07.2, the S-IVB auxiliary propulsion system executed a precise 54.2-second burn, guiding the spent rocket stage toward a targeted impact on the lunar surface near the Apollo 12 landing site. This deliberate collision was orchestrated to generate seismic vibrations, pivotal for studying the lunar interior's composition and structure.

The Apollo 16 LSM deployed on the Moon in the lunar highlands. Sensors are at the top ends of the booms and approximately 75 cm above the lunar surface.

Despite minor malfunctions in the launch vehicle's systems, which hindered a planned trajectory refinement, the S-IVB successfully struck the Moon within the designated impact zone. However, the exact moment of impact could not be pinpointed due to telemetry loss, introducing some uncertainty in the seismic data analysis.

Estimates suggest that the S-IVB made contact with the lunar surface at 075:08:04.0. The impact site was calculated to be at latitude 2.1° north and longitude 22.1° west, approximately 173 nautical miles from the intended target point. Notably, it landed 86 nautical miles away from the Apollo 12 seismometer, 121 nautical miles from Apollo 14's, and 585 nautical miles from Apollo 15's—each strategically positioned to capture and analyze the resulting seismic waves.

At the moment of impact, the S-IVB weighed approximately 30,805 pounds and hurtled through space at 8,202 feet per second, underscoring the precision and force required for this scientific endeavor. This event underscored Apollo's dual role in lunar exploration: not just landing astronauts but also conducting invaluable scientific experiments that expanded our understanding of Earth's nearest celestial neighbor.

Apollo 16's Passive Seismic Experiment

During the Apollo mission's crucial CSM/LM docking phase, an unexpected observation caught the crew's attention: light-colored particles emanated from the LM. At 007:18, astronauts reported seeing these particles streaming from an area near aluminum close-out panel 51, which covered the Mylar insulation for the reaction control system A. Positioned beneath the docking target on the LM ascent stage's +Z face, this panel became the focal point of investigation.

To assess the situation, the crew entered the LM at 008:17 and powered up its systems. Fortunately, all diagnostics returned normal results, reassuring the astronauts that operational integrity was maintained. The LM was subsequently powered down at 008:52.

To provide visual confirmation and aid mission control's assessment, the CM television was activated at 008:45, capturing footage of the particle emission. Notably, adjustments were made to the high gain antenna by rotating panel 51 out of direct sunlight, which resulted in a noticeable reduction in particle release.

On closer inspection via television feed, the source of these particles appeared to be a cluster resembling grass-like growths located at the base of panel 51. This incident underscored the meticulous monitoring and problem-solving capabilities inherent in Apollo missions, where even seemingly minor anomalies prompted thorough investigation and response.

Following the investigation, it was determined that the observed particles were shredded thermal paint. Importantly, this degradation in thermal protection posed no significant impact on future LM operations. With this issue addressed, the mission continued seamlessly.

At 020:05, the crew commenced a scheduled 45-minute in-flight electrophoresis demonstration, which proceeded successfully. They also achieved planned objectives in ultraviolet photography of the Earth from 58,000 and 117,000 nautical miles.

A midcourse correction was executed at 030:39:00.66, lasting 2.01 seconds, ensuring precise lunar orbit insertion as planned.

Later, at 038:18:56, the command module computer received an indication of an inertial measurement unit gimbal lock. Promptly responding, the computer transitioned the IMU to "coarse align" mode and activated necessary alarms, ensuring continued navigation and operational safety throughout the mission. These events highlighted the crew's capability to manage unexpected challenges and maintain mission objectives on course during the Apollo program's rigorous exploratory missions.

During the Apollo mission, the Command Module Pilot faced visibility challenges due to a significant number of particles from the LM panels floating near the spacecraft, obstructing views of the stars. To realign the platform, the crew utilized the Sun and Moon as reference points, ensuring navigation precision despite the visual obstruction.

The gimbal lock indication, initially suspected to be caused by an electrical transient from the thrust vector control enable relay during IMU alignment program exit, prompted a swift response. An erasable software program was uplinked to the crew and implemented on the computer. This program instructed the computer to disregard gimbal lock indications during critical operational phases, safeguarding against potential navigational disruptions.

The visual light flash phenomena experiment commenced at 049:10. The crew reported numerous instantaneous white flashes, which ceased promptly when the experiment concluded at 050:16. Notably, these flashes left no residual glow, highlighting their brief and intense nature. This experiment added to the mission's scientific observations, contributing valuable data despite the earlier operational challenges encountered.

At approximately 053:30, the second round of LM housekeeping tasks commenced, ensuring all systems were operating within normal parameters. These checks were completed by 055:11, affirming the spacecraft's readiness for further operations.

At 069:59:01, the scientific instrumentation module door was jettisoned as part of mission procedures.

During its descent to the Moon, at an altitude of 92.9 nautical miles, the service propulsion engine fired for 374.90 seconds. This maneuver was crucial for placing the spacecraft into a precise lunar orbit, achieving parameters of 170.3 by 58.1 nautical miles.

The translunar coast phase lasted 71 hours, 55 minutes, and 14.35 seconds, marking a significant journey toward lunar exploration.

Throughout these operations, the crew captured compelling images of crater patterns and the lunar horizon from the Command

Module, offering valuable visual insights into the lunar surface.

Significant maneuvers and challenges shaped the Apollo 16 mission's narrative during the lunar orbit and surface phase. At 078:33:45.04, a critical 24.35-second service propulsion system (SPS) maneuver was executed to position the spacecraft into a descent orbit measuring 58.5 by 10.9 nautical miles. This maneuver set the stage for subsequent operations.

The Lunar Module (LM) activation began at 093:34, slightly ahead of schedule, with all systems indicating normal functionality. Following undocking from the Command and Service Module (CSM), the LM underwent a thorough visual inspection by the command module pilot, ensuring readiness for its upcoming lunar descent.

The LM's undocking from the CSM occurred at 096:13:31 during the 12th revolution, marking a pivotal transition toward lunar surface operations. At 103:21:43.08, a brief 4.66-second firing of the service propulsion system placed the CSM into a near-circular lunar orbit measuring 68.0 by 53.1 nautical miles, priming it for scientific data collection activities.

However, during the 13th lunar revolution, planned circularization maneuvers were briefly interrupted by detected oscillations in a secondary system controlling the SPS engine's thrust direction. Flight controllers promptly maneuvered the CSM into a stationkeeping configuration with the LM, evaluating whether to proceed or consider re-docking in case of mission abort considerations.

After extensive analysis lasting 5 hours and 45 minutes, engineers determined that the system could be safely utilized, affirming the mission's continuity. Subsequently, the spacecraft was separated again at 102:30:00, and the CSM successfully performed the planned 4.66-second circularization maneuver using the primary system at 103:21:43.08.

The powered descent to the lunar surface during Apollo 16 encountered delays, pushing the start about six hours behind schedule. Consequently, Command Module Pilot Young and Lunar Module Pilot Duke initiated their descent from a higher altitude than any previous mission, beginning at 20.1 kilometers (10.9 nautical miles). As they descended to approximately 13,000 feet (4,000 meters), Young gained a comprehensive view of the landing site. The throttle-down of the LM's landing engine occurred as planned, and the spacecraft smoothly transitioned to its landing orientation at 7,200 feet (2,200 meters).

The LM touched down at 104 hours, 29 minutes, and 35 seconds into the mission, precisely at 02:23:35 GMT on April 21 (8:23:35 pm on April 20 in Houston). The landing site was the Plain of Descartes, positioned at latitude 8.97301° south and longitude 15.50019° east, approximately 886 feet northwest of the intended target. Despite the slight deviation, the availability of the Lunar Roving Vehicle (LRV) minimized the significance of this distance.

The powered descent engine firing lasted 734.0 seconds, beginning at 104:17:25 at an altitude of 10.9 nautical miles. Approximately 102 seconds of engine firing time remained at the moment of landing. Visual observations from Apollo 16's pan camera frame 4623 captured the LM's landing site, highlighting landmarks like Palmetto Crater, Spook Crater, and Flag Crater.

Due to the extended lunar orbit duration, the LM's systems were powered down to conserve electrical resources. Consequently, the crew's first extravehicular activity (EVA) was postponed to ensure adequate rest. The LM cabin was depressurized at 118:53:38 for the commencement of the EVA. Television coverage of surface activities was delayed until the LRV systems were operational, as the LM's steerable antenna remained locked and unusable for initial transmissions.

During initial setup on the lunar surface, the lunar experiments package was successfully deployed. However, an accidental mishap occurred when the commander tripped over an electronics cable, damaging the heat flow experiment and rendering it inoperative. These early moments on the lunar surface illustrated both the triumphs and challenges inherent in Apollo missions, where meticulous planning and adaptability were crucial for achieving scientific objectives despite unforeseen obstacles.

Upon safely landing on the lunar surface, Young and Duke immediately began conserving battery power by powering down non-essential Lunar Module (LM) systems. With initial procedures completed, they proceeded to configure Orion, their living quarters, for the next three days while removing their space suits to conduct initial geological observations of the landing site. Their first activity after this was settling down for their inaugural meal on the lunar surface. Following their meal, they prepared the cabin for sleep, marking their initial rest period on the Moon.

The delay caused by the CSM's engine malfunction necessitated significant adjustments to the mission schedule. To accommodate potential issues, Apollo 16 was planned to spend one less day in lunar orbit after completing surface exploration. Additionally, to optimize the astronauts' sleep schedule, the third and final moonwalk duration was reduced from seven to five hours. These adaptations were crucial in maintaining operational flexibility and ensuring the crew's safety and well-being during their historic exploration of the lunar surface.

On the morning of April 21, 1972, astronauts John Young and Charles Duke embarked on their historic day with the meticulous routine of a lunar extravehicular activity, commonly known as a moonwalk. After a nourishing breakfast, they began their preparations inside the Lunar Module. Donning their spacesuits and ensuring they were pressurized, they depressurized the cabin, readying themselves for the journey outside.

John Young was the first to step onto the LM's external platform, a small area above the ladder known as the "porch." Charles Duke, his partner in this remarkable venture, handed Young a jettison bag containing discarded items that would be left behind on the lunar surface. With precision, Young carefully lowered the equipment transfer bag (ETB), which carried essential tools and scientific gear for their activities during the EVA.

As Young descended the ladder, each step brought him closer to making history. With the touch of his boot, he became the ninth human ever to set foot on the Moon, marking a milestone in mankind's space exploration.

Upon setting foot on the lunar surface, John Young shared his profound sentiments, remarking, "There you are: Mysterious and unknown Descartes. Highland plains. Apollo 16 is gonna change your image. I'm sure glad they got ol' Brer Rabbit here, back in the briar patch where he belongs." His words captured the awe and the scientific curiosity driving their mission to explore the Descartes Highlands.

Shortly after, Charles Duke descended the ladder to join Young, becoming the tenth person to walk on the Moon at 36, a record that stood as the youngest lunar explorer. Overwhelmed with excitement, Duke expressed his exhilaration to CAPCOM Anthony England, saying, "Fantastic! Oh, that first foot on the lunar surface is super, Tony!"

Their initial tasks on the moonwalk included unloading critical equipment such as the Lunar Roving Vehicle (LRV), the Far Ultraviolet Camera/Spectrograph, and other scientific instruments, all accomplished smoothly.

Commander John Young drove the rover and navigated the rugged lunar terrain during their lunar surface operations. At the same time, Lunar Module Pilot Charles Duke assisted with navigation and other tasks, adhering to the carefully planned division of responsibilities that characterized the operational efficiency of Apollo's J missions.

Upon landing on the lunar surface, their initial objective was to unload crucial equipment, including the Lunar Roving Vehicle (LRV), essential scientific instruments like the Far Ultraviolet Camera/Spectrograph, and other necessary gear. This process proceeded smoothly and without complications.

However, during their first attempt to drive the lunar rover, John Young encountered an issue: the rear steering mechanism was not functioning properly. Recognizing the importance of swift action, Young promptly alerted Mission Control about the technical glitch. Despite this setback, he continued efficiently setting up the television camera, a vital component for documenting their activities on the Moon.

Meanwhile, Charles Duke undertook the significant task of erecting the United States flag, symbolizing their nation's achievement in space exploration.

Following the deployment of the Lunar Roving Vehicle (LRV) and other equipment, the next task on their agenda for the day was to set up the Apollo Lunar Surface Experiments Package (ALSEP). As they parked the rover, which also housed the TV camera used to observe the deployment, they encountered a fortunate turn of events—the previously malfunctioning rear steering of the rover began to function once again.

With the ALSEP successfully deployed, John Young and Charles Duke collected geological samples from the surrounding area. Approximately four hours into their Extravehicular Activity (EVA-1), they mounted the rover and embarked on their journey to the first geological site, Plum Crater. Situated on the rim of Flag Crater, Plum Crater spanned 118 feet (36 meters) in diameter, with Flag Crater itself measuring about 790 feet (240 meters) across.

At Plum Crater, located approximately 1.4 kilometers (0.87 miles) from the Lunar Module (LM), Young and Duke conducted sampling activities. Scientists believed the materials collected here had penetrated through the upper regolith layer to reach the underlying Cayley Formation, offering valuable insights into the Moon's geological history.

During this stop, at the request of Mission Control, Charles Duke retrieved a significant find—the largest rock ever returned by an Apollo mission. This breccia, later nicknamed "Big Muley" after mission geology principal investigator William R. Muehlberger, became a pivotal specimen in the scientific examination of lunar geology.

John W. Young on the Moon during Apollo 16 mission jumping about 42 Centimeters high. Charles M. Duke Jr. took this picture. The LM Orion is on the left. April 21, 1972

After concluding their activities at the experiment site, the crew of Apollo 16 embarked on a traverse westward in their lunar roving vehicle (LRV-2) toward Flag Crater. They conducted visual observations, photographed significant features, and gathered lunar samples to further scientific understanding.

The return route followed a path slightly south of their outbound journey, leading them to their next stop at Spook Crater. They returned to the experiment station near the LM, where they deployed the solar wind composition experiment. This experiment marked a pivotal moment in their scientific endeavors on the lunar surface.

One of the most iconic moments of Apollo 16 was captured during EVA-1, when John Young saluted the U.S. flag, seemingly "hanging in the air" due to the Moon's lower gravity than Earth. This memorable photograph was taken at the Descartes landing site, symbolizing the culmination of their mission.

Despite the successes, EVA-1 encountered several challenges with the LRV. As they navigated steep ridges and rocky terrain, the vehicle's rear wheels failed to respond when full throttle was applied. Despite this setback, Young later conducted a "grand prix" test ride to assess the vehicle's capabilities under more controlled conditions.

The crew re-entered the LM, repressurizing the cabin at 126 hours, 4 minutes, and 40 seconds into their mission. The first extravehicular activity lasted 7 hours, 11 minutes, and 2 seconds. During this time, they covered a distance of 13,800 feet (approximately 4.2 kilometers) in the lunar rover, spending 43 minutes actively driving and parking the vehicle for 3 hours and 39 minutes. They collected an estimated 65.9 pounds (29.9 kilograms) of lunar samples, contributing significantly to scientific research back on Earth.

During EVA-1 of Apollo 16, several issues arose with the Lunar Roving Vehicle (LRV). While navigating steep ridges and rough, rocky terrain, John Young encountered a problem where the rear wheels failed to respond when he applied full throttle. Despite this setback, the vehicle continued moving forward, though the front wheels dug into the lunar surface, complicating their progress.

In response to these challenges, Young later took the LRV for a "grand prix" test ride, a maneuver to evaluate its performance under more controlled conditions. This test ride, captured in mission imagery, allowed him to assess the vehicle's capabilities in different lunar terrain scenarios.

During EVA-1, the crew deployed various scientific instruments, including the ultraviolet camera, which was positioned in the shadow of the Lunar Module (LM).

Following their extensive extravehicular activities, the crew re-entered the LM, and at 126 hours, 4 minutes, and 40 seconds into their mission, the cabin was repressurized. The total duration of EVA-1 lasted 7 hours, 11 minutes, and 2 seconds. During this time, they covered a distance of 13,800 feet (approximately 4.2 kilometers) in the lunar rover. The vehicle was actively driven for 43 minutes, while the remainder of the time—3 hours and 39 minutes—was spent parked as they conducted experiments and collected samples.

During their first extravehicular activity, the astronauts gathered an estimated 65.9 pounds (29.9 kilograms) of lunar samples, contributing significantly to scientific research on Earth about the Moon's geology and composition.

During their second extravehicular activity on the lunar surface, the Apollo 16 crew embarked on a meticulously planned journey that unfolded over several hours within the confines of their lunar module. Emerging into the stark lunar daylight, they depressurized the cabin at precisely 142 hours, 39 minutes, and 35 seconds of mission time. Their objective: traverse the rugged terrain surrounding their landing site.

Equipped with the Lunar Roving Vehicle (LRV), the astronauts set out on a southerly course toward a region of interest nestled near the Cinco Craters on the northern slope of Stone Mountain. This lunar journey wasn't just about exploration; it was a scientific endeavor to gather samples from unexplored geological formations. As they navigated the undulating lunar landscape, they made strategic stops near notable craters like Stubby and Wreck, each location carefully selected for its geological significance. The final leg of their traverse led them northward to an experimental station and ultimately back to the safety of their lunar module.

However, the mission had its challenges. At station 8, the crew encountered a technical issue with the LRV's rear drive, prompting a troubleshooting procedure. They identified a power mode switching discrepancy as the culprit and swiftly reconfigured the switches, restoring the vehicle to full operational status.

Further into their journey, they faced another setback at stations 9 and 10 when the LRV's range and distance measurement systems malfunctioned. Quick thinking and procedural resets revived the navigation system, ensuring they could continue their exploration unhindered.

Despite these technical hurdles, the astronauts remained focused on their scientific objectives. Their visit to Stone Mountain's steep slopes, reaching elevations of 152 meters above the valley floor, provided unparalleled views of the lunar landscape. Duke, awestruck by the vista that included South Ray Crater, described it as nothing short of spectacular.

Astronaut John W. Young, Commander of the Apollo 16 mission, replaces tools in the hand tool carrier at the aft end of the "Rover" Lunar Roving Vehicle (LRV) during the second Apollo 16 extravehicular activity (EVA-2) at the Descartes landing site.

Their meticulous sampling efforts at each stop, particularly at Station 5 near a 20-meter crater, aimed to collect pristine Descartes material untainted by nearby ejecta from South Ray. While initially of uncertain origin, these samples were considered by geologists like Wilhelms as compelling candidates from the Descartes region.

Following their exploration at Station 5, the Apollo 16 astronauts continued their traverse across the lunar surface with a keen focus on scientific discovery. Station 6 beckoned—a blocky crater measuring 10 meters wide, promising samples from the Cayley Formation, distinguished by its firmer soil. Skipping Station 7 to optimize their schedule, they proceeded to Station 8, strategically located on the lower flank of Stone Mountain.

At Station 8, nestled within a ray emanating from South Ray Crater, they engaged in intensive sampling for over an hour. Here, amidst the lunar regolith, they meticulously collected specimens: black and white breccias alongside crystalline rocks rich in plagioclase—a bounty that would yield valuable insights into lunar geology.

Their journey continued to Station 9, dubbed the "Vacant Lot," a pristine area believed to be devoid of South Ray ejecta. They spent approximately 40 minutes here, methodically gathering additional samples to complement their growing collection.

They embarked on their final scientific endeavor as lunar daylight waned, arriving halfway between the Apollo Lunar Surface Experiments Package (ALSEP) site and their Lunar Module (LM). They conducted a series of core samplings and penetrometer tests, stretching over a 50-meter line, to further probe the lunar surface's composition.

Driven by their relentless pursuit of scientific data, astronauts Young and Duke petitioned for ten minutes to extend their moonwalk. This dedication culminated in a record-breaking Extravehicular Activity (EVA), 7 hours, 23 minutes, and 26 seconds—a testament to their endurance and the mission's success in pushing the boundaries of lunar exploration.

Returning to the LM, they sealed the cabin, pressurized the interior, and commenced their post-EVA routines. After a well-deserved meal, they debriefed with Mission Control, reviewing the day's activities and outcomes. With meticulous care, they reconfigured the LM cabin for their sleep period, marking the end of another momentous day on the lunar surface for Apollo 16.

After completing their exploration at Station 10, near both the Lunar Module (LM) and the Apollo Lunar Surface Experiments Package (ALSEP) area, the Apollo 16 crew found themselves ahead of schedule with their consumables usage. Capitalizing on this unexpected surplus, they extended their surface activity by 20 minutes, allowing for more detailed inspections and experiments.

During this extended period, the Lunar Module Pilot (LMP) thoroughly examined the damaged heat flow experiment. Visual inspection revealed a critical issue: a cable detached at the connector. Troubleshooting efforts at Mission Control suggested a potential solution. Still, the decision was made to wait to attempt repairs, as the time required could jeopardize plans for the upcoming third Extravehicular Activity (EVA).

As the period drew to a close, the crew re-entered the LM cabin and began the process of repressurization at 150 hours, 2 minutes, and 44 seconds into the mission. During this ingress, a two-inch portion of the Commander's antenna broke off, resulting in a significant drop in signal strength—measured at 15 to 18 decibels. This posed a critical communication challenge, as the Commander's backpack radio relayed crucial information from the LMP to both the LM and the Lunar Communications Relay Unit for transmission to ground stations.

To mitigate the antenna issue, the Commander later decided to utilize the LMP's oxygen purge system, which supported the damaged antenna and ensured vital communications remained operational.

Reflecting on their achievements, the second Extravehicular Activity (EVA) of Apollo 16 spanned an impressive 7 hours, 23 minutes, and 9 seconds. During this time, the crew covered 37,100 feet (approximately 11.3 kilometers) in the Lunar Rover Vehicle (LRV). This included 1 hour and 31 minutes of actual driving time and 3 hours and 56 minutes spent at various stops for scientific investigations and sample collection. Their efforts yielded an estimated 63.9 pounds (29.0 kilograms) of lunar samples—each sample a testament to their meticulous planning and dedication to advancing our understanding of the Moon's geological history.

On their third and final day on the lunar surface, the Apollo 16 astronauts embarked on their last Extravehicular Activity (EVA), exploring North Ray Crater—a monumental geological feature and the largest crater ever visited by any Apollo mission. Exiting their lunar module Orion, they navigated the smoother terrain north of their landing site, where craters were shallower and boulders less obstructive compared to previous days.

Their journey to North Ray Crater took them past Palmetto Crater, where the lunar landscape gradually transformed with larger, more abundant boulders as they approached their destination in the Lunar Rover Vehicle (LRV). Arriving at the crater's rim, they marveled at its vast dimensions: 1 kilometer wide and 230 meters deep. They encountered 'House Rock,' a towering boulder taller than a four-story building. Samples collected from this landmark proved pivotal in dispelling the pre-mission hypothesis of lunar volcanic origins, showcasing numerous micrometeoroid impact marks akin to bullet holes.

The third EVA began 30 minutes ahead of schedule as the cabin was depressurized at 165 hours, 31 minutes, and 28 seconds into the mission. Despite time constraints leading to the deletion of four planned stations, the astronauts focused their efforts on North Ray Crater. They meticulously photographed the terrain, gathered samples, and conducted scientific experiments, including using the lunar rake and obtaining a double core tube sample at the ALSEP site.

Their traverse continued southeast to Shadow Rock, the second sampling area, before retracing their route back to the LM. The day's activities culminated in an EVA lasting 5 hours, 40 minutes, and 3 seconds, during which they covered a distance of 37,400 feet (approximately 11.4 kilometers) in the LRV. This included 1 hour and 12 minutes of driving time and 2 hours and 26 minutes at various scientific stops. They collected an estimated 78.0 pounds (35.4 kilograms) of lunar samples, further enriching our understanding of the Moon's geological and environmental history.

After intensive exploration of the North Ray crater, the Apollo 16 astronauts embarked on their final scientific endeavors on the lunar surface. Departing from the crater about 1 hour and 22 minutes later, they reached Station 13, a significant boulder field approximately 0.5 kilometers away. During their descent, they achieved a lunar speed record, traveling downhill at a remarkable speed of 17.1 kilometers per hour (10.6 mph).

John Young adjusting the LRV's antenna near Shadow Rock

At Station 13, they encountered "Shadow Rock," a prominent 3-meter-high boulder where they conducted sampling of permanently shadowed soil, adding to their diverse collection of lunar specimens. Meanwhile, in orbit, Command Module Pilot Mattingly prepared the Command and Service Module (CSM) for their return, anticipating their arrival approximately six hours later.

After three hours and six minutes of further scientific activities and experiments near the LM, the crew completed their tasks. Before leaving the lunar surface, Duke placed a photograph of his family and an Air Force commemorative medallion—a personal tribute to their historic mission.

Before departure, Young drove the rover to a designated point 90 meters east of the LM, known as the 'VIP site,' where a remotely controlled television camera installed by Mission Control captured Apollo 16's liftoff from the Moon.

With their final excursion completed, the crew re-entered the LM after 5 hours and 40 minutes spent outside during this last EVA. The cabin was repressurized at 171 hours, 11 minutes, and 31 seconds into the mission, marking the conclusion of the Apollo program's fifth and final human exploration of the Moon.

Reflecting on their achievements, Apollo 16 set impressive records: 20 hours, 14 minutes, and 14 seconds spent outside the LM, covering a distance of 88,300 feet (approximately 26.9 kilometers) in the Lunar Rover Vehicle (LRV). Driving time totaled 3 hours and 26 minutes, with the vehicle parked for 10 hours and 1 minute during extravehicular activities. They collected a remarkable 211.00 pounds (95.71 kilograms) of lunar samples, contributing invaluable data to scientific research.

Their farthest point from the LM reached an impressive 15,092 feet, encapsulating the extent of their exploration and the enduring legacy of Apollo 16's mission to unlock the mysteries of the Moon.

During the crew's lunar surface operations, the Command Module Pilot (CMP), Thomas Mattingly, conducted a series of critical tasks from orbit aboard the Command and Service Module (CSM). His responsibilities included capturing photographs of the Moon's surface, measuring its physical properties, and making visual observations. These observations and measurements provided valuable scientific data that complemented and validated findings from previous Apollo missions, notably Apollo 15.

At 169 hours, 5 minutes, and 52.14 seconds into the mission, a pivotal CSM plane change maneuver lasting 7.14 seconds was executed. This adjustment altered the CSM's orbit to 64.6 by 55.0 nautical miles.

The climactic moment of the lunar module's liftoff from the lunar surface occurred at 175 hours, 31 minutes, and 47.9 seconds into the mission, precisely at 01:25:47 GMT on April 24 (08:25:47 p.m. EST on April 23). This historic event, televised to audiences worldwide, marked the culmination of their 71 hours, 2 minutes, and 13 seconds on the lunar surface.

The ascent stage engine fired 427.7 seconds, propelling the lunar module into a new orbit measuring 40.2 by 7.9 nautical miles. Following liftoff, several rendezvous sequence maneuvers were necessary to facilitate docking with the orbiting CSM. First, a vernier adjustment was performed at 175 hours, 42 minutes, and 18 seconds at 11.2 nautical miles, setting the stage for the complex dance of maneuvers required for a successful reunion in lunar orbit.

Eight minutes before the scheduled departure from the lunar surface, CAPCOM James Irwin relayed confirmation from Mission Control to astronauts Young and Duke that all systems were ready for liftoff. This critical communication signaled the final preparations for their ascent back to the Command Module.

Two minutes before liftoff, as part of the launch sequence, Young and Duke activated the "Master Arm" switch followed by the "Abort Stage" button. These actions initiated a series of small explosive charges designed to sever the ascent stage from the Lunar Module's descent stage (LM). The separation was facilitated by a guillotine-like mechanism that cleanly cut the cables connecting the two stages, ensuring a smooth transition for the ascent stage as it prepared to leave the lunar surface.

After successfully blasting off from the Moon's surface, the ascent stage of Apollo 16's Lunar Module (LM) quickly gained speed, reaching approximately 5,000 kilometers per hour (3,100 mph) as it ascended toward lunar orbit. The Lunar Roving Vehicle (LRV) camera captured the initial moments of the flight, documenting this historic departure.

Six minutes after liftoff, John Young and Charles Duke maneuvered the ascent stage into lunar orbit, where Thomas Mattingly awaited aboard the Command and Service Module (CSM). The rendezvous and redocking with the CSM were executed flawlessly, marking a critical milestone in the mission's return phase.

To ensure minimal transfer of lunar dust from the LM cabin to the pristine environment of the CSM, Young, and Duke thoroughly cleaned the cabin before opening the hatch that separated the two spacecraft. Upon reuniting with Mattingly, they began the meticulous process of transferring the precious lunar samples, film, and equipment collected during their surface mission into the CSM.

Once all transfers were completed, the crew prepared for the next phase of the mission. They scheduled a rest period before intentionally jettisoning the now-empty Lunar Module ascent stage the following day. This deliberate crash into the lunar surface was planned to calibrate the seismometer left behind by Young and Duke, an important scientific objective of the mission.

However, following its jettison, the Lunar Module (LM) experienced unexpected instability, tumbling at a rate of approximately 3 degrees per second. This anomaly was attributed to a guidance circuit breaker that may have inadvertently been left open, highlighting the complexities and challenges even during the mission's final phases.

Following the separation of the Command and Service Module (CSM) from the ascent stage at 195 hours, 03 minutes, and 13 seconds into the mission, subsequent maneuvers finalized Apollo 16's lunar operations. Due to circumstances, no deorbit burn maneuver was feasible for the ascent stage, resulting in its retention in lunar orbit for approximately one year.

Shortly after, at 195 hours, 23 minutes, and 12 seconds, the mass spectrometer deployment boom encountered a technical issue during its retract cycle, leading to its jettison.

Before departing lunar orbit, at 196 hours, 02 minutes, and 09 seconds into the mission's 62nd revolution, the CSM deployed a particles and fields subsatellite. This deployment took place into an orbit measuring 66 by 52 nautical miles at an inclination of -11 degrees. The subsatellite was slated for release during the 73rd revolution into a higher orbit measuring 170 by 58 nautical miles. This strategic deployment aimed to gather scientific data crucial for understanding lunar particles and fields, akin to the subsatellite mission accomplished during Apollo 15.

The subsatellite deployed during Apollo 16 had a specific mission to measure plasma and energetic-particle fluxes, vector magnetic fields, and lunar gravitational anomalies. Unfortunately, a planned orbit-shaping maneuver for the Command and Service Module (CSM) had not been executed before the subsatellite's ejection due to an engine gimbal anomaly earlier in the mission. As a result, the subsatellite was placed into an orbit with a significantly shorter lifetime than originally intended.

Communications interference delayed the subsatellite's activation for approximately 20 hours after its launch. This delay was caused by the ascent stage's failure to deorbit as planned but did not affect the subsatellite's operational systems.

After the unfortunate loss of tracking and telemetry data, the subsatellite deployed during the Apollo 16 mission encountered significant challenges that shortened its operational lifespan. Attempts to regain communication with the subsatellite were unsuccessful, leading to the belief that it impacted the lunar far side during its 425th revolution at a longitude of 110 degrees east. The subsatellite's lower-than-desired orbit, influenced by lunar gravitational anomalies and mass concentrations, contributed significantly to its shortened orbital duration.

Following this event, the crew proceeded with their planned activities. They jettisoned the expended Lunar Module (LM) ascent stage, which encountered issues likely due to a crew oversight in activating a necessary switch before sealing. As a result, the ascent stage tumbled uncontrollably and did not perform the required rocket burn for intentional de-orbiting. Approximately a year later, it eventually crashed into the lunar surface.

Despite these challenges, the crew successfully released a subsatellite from the Command and Service Module's (CSM) scientific instrument bay into lunar orbit. Although a planned orbit-shaping burn to optimize the CSM's orbit for the subsatellite deployment had been canceled, the subsatellite remained operational for just over a month instead of the intended year-long mission. Following its deployment, the CSM's Service Propulsion System (SPS) engine was successfully reignited during the 65th orbit around the Moon to initiate the spacecraft's return trajectory to Earth. This critical burn was executed flawlessly despite previous technical setbacks during the mission.

The Apollo 16 mission adjusted its schedule to accommodate an earlier transearth injection by deleting the second plane-change maneuver and some orbital science photography. This decision stemmed from an engine issue during the lunar orbit circularization maneuver.

The transearth injection maneuver occurred 200 hours, 24 minutes, and 15.36 seconds into the mission, following a 162.29-second burn at 52.2 nautical miles. This critical maneuver initiated the return journey after completing 64 lunar orbits over 125 hours, 49 minutes, and 32.59 seconds, achieving a velocity of 8,663.0 feet per second.

During the transearth phase, significant moments included transmitting high-quality television pictures inside the Command Module (CM) between 202 hours, 57 minutes, and 203 hours, 12 minutes. Additionally, pictures from the Lunar Roving Vehicle (LRV) camera on the lunar surface were broadcast from 203 hours, 29 minutes, to 204 hours, 12 minutes.

To ensure optimal entry into Earth's atmosphere, the first of two midcourse corrections was performed at 214 hours, 35 minutes, and 02.8 seconds. This maneuver lasted 22.6 seconds and adjusted the spacecraft's velocity by 3.4 feet per second to refine the trajectory toward Earth.

During the Apollo mission, at 218 hours, 39 minutes, and 46 seconds into the flight, the command module pilot initiated a critical extravehicular activity known as a transearth coast EVA. This event was broadcast live for 1 hour, 23 minutes, and 42 seconds, capturing audiences' attention back on Earth. Mattingly meticulously retrieved film cassettes from the scientific instrument module cameras inside the vast expanse of space. These devices were crucial for documenting the lunar journey, capturing stunning images of the Moon's surface and the spacecraft's surroundings.

Amid the EVA, Mattingly visually inspected the spacecraft's exterior equipment, ensuring all components functioned optimally under the extreme conditions of space. He also conducted a groundbreaking experiment, exposing a microbial sample to the cosmic environment for ten minutes. This scientific endeavor aimed to study how microorganisms respond to the harsh realities of space, providing invaluable data for future missions and scientific research.

By the conclusion of this spacewalk, the total accumulated time spent outside the spacecraft during the mission reached an impressive 22 hours, 17 minutes, and 36 seconds, underscoring the crew's dedication to scientific exploration and discovery.

Later in the mission timeline, at 243 hours and 35 minutes, a scheduled television press conference offered the crew an opportunity to communicate their experiences directly to the world. Lasting 18 minutes, this event allowed them to share a vivid account of their observations of the Moon's far side. Of particular interest was their detailed description of Guyot Crater, a geological formation that captivated both scientists and the public alike. The crater appeared brimming with material, overflowing and cascading down its slopes—a scene reminiscent of similar geological phenomena observed on Earth, particularly in Hawaii.

Additional scientific activities ensued during the tranquil transearth coast phase of the mission. These included comprehensive photography sessions, vital for the Skylab program's study of how particles emitted from the spacecraft behave and impact the surrounding environment. Additionally, the crew conducted a second session of light-flash observations, documenting peculiar optical phenomena observed during their journey through space.

During the Apollo mission's return journey to Earth, a pivotal moment unfolded when Mattingly embarked on an 83-minute extravehicular activity (EVA). This mission-critical maneuver aimed to retrieve essential film cassettes from the Scientific Instrument Module (SIM) bay cameras. Duke, remaining at the command module's hatch, provided crucial support throughout the operation, ensuring the safety and success of the task.

This EVA marked a historic milestone, occurring approximately 173,000 nautical miles (199,000 miles or 320,000 kilometers) from Earth. It stood as only the second "deep space" EVA in history, performed far from any planetary body—a testament to the precision and advanced capabilities of the Apollo missions. As of 2024, this remains one of just three EVAs ever conducted. All were carried out during Apollo's J-missions under similarly challenging circumstances.

Amidst the retrieval mission, Mattingly also set up a pioneering biological experiment known as the Microbial Ecology Evaluation Device (MEED). This unique experiment, exclusive to Apollo 16, sought to explore how microorganisms respond to the rigors of the space environment. The data gathered would provide critical insights into the feasibility of sustaining biological systems beyond Earth's atmosphere—a cornerstone of future space exploration endeavors.

Throughout the day, the crew diligently attended to housekeeping and maintenance tasks aboard the spacecraft, ensuring all systems operated efficiently during the return journey. Amidst their duties, they paused to share a meal—a moment of camaraderie and respite amidst the vastness of space.

On the penultimate day of their journey, the crew of the Apollo mission dedicated much of their time to conducting a series of rigorous experiments. These scientific endeavors culminated weeks of meticulous planning and preparation, aimed at gathering invaluable data from the far reaches of space.

Amidst their scientific duties, the astronauts took a brief hiatus for a twenty-minute press conference, a pivotal opportunity to communicate with journalists back on Earth. Prepared questions prioritized by the Manned Spacecraft Center in Houston covered both technical intricacies and broader aspects of the mission. This interaction provided insights into the challenges and triumphs experienced during their extraordinary voyage.

Throughout the day, the crew also tended to numerous housekeeping tasks aboard the spacecraft, ensuring that all systems remained in optimal condition as they prepared for the critical phase of atmospheric reentry scheduled for the following day. These preparations included meticulous checks and adjustments to guarantee a safe return to Earth's atmosphere after their extensive journey through space.

As the crew approached the end of their final full day in space, the spacecraft was approximately 143,000 kilometers (about 77,000 nautical miles) from Earth, steadily closing in at approximately 2.1 kilometers per second (7,000 feet per second). This phase marked the culmination of their mission, blending scientific achievement with meticulous preparation for a safe return home.

As the crew received their final wake-up call in space from CAPCOM England, the Command and Service Module (CSM) of Apollo 16 was approximately 45,000 nautical miles (83,000 kilometers) away from Earth, hurtling through space at a speed of over 2.7 kilometers per second (9,000 feet per second). With only hours remaining before their scheduled splashdown in the Pacific Ocean, the crew executed a crucial course correction burn, fine-tuning their trajectory using the spacecraft's thrusters to adjust their velocity by 0.43 meters per second (1.4 feet per second).

A pivotal moment arrived just minutes before atmospheric reentry: the command module, resembling a cone housing the three crewmembers, separated from the service module, which would later disintegrate upon reentry into Earth's atmosphere. At precisely 265 hours and 37 minutes into the mission, traveling at a staggering velocity of about 11 kilometers per second (36,000 feet per second), Apollo 16 dramatically descended into Earth's atmosphere.

During the final phase of Apollo 16's return journey, the crew executed their second midcourse correction, a precise 6.4-second maneuver that adjusted their velocity by 1.4 feet per second, ensuring they maintained their intended trajectory. This critical adjustment occurred at 262 hours, 37 minutes, and 20.7 seconds into the mission.

As they neared Earth, another pivotal moment occurred when the service module was jettisoned at 265 hours, 22 minutes, and 33 seconds. This marked the separation of the module that had supported them throughout their mission, clearing the way for the command module's solitary descent back through Earth's atmosphere.

At 265 hours, 37 minutes, and 31 seconds, the command module began its reentry into Earth's atmosphere at an altitude of 400,000 feet, hurtling through the sky at a velocity of 36,090 feet per second. This marked the culmination of a transearth coast spanning 65 hours, 13 minutes, and 16 seconds since their departure from the Moon.

The reentry followed a meticulously planned trajectory, adhering to a normal profile to ensure the crew's safe return. The command module's heat shield, exposed to temperatures ranging between 2,200 and 2,480 degrees Celsius (4,000 and 4,500 degrees Fahrenheit), withstood the intense heat of atmospheric friction.

Following a flawless descent, parachute deployment slowed the module's speed, guiding it gently to a precise splashdown in the Pacific Ocean, approximately 350 kilometers (189 nautical miles) southeast of Kiritimati Island. Within moments, swift recovery operations were launched by the nearby USS Ticonderoga, swiftly retrieving the astronauts and marking the successful conclusion of Apollo 16's pioneering mission in space exploration.

After the conclusion of Apollo 16's mission, later analysis suggested that the ascent stage had impacted the lunar surface shortly before the start of Apollo 17, although definitive data to support this were unavailable.

During the return to Earth, the Command Module (CM) descended under its drogue parachutes, which were visible to television viewers who followed the continuous coverage of the crew's recovery. At 19:45:05 GMT (02:45:05 p.m. EST) on April 27, the CM splashed down in the Pacific Ocean, marking the end of the mission after a total duration of 265 hours, 51 minutes, and 5 seconds.

The splashdown occurred approximately 3.0 nautical miles from the intended target point and 2.7 nautical miles from the recovery ship, USS Ticonderoga. The precise coordinates of the splashdown site were estimated to be latitude 0.70° south and longitude 156.22° west.

Upon impact with the water, the CM initially assumed an apexdown flotation attitude. However, within a remarkably swift 4 minutes and 30 seconds, it was successfully returned to its normal flotation position by the inflatable bag uprighting system. This rapid recovery process ensured the safety and comfort of the astronauts as they were retrieved from their capsule in the central Pacific Ocean, culminating in the successful conclusion of Apollo 16's momentous journey of exploration and scientific discovery.

After splashdown, the crew of Apollo 16 was swiftly retrieved by helicopter and safely aboard the recovery ship, USS Ticonderoga, a mere 37 minutes later. Following their recovery, the Command Module (CM) was recovered from the ocean waters just 62 minutes after splashdown.

At splashdown, the CM was estimated to weigh 11,995 pounds and have traveled an estimated 1,208,746 nautical miles throughout the mission's duration.

The crew remained aboard the USS Ticonderoga until 17:30 GMT on April 29th, after which they were flown to Hickam Air Force Base in Hawaii, arriving at 19:21 GMT. From there, they departed aboard a C-141 aircraft for Ellington Air Force Base in Houston at 20:07 GMT, arriving at their destination at 03:40 GMT on April 30th. Concurrently, the CM also arrived in Hawaii at 03:30 GMT on April 30th.

On May 1st at 18:00 GMT, the CM departed for North Island Naval Air Station in San Diego for deactivation, completing its journey and finalizing the post-mission activities. The crew's return and the CM's deactivation were met with welcome ceremonies aboard the USS Ticonderoga, marking the conclusion of Apollo 16's successful mission to explore the Moon and expand our understanding of the universe.

On May 7th, while removing propellants from the Command Module (CM) of Apollo 16, a tank cart incident occurred due to overpressurization, resulting in an explosion. Fortunately, although forty-six individuals were initially hospitalized on suspicion of inhaling toxic fumes, subsequent examinations revealed no symptoms of inhalation, and all affected personnel recovered safely. Importantly, the CM itself sustained no damage from the incident.

An investigation board later determined that the overpressurization was caused by a low ratio of neutralizer to oxidizer during the detanking process. This imbalance was attributed to extra oxidizer remaining in the CM tanks, a residual effect from an anomaly observed during the Apollo 15 parachute deployment. After this incident, significant changes were implemented in ground support equipment and detanking procedures to mitigate the risk of future overpressurization incidents.

Following the incident and its aftermath, deactivation procedures for the CM were completed at 00:00 GMT on May 11th. Subsequently, at 03:00 GMT on May 12th, the CM departed from North Island Naval Air Station and was transferred to the North American Rockwell Space Division facility in Downey, California, for thorough postflight analysis. It arrived at its destination at 10:30 GMT on the same day.

The Apollo 16 mission, despite ending a day earlier than scheduled, achieved remarkable success in fulfilling its primary and detailed objectives. Throughout its journey, data collection spanned lunar orbit, surface operations, and both the journey to and from the Moon. Notably, the mission delivered groundbreaking scientific discoveries, capturing unprecedented photographs of Earth's geocorona in the hydrogen (Lyman alpha) wavelength from beyond the confines of our atmosphere. Additionally, Apollo 16 expanded our understanding of Earth's magnetosphere, identifying two previously unknown auroral belts encircling our planet. Though some experiments, such as subsatellite tracking and the heat flow investigation, faced challenges, the mission's overall performance underscored its pivotal role in advancing lunar exploration and scientific inquiry during the Apollo era.

Upon analyzing post-mission data, several key conclusions emerged:

Lunar dust and soil continued to pose challenges, affecting equipment performance. Procedural adjustments were implemented, alongside equipment modifications, to mitigate these issues effectively.

The loss of the heat flow experiment highlighted the need for hardware designs that account for inadvertent loads caused by crew movements influenced by limited vision and mobility in pressurized suits.

The reliability of the S-band omnidirectional antenna system was proven following the failure of the S-band steerable antenna, demonstrating its crucial role in supporting lunar module operations.

The Apollo 16 particles and fields' subsatellite performance underscored deficiencies in the lunar gravitational model's accuracy for predicting orbital conditions and impact times.

The absence of cardiac arrhythmias during the mission was partly attributed to improved physiological balance through increased potassium intake and optimized rest-work cycles, enhancing crew sleep quality.

The mission successfully demonstrated the crew's capability to land safely in rugged lunar highlands terrain without prior high-resolution photography. Moreover, the lunar roving vehicle showcased its ability to navigate slopes of up to 20° under these challenging conditions.

Apollo 16 embarked on a multifaceted mission designed to push the boundaries of lunar exploration and scientific discovery. From the outset, its primary goals were clear: to thoroughly inspect and sample the Descartes region of the Moon, deploy and activate a series of pioneering surface experiments, and conduct a wide array of inflight experiments and photographic tasks.

The mission encountered triumphs and challenges across its detailed objectives during its journey. Photographic tasks from lunar orbit and visual observations were successfully completed, providing invaluable visual data of the lunar surface. However, timeline adjustments impacted the completeness of command module photographic tasks and the Skylab contamination study, affecting data acquisition.

Technological innovations and adaptations were also a focal point. While the crew successfully deployed an improved fecal collection bag and evaluated the lunar rover vehicle's capabilities in challenging terrain, the failure of the gas/water separator and the cable breakage in the heat flow experiment highlighted the ongoing need for robust equipment designs.

Scientific inquiry reached new heights with experiments ranging from seismic studies to cosmic ray detection and soil mechanics. These investigations yielded critical insights into lunar geology, solar wind composition, and the behavior of microbial organisms in space.

The mission's operational tests and demonstrations further underscored its significance, from testing the reliability of communication systems with the S-band transponder to demonstrating fluid electrophoresis in microgravity, and Apollo 16 set benchmarks for future manned space missions.

Despite some objectives not achieving full success, such as the accuracy of impact targeting and the completion of certain biological experiments, Apollo 16's legacy remains profound. Its achievements in lunar exploration and scientific research continue to inform and inspire our ongoing endeavors in space exploration.

Apollo 17

The Eleventh Crewed Mission

Apollo 17, the final piloted lunar landing mission of the Apollo program, flew 1,484,933.8 miles and marked a culmination of scientific exploration on the Moon. Launched on December 7, 1972, it was designated as a "J-type" mission, emphasizing extended surface operations and advanced scientific instrumentation. Commanded by Eugene Cernan, the crew included lunar module pilot Harrison Schmitt and command module pilot Ronald Evans.

The mission's ambitious objectives were to conduct a comprehensive selenological survey and sampling of the Taurus-Littrow region, a site chosen for its geological diversity and operational advantages. This included deploying the third Lunar Roving Vehicle (LRV), enhancing mobility and the range of exploration. Unique to Apollo 17 were experiments aimed at better understanding lunar geology and surface conditions and conducting experiments both on the lunar surface and from lunar orbit.

Upon landing, the astronauts began their three-day exploration, conducting geological inspections, collecting samples, and setting up various scientific instruments. The mission also involved in-flight experiments and extensive photographic documentation of the lunar surface and Earth.

In 1969, NASA announced the backup crew for Apollo 14, which included Gene Cernan, Ronald Evans, and Joe Engle. This assignment positioned them as the prime crew for Apollo 17, following the program's crew rotation protocol where backup crews typically flew as prime crew three missions later. Harrison Schmitt, a professional geologist and astronaut, had served on the backup crew of Apollo 15, making him next in line to fly as the lunar module pilot for Apollo 18.

The prime crew for the Apollo 17 lunar landing mission are: Commander, Eugene A. Cernan (seated), Command Module pilot Ronald E. Evans (standing on right), and Lunar Module pilot, Harrison H. Schmitt.

The prime crew of Apollo 17 consisted of Captain Eugene Cernan as commander, Commander Ronald Evans as command module pilot, and Dr. Harrison Schmitt as lunar module pilot. Cernan, born on March 14, 1934, in Chicago, Illinois, was 38 years old at the time of the mission. He had previously flown as pilot of Gemini 9-A and lunar module pilot of Apollo 10, the mission that tested the lunar module in lunar orbit and served as a rehearsal for the first piloted lunar landing.

Ronald Evans, born on November 10, 1933, in St. Francis, Kansas, was 39 years old during Apollo 17. He had a background in electrical engineering, earning a B.S. from the University of Kansas in 1956 and an M.S. from the U.S. Naval Postgraduate School in 1964. Evans was selected as an astronaut in 1966 and made his first spaceflight on Apollo 17.

Harrison Schmitt, born on July 3, 1935, in Santa Rita, New Mexico, was 37 years old during the mission. As the first professional scientist to explore the Moon, he brought a unique perspective. Schmitt held a B.S. from the California Institute of Technology and a Ph.D. in geology from Harvard University, and he was selected as an astronaut in 1965.

During the Apollo 17 mission, a diverse team of support personnel played crucial roles in ensuring its success. Serving as capsule communicators (CAPCOMs) were Major Charles Gordon Fullerton (USAF), Lt. Colonel Robert Franklyn Overmyer (USMC), Robert Alan Ridley Parker, Ph.D., Joseph Percival Allen IV, Ph.D., Captain Alan Bartlett Shepard, Jr. (USN), Commander Thomas Kenneth "Ken" Mattingly II (USN), Charles Moss Duke Jr., Stuart Allen Roosa, and John Watts Young.

The support crew, consisting of Overmyer, Parker, and Fullerton, provided essential assistance in mission preparation and support tasks. Their role included attending meetings and ensuring that all mission procedures and documentation were meticulously prepared and updated, supporting the prime crew during their training and eventual mission execution.

Behind the scenes, the flight directors orchestrated the mission from Mission Control. Gerald D. Griffin led the first shift, while Eugene F. Kranz and Neil B. Hutchinson manned the second shift, and M.P. "Pete" Frank and Charles R. Lewis oversaw the third shift. Their responsibilities were critical; they made real-time decisions and managed the mission operations to ensure crew safety and mission objectives were met.

In the context of crew rotations and backups for Apollo missions, NASA strategically selected experienced astronauts for backup roles to effectively leverage their prior mission knowledge. Originally, the backup crew for Apollo 17 was slated to be the crew of Apollo 15—David Scott, Alfred Worden, and James Irwin. However, due to their involvement in the Apollo 15 postal covers incident, they were replaced by John W. Young, Charles Duke, and Stuart Roosa from Apollo 16. This decision aimed to capitalize on experienced personnel while maintaining mission readiness and continuity.

In September 1970, NASA canceled the planned Apollo 18 mission. This cancellation followed strong recommendations from the scientific community urging NASA to include a trained geologist on a lunar landing mission rather than a pilot without professional geological training. As a result, NASA reassigned Harrison Schmitt, a professional geologist and astronaut, to serve as the lunar module pilot for Apollo 17.

Following this decision, Deke Slayton, NASA's director of flight crew operations, was tasked with selecting the other two members of the Apollo 17 crew. The options considered were whether to include the remaining members of the Apollo 15 backup crew, Dick Gordon and Vance Brand or to choose Gene Cernan and Ronald Evans from the Apollo 14 backup crew. Ultimately, Slayton opted for Cernan and Evans.

However, the decision to assign Cernan was not without controversy within NASA. Earlier in January 1971, Cernan had been involved in a helicopter crash where he misjudged his altitude during a training exercise near Cape Kennedy, resulting in the helicopter crashing into the Indian River. Jim McDivitt, manager of the Apollo Spacecraft Program Office at the time, objected to Cernan's selection due to this incident. Despite McDivitt's concerns, Slayton decided to appoint Cernan to the mission.

After being offered the command of Apollo 17, Cernan initially advocated for Joe Engle to join him on the mission. However, it was clear to him that Harrison Schmitt had already been assigned as the lunar module pilot, regardless of Cernan's preference. Consequently, Cernan accepted the crew assignment with Evans and Schmitt.

When Gene Cernan was assigned to Apollo 17, he was a 38-year-old captain in the United States Navy. He had been selected as an astronaut in NASA's third group in 1963. Cernan's previous spaceflight experience included serving as the pilot of Gemini 9A in 1966 and as the lunar module pilot of Apollo 10 in 1969. Before his assignment to Apollo 17, he had also served on the backup crew for Apollo 14.

Ronald Evans, assigned to Apollo 17 at the age of 39, had been selected as part of NASA's fifth group of astronauts in 1966. At the time, he was lieutenant commander in the United States Navy. Apollo 17 marked Evans' first spaceflight experience.

Harrison Schmitt, a civilian, was 37 years old when assigned to Apollo 17. He held a doctorate in geology from Harvard University and had been selected as an astronaut in NASA's fourth group in 1965. Schmitt's background as a professional geologist made him uniquely qualified for his role as the lunar module pilot on Apollo 17. Similar to Evans, Apollo 17 was Schmitt's first journey into space.

Before the cancellation of Apollo 18 through 20, Apollo 17 had been initially planned to launch in September 1971, as outlined in NASA's tentative schedule established in 1969. However, the in-flight abort of Apollo 13 and subsequent modifications to the Apollo spacecraft caused delays that pushed back subsequent missions.

By early 1970, with Apollo 20 canceled, NASA decided to limit the number of Apollo missions to up to two per year. This decision was influenced by logistical considerations and the need to ensure the safety and success of each mission.

A significant factor in scheduling Apollo 17 for December 1972 was political. The mission's timing was deliberately set after the November presidential election to avoid any potential impact on President Richard Nixon's re-election campaign. Nixon, who had been deeply concerned about the safety of the Apollo 13 astronauts during their crisis, initially considered omitting funding for Apollo 17 from the budget. However, he was eventually persuaded to approve the December 1972 launch date for the mission.

Apollo 17, like its predecessors Apollo 15 and Apollo 16, was designated as a "J-mission." These missions were characterized by extended lunar surface stays lasting up to three days, enhanced scientific capabilities, and the use of the Lunar Roving Vehicle (LRV) to explore greater distances on the Moon's surface.

As the Apollo program's final manned lunar landing mission, Apollo 17 prioritized visiting high-value landing sites that had not been previously explored. Several potential landing sites were considered and rejected during the site selection process. For instance, landing in the crater Copernicus was dismissed because Apollo 12 had already gathered samples from that area, and other Apollo missions had also visited nearby regions near Mare Imbrium, where Copernicus is situated.

The lunar highlands near the crater Tycho were also ruled out due to the rugged terrain that would have posed challenges for the astronauts. Similarly, a site on the far side of the Moon within the crater Tsiolkovskiy was deemed impractical due to technical difficulties in maintaining communication with Earth during surface operations and the associated operational costs.

Another potential landing site southwest of Mare Crisium was rejected over concerns that a Soviet spacecraft could easily access the area and retrieve lunar samples. Indeed, shortly after Apollo 17's site selection, the Soviet Luna 20 mission accomplished this task.

Harrison Schmitt, the lunar module pilot for Apollo 17 and a professional geologist, initially advocated for a landing on the far side of the Moon. However, Director of Flight Operations Christopher C. Kraft informed him that such a mission was not feasible due to NASA's budget constraints, particularly the lack of funds for necessary communication satellites to maintain contact with astronauts on the lunar far side.

For Apollo 17, the final selection of the landing site involved careful consideration of several factors aligned with the mission's primary scientific objectives. Among the three sites that were seriously considered—Alphonsus Crater, Gassendi Crater, and the Taurus-Littrow Valley—the Taurus-Littrow Valley emerged as the chosen destination.

The decision to select Taurus-Littrow was driven by several key factors crucial to the mission's scientific goals. First, the site offered access to ancient highlands material, located significantly distant from Mare Imbrium. This ensured that the samples collected would provide valuable insights into geological processes that occurred in lunar history.

Landing site and surrounding area, as imaged from the Apollo 17 command module, 1972

Secondly, Taurus-Littrow presented opportunities to study materials from relatively young volcanic activities, specifically those less than three billion years old. This focus on volcanic samples aimed to enhance our understanding of the Moon's geological evolution and its volcanic history.

A critical consideration in site selection was to minimize overlap with the orbital paths previously traversed by Apollo 15 and Apollo 16. By choosing Taurus-Littrow, mission planners aimed to maximize the coverage of new terrain and geological features not previously explored by manned missions. This strategic approach was intended to optimize acquiring fresh scientific data and discoveries during the final Apollo lunar landing.

One influential factor in favor of Taurus-Littrow was the observations made by Al Worden, the Command Module Pilot (CMP) of Apollo 15, during his orbital passes over the site. Worden had noted geological features that he suspected to be of volcanic origin, adding weight to the site's scientific potential and influencing its final selection for Apollo 17.

NASA's Apollo Site Selection Board made the final decision to select Taurus-

Littrow as the landing site for Apollo 17 after evaluating various factors. Gassendi crater was eliminated from consideration due to concerns about the difficulty of reaching its central peak, which was deemed challenging due to the rough terrain surrounding it. On the other hand, although operationally more feasible than Taurus-Littrow, Alphonsus crater was judged to be of lesser scientific interest than the potential discoveries expected at Taurus-Littrow.

At Taurus-Littrow, several compelling scientific opportunities influenced its selection. The site offered the possibility of obtaining samples of old highland material, particularly from the remnants of a landslide event on the valley's south wall. Additionally, there was the potential to investigate relatively young volcanic activity, including explosive volcanic features. Despite Taurus-Littrow's similarity to Apollo 15's landing site in being on the border of a lunar mare, its scientific advantages outweighed any similarities or potential drawbacks.

The Apollo Site Selection Board, composed of NASA personnel and scientists tasked with setting scientific objectives and choosing landing sites for the Apollo missions, unanimously recommended Taurus-Littrow at its final meeting in February 1972. Based on this strong recommendation, NASA officially selected Taurus-Littrow as the landing site for Apollo 17, affirming its potential to yield significant scientific discoveries and fulfill the mission's ambitious exploration goals on the lunar surface.

During their preparation for the Apollo 17 mission, astronauts engaged in an exhaustive training regimen to equip them with the skills necessary for lunar exploration. Central to their training was mastering the techniques of lunar sample collection under simulated lunar gravity conditions. This involved extensive practice with the specialized equipment, including the intricacies of their spacesuits tailored for lunar conditions.

Additionally, astronauts familiarized themselves with the Lunar Roving Vehicle (LRV) operations, a critical component enabling mobility across the lunar surface. This included navigation exercises that mirrored the challenges they would face on the Moon's rugged terrain.

Geological field training was a pivotal aspect of their preparation, designed to simulate lunar geology missions as realistically as possible. Astronauts were briefed on the terrain using aerial imagery and maps, acquainting themselves with the topographical features and planned routes. Subsequently, they conducted field trips, following these routes, executing assigned tasks and making scientific observations at designated stops.

The training also encompassed survival skills essential for lunar missions, ensuring astronauts were prepared for any contingencies that might arise during their exploration activities. Moreover, they underwent rigorous splashdown and recovery exercises, rehearsing procedures for a safe return to Earth and retrieval by recovery teams.

In October 1971, the geology field trips commenced with a visit to Big Bend National Park in Texas as part of the astronauts' rigorous training regimen for the upcoming Apollo missions. Initially, these excursions were not tailored specifically for the Taurus-Littrow landing site, finalized in February 1972. However, by June of that year, the focus shifted, and the astronauts began training at locations carefully chosen to simulate conditions they would encounter at Apollo 17's designated landing site.

Gene Cernan participates in geology training in Sudbury, Ontario, in May 1972

In October 1971, the geology field trips commenced with a visit to Big Bend National Park in Texas as part of the astronauts' rigorous training regimen for the upcoming Apollo missions. Initially, these excursions were not tailored specifically for the Taurus-Littrow landing site, finalized in February 1972. However, by June of that year, the focus shifted, and the astronauts began training at locations carefully chosen to simulate conditions they would encounter at Apollo 17's designated landing site.

Eugene Cernan and Harrison Schmitt, the Apollo 17 astronauts, brought valuable experience to their roles. Having served on backup crews for previous Apollo missions, they were familiar with mission procedures and protocols. However, concerns arose among their trainers, including Gordon Swann, about potential dynamics within the team. There was apprehension that Cernan might defer too much to Schmitt, a professional geologist, regarding geological matters.

Moreover, Cernan had to adjust to the absence of his original crewmate, Ronald Evans, with whom he had trained extensively for Apollo 14. Despite these challenges, Cernan and Schmitt developed a cohesive working relationship. Cernan demonstrated a growing proficiency in describing geological features during field trips and learned to work independently when necessary, showcasing his adaptability and commitment to mastering the scientific aspects of their mission.

The Apollo 17 landing crew meticulously planned their tasks to optimize efficiency upon arriving at new lunar sites. Eugene Cernan's responsibilities included technical operations like adjusting the Lunar Roving Vehicle's antenna for Earth transmissions, ensuring crucial communication channels remained open. Meanwhile, Harrison Schmitt took on the role of providing detailed geological assessments of each site they explored.

Back on Earth, scientists in the geology "backroom" relied heavily on Schmitt's reports to refine and adjust the planned activities for each location. These adjustments were then relayed through mission control to Cernan and Schmitt via the Capsule Communicator (CapCom), ensuring their activities aligned with the evolving geological objectives.

William R. Muehlberger, a scientist involved in training the astronauts, noted Schmitt's significant role in directing the mission's scientific activities from the lunar surface. He observed, "In effect [Schmitt] was running the mission from the Moon." This division of labor was deliberately structured to leverage Schmitt's expertise in geology, acknowledging his pivotal role in shaping the mission's scientific outcomes.

While the arrangement may have been unconventional in some respects, Muehlberger indicated that it was a practical approach that effectively utilized Schmitt's specialized knowledge.

During the Apollo 17 mission preparations, the involvement of backup crew members added depth to the training exercises, particularly in geology. Before the Apollo 15 astronauts David Scott and James Irwin were assigned as the backup crew for Apollo 17 in February 1972, initial field trips were already underway. Scott and Irwin participated in four of these trips, with their joint presence noted in two instances. Following their removal from the backup crew, the new backup commander and lunar module pilot (LMP), John Young and Charles Duke, respectively, took part in the remaining four field trips.

During these geological excursions, the backup crew followed the prime crew's activities with a half-hour delay, mirroring their tasks and scenarios. They operated under simulated conditions, complete with their own dedicated Capsule Communicator (CapCom) and Mission Control support, ensuring readiness for any role they might need to assume in the event of a crew change or backup mission activation.

In stark contrast to the extensive preparation of Apollo 17, the Apollo 11 crew, pioneers of lunar exploration, undertook only one field trip as part of their training regimen.

Ronald Evans, the Command Module Pilot (CMP) for Apollo 17, underwent a specialized training regimen tailored for his role, distinct from the geological field trips undertaken by the mission's lunar landing crew. Instead of participating in field excursions, Evans received dedicated training sessions facilitated by NASA geologist and pilot Dick Laidley.

Evans' training involved aerial surveys conducted at varying altitudes to simulate lunar observation conditions. At 40,000 feet (12,000 m), the exercises provided a perspective akin to viewing lunar features from the planned orbit distance of approximately 60 nautical miles, enhancing his ability to identify and analyze geological formations. Lower altitude flights, ranging from 1,000 feet (300 m) to 5,000 feet (1,500 m), allowed for more detailed examination and photography practice, akin to the lunar surface close-ups expected during the mission.

Each training session was preceded by extensive briefings lasting several hours, supplemented with comprehensive study guides. Following the exercises, thorough debriefings and evaluations were conducted to refine Evans' observational skills and prepare him for the photographic tasks he would undertake from the Command and Service Module (CSM) orbiting the Moon.

Later in the training cycle, Evans received further instruction in lunar geology from Farouk El-Baz, ensuring he remained well-versed in the geological features he would encounter and document during the mission. This comprehensive training approach equipped Evans to fulfill his critical role in the Apollo 17 mission, contributing to the overall scientific objectives of lunar exploration.

The Apollo 17 spacecraft comprised several critical components essential for its lunar mission. It featured the Command and Service Module (CSM-114), consisting of Command Module 114 (CM-114) and Service Module 114 (SM-114). Alongside this was Lunar Module 12 (LM-12), housed within the Spacecraft-Lunar Module Adapter (SLA-21). The Launch Escape System (LES) was also integral, equipped with a rocket motor designed to propel the Command Module to safety in an emergency during launch.

Cernan (seated, right) and Schmitt in the training Lunar Roving Vehicle, with the mockup Lunar Module in the background, August 1972

During liftoff, the LES was crucial for ensuring the crew's safety in the initial moments post-launch. It was jettisoned once the launch vehicle ascended beyond the point where its protection was necessary. Meanwhile, the SLA remained atop the S-IVB third stage of the Saturn V rocket after separating the CSM and LM.

Apollo 17 was launched using the Saturn V rocket designated SA-512, one of fifteen Saturn V rockets manufactured. It was the twelfth to be launched in the series. At liftoff, the entire assembly weighed approximately 6,529,784 pounds (2,961,860 kg), with the spacecraft accounting for 116,269 pounds (52,739 kg). This made Apollo 17's launch vehicle slightly lighter than Apollo 16 but heavier than any other crewed Apollo mission, highlighting the scale and complexity of the equipment required for lunar exploration.

The Apollo 17 launch vehicle assembly at Kennedy Space Center proceeded with a carefully orchestrated timeline of component arrivals and integration. The process began with the arrival of the S-II second stage on October 27, 1970, followed by the S-IVB stage on December 21 of the same year. However, the S-IC first stage, a pivotal component, reached the Kennedy Space Center in May 11, 1972, marking a significant milestone in the assembly timeline.

After the S-IC's arrival, the Instrument Unit, critical for navigation and guidance, was delivered on June 7, 1972. Meanwhile, the Lunar Module (LM-12) components were also reaching the Kennedy Space Center. The ascent stage arrived on June 16, 1971, followed by the descent stage the next day. These two components were mated in May 18, 1972, ensuring meticulous preparation and integration.

On March 24, 1972, the Command Module (CM-114), Service Module (SM-114), and Spacecraft-Lunar Module Adapter (SLA-21) all arrived at Kennedy Space Center, consolidating the core elements of the spacecraft assembly. Finally, the Lunar Roving Vehicle (rover), crucial for surface mobility during the lunar mission, reached Kennedy Space Center on June 2, 1972, completing the logistical puzzle necessary for the Apollo 17 mission.

Following the Command Module (CM) and Service Module (SM) mating on March 28, 1972, extensive testing of the Apollo 17 spacecraft commenced at Kennedy Space Center. The CM and Lunar Module (LM) were subjected to rigorous testing in vacuum chambers to simulate space conditions. These tests were crucial for verifying the spacecraft's performance and ensuring its readiness for the lunar mission.

During testing, issues surfaced with the LM's rendezvous radar assembly, which had received excessive voltage during earlier tests. Grumman, the manufacturer, replaced the faulty assembly. The LM's landing radar experienced intermittent malfunctions and was replaced to ensure reliability during lunar descent operations.

SA-512, Apollo 17's Saturn V rocket, on the launch pad awaiting liftoff, November 1972

The Lunar Roving Vehicle (LRV) underwent its own challenges during testing. The front and rear steering motors required replacement, necessitating several modifications to improve their functionality and reliability.

By July 1972, following testing in the vacuum chambers, the LM's landing gear was installed and mated with the CM and Spacecraft-Lunar Module Adapter (SLA). This combined spacecraft unit was then transferred to the Vehicle Assembly Building in August for further comprehensive testing.

After completing all necessary tests, including a simulated mission scenario, the Lunar Roving Vehicle (LRV) was integrated into the LM on August 13, marking a significant milestone in the assembly and readiness process for the Apollo 17 mission.

The assembly and preparation of the Saturn V launch vehicle for Apollo 17 marked a significant milestone in NASA's space exploration efforts. Beginning on May 15, 1972, the erection of the stages took place in High Bay 3 of the Vehicle Assembly Building at Kennedy Space Center. This complex process, involving the stacking of stages, was completed by June 27, coinciding with the processing of launch vehicles for Skylab 1 and Skylab 2, making it the first time since 1969 that three launch vehicles were concurrently processed in the same facility.

After the spacecraft was mounted onto the launch vehicle on August 24, 1972, the assembled unit was rolled out to Pad 39-A on August 28. This event drew considerable attention from local residents, with around 5,000 spectators observing the rollout. The prime and operating crews from Bend joined atop the crawler, underscoring the collaborative effort and public anticipation surrounding the mission.

At Pad 39-A, rigorous testing and preparations continued. The Command and Service Module (CSM) was electrically mated to the launch vehicle on October 11, 1972, marking another critical step in readiness assessments. Countdown demonstration tests were successfully conducted on November 20 and 21, confirming the operational readiness of systems and personnel.

Finally, on December 5, 1972, at 7:53 a.m. local time (12:53 UTC), the countdown to launch commenced. This marked the beginning of the final preparations for the historic Apollo 17 mission, the last crewed lunar landing of the Apollo program, as the world watched with anticipation for the culmination of years of meticulous planning and preparation.

One of the critical scientific endeavors during the Apollo missions was the deployment of the Apollo Lunar Surface Experiments Package (ALSEP) on the Moon. Designed to operate autonomously after the astronauts' departure, ALSEP was a suite of sophisticated instruments powered by nuclear sources deployed from Apollo 12 onwards.

Each ALSEP package was meticulously crafted to unravel the mysteries of the lunar environment. On Apollo 17, the program's final mission, the array of experiments included several groundbreaking instruments. The Heat Flow Experiment (HFE) aimed to measure the Moon's internal heat flow, offering insights into its geological processes. Accompanying this was the Lunar Surface Gravimeter (LSG), tasked with mapping variations in the lunar gravity field at the landing site, enhancing our understanding of the Moon's interior structure.

The Lunar Atmospheric Composition Experiment (LACE) was included to probe the lunar atmosphere, analyzing the composition of gases surrounding the Moon. Meanwhile, the Lunar Seismic Profiling Experiment (LSPE) sought to detect and study moonquakes and other seismic activities, providing crucial data for lunar geophysics. Lastly, the Lunar Ejecta and Meteorites Experiment (LEME) measured the velocity and energy of dust particles ejected from the lunar surface, shedding light on lunar impacts and environmental dynamics.

While the Heat Flow Experiment had been tested on earlier missions, the remaining instruments introduced on Apollo 17 marked their inaugural journey to the Moon, expanding the scope of lunar exploration. These experiments were pivotal in advancing our knowledge of the Moon's geophysical properties, atmospheric composition, and geological history, cementing ALSEP's legacy as a cornerstone of scientific exploration during the Apollo era.

The history of the Apollo Lunar Surface Experiments Package (ALSEP) is marked by its pivotal role in advancing our understanding of the Moon's environment during the Apollo missions. Among its various components, the Heat Flow Experiment (HFE) had previously flown on Apollo 13, 15, and 16, but it was only successfully deployed on Apollo 15. The results from this mission prompted a careful reattempt on Apollo 17, which proved successful, allaying scientists' concerns and providing valuable data on lunar heat flow dynamics.

Meanwhile, the Lunar Surface Gravimeter (LSG) aimed to detect fluctuations in lunar gravity, supporting Albert Einstein's general theory of relativity. However, it encountered operational issues and did not function as intended.

The Lunar Atmospheric Composition Experiment (LACE), equipped with a mass spectrometer, analyzed the composition of the lunar atmosphere. Unlike its predecessor, the Code Cathode Gauge, which measured particle quantities, LACE identified specific gases such as neon, helium, and hydrogen.

The Lunar Seismic Profiling Experiment (LSPE) utilized geophones to detect seismic activity, including explosions set off remotely after the astronauts' departure. Operating at high bit rates, it provided critical data during significant events like the ascent stage liftoff and impact.

The Lunar Ejecta and Meteorites Experiment (LEME) measured characteristics of lunar dust particles, primarily detecting slow-moving dust across the lunar surface rather than interstellar debris or comet impacts as initially hoped.

Budget constraints led to the deactivation of all powered ALSEP experiments by September 30, 1977. Despite the challenges and limitations, ALSEP played a crucial role in expanding our knowledge of the Moon's geological, atmospheric, and seismological aspects, leaving a lasting legacy in lunar exploration.

Apollo 17, like its predecessors Apollo 15 and 16, was equipped with a Lunar Roving Vehicle (LRV), a revolutionary addition that significantly expanded the astronauts' mobility and operational capabilities on the lunar surface. Throughout the mission's three moonwalks, the LRV served as a vital means of transport between different exploration sites, enabling astronauts Eugene Cernan and Harrison Schmitt to efficiently navigate the rugged lunar terrain.

Beyond transportation, the LRV played a crucial role in carrying essential equipment, including tools, communication gear, and the lunar samples collected during their expeditions. Additionally, it transported scientific instruments such as the Traverse Gravimeter Experiment (TGE) and the Surface Electrical Properties (SEP) experiment, enhancing the mission's scientific objectives.

Throughout its operational life on Apollo 17, the LRV covered a remarkable cumulative distance of approximately 35.7 kilometers (22.2 miles). The longest single traverse taken by Cernan and Schmitt extended about 7.6 kilometers (4.7 miles) from their lunar module, demonstrating the vehicle's capability to significantly extend the range of exploration during their lunar stay.

The Surface Electrical Properties (SEP) experiment conducted during Apollo 17 represented a pioneering effort to understand the electrical characteristics of the lunar regolith. This experiment was unique to Apollo 17 and involved two primary components: a transmitting antenna deployed near the lunar module and a receiver mounted on the Lunar Roving Vehicle (LRV).

Electrical signals were transmitted from the antenna into the lunar regolith throughout the mission's traverses across the lunar surface. These signals traveled through the lunar soil and were received by the antenna on the LRV at various stops along their exploration routes. By comparing the transmitted and received electrical signals, scientists could analyze and interpret the electrical properties of the lunar regolith.

The findings from the SEP experiment yielded valuable insights into the composition of the lunar surface. Specifically, the data indicated that the electrical properties observed were consistent with those expected from lunar rock composition. One significant conclusion drawn from the experiment was the absence of detectable water in the area of the Moon where Apollo 17 landed, extending to a depth of 2 kilometers (1.2 miles).

The Lunar Neutron Probe deployed during Apollo missions was a specialized device measuring 2.4 meters (7.9 feet) long and 2 centimeters (0.79 inches) in diameter. Inserted into a drilled hole on the lunar surface during the first Extravehicular Activity (EVA), its purpose was to detect neutron flux penetrating its length, providing insights into the "gardening" process affecting the lunar regolith. This process involves gradually mixing and burying surface materials due to impacts from micrometeorites and other lunar events.

Operational throughout the mission, the probe collected crucial data until its retrieval during the third and final EVA. Astronauts transported it back to Earth with core samples extracted from the same site. Comparing the neutron measurements from the probe with those obtained from the extracted cores enabled scientists to deduce significant findings about lunar surface dynamics.

The results indicated that the top layer of the lunar regolith undergoes turnover approximately every million years, driven by surface "gardening" processes. In contrast, deeper layers, down to a meter, experience much slower turnover rates, estimated at a billion years. These findings have been instrumental in shaping current theories about lunar geological processes and the evolution of its surface over geological timescales.

On Apollo 17, the Command Module (CM) carried out a unique biological experiment known as BIOCORE, designed to study the effects of cosmic rays on living organisms. This experiment involved five pocket mice (Perognathus longimembris), each implanted with radiation monitors under their scalps to measure potential damage from cosmic radiation exposure during the mission.

The mice were housed individually in metal tubes within a sealed container that provided its oxygen supply, ensuring their isolation and survival in space. The choice of pocket mice for this experiment was deliberate: they were well-documented, small in size, and capable of enduring environmental stress without needing drinking water during the mission, thanks to their highly concentrated waste output.

Officially identified by the numbers A3326, A3400, A3305, A3356, and A3352, the mice were affectionately nicknamed by the Apollo 17 crew as Fe, Fi, Fo, Fum, and Phooey, according to Eugene Cernan. This lighthearted naming reflected the camaraderie and human touch amidst the mission's rigorous scientific pursuits.

BIOCORE represented a pioneering effort in space biology, seeking to understand how cosmic radiation impacts living organisms beyond Earth's protective atmosphere. The data collected from this experiment contributed valuable insights into the biological effects of space radiation exposure, essential for planning future long-duration space missions and ensuring astronaut health and safety.

The BIOCORE experiment aboard Apollo 17 yielded intriguing results regarding the effects of cosmic rays on living organisms in space. Four of the five pocket mice carried on the mission survived the flight, although only two showed signs of being healthy and active upon return. Unfortunately, the cause of death for the fifth mouse could not be determined. Upon examination, lesions were found in the scalps of the surviving mice, and in one case, lesions were also observed in the liver. Interestingly, these lesions did not appear to be correlated, and there was no clear evidence attributing them to cosmic ray exposure.

In addition to BIOCORE, the Biostack experiment on Apollo 17 was similar to one flown on Apollo 16. It aimed to investigate the effects of cosmic rays encountered during space travel on various biological samples, all housed in a sealed container. These included microorganisms, seeds, and the eggs of simple animals like brine shrimp and beetles.

Post-mission analysis revealed that while the microorganisms and seeds showed minimal effects from cosmic ray exposure, many eggs across all species failed to hatch or mature normally. A significant number also died or exhibited abnormalities, indicating a potential sensitivity to cosmic radiation.

The Scientific Instrument Module (SIM) bay aboard the Apollo 17 Service Module (SM) housed advanced scientific instruments to conduct detailed studies during lunar orbit. Among the new experiments included were the lunar sounder, infrared scanning radiometer, and far-ultraviolet spectrometer, complemented by previously carried instruments such as the mapping camera, panoramic camera, and laser altimeter.

The lunar sounder operated by emitting electromagnetic pulses toward the lunar surface. Its primary goal was to gather data essential for constructing a geological model of the Moon's interior down to approximately 1.3 kilometers (0.81 miles) beneath the surface. By analyzing the echoes and reflections of these pulses, scientists aimed to map the structure and composition of the lunar crust.

The infrared scanning radiometer was designed to generate a detailed temperature map of the lunar surface. This map was instrumental in identifying thermal variations across the Moon and discovering surface features like rock fields, structural differences in the crust, and volcanic activity. This data provided crucial insights into the Moon's dynamic thermal environment.

The far-ultraviolet spectrometer played a critical role in studying the lunar atmosphere. It measured the lunar exosphere's composition, density, and constituents, providing valuable information on gases and particles present in the Moon's tenuous atmosphere. Additionally, it detected far-ultraviolet radiation reflected off the lunar surface, contributing to our understanding of solar interactions with the Moon.

Complementing these instruments, the laser altimeter accurately measured the spacecraft's altitude above the lunar surface with a precision of approximately 2 meters

(6.6 feet). This altitude data was crucial for the operations of the panoramic and mapping cameras, which captured the lunar terrain's detailed imagery and topographical information.

These instruments in the SIM bay of Apollo 17's Service Module facilitated groundbreaking scientific investigations during lunar orbit, significantly advancing our knowledge of the Moon's geology, atmosphere, and surface characteristics. Their contributions marked a significant milestone in lunar exploration, paving the way for future missions and deeper insights into planetary science.

On December 5, 1972, under the expansive Florida sky at Cape Kennedy, preparations for the historic Apollo 17 mission were meticulously underway. The Saturn V rocket, designated SA-512, was towering on the launch pad, poised to carry the crew toward the moon. The Command and Service Module (CSM-114), affectionately named "America," and the Lunar Module (LM-12), known as "Challenger," were integral parts of this monumental endeavor.

The final countdown commenced at T-28 hours, marking the beginning of a meticulously orchestrated sequence. Scheduled holds punctuated the countdown, including a significant pause at T-9 hours for nine hours and another at T-3 hours 30 minutes for an hour. Progress was momentarily stalled at T-2 minutes 47 seconds when the Terminal Countdown Sequencer failed to initiate the S-IVB LOX tank pressurization command. This glitch prompted an automatic hold at T-30 seconds, extending for 1 hour and 5 minutes.

Undeterred, the team recycled the countdown to T-22 minutes, encountering another hold at T-8 minutes to address the sequencer issue, which extended for 1 hour and 13 minutes. Eventually, the countdown resumed at T-8 minutes and proceeded smoothly toward the final launch phase. In total, these delays amounted to 2 hours and 40 minutes, a testament to the resilience and precision of the mission control team.

As evening descended over Cape Kennedy, mild temperatures and gentle surface winds prevailed, courtesy of a warm, moist air mass enveloping Florida. This contrasted sharply with the frigid air sweeping across the southern United States, separated by a northeast-southwest oriented cold front traversing the Florida panhandle. Near the launch site, light northwesterly winds prevailed, with the most intense wind belt positioned to the north, resulting in favorable conditions aloft.

At the precise moment of launch, the sky revealed stratocumulus clouds covering 20 percent of the expanse at 2,600 feet, while cirrus clouds adorned 50 percent of the sky at a lofty 26,000 feet. The temperature stood at a comfortable 70.0°F, with a relative humidity of 93 percent and a barometric pressure of 14.785 lb/in². Anemometer readings at 60 feet above ground measured winds at 8.0 knots from the south-southeast, while at 330 feet, winds increased slightly to 10.5 knots from the north-northeast.

Originally scheduled for December 6, 1972, at 9:53 p.m. EST (2:53 a.m. UTC on December 7), Apollo 17 marked the final crewed launch of the Saturn V rocket and the only one to take place at night. However, the historic launch faced a delay of two hours and forty minutes due to a critical issue in the countdown sequence.

At T-30 seconds before liftoff, the launch sequencer failed to command the pressurization of the liquid oxygen tank in the third stage of the Saturn V. Although launch control manually initiated the pressurization, the sequencer did not register the correction, triggering an automatic hold in the countdown. Technicians quickly identified the problem and reset the countdown at T-22 minutes, working diligently to rectify the malfunction.

This unexpected delay, caused by a rare hardware issue, represented the only setback in the Apollo program. After resolving the sequencer malfunction, the countdown resumed smoothly. Finally, at 12:33 a.m. EST on December 7, 1972, Apollo 17 thundered skyward, commencing its journey toward the Moon.

The launch window, originally scheduled for 9:53 p.m. EST on December 6, remained open until 1:31 a.m., providing a narrow but crucial timeframe for the mission to begin within the designated period.

Apollo 17 embarked on its journey to the Moon from Kennedy Space Center Launch Complex 39, Pad A, precisely at 12:33:00 a.m. EST on December 7, 1972. The launch was meticulously timed within a window spanning from 02:53:00 GMT to 06:31:00 GMT on the same day, leveraging optimal conditions with a sun elevation angle of 13.3° over the lunar surface.

During the initial moments of ascent, between 000:00:12.9 and 000:00:14.3, the vehicle executed a controlled roll maneuver, transitioning from a launch pad azimuth of 90° to a flight azimuth of 91.504°. The powerful S-IC engine powered the rocket until shutdown at 000:02:41.20, followed swiftly by separating the S-IC stage and igniting the S-II stage engine (S-II).

Nine minutes and nineteen seconds into flight, the S-II engine ceased operation at 000:09:19.66, separating from the S-IVB stage. Shortly after 000:09:23.80, the S-IVB stage ignited, continuing the propulsion toward the Moon. The first cutoff of the S-IVB engine occurred at 000:11:42.65, with deviations from the planned trajectory minimal: just +1.0 ft/sec in velocity and -0.1 nautical miles in altitude.

Throughout the ascent, the spacecraft encountered maximum wind speeds of 87.6 knots at 311° true north at 38,945 feet. The maximum wind shear experienced was 0.0177 seconds^-1 at 26,164 feet, highlighting the precise navigation and control required during this critical mission phase.

Approximately half a million spectators gathered near Kennedy Space Center to witness the early-morning launch of Apollo 17. Despite the hour, the spectacle was visible as far as 800 kilometers (500 miles) away, with observers in Miami, Florida, catching sight of a distinctive "red streak" crossing the northern sky.

Among the notable attendees at this historic event were astronauts Neil Armstrong and Dick Gordon, distinguished figures who had themselves ventured into space. Also present was centenarian Charlie Smith, who claimed an impressive age of 130 years at the time, adding a touch of extraordinary longevity to the occasion.

Following insertion into a parking orbit, recorded at 000:11:42.65 (S-IVB cutoff plus 10 seconds for engine tailoff and transient effects), Apollo 17 exhibited specific orbital characteristics. The spacecraft's trajectory displayed an apogee and perigee of 90.3 by 90.0 nautical miles, orbiting at an inclination of 28.526°. The orbital period was measured at 87.83 minutes, maintaining a velocity of 25,604.0 feet per second.

Apollo 17 launches on December 7, 1972

These parameters were calculated based on a spherical Earth model with a radius of 3,443.934 nautical miles. The International Designator assigned to the Command and Service Module (CSM) upon achieving orbit was 1972-096A, while the S-IVB stage was designated 1972-096B.

Upon undocking at the Moon's surface, the Lunar Module's ascent stage would assume the designation 1972-096C, with the descent stage designated 1972-096D. This would mark each phase of Apollo 17's journey with precision and clarity in international space tracking systems.

After confirming the inflight systems checks, Apollo 17 executed its crucial translunar injection maneuver, known as the second firing of the S-IVB engine, precisely at 003:12:36.60 into the mission. This maneuver was timed meticulously to achieve a velocity of 35,579.4 feet per second, setting the spacecraft on its trajectory toward the Moon after completing two Earth orbits spanning 3 hours, 6 minutes, and 44.99 seconds.

The S-IVB engine shutdown occurred at 003:18:27.64, followed by the translunar injection ten seconds later, ensuring that Apollo 17 maintained its course to arrive at the Moon as originally scheduled. To optimize their trajectory, the crew performed a minor midcourse correction, a 1.73-second burn, adjusting the velocity by 10.5 feet per second at 035:29:59.91 into the mission.

Following these critical maneuvers, Apollo 17 settled into its trajectory, orbiting the Earth as the crew meticulously monitored and verified the spacecraft's systems to ensure readiness for departure from Earth orbit. At 3:46 a.m. EST, the S-IVB third stage reignited for the pivotal 351-second trans-lunar injection burn, propelling the spacecraft on its definitive journey toward the Moon.

During Apollo 17, critical maneuvers and operations unfolded meticulously, reflecting the precision and complexity of lunar missions. At 003:42:27.6, the Command and Service Module (CSM) separated from the S-IVB stage, a pivotal step before transposition and docking occurred at 003:57:10.7. This maneuver, however, was not without its challenges, as indications of a ring latch malfunction during docking required immediate attention. After pressurizing the Lunar Module (LM) and removing the hatch, troubleshooting revealed issues with latches 7, 9, and 10, which were manually set to ensure secure docking.

Subsequent maneuvers continued to shape the trajectory of the mission. The docked spacecraft were ejected from the S-IVB at 004:45:02.3, followed by a precise 79.9-second separation maneuver at 005:03:01.1. Later, at 006:09:59.8, the S-IVB tanks were vented, and the auxiliary propulsion system fired for 98.2 seconds to target the S-IVB for lunar impact. A second critical maneuver lasting 102.4 seconds was performed at 011:14:59.8, further refining the spacecraft's trajectory.

The culmination of these maneuvers came with the deliberate lunar impact of the S-IVB at 086:59:40.99. The impact site, located at latitude 4.33° south and longitude 12.37° west, was meticulously planned despite being 84 nautical miles from the intended target point. This event, significant in its scientific implications, was monitored closely by seismometers left by previous Apollo missions—Apollo 12, 14, 15, and 16—all of which recorded the impact. At the moment of impact, the S-IVB carried a weight of 30,712 pounds and was traveling at a speed of 8,346 feet per second, underscoring the force involved in this controlled lunar event.

Complications earlier in the mission, including a 2-hour 40-minute launch delay, necessitated ground controllers to adjust Apollo 17's trajectory. Opting for a faster path ensured the spacecraft would reach lunar orbit as scheduled despite the setback. Approximately half an hour after the S-IVB's trans-lunar injection burn, the CSM separated from the S-IVB, aligning itself to dock with the LM still attached to the spent stage. Following the successful extraction of the LM, Mission Control programmed the now-redundant S-IVB to impact the Moon, activating seismometers left by previous missions.

As Apollo 17 approached the conclusion of its first day, approximately nine hours post-launch, the crew entered a scheduled sleep period, preparing for the challenges of the mission's second day.

At 040:10 into Apollo 17's mission, the commander and lunar module pilot began transferring into the Lunar Module (LM). Upon ingress, they encountered an issue with the #4 docking latch, which was not properly latched. Acting swiftly, the command module pilot adjusted the latch handle between 30° and 45°, successfully disengaging the hook from the docking ring. After consulting with ground control, it was decided to defer further action on the latch until the second LM activation.

The subsequent tasks proceeded smoothly, with the remainder of the LM housekeeping proceeding nominally, concluding with the closure of the LM at 042:11. Planned experiments, including heat flow and convection demonstrations, commenced promptly. The first demonstration, initiated at 042:55, was conducted with the spacecraft in attitude hold, while the second run utilized the passive thermal control mode. Both demonstrations yielded satisfactory results and were completed by 046:00.

Following these experiments, the second LM housekeeping session commenced at 059:59 and wrapped up at 062:16, during which all LM systems checks returned as nominal. Concurrently, the command module pilot continued troubleshooting the #4 docking latch issue encountered earlier. Following ground control instructions, he manipulated the latch handle, successfully cocking the latch into position. This adjustment ensured readiness for the mission's upcoming CSM/LM rendezvous phase.

Beginning with Apollo 11, astronauts reported observing peculiar light flashes that penetrated their closed eyelids during spacecraft darkness, typically while in sleep mode. These flashes, often described as "streaks" or "specks" of light, occurred at an average rate of about two per minute. Crew members observed them during the journey to the Moon, back to Earth, and in lunar orbit. Notably, these phenomena were not observed while on the lunar surface.

To investigate these light flashes, the Apollo 17 crew repeated an experiment conducted during Apollo 16. Command Module Pilot Ronald Evans wore a specialized device over his eyes that recorded the timing, intensity, and trajectory of high-energy atomic particles that penetrated the device. Meanwhile, the other two crew members wore blindfolds to prevent any external light from reaching their eyes. Data analysis led investigators to conclude that these flashes likely result from charged particles passing through the astronauts' retinas.

Apollo 17 also carried a sodium-iodide crystal identical to those in the gamma-ray spectrometer flown on Apollo 15 and 16. This crystal was intended to gather data on gamma-ray emissions, which, upon examination back on Earth, would help establish a baseline for subtracting contributions from the Command Module (CM) or cosmic radiation. This subtraction aimed to refine the interpretation of earlier gamma-ray spectrometer results and enhance our understanding of lunar surface composition.

Additionally, Apollo 17's mission included investigating lunar gravity variations using S-band transponders. These instruments were directed toward the Moon to collect data on its gravitational field. Previous data from Lunar Orbiter probes had revealed that lunar gravity exhibits slight variations due to mass concentrations known as "mascons". Data gathered during the Apollo missions and information from lunar subsatellites left by Apollo 15 and 16 contributed to mapping these variations and deepening our knowledge of lunar gravitational anomalies.

At 068:19 into the Apollo 17 mission, the crew engaged in a significant observational task—a one-hour study of visual light flashes. During this period, they reported witnessing a spectrum of light flashes, varying in intensity from bright to faint, contributing valuable data to ongoing scientific investigations.

Later, at 081:32:40, another operational milestone occurred with the jettisoning of the scientific instrument module bay door. This action marked a step toward optimizing the spacecraft's configuration for upcoming maneuvers.

One of the mission's critical maneuvers occurred at 086:14:22.60 when the service propulsion engine was ignited. This burn lasted 393.16 seconds, strategically positioning the spacecraft into a lunar orbit characterized by parameters of 170.0 by 52.6 nautical miles. This precise insertion maneuver was conducted at 76.8 nautical miles above the lunar surface, culminating an 83-hour, 2-minute, and 18.11-second translunar coast phase.

As Apollo 17 progressed toward its lunar objectives, adjustments, and critical maneuvers defined the mission's unfolding narrative. Following the launch delay, which extended the first day of the mission, Mission Control and the crew made strategic decisions to optimize the crew's sleep schedule. This adjustment aimed to align wake-up times for subsequent days, crucially anticipating an early morning (EST) wake-up on the day of the planned lunar landing, originally scheduled for early afternoon (EST).

On the mission's third day, after a necessary rest period, the crew executed their first mid-course correction. This brief two-second burn of the Command and Service Module's (CSM) service propulsion engine refined the spacecraft's trajectory toward the Moon. Subsequently, they thoroughly checked the Lunar Module (LM)'s systems upon opening the hatch separating the CSM and LM, confirming all systems were operating nominally.

To ensure synchronization with the flight plan, mission clocks were adjusted forward by 2 hours and 40 minutes, matching the duration of the launch delay. One hour of this adjustment occurred at 45:00:00 into the mission, with the remaining hour at 65:00:00.

Entering the Lunar Orbit/Lunar Surface Phase marked a pivotal stage in the mission. At 090:31:37.43, a 22.27-second maneuver of the service propulsion system lowered the spacecraft into a descent orbit measuring 59.0 by 14.5 nautical miles, preparing for the forthcoming undocking of the LM. The CSM and LM remained in this orbit for 17 hours until they were undocked through a precise

3.4-second maneuver at 107:47:56, executed at 47.2 nautical miles and in an orbit of 61.5 by 11.5 nautical miles.

Following undocking, a 3.80-second maneuver at 109:17:28.92 circularized the CSM's orbit to 70.0 by 54.0 nautical miles. Shortly after, the second LM descent orbit insertion maneuver, a 21.5-second burn at 109:22:42, further adjusted the LM's orbit to 59.6 by 6.2 nautical miles. The highlight of this phase was the extensive 725-second powered descent maneuver initiated at 110:09:53 from an altitude of 8.7 nautical miles, paving the way for the lunar landing.

During their journey toward the Moon, the Apollo 17 crew was immersed in critical tasks and scientific observations. As they ventured farther from Earth, they focused on documenting their celestial voyage, capturing iconic images of our planet, including the renowned photograph now immortalized as "The Blue Marble."

Amidst their duties, they encountered technical challenges typical of space travel. One such hurdle involved a latch that had unexpectedly come undone, requiring astronaut Ron Evans to meticulously address the issue while his fellow crew members, Harrison Schmitt, and Eugene Cernan, were engaged in the meticulous housekeeping duties essential for their mission's success. Evans's expertise prevailed, ensuring the latch was properly secured for the anticipated docking of the Command and Service Module (CSM) with the Lunar Module (LM) upon their return from the lunar surface.

Their outbound journey also included conducting vital scientific experiments. They demonstrated heat flow and convection principles, crucial for understanding thermal dynamics in space, and participated in the Apollo light-flash experiment, which aimed to study the mysterious phenomenon of light flashes observed by astronauts during spaceflight.

As their spacecraft approached lunar orbit, another pivotal moment arrived when the SIM door on the Service Module (SM) was jettisoned, marking a procedural step before their critical maneuver to enter lunar orbit. At precisely 2:47 p.m. EST on December 10th, the CSM's powerful service propulsion system engine fired, decelerating their combined spacecraft stack and initiating their descent into lunar orbit. Once stabilized in orbit around the Moon, the crew focused intently on readying themselves for the upcoming landing at the designated site, Taurus-Littrow.

"The Blue Marble" is a famous photograph of the Earth taken on December 7, 1972, by the crew of the Apollo 17 spacecraft en route to the Moon at a distance of about 29,400 kilometres (18,300 mi). It shows Africa, Antarctica, and the Arabian Peninsula.

On the pivotal day of their lunar landing, the Apollo 17 mission commenced with a thorough examination of the Lunar Module's systems, confirming everything was operational for the next phase. Astronauts Eugene Cernan, Ronald Evans, and Harrison Schmitt, suited up in their specialized gear, preparing for the intricate maneuvers ahead. Cernan and Schmitt entered the Lunar Module (LM) Challenger, while Evans remained aboard the Command and Service Module (CSM) America in orbit around the Moon.

With precise coordination, the LM disengaged from the CSM, and for the next hour and a half, the two spacecraft maintained close proximity. During this period, Cernan and Schmitt meticulously inspected their surroundings and conducted final checks to ensure all systems were primed for the descent to Taurus-Littrow.

Apollo 17 SIM bay on the service module America, seen from the Lunar Module Challenger in orbit around the Moon

Following separation from the CSM, Challenger adjusted its orbit, aligning its trajectory so that its lowest point would be approximately 10.5 miles (16.9 kilometers) above the chosen landing site. As they readied for descent, Cernan and Schmitt focused on the complex procedures necessary to land on the lunar surface safely. At the same time, Evans, orbiting above, conducted scientific observations and experiments crucial to the mission's broader objectives.

Less than three hours after the Lunar Module (LM) Challenger separated from the Command and Service Module (CSM), astronauts Eugene Cernan and Harrison Schmitt initiated their descent toward the Taurus-Littrow valley on the lunar surface. With the precise ignition of the LM's descent propulsion system (DPS) engine, the spacecraft began its controlled journey downwards. Approximately ten minutes into the descent, the LM pitched forward, providing Cernan and Schmitt with their first panoramic view of the landing site. Cernan skillfully guided the LM toward a designated landing spot while Schmitt fed crucial flight data to the onboard computer.

At 2:55 p.m. EST on December 11, just over twelve minutes after the DPS ignition, Challenger gently touched down on the Moon's surface. The landing occurred slightly east of the intended target, approximately 656 feet (200 meters) away, within the awe-inspiring Taurus-Littrow region.

Immediately following touchdown, Cernan and Schmitt swiftly transitioned into configuring the LM for their lunar stay. They prepared meticulously for their first extravehicular activity (EVA-1), donning their spacesuits and depressurizing the LM cabin at 114:21:49 GMT. Stepping onto the lunar surface, they unloaded the lunar roving vehicle (LRV-3) at 114:51:10 GMT, setting the stage for groundbreaking exploration and scientific endeavor in humanity's ongoing quest beyond Earth.

During their extended 75-hour stay on the lunar surface, astronauts Eugene Cernan and Harrison Schmitt conducted a series of meticulously planned activities and explorations. Throughout three exhilarating moonwalks (EVAs), they executed a sequence of scientific tasks to further humanity's understanding of our celestial neighbor.

Their priority was deploying the lunar roving vehicle (LRV), a crucial tool that would greatly enhance their mobility across the rugged lunar terrain. With the LRV in operation, they embarked on a systematic journey to nine predetermined geological survey stations. They meticulously collected samples at each station and conducted detailed observations, gathering invaluable data about the Moon's geology and composition.

In addition to these planned stops, Schmitt, as the Lunar Module Pilot, had the flexibility to make twelve short sampling stops while riding the rover. Using a handheld scoop, these impromptu collections allowed them to gather samples efficiently without needing to dismount.

Throughout their lunar surface operations, Commander Cernan took the helm as the rover driver, navigating the lunar landscape precisely, while Schmitt served as navigator and assisted with operational tasks. This division of roles was standard practice during Apollo's J-missions, ensuring efficient execution of mission objectives.

The inaugural lunar excursion of Apollo 17 commenced just four hours after touchdown, at 6:54 p.m. EST on December 11th. Commander Eugene Cernan, filled with a sense of historical significance, descended the ladder of the Lunar Module (LM) Challenger and uttered his memorable dedication: "I'm on the footpad. And, Houston, as I step off at the surface at Taurus-Littrow, we'd like to dedicate the first step of Apollo 17 to all those who made it possible."

Stepping onto the lunar surface, Cernan conducted an initial survey around the LM, remarking on the immediate landing site before being joined by Lunar Module Pilot Harrison Schmitt. Their first task was to unload the lunar roving vehicle (LRV) and essential equipment from the LM to facilitate their exploration.

During these early moments on the Moon, a minor mishap occurred when Cernan inadvertently caught his hammer under the right-rear fender extension of the rover, breaking it off. This incident, reminiscent of a similar occurrence on Apollo 16, was resolved temporarily with duct tape and a paper map attached to the damaged fender during the second EVA. However, lunar dust interfered with the tape's adhesive, limiting its effectiveness.

Cernan on the lunar surface, December 13, 1972

Following the deployment of the LRV and preparations for traversing to the Apollo Lunar Surface Experiments Package (ALSEP) site, another issue arose when Cernan inadvertently dislodged the right rear fender extension of the LRV. The extension was initially secured with tape but later detached during EVA-1, resulting in a shower of lunar dust over the crew and the rover.

Undeterred by these challenges, Cernan and Schmitt proceeded with their mission objectives, conducting a test drive of the LRV, gathering samples, and capturing panoramic photographs of the lunar landscape.

Following the deployment and successful testing of the lunar roving vehicle (LRV), the crew of Apollo 17 proceeded with deploying the Apollo Lunar Surface Experiments Package (ALSEP), situated just west of their landing site in the Taurus-Littrow valley. However, this operation encountered unexpected challenges, particularly with the drilling of core holes, which proved more difficult than anticipated. As a result, the geological portion of their first Extravehicular Activity (EVA) had to be shortened, leading to the cancellation of a planned visit to Emory crater.

Undeterred, Commander Eugene Cernan and Lunar Module Pilot Harrison Schmitt redirected their efforts to the Steno Crater, located south of the landing site. Their objective at Steno was to sample subsurface materials exposed by the crater's impact. They collected 14 kilograms (31 pounds) of lunar samples, conducted seven gravimeter measurements to study lunar gravity variations, and deployed two explosive packages throughout their exploration. These packages were later detonated remotely, with the resulting seismic signals captured by geophones and seismometers left behind by previous missions, contributing to ongoing lunar geophysical studies.

The first EVA concluded after seven hours and twelve minutes, marking a milestone in lunar exploration. Following their extensive extravehicular activities, Cernan and Schmitt retreated to the pressurized safety of the Lunar Module (LM), where they would spend the next 17 hours.

During their EVA, notable milestones included the deployment of the American flag at 115:40:58 and the precise offloading of the ALSEP package at 115:58:30. Further activities included traverse gravimeter readings and the drilling of two holes for heat flow experiment probes and a deep core sample at the ALSEP site. Their scientific endeavors also encompassed conducting the Surface Electrical Properties experiment and deploying another seismic profiling explosive charge.

The crew re-entered the LM at 121:33:42, concluding a productive first EVA encompassing approximately 3.3 kilometers (10,800 feet) of lunar travel in the rover, with a total vehicle drive time of 33 minutes.

On December 12, awoken by the dramatic strains of "Ride of the Valkyries" playing from Mission Control, Eugene Cernan and Harrison Schmitt embarked on their second lunar excursion. Their first task was to address the rover's damaged fender, which had become a concern after being dislodged during the previous day's activities. Overnight, flight controllers devised a makeshift solution communicated by John Young: they taped together four stiff paper maps to form a "replacement fender extension" and secured it with clamps onto the rover's fender. This improvisation proved effective, holding up until near the end of their third excursion.

Setting out for their destination at Nansen Crater, located at the base of the imposing South Massif, Cernan and Schmitt ventured to a record-setting distance of 7.6 kilometers (4.7 miles, 25,029 feet) from the Lunar Module Challenger. This remains the furthest any humans have traveled from the safety of a pressurized spacecraft on a planetary body, a testament to both their courage and the capabilities of the lunar rover. They carefully monitored their "walkback limit," a safety measure ensuring they could return to the LM on foot if the rover experienced mechanical issues.

Their second Extravehicular Activity (EVA-2) commenced 80 minutes behind schedule, with the LM cabin depressurizing at 137:55:06 GMT. Before embarking on their traverse, ground controllers transmitted instructions to improvise another replacement for the compromised fender extension. This time, the astronauts crafted a rig from four maps taped together and secured with clamps from portable utility lights, effectively addressing the issue.

During their traverse to the Surface Electrical Properties experiment site, the mission plan was adapted to allow more time at points of geological interest. This flexibility enabled Cernan and Schmitt to conduct thorough scientific observations and collect valuable data, further enhancing our understanding of the Moon's composition and geological history.

During their second lunar excursion on December 12, Eugene Cernan and Harrison Schmitt encountered several notable events and discoveries. At Station 3, Schmitt had a memorable moment when he stumbled awkwardly while working, prompting a humorous remark from CAPCOM Bob Parker about NASA needing his services for Houston's ballet group. This incident led to Station 3 being humorously renamed Ballet Crater in 2019. Meanwhile, Cernan collected a sample at Station 3 intended for future analysis under better conditions, playfully suggesting to Parker about leaving a note inside the sample container—a whimsical gesture that remained unopened until 2022.

Continuing to Station 4—Shorty Crater—the astronauts made a remarkable discovery: orange soil composed of tiny beads of volcanic glass, estimated to have formed over 3.5 billion years ago. This find sparked excitement at Mission Control, initially suggesting a potential volcanic vent discovery. However, post-mission analysis clarified that Shorty Crater is an impact crater. The orange soil, it was determined, originated from a lava fountain that ejected molten lava droplets high into the lunar sky during the Moon's early history. These droplets solidified and were later exposed by the impact that formed Shorty Crater less than 20 million years ago.

During their exploration at Station 4, the crew deployed three explosive packages as part of the lunar seismic profiling experiment, conducted seven traverse gravimeter measurements, gathered numerous samples, and completed their assigned photographic tasks. After a productive outing lasting 7 hours and 36 minutes, they returned to the LM.

After a series of productive stops exploring the lunar surface, Eugene Cernan and Harrison Schmitt made their final excursion to Camelot Crater before returning to the safety of the Lunar Module. They accumulated 34 kilograms (75 lb) of samples throughout their journey, conducted seven gravimeter measurements, and strategically deployed three additional explosive packages for scientific experiments.

The final moonwalk of the Apollo program commenced at 5:25 p.m. EST on December 13th, marking a poignant moment in lunar exploration. Astronauts Eugene Cernan and Harrison Schmitt embarked on a journey northeast of their landing site aboard the lunar rover. Their route took them to the base of the North Massif and the intriguing Sculptured Hills, where they paused at various stations to conduct scientific examinations.

Harrison Schmitt working next to Tracy's Rock during EVA-3

One notable stop was at station 6, where they encountered a massive split boulder named Tracy's Rock, affectionately named after Cernan's daughter. This house-sized geological formation became a focal point of their exploration, offering insights into the Moon's ancient history and geological processes. As they pressed on to their ninth and final planned station at Van Serg crater, they meticulously gathered lunar samples and performed gravimeter measurements, adding to their scientific haul.

Schmitt identified an unusual fine-grained rock during their traverse, unlike the surrounding terrain. Intrigued, he placed it upright earlier in the mission and made a point to retrieve it before concluding their Extravehicular Activity (EVA-3). This rock, designated as Sample 70215, weighed a substantial 17.7 pounds (8.0 kg) and remains one of the largest lunar rocks returned by any Apollo mission. A fragment of Sample 70215 is exhibited at the Smithsonian Institution, offering a rare opportunity for the public to touch a piece of the Moon.

In a scientific coup, Schmitt also collected Sample 76535 near the base of the North Massif, which was identified as a troctolite. This sample gained renown as the oldest known "unshocked" lunar rock, providing crucial insights into lunar formation theories, including a potential metallic core or a core dynamo.

Their EVA-3 achieved unprecedented milestones, setting records for both duration and distance traveled from a spacecraft on another celestial body. The astronauts spent seven hours and thirty-seven minutes outside their lunar module, exploring and conducting experiments. Their journey extended the boundaries of human exploration and showcased their resourcefulness; a makeshift fender fix, crafted from taped-together maps, ingeniously protected their rover throughout the entire EVA. This innovative solution earned them unexpected accolades, including an honorary lifetime membership from the "Auto Body Association of America president," highlighting their ingenuity under the demanding lunar conditions.

As their momentous moonwalk drew to a close, the Apollo 17 crew made a symbolic gesture to the nations of Earth. Just before re-entering the Lunar Module (LM), Eugene Cernan and Harrison Schmitt collected a special breccia rock, dedicating it to the 70 nations represented by students visiting the Mission Control Center in Houston, Texas. This rock, dubbed the Friendship Rock, conveyed unity and exploration. Portions of it were later distributed to these nations as a testament to international cooperation in space exploration.

Before leaving the lunar surface for the final time, the astronauts unveiled a plaque on the LM commemorating the achievements of the Apollo program. Eugene Cernan, the last person to step onto the Moon during Apollo missions, reflected solemnly: "I'm on the surface; and, as I take man's last step from the surface, back home for some time to come – but we believe not too long into the future – I'd like to just [say] what I believe history will record. That America's challenge of today has forged man's destiny of tomorrow. And, as we leave the Moon at Taurus–Littrow, we leave as we came and, God willing, as we shall return, with peace and hope for all mankind. 'Godspeed the crew of Apollo 17.'"

Following their poignant moment of reflection, Cernan and Schmitt re-entered the LM, concluding their final lunar excursion after seven hours and fifteen minutes. Inside the LM, they removed their spacesuits, reconfigured the cabin for rest, and communicated their geological findings to mission control, a ritual they had followed after each of their successful EVAs

During their third Extravehicular Activity (EVA-3), the Apollo 17 astronauts achieved specific scientific objectives with precision and determination. Their tasks included conducting nine traverse gravimeter measurements and capturing panoramic photography with a 500 mm lens. However, the surface electrical properties experiment had to be terminated due to increasing receiver temperatures that threatened the integrity of the data tape. As a precaution, the tape recorder was removed on their return journey to the Lunar Module (LM).

In addition to these experiments, the crew successfully retrieved the cosmic ray experiment and the lunar neutron probe experiment at 161 hours, 20 minutes, and 17 seconds into their mission timeline. They also deployed several seismic profiling charges to study the Moon's subsurface structure further.

Throughout EVA-3, the astronauts documented their journey with numerous 500 mm panoramic images, capturing iconic scenes like the LM positioned prominently amidst the rolling lunar hills at Station 6.

The third EVA was a record-setting endeavor, lasting 7 hours, 15 minutes, and 8 seconds. During this time, the astronauts covered an impressive distance of 39,700 feet (approximately 12.1 kilometers) in the lunar rover, spending 1 hour and 31 minutes actively driving the vehicle. Their efforts culminated in collecting an estimated 136.7 pounds (62.0 kg) of lunar samples, further enriching our understanding of the Moon's geological history and composition.

While Eugene Cernan and Harrison Schmitt explored the lunar surface, Ronald Evans remained aboard the Command and Service Module (CSM), carrying out a series of vital scientific tasks in lunar orbit. Tasked with operating the CSM's scientific instruments housed in the Scientific Instrument Module (SIM) bay, Evans meticulously observed and photographed various surface features from his unique vantage point.

One of Evans' primary responsibilities was to adjust the CSM's orbit from its elliptical trajectory to a circular one. This adjustment ensured the CSM maintained a consistent altitude above the lunar surface throughout its orbit, facilitating continuous observation and data collection.

Using handheld cameras, Evans conducted visual and photographic surveys of geological formations and other designated targets on the lunar surface. He mainly focused on capturing images during the transition from lunar night to day, known as "sunrise," when the lunar surface emerges from darkness into sunlight—this period provided optimal lighting conditions for detailed photography of features such as the craters Eratosthenes and Copernicus and the Mare Orientale region.

Evans enhanced his photographic capabilities by using exposure techniques and Earthlight illumination to photograph areas not directly lit by the sun during his orbital passes. This method allowed him to capture clear images of lunar landscapes under varying lighting conditions.

According to the Apollo 17 Mission Report, Evans completed all scientific photographic tasks assigned to him, demonstrating his skill and precision in documenting critical lunar features for scientific analysis.

Similarly to the experiences of the Apollo 16 crew, Ronald Evans aboard the Apollo 17 Command and Service Module (CSM) and Harrison Schmitt, while in lunar orbit, reported observing transient lunar phenomena (TLP). Evans noted these phenomena, characterized by sudden light flashes or glows on the lunar surface, in the vicinity of Grimaldi crater and Mare Orientale.

An oblique, black-and-white view of a portion of Mare Orientale from the CSM, illustrating the illuminating effect of Earthlight on the lunar terrain below during local nighttime; Evans reported seeing a light "flash" apparently originating from the surface in this area

Transient lunar phenomena are intriguing yet not fully understood occurrences. They have been observed intermittently throughout history and can potentially originate from several sources. One proposed explanation suggests they might arise from outgassing events within the Moon's interior. Locations where Evans observed these flashes coincide with known regions of lunar outgassing, although this correlation remains speculative.

Another plausible cause of transient lunar phenomena is meteorite impacts. Collisions between meteoroids and the Moon's surface can release energy, producing brief light flashes visible from Earth or lunar orbit.

Ronald Evans played a pivotal role aboard the Apollo 17 Command and Service Module (CSM), managing a demanding flight plan that kept him extensively occupied throughout the mission. His responsibilities included piloting the CSM during its orbital phase, adjusting its trajectory, and overseeing the operation of instruments housed in the Scientific Instrument Module (SIM) bay.

During his approximately 148 hours in lunar orbit, Evans encountered a few challenges typical of space missions. Despite the meticulous planning, fatigue caught up with him on one occasion, causing him to oversleep by an hour despite efforts from Mission Control to wake him. In another instance, he realized he had misplaced his scissors, essential for opening food packets, and was generously loaned a pair by his fellow crewmates, Eugene Cernan and Harrison Schmitt.

Throughout his orbital duties, Evans ensured that the instruments in the SIM bay operated smoothly, providing valuable scientific data. At the same time, minor issues with the lunar sounder and mapping camera did not significantly hinder the mission's objectives.

One of Evans' critical solo tasks was performing a plane change maneuver to align the CSM's orbital plane with that of the Lunar Module (LM) as it prepared to rendezvous for their return journey to Earth. This maneuver, executed by firing the Service Propulsion System (SPS) engine for about 20 seconds, successfully adjusted the CSM's trajectory to synchronize with the LM's path.

After completing their mission objectives, the Apollo 17 crew reentered the Lunar Module (LM). Following the equipment jettison, the cabin was repressurized at 168 hours, 7 minutes, and 56 seconds into the mission timeline, marking the conclusion of the Apollo program's sixth and final human exploration of the Moon.

The astronauts spent 22 hours, 3 minutes, and 57 seconds outside the LM conducting scientific activities on the lunar surface. In the lunar rover vehicle, they covered a remarkable distance of 117,000 feet (approximately 35.7 kilometers), with a total drive time of 4 hours and 29 minutes. The crew collected substantial lunar samples totaling 243.65 pounds (110.52 kilograms, as officially determined by the Lunar Receiving Laboratory in Houston).

The farthest point reached from the LM during their explorations was 24,180 feet. The mission also achieved continuous high-quality television transmissions during all three Extravehicular Activities (EVAs), providing unprecedented live coverage of their lunar explorations to audiences on Earth.

During the Apollo 17 mission, while the astronauts explored the lunar surface, a range of scientific activities continued in lunar orbit aboard the Command and Service Module (CSM). Building on instruments used in previous missions, several new experiments were deployed in the Service Module to enhance our understanding of the Moon's environment and structure.

Among the existing instruments were the panoramic camera, mapping camera, and laser altimeter, which had proven essential for geological mapping and surface topography assessment. Additionally, three new experiments were introduced:

Ultraviolet Spectrometer: This instrument measured lunar atmospheric density and composition using ultraviolet light, providing insights into the Moon's tenuous atmosphere and its interactions with solar radiation.

Infrared Radiometer: Designed to map the thermal characteristics of the lunar surface, the infrared radiometer measured surface temperatures across different regions, helping scientists understand the Moon's heat distribution and thermal properties.

Lunar Sounder: This experiment collected data on the Moon's subsurface structure, transmitting signals that penetrated beneath the lunar surface to reveal geological layers and potentially detect subsurface structures like lava tubes or buried impact craters.

While conducting these scientific observations, mission controllers noted that the CSM's orbit did not decay as predicted during the Lunar Module's stay on the Moon's surface. A precise 37.50-second orbital trim maneuver was executed at 178 hours, 54 minutes, and 5.45 seconds into the mission to correct the orbit. This maneuver lowered the CSM's orbit to approximately 67.3 by 62.5 nautical miles (124.6 by 115.8 kilometers).

Furthermore, a planned 20.05-second plane change maneuver was performed at 179 hours, 53 minutes, and 53.83 seconds. This adjustment prepared the CSM's trajectory for rendezvous with the Lunar Module and resulted in a final orbital configuration of 62.8 by 62.5 nautical miles (116.3 by 115.8 kilometers).

On December 14th, at 5:54 p.m. EST, astronauts Eugene Cernan and Harrison Schmitt achieved a critical milestone in the Apollo program. After completing a successful exploration mission on the lunar surface, their lunar module lifted off, beginning the journey back to lunar orbit. The ascent was swift, taking just over seven minutes, powered by the ascent stage's engine igniting precisely at 05:54:37 GMT. At that moment, the lunar module had spent 74 hours, 59 minutes, and 40 seconds on the moon.

Cernan guided the ascent stage to achieve an initial lunar orbit 48.5 by 9.1 nautical miles. This maneuver was followed by several precise rendezvous adjustments, including a 10-second vernier adjustment at 185:32:12 GMT, which refined the orbit to 48.5 by 9.4 nautical miles. Finally, at 186:15:58 GMT, a 3.2-second terminal phase initiation placed the ascent stage into a stable orbit of 64.7 by 48.5 nautical miles.

Approximately two hours after liftoff, at 187:37:15 GMT and at an altitude of 60.6 nautical miles, the ascent stage successfully docked with the Command and Service Module (CSM), piloted by Ronald Evans. The crew then transferred crucial equipment and lunar samples from the lunar module to the CSM in preparation for their return to Earth.

On December 14th at 5:54 p.m. EST, astronauts Eugene Cernan and Harrison Schmitt successfully launched from the lunar surface aboard the lunar module's ascent stage. Their ascent back to lunar orbit was swift, taking just over seven minutes. Once in orbit, the lunar module, under Cernan's command, and the Command and Service Module (CSM), piloted by Ronald Evans, executed a precise rendezvous and redocking maneuver approximately two hours after liftoff from the lunar surface.

Following a successful docking, the crew transferred critical equipment and lunar samples from the lunar module to the CSM in preparation for their journey back to Earth. With their tasks completed, the ascent stage was deliberately jettisoned at 191:18:31 GMT and remotely directed to impact the lunar surface. A 12-second separation maneuver at 191:23:31 GMT positioned the CSM into a stable orbit of 63.9 by 61.2 nautical miles.

At 193:00:10 GMT, a 116-second deorbit firing was initiated, depleting the ascent stage's propellant and ensuring its collision with the moon. The final impact occurred at 193:17:21 GMT, concluding Apollo 17's lunar module mission. Meanwhile, aboard the CSM, the astronauts prepared for their transearth injection, marking the beginning of their return journey home.

After impacting the lunar surface, the Apollo 17 ascent stage landed approximately 0.94 nautical miles (1.75 km) from its intended target and 5.35 nautical miles (9.9 km) southwest of the Apollo 17 landing site. Seismic stations monitored this event from previous missions—Apollo 12, 14, 15, and 16—which recorded the impact.

The Apollo 17 crew conducted scientific experiments following the landing, including detonating explosive packages on the lunar surface. These detonations occurred at precise times: 210:15:35 and 212:45:01 GMT. Both events were captured by lunar seismic profiling geophones, with the resulting flashes and dust from the second explosion visible on television screens.

However, attempts to operate the television assembly and lunar communications relay unit were unsuccessful at various times (218:20 GMT, 235:04 GMT, and 235:13 GMT) due to an overtemperature failure in the relay unit.

After completing their scientific tasks, the crew initiated their return journey to Earth. A critical 143.69-second maneuver at 234:02:09.18 GMT, at 62.1 nautical miles, positioned the spacecraft for transearth injection. Transearth injection was achieved precisely at 234:04:32.87 GMT, propelling the spacecraft at 8,374.3 feet per second. This marked the culmination of 75 lunar orbits over 147 hours, 43 minutes, and 37.11 seconds of mission time, including an additional day spent in lunar orbit conducting scientific research.

During the transearth phase of Apollo 17, additional explosive packages were detonated on the lunar surface at precise times: 235:09:52 GMT and 238:12:50 GMT. Both generated strong signals received by the lunar geophones, contributing to ongoing scientific observations of lunar seismic activity.

At 254:54:40 GMT, the command module pilot initiated a significant extravehicular activity (EVA) during the transearth coast phase. This televised event lasted 1 hour, 5 minutes, and 44 seconds, during which he retrieved important scientific instruments, including the lunar sounder film, panoramic camera, and mapping camera cassettes from the scientific instrument module bay. This brought the total EVA time for the mission to 23 hours, 9 minutes, and 41 seconds. The retrieval of these instruments provided crucial data and imagery from the lunar surface.

Further explosive packages were detonated on the lunar surface at 257:43:56 GMT, 259:12:02 GMT, and 262:34:29 GMT, all of which were detected by lunar surface geophones. This continued to enhance NASA's understanding of lunar seismic activity.

Throughout the remainder of the transearth flight, the crew engaged in scientific experiments, including another light-flash experiment, and operated instruments such as the infrared radiometer and ultraviolet spectrometer to gather additional data en route to Earth. One midcourse correction maneuver was performed, a precise 9-second, 2.1-feet-per-second adjustment at 298:38:01 GMT, ensuring optimal trajectory toward Earth.

During the return journey, at approximately 160,000 nautical miles (184,000 miles; 296,000 km) from Earth, Eugene Cernan performed a historic extravehicular activity lasting 65 minutes. Assisted by Harrison Schmitt, who remained at the command module's hatch, Cernan retrieved film cassettes from the service module's Scientific Instrument Module (SIM) bay. This marked the third "deep space" EVA in history, a remarkable achievement conducted far from any planetary body. As of 2024, this remains one of only three such EVAs, all performed during Apollo's J-missions under similar circumstances, and notably, it was the final EVA of the Apollo program.

During the return journey to Earth aboard Apollo 17, the crew continued their scientific operations, utilizing instruments such as the infrared radiometer and ultraviolet spectrometer in the Service Module (SM). A minor midcourse correction lasting 9 seconds was performed to refine their trajectory toward Earth.

On December 19th, the crew jettisoned the Service Module (SM) at 301:23:49 GMT, leaving only the Command Module (CM) to reenter Earth's atmosphere. The CM followed a standard entry profile, reentering the Earth's atmosphere at an altitude of 400,000 feet (121,920 meters) and a velocity of 36,090.3

feet per second (11,000 meters per second), after a transearth coast duration of 67 hours, 34 minutes, and 8 seconds.

The parachute system deployed as planned, guiding the Command Module to a splashdown in the Pacific Ocean at 19:24:59 GMT (02:24:59 p.m. EST) on December 19th. The splashdown occurred approximately 6.4 kilometers (4.0 miles) from the recovery ship, USS Ticonderoga, at a location estimated to be at latitude 17.88° south and longitude 166.11° west. The Command Module assumed an apex-up flotation attitude upon landing.

Within 52 minutes after splashdown, the crew—Eugene Cernan, Ronald Evans, and Harrison Schmitt—were safely retrieved from the water by a recovery helicopter piloted by Commander Edward E. Dahill III and brought aboard the USS Ticonderoga. The Command Module itself was recovered 71 minutes after splashdown.

After their successful recovery from the USS Ticonderoga, the crew of Apollo 17 departed the ship at 00:38 GMT on December 21st and arrived in Houston at 15:50 GMT the same day. Their return marked the conclusion of their mission, which had achieved significant scientific and exploratory milestones on the lunar surface.

Following the crew's return, the Command Module (CM) was transported to North Island Naval Air Station in San Diego, reaching its destination at 19:30 GMT on December 27th. Deactivation procedures began immediately and were completed by 22:00 GMT on December 30th.

Subsequently, at 19:00 GMT on January 2nd, the CM was transferred to the North American Rockwell Space Division facility in Downey, California, for detailed postflight analysis. It arrived at its destination by 22:00 GMT the same day, where engineers and scientists would conduct thorough examinations to glean insights and data from the mission.

Apollo 17 stands as a pinnacle of achievement in the storied history of NASA's Apollo program, marking the culmination of years of meticulous preparation and technological advancement. Every aspect of the mission, from launch to return, exemplified the skill and precision of its experienced personnel and the robust performance of its cutting-edge equipment.

An analysis of post-mission data revealed several key conclusions. Firstly, Apollo 17 was hailed as the most productive and trouble-free piloted mission of its time. It underscored the evolution and refinement of spacecraft hardware, operational procedures, training protocols, meticulous planning, and the execution of scientific experiments in the unforgiving environment of space.

A groundbreaking aspect of Apollo 17 was its success in integrating highly trained scientists into the astronaut corps. This mission validated the practicality of training scientists to perform as qualified astronauts while leveraging their specialized expertise and scientific acumen to enhance mission objectives.

A unique challenge arose during night launches: the absence of visible stars and the horizon deprived the crew of traditional out-of-the-window alignment techniques for attitude reference. This necessitated a heavy reliance on ground control and automated systems for precise navigation and alignment, showcasing the adaptability and foresight of mission planners and engineers.

A U.S. Navy Sikorsky SH-3G Sea King (BuNo 149930) of Helicopter Combat Support Squadron 1 (HC-1) "Pacific Fleet Angels" recovers an Apollo 17 astronaut on 19 December 1972, with the aircraft carrier USS Ticonderoga (CVS-14) in the background.

The spacecraft's dynamic conditions during the initial launch phases posed additional challenges. The intense vibrations and G-forces rendered in-flight troubleshooting and corrective actions by the crew impractical. As a result, the mission highlighted the critical role of ground control and automation in swiftly diagnosing and addressing system anomalies.

Moreover, issues encountered during Apollo 17 and preceding missions underscored the need for further research and development. Specifically, mechanisms designed to extend and retract equipment in the harsh vacuum of space required enhancements to ensure reliability and performance over repeated operations.

In essence, Apollo 17 achieved its scientific and exploratory objectives on the lunar surface and set a standard for future space missions. It remains a testament to human ingenuity, perseverance, and collaboration, reflecting NASA's unwavering commitment to pushing the boundaries of exploration and knowledge in the vastness of space.

Apollo 17, NASA's final mission, achieved a remarkable array of scientific and operational objectives during its journey to the Moon. Tasked with exploring the Taurus-Littrow region, the mission's primary objectives were multifaceted and included selenological inspection, surveying, and sampling of lunar materials and surface features. These goals were achieved and exceeded expectations through meticulous planning and the deployment of advanced scientific instrumentation.

One of the mission's pivotal accomplishments was the successful emplacement and activation of surface experiments, which provided invaluable data on lunar geology and atmospheric conditions. In addition to these surface activities, Apollo 17 conducted numerous in-flight experiments and photographic tasks, capturing iconic images such as the micrograph of orange soil particles and detailed views of lunar rock samples.

Detailed objectives spanned a wide spectrum of scientific inquiries. The mission obtained crucial photographs and altitude data from lunar orbit, shedding light on lunar topography and geological formations. Visual light flash phenomena were documented, providing insights into lunar surface interactions. Meanwhile, the command module's photographic surveys captured scientific features of interest and celestial sources of low brightness.

Apollo 17's scientific payload, including the Apollo Lunar Surface Experiments Package (ALSEP V), conducted groundbreaking experiments. These ranged from measuring lunar heat flow and seismic activity to analyzing the moon's atmospheric composition and gravitational field. Despite some operational challenges, such as overheating affecting the lunar ejecta and meteorites experiment, the mission successfully gathered significant scientific data.

Beyond its scientific endeavors, Apollo 17 also fulfilled passive objectives, including long-term lunar surface exposure experiments and tests related to gamma-ray spectrometry and soil mechanics. Classified experiments conducted with the Department of Defense underscored the mission's dual-purpose nature, advancing both civilian and military research objectives.

The launch vehicle objectives were equally precise and achieved with accuracy. The Saturn V rocket flawlessly propelled the spacecraft into its planned Earth parking orbit before guiding it onto a precise translunar trajectory. Even the controlled impact of the S-IVB stage on the lunar surface, within specified coordinates, demonstrated NASA's mastery of orbital mechanics and precision targeting.

In conclusion, Apollo 17 stands as a testament to human ingenuity and exploration. Its successful execution of diverse scientific tasks and operational tests expanded our understanding of the Moon's geology, atmosphere, and cosmic environment. More importantly, it marked the capstone of an era of lunar exploration, leaving a lasting legacy of scientific achievement and paving the way for future missions into the cosmos.

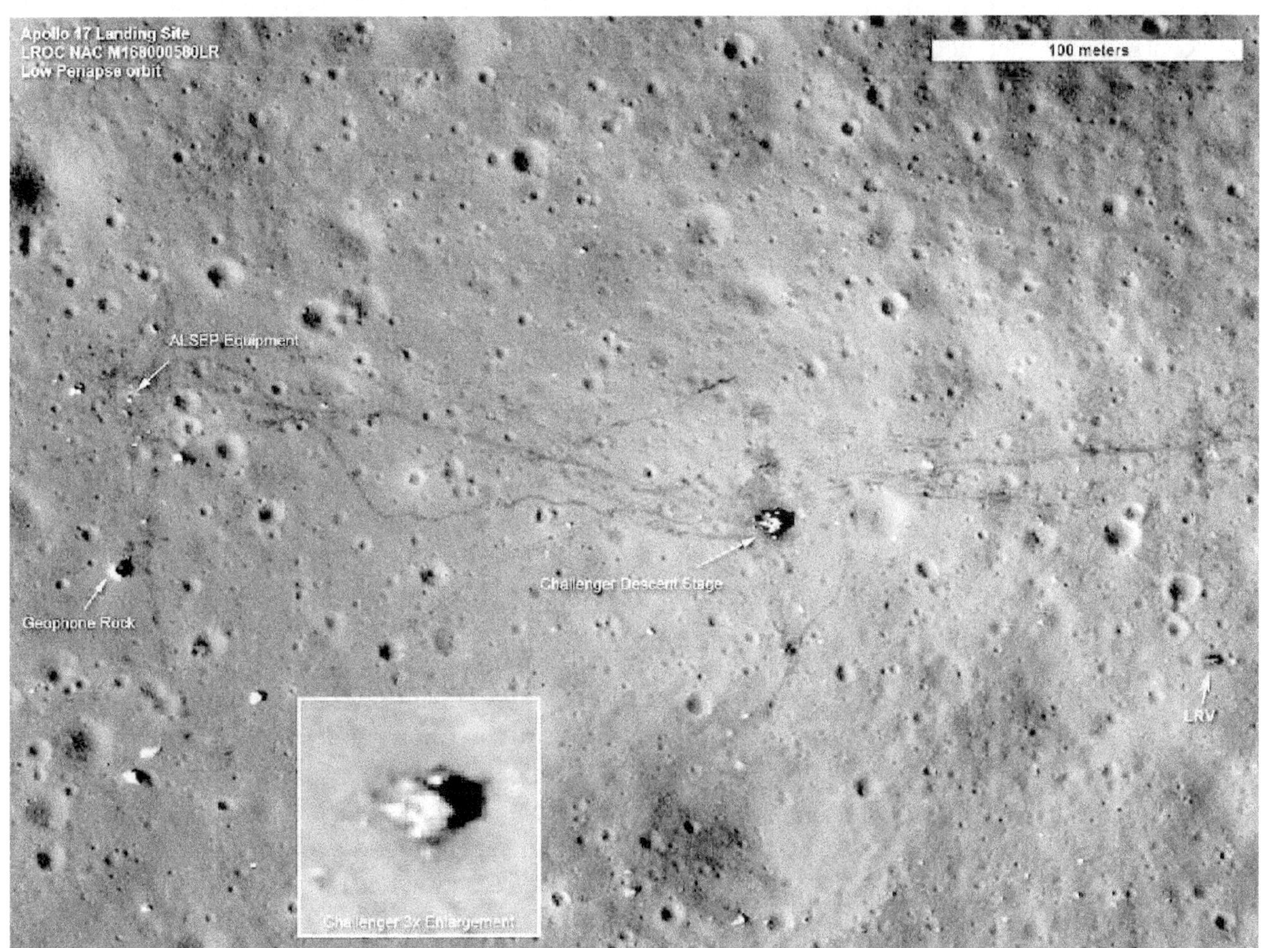

Epilogue

After their historic mission, the Apollo 17 crew embarked on extensive domestic and international tours, spanning 29 states and 11 countries. Their journey began with a memorable appearance at Super Bowl VII, where they led the crowd in reciting the Pledge of Allegiance and showcased the Command Module America during pregame festivities.

After returning to Earth, none of the Apollo 17 astronauts flew in space again. Gene Cernan retired from both NASA and the Navy in 1976, leaving behind a distinguished career marked by his role as the last man to walk on the moon. He passed away in 2017, leaving a lasting legacy in space exploration.

Ron Evans concluded his service with the Navy in 1976 and NASA in 1977, transitioning to the private sector. His contributions to spaceflight continued to inspire until he died in 1990.

Harrison Schmitt departed NASA in 1975, preceding a successful venture into politics. He was elected as a United States Senator from New Mexico in 1976, serving a single six-year term. His unique perspective as a scientist-astronaut enriched his legislative work and furthered the legacy of Apollo's scientific endeavors.

The Command Module America, a testament to the Apollo 17 mission's achievements, is prominently exhibited at Space Center Houston, located within the Lyndon B. Johnson Space Center in Houston, Texas. This artifact is a tangible reminder of humanity's historic journeys to the moon and back.

Meanwhile, on the lunar surface, the ascent stage of the Lunar Module Challenger met its deliberate end on December 15, 1972. At 06:50:20.8 UTC (1:50 a.m. EST), it impacted the Moon at coordinates 19.96°N 30.50°E. This intentional maneuver was part of the mission's planned operations.

The Challenger's descent stage, however, remains where it first touched down, marking the exact landing site with coordinates 20.19080°N 30.77168°E. Decades later, in 2023, a study utilizing Apollo-era data from the Lunar Seismic Profiling Experiment revealed an intriguing phenomenon: the descent stage induces subtle tremors each lunar morning as its components react to the fluctuating lunar temperatures, providing ongoing scientific insights from this enduring lunar landmark.

Eugene Cernan's flown Apollo 17 spacesuit holds a place of honor in the Smithsonian's National Air and Space Museum (NASM) collection, where it was transferred in 1974. This iconic artifact preserves the legacy of Cernan's historic lunar mission, allowing visitors to connect with the realities of space exploration.

Meanwhile, Harrison Schmitt's Apollo 17 spacesuit is carefully stored at NASM's Paul E. Garber Facility. Remarkably well-preserved, it is recognized by Amanda Young of NASM as one of the best-maintained flown Apollo lunar spacesuits. Due to its exceptional condition, Schmitt's suit remains in storage rather than publicly displayed.

Ron Evans' spacesuit, also transferred from NASA to NASM in 1974, is similarly housed in storage at the museum. Though not currently on public display, these suits represent invaluable pieces of space history, awaiting potential future exhibitions to inspire and educate generations to come.

Since Apollo 17's historic mission, ongoing efforts have aimed to document and explore the lunar landing site where the Lunar Module's descent stage, the Lunar Roving Vehicle (LRV), and other mission hardware remain. In 2009 and subsequently, in 2011, the Lunar Reconnaissance Orbiter conducted photographic surveys of the site from progressively lower orbits, providing detailed visual data of the artifacts left behind by the astronauts.

Beyond orbital photography, ambitions to physically visit the site have persisted. In

2018, the German space company PTScientists announced plans to land two lunar rovers near the Apollo 17 landing site. This initiative underscores the enduring interest in lunar exploration and the potential for new missions to revisit and study the historic locations where humanity first ventured beyond Earth. These efforts expand our scientific understanding of the Moon and honor the legacy of the Apollo missions and pave the way for future lunar exploration endeavors.

Apart from the iconic moonwalks of the Apollo program and a distinctive trio of deep-space EVAs during the J-missions, nearly all other spacewalks have occurred within Low-Earth orbit. These EVAs typically involve astronauts tethered to their spacecraft by a safety tether, ensuring they remain within a short distance for safety.

However, notable exceptions occurred in 1984 and 1994, marked by a series of seven EVAs where astronauts engaged in untethered activities using innovative equipment like the Manned Maneuvering Unit (MMU) and the Simplified Aid For EVA Rescue Unit (SAFER). These missions represented significant milestones in spacewalk history, allowing astronauts to maneuver freely in space beyond tethered limitations.

Among these missions, a standout achievement was recorded during STS-41-B in 1984, when Bruce McCandless made history by venturing up to approximately 100 meters (320 feet) away from the spacecraft during the MMU's maiden flight. This daring feat demonstrated astronauts' capability to navigate independently in the vacuum of space, pushing the boundaries of human exploration beyond Earth's atmosphere.

A look back at the Apollo Program

The Apollo program, a monumental human exploration effort, comprised a sophisticated array of spacecraft and stages meticulously engineered to achieve the audacious goal of landing astronauts on the Moon. At its core was the Apollo Space Vehicle, a marvel of engineering consisting of several key stages and components.

The journey began with the S-IB stage, generating an impressive 1.64 million pounds of thrust at liftoff, propelling the entire vehicle to an astonishing speed of 7,620 feet per second within 2.5 minutes. Following this initial push, the S-IC stage took over, its powerful engines delivering 7.65 million pounds of thrust, accelerating the spacecraft to 7,880 feet per second and reaching an altitude of 58 nautical miles in just as swift a timeframe.

The interstage components bridged these stages, crucially facilitating transitions and preparations for subsequent phases. The S-II stage, housed within the interstage, further accelerated the vehicle to an impressive 22,850 feet per second over 370 seconds, culminating in an altitude of 101 nautical miles. This stage also provided essential support for the S-II engines and retro-rocket mounting.

As the spacecraft ascended, the S-IVB stage took over, increasing its velocity to 25,553 feet per second during Apollo 7 and 25,568 feet per second for subsequent missions, effectively placing the vehicle into orbit. This final push toward Trans-Lunar Injection (TLI) saw speeds peaking at 35,500 feet per second, crucial for escaping Earth's gravitational pull.

The evolution of space exploration after Project Apollo was a significant shift with the emergence of commercial astronauts alongside traditional government-sponsored astronauts. Historically, governmental space agencies like NASA or military programs exclusively selected and trained astronauts, focusing on missions crucial to national interests and scientific advancement.

The landmark moment arrived with the privately funded SpaceShipOne in 2004, which demonstrated the feasibility of suborbital spaceflight outside traditional government channels. This pivotal achievement opened the door to a new era in which private companies began developing their own spacecraft and training programs, leading to the creation of commercial astronauts.

Commercial astronauts differ from their governmental counterparts in several key aspects. They are typically trained and employed by private companies such as SpaceX, Blue Origin, or Virgin Galactic, which are actively involved in commercial space travel and exploration. These astronauts undergo rigorous training similar to government-sponsored astronauts, preparing them for space missions ranging from suborbital flights to potentially crewed missions to the International Space Station (ISS) and beyond.

Continuing into 1973, the Soviet Union established Civilian Specialist Group 5, welcoming Vladimir Aksyonov, Valeri Romanov, and others. This group played essential roles in various space missions, emphasizing the USSR's continued expansion of its space exploration capabilities.

January 1, 1974, marked the formation of the Physician Group in the USSR, with Zyyadin Abuzyarov selected to provide specialized medical support essential for the health and safety of cosmonauts during their missions.

In 1976, Air Force Group 6, known as the Space Shuttle Buran crew, was introduced. This crew comprised Leonid Ivanov, Leonid Kadenyuk, and others. This group's training focused on preparing for the Soviet Buran program, reflecting the USSR's pursuit of reusable spacecraft technology comparable to the American Space Shuttle.

Finally, on November 25, 1976, the Intercosmos Group was established, uniting cosmonauts from Poland, East Germany, and Czechoslovakia. Mirosław Hermaszewski, Sigmund Jähn, and Vladimír Remek were among the members, highlighting international cooperation in space exploration during the Cold War era.

In 1978, NASA ushered in a new era of space exploration by selecting Group 8 astronauts, aptly nicknamed TFNG – Thirty-Five New Guys. This cohort marked a significant turning point for NASA after a nine-year hiatus in astronaut selections following the conclusion of the Apollo missions. Among them were notable pioneers: Judith Resnik and Sally Ride, the first American women astronauts. Judith Resnik, additionally, made history as the first Jewish American in space, while Guion Bluford and Frederick D. Gregory became the first African Americans to fly in space. Ellison Onizuka broke ground as the first Asian-American astronaut, reflecting NASA's commitment to diversity and inclusivity.

This diverse group included pilots and mission specialists, each chosen for their unique skills and expertise. Alongside them, NASA's Shuttle Program introduced payload specialists, scientists selected for individual missions, and occasionally international astronauts, expanding the program's global reach and scientific collaboration.

From this group, Sally Ride achieved another milestone by becoming America's first woman in space aboard STS-7. She later joined Kathryn Sullivan on a mission where Sullivan became the first American woman to perform an extravehicular activity (EVA). Dr. Norman Thagard, who flew with Ride, became the first American launched aboard a Russian rocket to the Mir space station. Shannon Lucid, another member of Group 8, set records for American space endurance during her stint on Mir, surpassing previous milestones set during Skylab and Shuttle missions.

Tragically, the Challenger Disaster in 1986 claimed the lives of several from this group, including Scobee, Resnik, Onizuka, and McNair, leaving a profound impact on NASA and the entire astronaut community.

Over the years, many of these astronauts continued to serve with distinction. Anna Fisher holds the record for the longest active duty tenure, retiring in 2017 after a career marked by significant contributions to space exploration. Robert Gibson and Rhea Seddon made history by becoming the first active-duty astronauts to marry, while Shannon Lucid's tenure remained unbroken until her retirement announcement in 2012. Sally Ride's contributions extended beyond missions, as she served on investigative commissions following the Challenger and Columbia accidents.

Meanwhile, on March 1, 1978, the USSR's Intercosmos program welcomed a new cohort, reinforcing global cooperation in space with astronauts from Bulgaria, Romania, Cuba, Hungary, Mongolia, and other nations. This era of international collaboration underscored a shared ambition for scientific discovery beyond national borders.

Simultaneously, on May 1, 1978, the European Space Agency (ESA) launched its Spacelab Payload Specialists Group 1, including representatives from West Germany, Switzerland, the Netherlands, and Italy. These specialists brought a wealth of scientific expertise to shuttle missions, advancing microgravity and space technology research.

In August 1979, the USAF Manned Spaceflight Engineer Program introduced its inaugural Group 1, consisting of distinguished individuals such as Frank J. Casserino, Michael A. Hamel, and Gary E. Payton. This group was unique in its focus on integrating military expertise into manned space missions. Gary E. Payton stood out as the sole member to fly into space, serving as a Payload Specialist on a dedicated Department of Defense Shuttle mission, highlighting the program's specialized role in national security initiatives.

Meanwhile, on April 1, 1979, the USSR's Intercosmos program expanded with the addition of Tuân Pham and Thanh Liem Bui from Vietnam. This international collaboration further underscored the program's commitment to fostering global partnerships in space exploration, showcasing the diverse contributions from nations worldwide.

On May 29, 1980, NASA unveiled its highly anticipated Group 9 of astronauts, a cohort comprising pilots and mission specialists. Among the pilots were notable names like John Blaha, Charles Bolden, and Michael J. Smith. The mission specialist roster included individuals such as Franklin Chang-Diaz, Mary Cleave, and Bonnie Dunbar, reflecting a diverse range of expertise crucial for NASA's evolving missions. Franklin Chang-Diaz, a member of this group, made history as the first Hispanic-American astronaut in space. Tragically, Michael J. Smith later perished in the Challenger disaster, a pivotal event in space exploration history.

John Blaha's career extended beyond NASA's Shuttle program; he ventured aboard the Mir space station, underscoring NASA's collaborative efforts in international space ventures. Jerry Ross and Franklin Chang-Diaz notably hold the record for the most crewed spaceflights, each having flown seven missions.

Charles Bolden, another prominent member of Group 9, went on to achieve significant milestones beyond his space missions. In 2009, Bolden became NASA's second astronaut and the first African-American Administrator, a role he assumed full-time. Frederick Gregory, also African-American and a former Shuttle commander, had previously held this position temporarily between other appointments.

In contrast, on July 30, 1980, the Soviet Union introduced the LII–1/IMBP–3/MAP/NPOE-5/AN–2 Cosmonaut Group, a diverse assembly of specialists from various Soviet organizations. Svetlana Savitskaya, among the distinguished members, later became the second woman to venture into space and the first woman to conduct a spacewalk.

Simultaneously, France launched its inaugural CNES Group 1, featuring Patrick Baudry and Jean-Loup Chrétien. Chrétien made history by becoming one of the first Frenchmen in space, journeying aboard the Soviet Union's Salyut 7 space station in 1982. Patrick Baudry followed suit on the Space Shuttle STS-51-G mission in 1985, marking France's continued engagement in international space missions.

In August 1982, the USAF Manned Spaceflight Engineer Program introduced its Group 2, comprising James B. Armor Jr., Livingston L. Holder Jr., and Maureen C. LaComb. This group represented a continuation of the program's mission to integrate military expertise into space missions. Sadly, Charles E. Jones, a member of this group, tragically lost his life in the September 11 attacks aboard American Airlines Flight 11. William A. Pailes was the only member of Group 2 to fly into space, serving as a Payload Specialist on a dedicated Department of Defense Shuttle mission.

On September 11, 1982, India made its mark in space exploration by selecting Ravish Malhotra and Rakesh Sharma for the Intercosmos program. Rakesh Sharma became the first Indian citizen to travel to space, achieving this milestone aboard the Soviet spacecraft Soyuz T-11 in 1984.

Later that year, on December 1, Germany contributed to the Spacelab Payload Specialists Group, including Reinhard Furrer and Ernst Messerschmid. This group bolstered European participation in Shuttle missions, emphasizing scientific research and international collaboration in microgravity studies.

In April 1983, the Soviet Union marked a significant step forward in its space exploration efforts by selecting Ural Sultanov and Magomed Tolboev as the second group of test pilots for the "Buran" project, based at the Gromov Flight Research Institute. This selection underscored the USSR's commitment to developing its space shuttle program to rival the United States in manned spaceflight capabilities.

Later in December 1983, Canada entered into space exploration with the selection of its first astronaut group by the National Research Council (NRC). This pioneering group included Roberta Bondar, Marc Garneau, Steve MacLean, Ken Money, Robert Thirsk, and Bjarni Tryggvason. They were instrumental in laying the groundwork for Canada's involvement in space, initially flying aboard US Space Shuttles.

By 1989, with the establishment of the Canadian Space Agency (CSA), these astronauts were formally transferred to the CSA, solidifying Canada's role in international space missions. All members of this group, except Ken Money, who resigned in 1992, flew aboard US Space Shuttles by 1997. Their contributions marked a significant chapter in Canadian space history, highlighting the country's growing presence and achievements in space exploration.

In February 1984, the Soviet Union expanded its cosmonaut roster by selecting Aleksandr Kaleri and Sergei Yemelyanov for the NPOE–6 Cosmonaut Group. This group played a crucial role in advancing Soviet space missions, contributing to ongoing exploration and scientific research efforts.

On May 23, 1984, NASA introduced its Group 10, colloquially known as "The Maggots," comprising pilots and mission specialists. Notable pilots included Kenneth Cameron, John Casper, and Frank Culbertson, while the mission specialist lineup featured prominent figures such as Ellen Baker, Marsha Ivins, and Kathryn Thornton. This diverse group of astronauts brought a wealth of expertise to NASA's missions, supporting a wide range of scientific and operational objectives.

William Shepherd, a member of Group 10, later made history as the commander of the first crew aboard the International Space Station during Expedition 1. Another standout member, James Wetherbee, achieved a unique distinction as the only individual to command five spaceflight missions. However, the group also faced tragedy with the loss of Sonny Carter in 1991, who died in a plane crash while on NASA business, highlighting the risks inherent in space exploration and aviation.

In June 1984, the Soviet Union continued to bolster its space capabilities by selecting Victor Zabolotski for the third group of test pilots for the "Buran" project at the Gromov Flight Research Institute. This ongoing expansion of the cosmonaut corps underscored the USSR's commitment to developing its space shuttle program, reflecting a period of intense competition and advancement in global space exploration efforts.

In 1985, the global landscape of space exploration continued to evolve with significant selections and developments across several nations:

India (May): The ISRO Insat Group, consisting of Nagapathi Chidambar Bhat and Paramaswaren Radhakrishnan Nair, were chosen for Space Shuttle missions. However, none flew due to the tragic Challenger disaster in 1986, which led to the cancellation of Bhat's scheduled flight.

Mexico (June): Rodolfo Neri Vela and Ricardo Peralta y Fabi represented Mexico's space aspirations. Neri Vela achieved a milestone by flying on Shuttle mission STS-61-B in November 1985.

USA (June 4): NASA Group 11, comprising pilots like Michael A. Baker and Robert D. Cabana, and mission specialists including Jerome Apt and Linda Godwin, joined NASA's ranks. Tragically, Stephen Thorne lost his life in a private airplane crash before his first space assignment.

USA (July 19): The NASA Teacher in Space Program selected Christa McAuliffe as the prime and Barbara Morgan as the backup Payload Specialists for STS-51-L. McAuliffe tragically perished in the Challenger disaster, while Morgan later joined the NASA Astronaut Corps and flew on STS-118 in 2007.

Japan (August 1): The 1985 NASDA Group introduced Mamoru Mohri, Chiaki Mukai, and Takao Doi, enhancing Japan's presence in global space exploration efforts.

USA (August): The USAF Manned Spaceflight Engineer Program – Group 3 expanded the program with astronauts like Joseph J. Caretto and Teresa M. Stevens.

USSR (September 2): Viktor Afanasyev and Sergei Krikalyov were members of the GKNII–2/NPOE–7 Cosmonaut Group, reinforcing Soviet capabilities in manned spaceflight.

France (September 18): CNES Group 2, featuring Claudie André-Deshays and Jean-François Clervoy, exemplified France's commitment to space exploration and scientific research.

Syria (September 30): Muhammed Ahmed Faris and Munir Habib Habib were selected for the 1985 Intercosmos Group, marking Syria's involvement in international space missions.

Indonesia (October): The Indonesian Palapa Group, consisting of Taufik Akbar and Pratiwi Sudarmono, prepared for Shuttle missions, though none flew due to the Challenger disaster.

ESA (December 27): Dirk D. Frimout from Belgium participated in the ATLAS-1 mission, contributing to the European Space Agency's scientific endeavors in space.

In 1986, the Soviet Union continued to advance its space capabilities by selecting the fourth group of test pilots for the "Buran" project at the Gromov Flight Research Institute. This group included Sergey Tresvyatski and Yuri Schaeffer, who underwent rigorous training and testing as part of the Buran program.

The significance of their selection was underscored by the Interdepartmental Qualification Committee (IAC) decision on June 5, 1987. Following this decision, all Buran test pilots, including Tresvyatski and Schaeffer, were officially awarded the qualification of test cosmonauts. This recognition affirmed their readiness and competence to pilot and operate the Buran spacecraft, marking a crucial milestone in the Soviet Union's efforts to develop and deploy advanced space technologies.

In 1987, significant astronaut and cosmonaut groups were selected, marking milestones in international space exploration.

On January 5th, Bulgaria's Shipka Group, consisting of Aleksandr Aleksandrov and Krasimir Stoyanov, entered training for future space missions, contributing to Bulgaria's space program.

March 26th saw the selection of the TsPK-8/NPOE-8 Cosmonaut Group in the Soviet Union. This group included Valery Korzun, Vladimir Dezhurov, Yuri Gidzenko, Yuri Malenchenko from TsPK, and Sergei Avdeyev from NPOE. They undertook rigorous training at the Yuri Gagarin Cosmonaut Training Center (TsPK) and the Rocket Space Corporation Energia (NPOE), preparing for missions aboard the Soyuz spacecraft and future space stations.

On June 5th, NASA announced Group 12 in the United States, affectionately known as "The GAFFers" or "George Abbey Final Fifteen." This group included pilots such as Andrew M. Allen, Kenneth Bowersox, Curtis Brown, and mission specialists like Mae Jemison, the first African-American woman in space, and Michael Foale, who later participated in extended missions on both the Mir space station and the International Space Station (ISS). William Readdy and Kenneth Bowersox played critical roles in NASA leadership and space missions in subsequent years.

Later in August, Germany's 1987 astronaut group, featuring Renate Brümmer, Hans Schlegel, Gerhard Thiele, Heike Walpot, and Ulrich Walter, was announced. They contributed to European Space Agency (ESA) missions, expanding international collaboration in space exploration.

On February 12, 1988, the OS "Mir" Group from Afghanistan was formed. This group consisted of Mohammad Dauran Ghulam Masum and Abdul Ahad Mohmand. Their selection marked Afghanistan's entry into the Soviet space program, contributing to international collaboration aboard the Mir space station.

In 1989, several significant developments unfolded in space exploration, reflecting the era's international fervor and technological prowess.

On January 25th, the Soviet Union's Cosmonaut Group saw a new cohort of astronauts selected across multiple institutes. Among them were Vladimir Karashtin and Vasili Lukiyanyuk from IMBP, Anatoli Polonsky and Valeri Tokarev from GNKII, Aleksandr Yablontsev from NPOE, and others from TsPK, including Sergei Kirchevsky and Gennady Padalka. These individuals were chosen for their exceptional training and expertise, marking another stride in the Soviet space program's ongoing quest for supremacy.

March 22nd marked a pivotal moment for the Buran project as the Gromov Flight Research Institute concluded its final round of test pilot selections. This project, emblematic of Soviet ambition in reusable spaceflight, had seen sporadic activity until its closure in 1993. Among the few who journeyed into space from this cadre were Igor Volk and Anatoly Levchenko, exemplifying the resilience and selective nature of the program.

In May, Italy made its mark with the selection of Franco Malerba, Franco Rossitto, Umberto Guidoni, and Cristiano Batalli Cosmovici, highlighting the growing international participation in space missions and fostering collaborative efforts across borders.

On September 29th, NASA appointed Charles R. Chappell, Michael Lampton, and Byron K. Lichtenberg as ATLAS Payload Specialists. Their roles underscored NASA's commitment to advancing scientific exploration through specialized payload missions, enriching our understanding of outer space's mysteries.

By November 25th, the United Kingdom's Project Juno added a new chapter to international space cooperation. Helen Sharman and Timothy Mace were chosen to represent Britain, with Sharman later making history as the first British-born person in space aboard Soyuz TM-12 in May 1991. This milestone affirmed the UK's role in space exploration and cemented collaborative efforts with the Soviet Union.

In early 1990, the landscape of space exploration continued to evolve with the addition of several notable astronaut groups worldwide.

On January 17th, NASA introduced Group 13, affectionately known as "The Hairballs," a diverse cohort comprising pilots and mission specialists. Among the pilots were Kenneth Cockrell, Eileen Collins (who later became the first female shuttle pilot and commander), and William G. Gregory. The group's mission specialists included notable figures such as Bernard Harris, the first African American to perform a spacewalk, and Ellen Ochoa, who later became the first Hispanic woman in space. Their nickname, as recounted by Thomas David Jones in "Sky Walking," stemmed from their group patch featuring a black cat.

February saw France's CNES Group 3 selection, including astronauts like Léopold Eyharts and Jean-Marc Gasparini. This marked the final wave of CNES astronauts chosen independently before integrating into the broader European Space Agency (ESA) Astronaut Corps in 1999, highlighting France's commitment to international collaboration in space exploration.

By May 11th, the Soviet Union's TsPK–11 Cosmonaut Group added new members such as Talgat Musabayev and Salizhan Sharipov, bolstering the nation's space program amidst ongoing geopolitical shifts.

Germany joined the ranks on October 8th with its 1990 astronaut group, featuring pioneers like Reinhold Ewald and Klaus-Dietrich Flade. This underscored the country's re-emergence as a key player in European space endeavors.

1992 marked a pivotal moment in global space exploration, with several countries making significant strides in their astronaut programs.

In March, Russia's NPOE-10 Cosmonaut Group welcomed Aleksandr Lazutkin, Sergei Treshchov, and Pavel Vinogradov, reinforcing Russia's commitment to manned space missions amidst the post-Soviet era.

On March 31st, NASA introduced Group 14, famously dubbed "The Hogs," comprising a diverse array of pilots and mission specialists. Notable members included Scott Horowitz, Brent Jett, and Kevin Kregel among the pilots, and a distinguished group of mission specialists such as Marc Garneau and Chris Hadfield from Canada, Maurizio Cheli from Italy, Jean-François Clervoy from France, and Koichi Wakata from Japan. This marked the first inclusion of international mission specialists, reflecting NASA's evolving collaboration with global partners in space exploration.

Japan furthered its space ambitions in April by selecting Koichi Wakata under the 1992 NASDA Group, highlighting its commitment to advancing its presence in manned space missions.

In June, Canada's CSA Group 2 was announced. It features astronauts like Dafydd Williams, Julie Payette, and Chris Hadfield and showcases Canada's ongoing contribution to international space endeavors, primarily through missions aboard the US Space Shuttle.

May 15th witnessed the European Space Agency (ESA) selecting its 1992 astronaut group, including Maurizio Cheli from Italy, Jean-François Clervoy from France, Pedro Duque from Spain, Christer Fuglesang from Sweden, Marianne Merchez from Belgium, and Thomas Reiter from Germany. This group underscored ESA's commitment to fostering European space exploration and scientific research collaboration.

In 1994, the global landscape of space exploration saw significant advancements and collaborations among various astronaut groups.

On April 1st, Russia's NPOE–11 Cosmonaut Group was established, welcoming new members Nadezhda Kuzhelnaya and Mikhail Tyurin. This addition underscored Russia's ongoing commitment to maintaining a robust cosmonaut corps during transition and reform.

December 12th marked the introduction of NASA's Group 15, affectionately known as "The Flying Escargot." This group included a cadre of talented pilots such as Scott Altman, Jeffrey Ashby, and Rick Husband, who tragically perished aboard the Space Shuttle Columbia in 2003. Among the mission specialists were notable figures like Kalpana Chawla, who became the first woman of Indian descent to fly in space, and Michael Anderson, who also lost his life in the Columbia disaster. The group's international contingent featured Jean–Loup Chrétien from France, Takao Doi from Japan, Michel Tognini from France, and Dafydd Williams from Canada, highlighting NASA's continued collaboration with global partners in manned space missions.

Jean-Loup Chrétien's contributions were particularly noteworthy. He had previously trained as a backup Spacelab crew member in the 1980s and became the first non-US or Soviet/Russian astronaut to perform a spacewalk. His experience bridging US and Soviet/Russian space programs symbolized the era's spirit of international cooperation and shared scientific exploration.

In 1996, the global landscape of space exploration expanded with new astronaut selections and international collaborations.

On February 9th, Russia's MKS/RKKE–12 Cosmonaut Group welcomed Oleg Kotov and Yuri Shargin from MKS, and Konstantin Kozeyev and Sergei Revin from RKKE. This group's formation highlighted Russia's ongoing commitment to manned space missions amidst restructuring and reforms.

On March 26th, Oleg Kononenko was added to Russia's MKS supplemental cosmonaut group, further bolstering the roster of skilled cosmonauts preparing for future missions.

May 1st marked the introduction of NASA's Group 16, nicknamed "The Sardines." This diverse group included pilots like Duane G. Carey, Charles O. Hobaugh, and James M. Kelly, as well as an extensive list of mission specialists such as Laurel Clark and Michael J. Massimino. International mission specialists from Spain, Sweden, Italy, Canada, Japan, and other nations underscored NASA's commitment to global collaboration in space exploration.

Notably, tragedy struck with members like David McDowell Brown, Laurel Clark, and William Cameron McCool, who were crewmembers on the final mission of the Space Shuttle Columbia in 2003.

June brought Japan's NASDA Group selection of Soichi Noguchi, highlighting Japan's ongoing advancements in its space program and its role in international space missions.

October marked a significant milestone with China's Group 1996, including Li Qinglong and Wu Jie, who trained at the Yuri Gagarin Cosmonaut Training Center as part of China's emerging astronaut program.

In November, Ukraine's Shuttle-97 Group added Leonid Kadeniuk and Yaroslav Pustovyi, contributing to international space endeavors amid Ukraine's growing presence in the aerospace community.

In 1997, space exploration continued to advance with significant astronaut selections and missions from various countries.

Israel's inaugural Shuttle Group welcomed Yitzhak Mayo and Ilan Ramon in April. Ilan Ramon made history as the first Israeli astronaut to venture into space. His mission as a Payload Specialist aboard Space Shuttle Columbia, STS-107, tragically ended in disaster upon re-entry in 2003, highlighting the risks inherent in space exploration.

July 28th marked the formation of Russia's TsPK–12/RKKE-13 Cosmonaut Group. This group included Dmitri Kondratyev, Yury Lonchakov, and Sergey Volkov from TsPK, along with Oleg Skripochka and Fyodor Yurchikhin from RKKE, among others. Their selection underscored Russia's ongoing commitment to maintaining a robust cosmonaut corps and participating in international space missions.

In 1998, significant developments in global space exploration programs highlighted expanding capabilities and international collaboration.

China's Group 1 of Astronauts was established in January, marking a significant milestone in China's space program. This group included notable names like Yang Liwei, who became the first Chinese astronaut to travel to space aboard Shenzhou 5 in October 2003, making China the third nation to achieve manned spaceflight independently.

On February 24th, Mikhail Korniyenko was included in the formation of Russia's RKKE-14 Cosmonaut Group, contributing to Russia's ongoing commitment to manned space missions.

March 2nd marked a historic moment for Slovakia with the establishment of the OS "Mir" Stefanik Group, comprising Ivan Bella and Michal Fulier. This group reflected Slovakia's emerging presence in international space endeavors.

June 4th brought NASA's Group 17, affectionately known as "The Penguins," into the fold. This diverse group included pilots like Christopher Ferguson and mission specialists such as Barbara Morgan, originally selected as the backup "Teacher-In-Space" for Christa McAuliffe. Their international counterparts from France, Italy, Brazil, Germany, Canada, and others underscored NASA's commitment to global collaboration in space missions.

In October, the European Space Agency (ESA) announced its 1998 astronaut group, featuring astronauts like Frank De Winne from Belgium, Léopold Eyharts from France, and Paolo Nespoli from Italy. This group reinforced ESA's role in advancing European space capabilities and participating in multinational space missions.

In 1999, the momentum in global space exploration continued with notable astronaut selections and collaborations.

In February, Japan's NASDA Group for 1999 was formed, welcoming Satoshi Furukawa, Akihiko Hoshide, and Naoko Sumino. This group represented Japan's ongoing commitment to advancing its space program and participating in international space missions.

On November 1st, the European Space Agency (ESA) expanded its astronaut corps with the addition of Claudie André-Deshays, Philippe Perrin, and Michel Tognini. This expansion marked a significant integration of the remaining CNES (France) astronauts into ESA's ranks, enhancing Europe's space exploration and research capabilities.

In July 2000, NASA announced the selection of its Group 18 astronauts, dubbed "The Bugs," for their diverse backgrounds and exceptional skills. Among the distinguished pilots chosen were Dominic A. Antonelli, Eric A. Boe, Kevin A. Ford, Ronald J. Garan Jr.,

Douglas G. Hurley, Terry W. Virts Jr., and Barry E. Wilmore. These seasoned aviators brought a wealth of flight experience and technical acumen to the team, each demonstrating remarkable proficiency in handling advanced aircraft and complex missions.

Joining them were a cadre of mission specialists, including Michael R. Barratt, Robert L. Behnken, Stephen M. Bowen, B. Alvin Drew, Andrew J. Feustel, Michael T. Good, Timothy L. Kopra, K. Megan McArthur, Karen L. Nyberg, and Nicole P. Stott. This group comprised experts in various scientific disciplines, engineering, and operational roles critical to NASA's ambitious exploration endeavors.

The selection of Group 18 marked a pivotal moment in NASA's ongoing commitment to human space exploration. It underscored the agency's rigorous standards and the caliber of individuals needed for future missions.

In 2003, significant developments in global space exploration highlighted diverse advancements across different programs and nations. Russia's TsPK-13/RKKE-15/IMBP-6 Cosmonaut Group saw the selection of Anatoli Ivanishin, Aleksandr Samokutyayev, Anton Shkaplerov, Evgeny Tarelkin, and Sergei Zhukov from TsPK; Oleg Artemyev, Andrei Borisenko, and Mark Serov from RKKE; and Sergey Ryazansky from IMBP. This cohort represented Russia's commitment to advancing its space capabilities, with members skilled in piloting, science, and engineering crucial for future missions.

Meanwhile, Kazakhstan's Group 1 included Aydyn Aimbetov and Mukhtar Aymakhanov, emphasizing the international collaboration in space exploration efforts.

September 11th marked a milestone in the United States with the first commercial astronauts selected for SpaceShipOne. Brian Binnie and Mike Melvill achieved spaceflight aboard SpaceShipOne, showcasing private sector advancements in aerospace technology and exploration ambition. Doug Shane and Peter Siebold, also part of the group, contributed to subsequent developments in commercial spaceflight, with Siebold notably piloting SpaceShipTwo in later missions.

On May 6, 2004, NASA introduced Group 19, aptly nicknamed "The Peacocks," marking a significant evolution in astronaut selection and training. Comprising a diverse array of talent, this group reflected NASA's strategic shift toward inclusivity and specialization as the space agency prepared for the final missions of the Space Shuttle era.

Among the distinguished pilots were Randolph Bresnik and James Dutton, both seasoned aviators bringing extensive flight experience to the team. Their roles would be crucial in piloting and maneuvering the Space Shuttle, ensuring precise execution of complex missions.

The mission specialists included Thomas Marshburn, Christopher Cassidy, R. Shane Kimbrough, José M. Hernández, Robert Satcher, and Shannon Walker. Each specialist brought unique expertise in fields ranging from medicine to engineering, essential for conducting scientific experiments and maintaining spacecraft operations during missions.

A groundbreaking addition to Group 19 included educator mission specialists Joseph M. Acaba, Richard R. Arnold, and Dorothy Metcalf-Lindenburger. Their selection underscored NASA's commitment to educational outreach, aiming to inspire future generations through direct involvement in space missions.

Furthermore, Group 19 welcomed international mission specialists Satoshi Furukawa, Akihiko Hoshide, and Naoko Yamazaki from Japan, highlighting NASA's collaborative efforts with global partners in advancing space exploration.

This cohort marked the end of an era as the last group trained specifically for Space

Shuttle missions, symbolizing a transition toward new horizons in space exploration methodologies and technologies.

In 2006, the landscape of space exploration expanded globally with notable advancements and selections across various astronaut programs.

Virgin Galactic's Astronaut Pilots Group from the UK, consisting of Steve Johnson, Alistair Hoy, David MacKay, and Alex Tai, signaled the emergence of commercial spaceflight ambitions. These pilots were pivotal in Virgin Galactic's efforts to pioneer private space travel, marking a significant step toward making space accessible to private individuals.

Malaysia's Angkasawan Group selection on September 4th highlighted the country's ambitious foray into space exploration. Sheikh Muszaphar Shukor, Faiz Khaleed, Siva Vanajah, and Mohammed Faiz Kamaludin were chosen to undergo training for a flight to the International Space Station (ISS). Sheikh Muszaphar Shukor achieved historical significance as the first Malaysian in space aboard Soyuz TMA-11, symbolizing Malaysia's aspirations for scientific and technological advancements.

Russia's TsPK-14/RKKE-16 Cosmonaut Group announcement on October 11th included Aleksandr Misurkin, Oleg Novitskiy, Aleksey Ovchinin, Maksim Ponomaryov, Sergey Ryzhikov from TsPK, and Yelena Serova, Nikolai Tikhonov from RKKE. This group represented Russia's ongoing commitment to advancing its space exploration capabilities through a new generation of skilled cosmonauts essential for missions aboard the Soyuz spacecraft and the ISS.

Closing out the year, South Korea's Astronaut Program Group introduced Yi So-yeon and Ko San in December. Ko San eventually emerged as the prime candidate in 2007, preceding Yi So-yeon's selection in 2008. Yi So-yeon made history as South Korea's first astronaut, embarking on a mission to the ISS, exemplifying the country's rapid strides in space exploration.

These selections underscored a global shift toward collaborative efforts in space exploration, blending governmental initiatives with the rising influence of commercial ventures, and inspiring diverse nations to reach for the stars in the pursuit of scientific discovery and technological innovation.

In July 2008, Virgin Galactic expanded its roster by adding three new pilots to its Astronaut Pilots Group from the UK. Robert Bendall, Rich Dancaster, and Brad Lambert joined the ranks of commercial astronauts selected to pioneer the next phase of private space exploration. Their expertise and dedication represented Virgin Galactic's ongoing commitment to pushing the boundaries of space tourism and advancing the accessibility of space travel beyond traditional governmental programs. These selections reinforced Virgin Galactic's position at the forefront of commercial spaceflight initiatives, marking a pivotal moment in the evolution of private-sector involvement in human space exploration.

In 2009, the global landscape of space exploration saw significant advancements and selections across multiple astronaut programs, underscoring diverse international collaboration and advancements in human spaceflight capabilities.

Japan's JAXA Group, announced on February 25th, introduced Takuya Onishi and Kimiya Yui. This selection highlighted Japan's continued commitment to expanding its presence in space exploration through skilled astronauts prepared for missions aboard the International Space Station (ISS).

On May 13th, Canada's CSA Group welcomed Jeremy Hansen and David Saint-Jacques, demonstrating Canada's ongoing participation in international space missions and contributions to scientific research aboard the ISS.

ESA's Group, aptly named "The Shenanigans," announced on May 20th, featured Samantha Cristoforetti (Italy), Alexander Gerst (Germany), Andreas Mogensen (Denmark), Luca Parmitano (Italy), Timothy Peake (United Kingdom), and Thomas Pesquet (France). This diverse cohort, selected from over 8,413 European applicants, underscored the ESA's role in fostering a unified approach to space exploration among member states.

NASA's Group 20, known informally as "Chumps," was unveiled on June 29th, marking a significant milestone as the first cohort of astronauts chosen for the post-Space Shuttle era. Comprising mission specialists Serena M. Auñón, Jeanette J. Epps, Jack D. Fischer, Michael S. Hopkins, Kjell N. Lindgren, Kathleen (Kate) Rubins, Scott D. Tingle, Mark T. Vande Hei, and Gregory R. (Reid) Wiseman, this group was selected from a pool of over 3,500 applicants. Notably, Fischer, Tingle, and Wiseman were initially designated as pilots, reflecting NASA's transition toward a unified mission specialist role in the absence of the Shuttle program.

On September 8th, JAXA added Norishige Kanai to its ranks, further expanding Japan's astronaut corps and capabilities in human spaceflight missions.

In 2010, the global landscape of space exploration continued to evolve with significant selections and advancements across various astronaut programs, reflecting a diverse range of national and commercial initiatives.

China's Group 2, announced in March, introduced Cai Xuzhe, Chen Dong, Liu Yang, Tang Hongbo, Wang Yaping, Ye Guangfu, and Zhang Lu. This cohort represented China's commitment to expanding its human spaceflight capabilities and preparing astronauts for future missions aboard Chinese spacecraft and space stations.

The Association of Spaceflight Professionals unveiled its Group 1 on April 12th, comprising Jim Crowell, Bruce Davis, Kristine Ferrone, Amnon Govrin, Chad Healy, Ryan Kobrick, Joseph Palaia, Luís Saraiva, Brian Shiro, Laura Stiles, and Veronica Ann Zabala-Aliberto. This group highlighted the growing influence of private sector involvement in space exploration, focusing on scientific research, technology development, and commercial spaceflight initiatives.

On June 7th, the Association of Spaceflight Professionals introduced its Group 2 of Commercial Astronauts, including Ben Corbin, José Miguel Hurtado Jr., Jason Reimuller, Todd Romberger, Erik Seedhouse, and Alli Taylor. This selection underscored the increasing role of commercial entities in advancing human presence in space and promoting innovation and accessibility in space travel.

Russia's TsPK-15/RKKE-17 Cosmonaut Group, announced on October 12th, featured Aleksey Khomenchuk, Denis Matveev, Sergey Prokopyev from TsPK, and Andrei Babkin, Ivan Vagner, Sergey Kud-Sverchkov, and Svyatoslav Morozov from RKKE. This group continued Russia's tradition of excellence in space exploration, preparing cosmonauts for missions aboard the Soyuz spacecraft and the ISS, emphasizing the country's ongoing commitment to scientific research and international collaboration.

In 2011, significant developments in global space exploration continued with new astronaut selections and advancements across different programs and initiatives.

In January and February, Roscosmos, the Russian space agency, enrolled a united squad of astronauts, marking a consolidation of talent within the Russian space program. This group included Oleg Artemyev, Andrei Babkin, Ivan Vagner, Andrei Borisenko, Sergei Zhukov, Oleg Kononenko, Mikhail Kornienko, Sergey Kud-Sverchkov, Svyatoslav Morozov, Sergei Revin, Sergey Ryazansky, Yelena Serova, and Nikolai Tikhonov. This move aimed to streamline training and preparation processes at the Y. A. Gagarin Cosmonaut Training Center, emphasizing Russia's commitment to maintaining and expanding its human spaceflight capabilities.

On February 28th, the Association of Spaceflight Professionals announced Group 3, comprising Christopher Altman, Jon-Erik Dahlin, Melania Guerra, Mindy Howard, Kris Lehnhardt, Abhishek Tripathi, Cosan Unuvar, Pavel Zagadailov, and Luis Zea. This group highlighted the growing diversity and specialization within the commercial space sector, focusing on scientific research, technology development, and expanding human presence beyond Earth's orbit.

In October, Virgin Galactic's Astronaut Pilots Group welcomed Keith Colmer, underscoring ongoing efforts in commercial spaceflight initiatives. This selection reinforced Virgin Galactic's role in advancing private sector involvement in space exploration, paving the way for future missions and space tourism endeavors.

In 2012, developments in space exploration continued with new additions to Russia's cosmonaut corps and advancements in training programs.

In February, Fyodor Yurchikhin joined the united squad of Roscosmos cosmonauts. Yurchikhin's enrollment underscored Russia's ongoing commitment to maintaining and expanding its cadre of experienced cosmonauts, essential for missions aboard the International Space Station (ISS) and future exploration endeavors.

On October 30th, the TsPK Addition Group in Russia introduced a cohort of finalists, including Oleg Blinov, Nikolay Chub, Pyotr Dubrov, Andrey Fedyaev, Ignat Ignatov, Anna Kikina, Sergey Korsakov, and Dmitriy Petelin. This selection highlighted Russia's efforts to identify and prepare the next generation of cosmonauts, emphasizing the country's strategic focus on training and developing skilled personnel for future space missions.

In 2013, significant developments unfolded across the aerospace landscape, reflecting space exploration ventures' diverse and dynamic nature. Among these milestones, the Virgin Galactic Astronaut Pilots Group, based in the UK, saw former NASA astronaut Frederick W. Sturckow and Michael "Sooch" Masucci join its ranks, marking a pivotal step in the burgeoning field of commercial space travel.

Simultaneously, the Association of Spaceflight Professionals welcomed its Group 4 members, including David Ballinger, Jessica Cherry, Michael Gallagher, Jamie Guined, Tanya Markow-Estes, and Aaron Persad. These individuals, each contributing unique expertise, highlighted the increasing specialization within the aerospace industry and its global reach.

Meanwhile, on June 17th, NASA introduced Group 21, affectionately dubbed the 8-Balls, underscoring the agency's ongoing commitment to fostering diverse talent pools. Comprising Josh A. Cassada, Victor J. Glover, Tyler N. Hague, Christina M. Hammock, Nicole Aunapu Mann, Anne C. McClain, Jessica U. Meir, and Andrew R. Morgan, this cohort symbolized NASA's dedication to advancing scientific discovery and space exploration capabilities.

These developments underscored a transformative period in space history, characterized by collaborative efforts between commercial entities and governmental agencies alike.

In 2014, the landscape of space exploration continued to evolve with notable additions to key astronaut groups. On July 24th, the Virgin Galactic Astronaut Pilots Group in the UK welcomed Todd Ericson into its ranks, further bolstering the cadre of commercial astronauts poised to pioneer new frontiers in space tourism and research.

Meanwhile, on August 14th, Mukhtar Aimakhanov was selected to join a united detachment of Roscosmos astronauts in

Russia, highlighting international collaborations that enrich space missions with diverse expertise and perspectives.

In 2015, pivotal developments continued to shape the landscape of space exploration, highlighting significant advancements and international collaborations. On January 23rd, Mark Stucky joined the Virgin Galactic Astronaut Pilots Group in the UK, adding to the roster of commercial astronauts poised to push the boundaries of space tourism and research.

Later in the year, on July 9th, NASA's Commercial Crew Program made strides with the inclusion of Robert Behnken, Sunita Williams, Eric Boe, and Douglas Hurley. This marked a crucial step toward NASA's goal of fostering commercial partnerships to transport astronauts to and from the International Space Station (ISS), enhancing accessibility to space and scientific research capabilities.

Simultaneously, the European Space Agency (ESA) expanded its Astronaut Corps with the addition of Matthias Maurer, reflecting Europe's commitment to advancing space exploration and scientific discovery.

In Denmark, Copenhagen Suborbitals made headlines in the commercial astronautics arena. Mads Stenfatt, Anna Olsen, and Carsten Olsen contributed to innovative suborbital missions, underscoring the diverse global efforts to explore space and push technological boundaries.

In 2017, the global space community witnessed significant advancements and new additions to astronaut programs worldwide, reflecting a diverse range of talents and aspirations in space exploration.

On June 7th, NASA introduced Group 22, aptly named The Turtles, comprising a diverse cohort of astronauts from the United States. This group included Kayla Barron, Zena Cardman, Raja Chari, Matthew Dominick, Robert Hines, Warren Hoburg, Jonny Kim, Jasmin Moghbeli, Loral O'Hara, Francisco Rubio, and Jessica Watkins. Their selection underscored NASA's commitment to fostering a new generation of explorers equipped to undertake missions to the International Space Station (ISS) and beyond. Originally part of Group 22, Robb Kulin resigned from NASA in August 2018 before completing his training, highlighting the rigorous nature and personal commitments required of astronaut candidates.

Meanwhile, on July 1st, the Canadian Space Agency (CSA) announced its 2017 astronaut selection, welcoming Jennifer Sidey and Joshua Kutryk. This expansion reflected Canada's ongoing contributions to international space missions and scientific research aboard the ISS.

In Germany, the 2017 Die Astronautin Selection program made waves by selecting Insa Thiele-Eich and later replacing Nicola Baumann with Suzanna Randall. This initiative aimed to propel women astronauts to the forefront of European space exploration efforts, emphasizing inclusivity and diversity in astronaut training and missions.

In 2018, significant advancements in global space exploration programs highlighted the diverse and expanding landscape of astronaut training and missions.

On August 10th, Russia's 17th Cosmonaut Group was announced, comprising Konstantin Borisov, Alexander Gorbunov, Alexander Grebenkin, Sergei Mikayev, Kirill Peskov, Oleg Platonov, Yevgeny Prokopyev, and Alexei Zubritsky. Following rigorous training, all members except Yevgeny Prokopyev passed the state exam in December 2020, qualifying them for future spaceflight assignments. Prokopyev, however, did not qualify initially and was reassigned to undergo further basic space training.

Meanwhile, on September 3rd, the United Arab Emirates (UAE) introduced its Emirati Astronaut Group, consisting of Hazza Al Mansouri and Sultan Al Neyadi. Al Mansouri and Al Neyadi were selected as candidates to fly to the International Space Station (ISS) aboard a Soyuz spacecraft. Al Mansouri completed his mission in 2019, while Al Neyadi served as his backup. In 2020, both astronauts were designated to train as full-fledged mission specialist astronauts in Houston, marking a significant milestone for the UAE's space program and integration into international space missions. Al Neyadi became the first Emirati astronaut to embark on a long-duration mission aboard SpaceX Crew-6 in March 2023.

In December 2019, India marked a significant milestone in its space exploration efforts by introducing the 1st Vyomnaut Group. Comprising Prashanth Nair, Angad Prathap, Ajit Krishnan, and Subhanshu Shukla, this group represented India's first cohort of astronauts selected for future missions under the Gaganyaan program.

The Gaganyaan H1 mission, scheduled for 2025, will see three members of the Vyomnaut Group undertake India's first crewed spaceflight, symbolizing the nation's ambitions to launch human missions into space independently. Additionally, the fourth astronaut will be a backup for the Gaganyaan H1 mission but is slated to embark on a short-duration mission to the International Space Station (ISS) aboard Ax-4 in late 2024.

In October 2020, China made significant strides in its space program with the announcement of the selection of 18 new Group 3 astronauts. This diverse cohort, consisting of 17 men and 1 woman, was selected across three specialized categories essential for advancing China's space exploration objectives.

Among the group are seven spacecraft pilots, including Tang Shengjie and Li Guangsu, tasked with commanding and navigating future space missions. Additionally, seven flight engineers were selected, such as Zhu Yangzhu, Jiang Xinlin, and Li Cong, who will play crucial roles in spacecraft operations and maintenance. Completing the team are four mission payload specialists, like Gui Haichao, who will focus on conducting scientific experiments and managing mission-specific equipment aboard the spacecraft.

This announcement highlighted China's commitment to expanding its human spaceflight capabilities and advancing scientific research in space. The selection of Group 3 astronauts underscored China's strategic vision and readiness to explore new frontiers, reinforcing its position as a key player in global space exploration endeavors.

In 2021, the global landscape of space exploration saw notable achievements and advancements across multiple astronaut programs, reflecting diverse achievements and contributions.

On January 27th, Russia announced its 18th Cosmonaut Group, which included Sergey Irtuganov, Alexander Kolyabin, Sergey Teteryatnikov, and Harutyun Kiviryan. These individuals represent Russia's ongoing commitment to expanding its cosmonaut corps and advancing its capabilities in space exploration.

March 30th marked a historic moment with the Inspiration4 mission in the USA. Sponsored by mission commander Jared Isaacman, Inspiration4 became the first all-civilian orbital spaceflight mission funded privately and not by a nation-state. The crew, comprising Jared Isaacman, Sian Proctor, Hayley Arceneaux, and Chris Sembroski, achieved the highest human orbit of the 21st century. Sian Proctor was the first female commercial astronaut spaceship pilot and African American female spacecraft pilot. At the same time, medical officer Hayley Arceneaux became the first astronaut to fly with a prosthesis.

In April, the United Arab Emirates (UAE) introduced its Emirati Astronaut Group 2, consisting of Nora Al Matrooshi and Mohammad Al Mulla. They began training alongside NASA's Astronaut Group 23 class, highlighting the UAE's continued investment in human spaceflight and international collaboration.

Later in the year, on December 6th, NASA announced its Group 23 astronauts, including Nichole Ayers, Marcos Berríos, Christina Birch, Deniz Burnham, Luke Delaney, Andre Douglas, Jack Hathaway, Anil Menon, Christopher Williams, and Jessica Wittner. This diverse group represents NASA's ongoing efforts to prepare for future space missions and scientific endeavors, underscoring the agency's commitment to exploration and discovery.

In 2022, the global space exploration community saw significant developments with the announcement of new astronaut selections and expansions in training programs.

On October 2nd, China unveiled its Group 4 astronauts, consisting of 12-14 new members categorized into spacecraft pilots, flight engineers, and mission payload specialists. The specific names of these astronauts were not disclosed, underscoring China's ongoing commitment to enhancing its human spaceflight capabilities and advancing its space exploration objectives.

Later in the year, on November 23rd, the European Space Agency (ESA) introduced its 2022 Astronaut Group. This diverse cohort included career astronauts such as Sophie Adenot from France, Pablo Álvarez Fernández from Spain, Rosemary Coogan from the UK, Raphaël Liégeois from Belgium, and Marco Alain Sieber from Switzerland. Additionally, the group featured reserve and project astronauts like Meganne Christian and John McFall from the UK, Anthea Comellini and Andrea Patassa from Italy, Sara García Alonso from Spain, Carmen Possnig from Austria, and others from across Europe. Notably, John McFall was included in a feasibility study focusing on astronauts with disabilities, highlighting ESA's commitment to inclusivity and accessibility in space exploration.

In 2023, several countries continued to advance their space exploration efforts by selecting and announcing new astronauts.

On February 12th, Saudi Arabia introduced its second astronaut group, including Rayyanah Barnawi, Ali AlQarni, Mariam Fardous, and Ali AlGhamdi. Two of these astronauts were selected to participate in Axiom Mission 2, marking Saudi Arabia's continued involvement in international space missions and scientific research.

Australia joined the ranks of spacefaring nations on March 8th by introducing Katherine Bennell-Pegg as part of its inaugural Astronaut Group 1. This marked a significant milestone for Australia's space program, highlighting its commitment to expanding its space exploration and research capabilities.

In April, Turkey launched its first Astronaut Group, comprising Alper Gezeravcı and Tuva Cihangir Atasever. This initiative represented Turkey's entry into human spaceflight, aiming to contribute to international collaborations and scientific endeavors in space.

Additionally, in June, Alysson Muotri was recognized for significant contributions to space research, reflecting advancements in scientific fields related to space exploration.

Looking ahead to 2024, Hungary announced its first astronaut mission, HUNOR 1, scheduled for May 27th. Tibor Kapu was selected as the primary astronaut, with Gyula Cserényi serving as the reserve astronaut. This mission marked Hungary's entry into crewed space missions, symbolizing its aspirations in scientific discovery and space exploration.

These developments in 2023 and 2024 underscored a global commitment to expanding human presence in space, fostering international cooperation, and pushing the boundaries of scientific knowledge beyond Earth's atmosphere.

The commercial space sector has evolved significantly, driven by a burgeoning market that has surpassed $330 billion and is projected to approach $3 trillion in the coming decades. Among the sectors poised for substantial growth, human spaceflight stands out prominently. Commercial astronauts are set to play a pivotal role in this transformative phase of space exploration.

The foundation for commercial astronauts was laid with the inception of the Ansari X PRIZE in 2004, which aimed to spur the development of the first privately-built, reusable crewed spacecraft. This milestone initiative saw the selection of pioneering individuals who would shape the future of commercial space travel.

Notable figures among the early commercial astronauts include Steve Bennett and Matt Shewbridge from Starchaser Industries, alongside esteemed former NASA astronauts such as John Bennett Herrington of Pioneer Rocketplane, Richard Searfoss, and pilot Dick Rutan of XCOR Aerospace. Canadian engineer Brian Feeney represented the da Vinci Project, contributing to the advancement of private space initiatives. Additionally, Wally Funk, a trailblazer from the Mercury 13 program, continued her legacy with Interorbital Systems, further enriching the commercial space landscape.

Boeing's commitment to advancing space exploration was underscored by the significant addition of former NASA astronaut Chris Ferguson to its Space Exploration Team. Ferguson's extensive experience and expertise from his NASA tenure make him a valuable asset to Boeing's ambitious space initiatives.

Boeing's astronaut corps included various candidates, ranging from former NASA astronauts to commercial scientist astronauts and test pilots who have yet to experience spaceflight. This eclectic mix highlights Boeing's strategic approach to assembling a team equipped to tackle the challenges and opportunities of the evolving space industry.

SpaceX, a commercial space exploration pioneer, has integrated former NASA astronauts into its team while maintaining distinct criteria for crewing its commercial vehicles to the International Space Station (ISS). However, specific citations are required to verify this information.

An example of SpaceX's integration with NASA is Anil Menon, formerly the medical director at SpaceX. He transitioned to become a NASA astronaut, selected in 2021 as part of NASA Astronaut Group 23. This transition exemplifies SpaceX and NASA's collaborative synergy, contributing to advancements in human spaceflight and exploration efforts.

The Association of Spaceflight Professionals (ASP) emerged as a pioneering initiative in commercial space exploration, purportedly forming the world's first commercial astronaut corps. This organization reportedly secured funding through the NASA Flight Opportunities Program in March 2012, marking a significant milestone in its efforts to conduct crewed spaceflight missions.

ASP's notable endeavors include several million dollars reportedly allocated for detailed spectroscopic analysis of high-altitude noctilucent cloud formations during suborbital flights. These missions utilize rapidly reusable, task-and-deploy spaceplanes, highlighting ASP's innovative approach to scientific research in space.

ASP's selection process for commercial astronauts mirrors that of NASA's prestigious Astronaut Corps, emphasizing rigorous training and qualification standards. Some ASP members are reported to serve as astronaut trainers themselves and have participated as finalists in National Space Agency astronaut candidate selection campaigns.

One prominent ASP member mentioned is Yi So-yeon, known for her historic orbital mission to the International Space Station. Her involvement underscores ASP's international reach and collaborative efforts in advancing commercial space ventures.

Virgin Galactic and its subsidiary Scaled Composites have been pivotal in advancing commercial spaceflight, and a roster of astronauts and key figures has contributed to their missions and developments.

Among the notable individuals associated with Scaled Composites and Virgin Galactic are:

Michael Alsbury, tragically killed in the 2014 Virgin Galactic crash, was instrumental in the development and testing phases.

Rob Bendall, representing Canada, a key figure in Virgin Galactic's early test flights and operations.

Richard Branson, the founder of Virgin Galactic, who played a pivotal role in promoting and supporting commercial space tourism.

Peter Kalogiannis, an integral member of the engineering team at Virgin Galactic.

Niki Lauda, the renowned Austrian Formula 1 champion and aviation enthusiast.

Brian Maisler, Clint Nichols, and Wes Persall contributed to various aspects of Virgin Galactic's operations.

Burt Rutan, the legendary aerospace engineer and founder of Scaled Composites, is known for his innovative design and contributions to spacecraft development.

Key pilots involved in test flights and operational missions are Peter Seiffert and Peter Siebold.

Mark Stucky and Dave Mackay are both renowned pilots who have flown Virgin Galactic's spacecraft on numerous test and operational flights.

The Teachers in Space program, launched in 2005, aimed to pioneer educational opportunities in space exploration. In 2012, the United States Rocket Academy announced a significant expansion, broadening the initiative to include a wider array of participants and rebranding it as Citizens in Space.

Under the Citizens in Space banner, the program's initial phase focused on selecting and training ten citizen astronaut candidates to serve as payload operators. This cohort included four astronaut candidates already undergoing training: Maureen Adams, Steve Heck, Michael Johnson, and Edward Wright. Among the notable participants was Gregory Kennedy, an informal educator and aerospace historian who contributed his expertise to the program's mission.

Copenhagen Suborbitals, founded in 2008 in Denmark, has set an ambitious goal to achieve human spaceflight beyond the Kármán line, aiming to make Denmark the fourth nation to accomplish this milestone. The organization represents a grassroots effort in amateur space exploration, driven by the passion and dedication of its members to push the boundaries of space technology.

Mars One, an ambitious private initiative announced in May 2012 by Dutch entrepreneur Bas Lansdorp, aimed to establish the first permanent human colony on Mars by 2023. The project garnered significant attention with its astronaut selection process beginning in April 2013, attracting over 200,000 applicants from around the globe by August of that year. Round Two of selections, announced in December 2013, identified 1,058 finalists from 107 countries eager to embark on the mission.

Despite early enthusiasm, Mars One faced substantial skepticism and criticism regarding its medical, technical, and financial viability. Concerns ranged from the feasibility of sustaining life on Mars to doubts about the project's funding model. Criticism intensified with unverified allegations suggesting Mars One might have been a fundraising scheme rather than a genuine space mission.

In a notable development, Mars One declared bankruptcy in a Swiss court on January 15, 2019, leading to the permanent dissolution of the company. This marked the end of the Mars One venture, highlighting the challenges and uncertainties inherent in pioneering efforts toward human settlement on Mars.

Inspiration Mars

The Inspiration Mars Foundation, established by Dennis Tito, envisioned a pioneering human mission: a flyby of Mars initially slated for January 2018, later rescheduled to 2021 after missing the 2018 target. This ambitious endeavor aimed to send astronauts on a historic journey around the Red Planet.

Among the notable figures involved were Jane Poynter and Taber MacCallum, a husband-and-wife team renowned for participating in the Biosphere 2 experiment. Their involvement underscored the foundation's commitment to assembling a capable, experienced crew for the challenging mission.

Waypoint2Space

Waypoint2Space achieved a significant milestone in 2014 when it received FAA safety approval for its spaceflight training services. Collaborating closely with NASA's Johnson Space Center in Houston, the company provides specialized training essential for astronauts preparing for missions beyond Earth's atmosphere. This partnership highlights Waypoint2Space's dedication to advancing human space exploration through comprehensive and rigorous training programs.

Truax Engineering

Truax Engineering holds a notable place in the history of private spaceflight initiatives. It was the first private firm to undertake the development of a suborbital space rocket. The company selected Jeana Yeager, an engineer and accomplished aviator, as its inaugural test pilot. However, despite early promise, the project faced financial challenges and was ultimately halted in 1991.

In preparation for the International Space Station's retirement in 2030, NASA is intensifying efforts to transition operations to future private space stations in low-Earth orbit. This pivotal shift aims to avoid disruptions in microgravity research and technology development, which have been pivotal on the ISS.

Key to this transition is the emergence of private sector initiatives, such as the ambitious plans of Houston-based Axiom Space to construct its own space station. NASA officials emphasize the critical need for continuity in space-based research, underscoring concerns over potential disruptions should there be any gap in operational capability post-2030.

"The impact of a gap would be disruptive," stated ISS director Robyn Gatens at a recent conference. To mitigate this risk, NASA is working closely with commercial partners to ensure that a viable successor to the ISS is operational by 2028. This initiative aligns with a strategy outlined by the White House Office of Science and Technology Policy, aimed at maintaining an uninterrupted American presence in low-Earth orbit and fostering a robust commercial space industry.

Despite these efforts, transitioning to commercial space stations poses significant challenges, including technical complexities and budgetary considerations. John Mulholland, Boeing's ISS program manager, highlighted the need for increased funding, particularly for projects like the United States Deorbit Vehicle (USDV) and upgrades to scientific instruments such as the Alpha Magnetic Spectrometer (AMS).

Looking ahead, NASA plans to leverage its expertise to support private space station providers through its Commercial Low Earth Orbit Destinations (CLD) program. This support, however, will gradually diminish as private entities assume greater responsibility.

Post-2030, NASA envisions operating a national laboratory, the LEO National Lab, designed to facilitate government-sponsored research across multiple commercial platforms. This initiative aims to complement rather than compete with private sector interests, ensuring continued scientific advancement in low-Earth orbit.

While international partners like Japan, Canada, and the European Space Agency are committed to supporting the ISS until its planned retirement, Russia intends to shift focus to its own orbital space station by 2028. This evolving landscape underscores the dynamic future of human space exploration beyond the ISS era.

In summary, NASA's proactive approach seeks to secure a seamless transition to private space stations, ensuring the continuity of groundbreaking research and technological innovation in the unique environment of low-Earth orbit.

About the Author

Thornton D. "TD" Barnes is a distinguished author, entrepreneur, and former military intelligence specialist. Born in Dalhart, Texas, and raised on a ranch near Clayton, New Mexico and Dalhart, Texas, he cultivated a passion for exploration. After high school in Oklahoma, Barnes embarked on a ten-year military journey, initially serving in Korea as an intelligence specialist. While in the Army, he also specialized in missile and radar electronics, defending against Soviet threats and later attending the Artillery Officer Candidate School. An injury ended his military career, but Barnes soon transitioned to aerospace endeavors. He worked on pivotal projects at NASA's High Range in Nevada, including the X-15 and the NERVA nuclear rocket project. Furthermore, he participated in the CIA's Mach 3 A-12 Project OXCART and stealth projects at Area 51.

Barnes founded and led an oil and gas exploration company outside the aerospace sphere for over 40 years, delving into uranium and gold mining ventures. He's dedicated to preserving Area 51's history, serving as president of Roadrunners Internationale and the Nevada Aerospace Hall of Fame Director Emeritus. His contributions have been spotlighted in documentaries on National Geographic, the History Channel, and other major networks. Barnes has authored several books, including "The Secret Genesis of Area 51," "The CIA Area 51 Chronicles," and a four-book series on the US Space Programs. He currently resides in Henderson, Nevada, continuing to influence aerospace, exploration, and literature, focusing on the formally highly classified of the CIA's era at Area 51.

Bibliography

Apollo Timeline: https://www.nasa.gov/wp-content/uploads/2023/04/sp-4029.pdf

https://www.britannica.com/story/timeline-of-the-apollo-space-missions

https://www.britannica.com/story/timeline-of-the-apollo-space-missions

https://www.nasa.gov/international-space-station/expedition-missions/

https://en.wikipedia.org/

ChatGPT. https://chatgpt.com

Index

Í

www.ingramcontent.com/pod-product-compliance
Lightning Source LLC
LaVergne TN
LVHW060820170826
845678LV00010B/1849

* 9 7 9 8 2 2 7 4 3 6 5 5 9 *